LOGIC FOR
COMPUTER SCIENCE

Harper & Row Computer Science and Technology Series

LOGIC FOR COMPUTER SCIENCE
Foundations of Automatic Theorem Proving

Jean H. Gallier
University of Pennsylvania

1817

HARPER & ROW, PUBLISHERS, New York
Cambridge, Philadelphia, San Francisco,
London, Mexico City, São Paulo, Singapore, Sydney

To Anne, my wife,
Philippe and Sylvie, my children,
and my mother, Simone

Sponsoring Editor: John Willig
Project Editor: Lucy Zakarian
Cover Design: ARCON
Production: Willie Lane
Compositor: Textset Inc
Printer and Binder: R. R. Donnelley & Sons Company

LOGIC FOR COMPUTER SCIENCE:
Foundations of Automatic Theorem Proving

Library of Congress Cataloging in Publication Data
Gallier, Jean H.
    Logic for computer science.

    Bibliography: p.
    Includes index.
    1. Automatic theorem proving. 2. Logic, Symbolic
and mathematical. I. Title.
QA76.9.A96G35 1986   511.3   85-14071
ISBN 0-06-042225-4

85 86 87 88 9 8 7 6 5 4 3 2 1

CONTENTS

PREFACE

This book is intended as an introduction to mathematical logic, with an emphasis on proof theory and procedures for constructing formal proofs of formulae algorithmically. The book is designed primarily for students, computer scientists, and, more generally, for mathematically inclined readers interested in the formalization of proofs and the foundations of automatic theorem proving.

The book is self-contained and the level corresponds to senior undergraduates and first-year graduate students. However, there is enough material for at least a two-semester course, and some chapters (Chapters 6, 7, 9, and 10) contain material that could form the basis of seminars. It would be helpful, but not indispensable, if the reader has had undergraduate-level courses in set theory and/or modern algebra.

Since the main emphasis of the text is on the study of proof systems and algorithmic methods for constructing proofs, it contains some features rarely found in other texts on logic. Four of these features are:

1. The use of Gentzen systems
2. A justification of the resolution method via a translation from a Gentzen system
3. A presentation of SLD-resolution and a presentation of the foundations of PROLOG
4. Fast decisions procedures based on congruence closures

A fruitful way to use this text is to teach PROLOG concurrently with the material in the book, and ask the student to implement in PROLOG some of the procedures given in the text, in order to design a simple theorem prover.

Even though the main emphasis of the book is on the design of procedures for constructing formal proofs, the treatment of the semantics is perfectly rigorous. The following paradigm has been followed: Having defined the syntax of the language, it is shown that the set of well-formed formulae is a freely generated inductive set. This is an important point, which is often glossed over. Then, the concepts of satisfaction and validity are defined by recursion over the freely generated inductive set (using the unique homomorphic extension theorem, which can be rigorously justified). Finally, the proof theory is developed, and procedures for constructing proofs are given. Particular attention is given to the complexity of such procedures.

In our opinion, the choice of Gentzen systems over other formal systems is pedagogically very advantageous. Gentzen-style rules reflect directly the semantics of the logical connectives, lead naturally to mechanical proof procedures, and have the advantage of having duality built in. Furthermore, in our opinion, Gentzen systems are more convenient than tableaux systems or natural deduction systems for proof-theoretical investigations (cut-free proofs in particular). In three years of teaching, I have found that Gentzen-like systems were very much appreciated by students.

Another good reason for using a formal system inspired from Gentzen (a sequent calculus) is that the completeness theorem is obtained in a natural and simple way. Furthermore, this approach even yields a program (the search procedure) for constructing a proof tree for a valid formula. In fact, in our presentation of the completeness theorem (inspired by Kleene, 1967), the search for a proof tree is described by a program written in pseudo-PASCAL. We also show how a proof procedure for first-order logic with equality can be developed incrementally, starting with the propositional case.

The contents of the book are now outlined.

Chapter 1 sets the goals of the book.

Chapter 2 has been included in order to make the book as self-contained as possible, and it covers the mathematical preliminaries needed in the text. It is recommended to refer to this chapter only when needed, as opposed to reading it entirely before proceeding to Chapter 3.

Propositional logic is studied in Chapter 3. This includes the syntax and semantics of propositional logic. Gentzen systems are introduced as a method for attempting to falsify a proposition. The completeness theorem is shown as well as some of its consequences (the conjunctive and disjunctive normal forms). By introducing infinite sequents, the extended completeness theorem and the compactness theorem are obtained. An informal exposition of the complexity classes P, NP, and of the concept of NP-completeness is given at the end of the chapter.

The resolution method for propositional logic is presented in Chapter 4. This chapter uses a new approach for proving the completeness of resolution. Completeness is obtained by defining a special Gentzen system whose completeness follows easily from the results of Chapter 3, and giving an algorithm for converting proofs in the special Gentzen systems into resolution refutations. Some complexity issues are also examined.

Chapter 5 is devoted to first-order logic. The syntax and semantics are presented. This includes the notions of first-order languages, structures, and models. Gentzen systems are extended to deal with quantifiers and equality. The concept of a Hintikka set is also introduced. It is shown that every Hintikka set is satisfiable in a model whose domain is (a quotient of) a term algebra. This result, together with a generalization of the search procedure, is used to derive the main theorems of first-order logic: completeness, compactness, model existence, and Löwenheim-Skolem theorems. One of the main themes in this chapter is that the search procedure is a Hintikka set builder.

Chapter 6 is devoted to Gentzen's cut elimination theorem and some of its applications. A simple semantic proof derived from the completeness theorem is given for the Gentzen system LK. An entirely proof-theoretic (constructive) argument is also given for a simpler system $G1^{nnf}$. This proof due to Schwichtenberg has the advantage that it also yields a precise upper bound on the length of cut-free proofs obtained from a proof with cut. This result is then extended to a system with equality.

A constructive proof of Craig's interpolation theorem is given, and Beth's definability theorem and Robinson's joint consistency theorem are also proved. This chapter contains more advanced material than the previous chapters.

Chapter 7 is devoted to Gentzen's sharpened Hauptsatz, Herbrand's theorem, and the Skolem-Herbrand-Gödel theorem. As in Chapter 6, this chapter contains more advanced material. Gentzen's sharpened Hauptsatz for prenex sequents is proved constructively, using proof transformation techniques. A version of the sharpened Hauptsatz is also proved constructively for sequents consisting of formulae in NNF. To prove this result, a new Gentzen system with quantifier rules applying to certain subformulae is defined. This version of the sharpened Hauptsatz for sequents in NNF appears to be new. Using these results, constructive versions of Herbrand's theorem are proved, as well as Andrews's version of the Skolem-Herbrand-Gödel theorem (Andrews, 1981). The class of primitive recursive functions and the class of recursive functions are also briefly introduced.

In Chapter 8, the resolution method for first-order logic is presented. A recursive unification algorithm inspired from Robinson's algorithm (Robinson, 1965) is presented. Using results from Chapter 4 and the Skolem-Herbrand-Gödel theorem, the completeness of first-order resolution is shown, using the lifting technique. Paramodulation is also briefly discussed.

Chapter 9 is devoted to SLD-resolution and the foundations of PROLOG. Using techniques from Chapter 4, the completeness of SLD-resolution is shown, by translating proofs in a certain Gentzen system into SLD-refutations. This approach appears to be new. Logic programs are defined, and a model-theoretic semantics is given. It is shown that SLD-resolution is a sound and complete computational proof procedure for logic programs. Most of this material can only be found in research papers, and should be useful to readers interested in logic programming.

In Chapter 10 (the last chapter), a brief presentation of many-sorted first-order logic is given. This presentation should be sufficient preparation for readers interested in the definition of abstract data types, or computing with rewrite rules. Finally, an extension of the congruence closure method of Nelson and Oppen (Nelson and Oppen, 1980) to the many-sorted case and its application to fast decision procedures for testing the validity of quantifier-free formulae are presented.

This book grew out of a number of class notes written for a graduate course in logic for computer scientists, taught at the University of Pennsylvania. The inspiration for writing the book came from Sheila Greibach (my advisor at UCLA) and Ronald Book (my "supervisor" at UCSB, while I was a "Post Doc"), who convinced me that there is no better way to really know a topic than to write about it.

I wish to thank my colleagues Saul Gorn, Dale Miller, and Alex Pelin for reading the manuscript very carefully and for their many helpful comments. I also wish to thank my students William Dowling, Thomas Isakowitz, Harry Kaplan, Larry Krablin, Francois Lang, Karl Schimpf, Jeff Stroomer, Stan Raatz, and Todd Rockoff for their help in "debugging" the manuscript. This includes reporting of typos, stylistic improvements, additional exercises, and correction of mistakes. In addition, my appreciation goes to the reviewers of the manuscript: Robert S. Boyer, University of Texas, Austin; Nathan Friedman, McGill University; Albert A. Grau, Northwestern University; Ray Gumb, New Mexico Tech; Kim King, Georgia Tech University; James Miller, Bradley University; J. Strother Moore, University of Texas, Austin; Andrzej Proskurowski, University of Oregon; William J. Rapaport, SUNY, Buffalo; Wilfred Sieg, Columbia University.

Jean H. Gallier

HOW TO USE THIS BOOK AS A TEXT

This book is written at the level appropriate to senior undergraduate and first-year graduate students in computer science or mathematics. The prerequesites are the equivalent of undergraduate-level courses in either set theory, abstract algebra, or discrete structures. All the mathematical background necessary for the text itself is contained in Chapter 2, and in the Appendix. Some of the most difficult exercises may require deeper knowledge of abstract algebra.

Most instructors will find it convenient to use Chapter 2 on a call by need basis, depending on the background of the students. However, to the author's experience, it is usually desirable to review the material contained in Sections 2.1, 2.2, and 2.3.

To help the instructor make up a course, we give below a graph showing the dependence of the sections and chapters. This graph only applies to the text itself and not to the exercises, which may depend on any earlier sections.

The core of the subject that, in the author's opinion, should be part of any course on logic for computer science, is composed of Sections 3.1, 3.2, 3.3 (excluding 3.3.5), 3.4, 3.5, 5.1, 5.2, 5.3, 5.4, and 5.5. The sections that are next in priority (as core sections) are 3.6, 5.6, 6.1, 6.2, 6.3, 7.1, 7.2, 7.3, and 7.5. More advanced topics suitable for seminars are covered in Sections 6.4, 6.5, 6.6, 6.7, 7.4, 7.6, and in Chapter 10. Sections marked with a star (*) give a glimpse of topics only sketched in this book. They can be omitted at first reading.

Some results from Section 2.4 are required in Chapter 5. However, in order to shorten Chapter 2, this material as well as the material on many-sorted algebras has been made into an appendix. Similarly, to be perfectly rigorous, Chapter 8 depends on Section 7.6 (since the Skolem-Herbrand-Gödel theorem proved in Section 7.6 is used to prove the completeness of resolution). However, if the emphasis of the course is on theorem-proving techniques rather than on foundations, it is possible to proceed directly from Section 5.5 to Chapter 8 after having covered Chapter 4. The instructor may simply quote the Herbrand-Skolem-Gödel theorem from Section 7.6, without proof.

Hence, depending on the time available and the level of the class, there is flexibility for focusing more on automatic theorem-proving methods, or more on foundations. A one-semester course emphasizing theorem-proving techniques may consist of the core, plus Chapters 4, 8, and possibly part of Chapter 9. A one-semester course emphasizing foundations may consist of the core, plus Chapters 6 and 7.

The ideal situation is to teach the course in two semesters, with automatic theorem-proving techniques first. The second semester covers the foundations and finishes with a more complete coverage of Chapter 9, Chapter 10, and possibly some material on decision procedures or on rewrite rules.

It is also possible to use Chapters 6 and 7 as the core of a seminar on analytic versus nonanalytic proofs.

Problems are usually found at the end of each section. The problems range from routine to very difficult. Difficult exercises or exercises requiring knowledge of material not covered in the text are marked with a star (*). Very difficult exercises are marked with a double star (**). A few programming assignments have been included.

Some historical remarks and suggestions for further reading are included at the end of each chapter. Finally the end of a proof is indicated by the symbol ☐ (box). The word *iff* is used as an abbreviation for *if and only if*.

DEPENDENCY OF SECTIONS

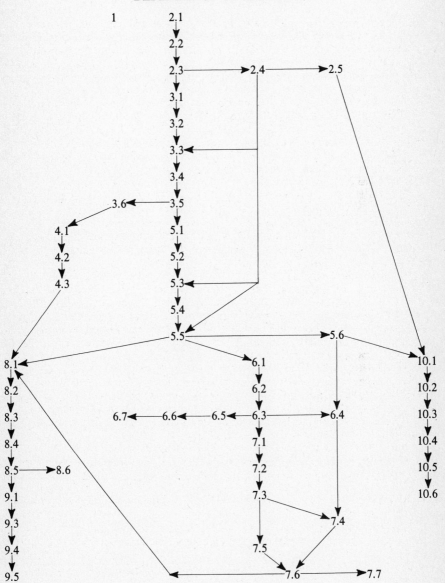

Chapter 1

Introduction

Logic is concerned mainly with two concepts: *truth* and *provability*. These concepts have been investigated extensively for centuries, by philosophers, linguists, and mathematicians. The purpose of this book is by no means to give a general account of such studies. Instead, the purpose of this book is to focus on a mathematically well defined logical system known as *first-order logic* (and, to some extent, *many-sorted logic*), and prove some basic properties of this system. In particular, we will focus on algorithmic methods for proving theorems (often referred to as *automatic theorem proving*).

Every logical system consists of a *language* used to write statements also called *propositions* or *formulae*. Normally, when one writes a formula, one has some intended *interpretation* of this formula in mind. For example, a formula may assert a true property about the natural numbers, or some property that must be true in a data base. This implies that a formula has a well-defined *meaning* or *semantics*. But how do we define this meaning precisely? In logic, we usually define the meaning of a formula as its *truth value*. A formula can be either true (or valid) or false.

Defining rigorously the notion of truth is actually not as obvious as it appears. We shall present a concept of truth due to Tarski. Roughly speaking, a formula is true if is is satisfied in all possible interpretations. So far, we have used the intuitive meaning of such words as *truth, interpretation*, etc. One of the objectives of this book is to define these terms rigorously, for the language of first-order logic (and many-sorted first-order logic). The branch of logic in which abstract structures and the properties true in these structures are studied is known as *model theory*.

Once the concept of truth has been defined rigorously, the next question is to investigate whether it is possible to find methods for deciding in a finite number of steps whether a formula is true (or valid). This is a very difficult task. In fact, by a theorem due to Church, there is no such general method for first-order logic.

However, there is another familiar method for testing whether a formula is true: to give a *proof* of this formula.

Of course, to be of any value, a proof system should be *sound*, which means that every provable formula is true.

We will also define rigorously the notion of proof, and proof system for first-order logic (and many-sorted first-order logic). The branch of logic concerned with the study of proof is known as *proof theory*.

Now, if we have a sound proof system, we know that every provable formula is true. Is the proof system strong enough that it is also possible to prove every true formula (of first-order logic)?

A major theorem of Gödel shows that there are logical proof systems in which every true formula is provable. This is referred to as the *completeness* of the proof system.

To summarize the situation, if one is interested in algorithmic methods for testing whether a formula of first-order logic is valid, there are two logical results of central importance: one positive (Gödel's completeness theorem), the other one negative (Church's undecidability of validity). Roughly speaking, Gödel's completeness theorem asserts that there are logical calculi in which every true formula is provable, and Church's theorem asserts that there is no decision procedure (procedure which always terminates) for deciding whether a formula is true (valid). Hence, any algorithmic procedure for testing whether a formula is true (or equivalently, by Gödel's completeness theorem, provable in a complete system) must run forever when given certain non-true formulae as input.

This book focuses on Gödel's positive result and its applications to automatic theorem proving. We have attempted to present a coherent approach to automatic theorem proving, following a main thread: Gentzen-like sequent calculi. The restriction to the positive result was dictated mostly by the lack of space. Indeed, it should be stressed that Church's negative result is also important, as well as other fundamental negative results due to Gödel. However, the omission of such topics should not be a severe inconvenience to the reader, since there are many texts covering such material (see the notes at the end of Chapter 5).

In spite of the theoretical limitation imposed by Church's result, the goal of automatic theorem proving (for short, atp) is to find *efficient* algorithmic methods for finding proofs of those formulae that are true.

A fairly intuitive method for finding such algorithms is the completeness proof for Gentzen-like sequent calculi. This approach yields a complete procedure (the *search* procedure) for proving valid formulae of first-order logic.

However, the *search* procedure usually requires an enormous amount of space and time and it is not practical. Hence, we will try improve it or find more efficient proof procedures.

For this, we will analyze the structure of proofs carefully. Fundamental results of Gentzen and Herbrand show that if a formula is provable, then it has a proof having a certain form, called a *normal form*.

The existence of such normal forms can be exploited to reduce the size of the search space that needs to be explored in trying to find a proof. Indeed, it is sufficient to look for proofs in normal form.

The existence of normal forms is also fundamental because it reduces the problem of finding a proof of a first-order formula to the problem of finding a proof of a simpler type of formula, called a proposition. Propositions are much simpler than first-order formulae. Indeed, there are algorithms for deciding truth. One of the methods based on this reduction technique is the *resolution method*, which will be investigated in Chapters 4 and 8.

Besides looking for general methods applying to the class of all true (first-order) formulae, it is interesting to consider subclasses for which simpler or more efficient proof procedures exist. Indeed, for certain subclasses there may be decision procedures. This is the case for propositions, and for quantifier-free formulae. Such cases are investigated in Chapters 3 and 10 respectively.

Unfortunately, even in cases in which algorithms exist, another difficulty emerges. A decision procedure may take too much time and space to be practical. For example, even testing whether a proposition is true may be very costly. This will be discussed in Chapter 3.

Automatic theorem proving techniques can be used by computer scientists to axiomatize structures and prove properties of programs working on these structures. Another recent and important role that logic plays in computer science, is its use as a *programming language* and as a *model of computation*. For example, in the programming language PROLOG, programs are specified by sets of assertions. In such a programming language, a computation is in fact a proof, and the output of a program is extracted from the proof. Promoters of such languages claim that since such programs are essentially logical formulae, establishing their correctness is trivial. This is not quite so, because the concept of correctness is relative, and the semantics of a PROLOG program needs to be expressed in a language other than PROLOG. However, using logic as a vehicle for programming is a very interesting idea and should be a selling point for skeptics. This use of logic will be investigated in Chapter 9.

Chapter 2

Mathematical Preliminaries

This chapter is devoted to mathematical preliminaries. This fairly lengthy chapter has been included in order to make this book as self-contained as possible. Readers with a firm mathematical background may skim or even skip this chapter entirely. Classroom experience shows that anyone who is not acquainted with the material included in Section 2.3 should probably spend some time reading Sections 2.1 to 2.3. In any case, this chapter can be used as a library of useful facts and may be consulted whenever necessary.

Since trees, inductive definitions and the definition of functions by recursion play an important role in logic, they will be defined carefully. First, we review some basic concepts and establish the terminology and notation used in this book. It is assumed that the reader is familiar with the basic properties of sets. For more details, the reader may consult Enderton, 1972; Enderton, 1977; Lewis and Papadimitriou, 1981; or Suppes, 1972.

2.1 Relations, Functions, Partial Orders, Induction

First, we review the concepts of Cartesian product, tuple and relation.

2.1.1 Relations

Given two sets A and B (possibly empty), their *Cartesian product* denoted by $A \times B$ is the set of ordered pairs

$$\{<a,b> \mid a \in A,\ b \in B\}.$$

Given any finite number of sets $A_1,...,A_n$, the *Cartesian product* $A_1 \times ... \times A_n$ is the set of ordered n-tuples

$$\{< a_1, ..., a_n > \ | \ a_i \in A_i, 1 \leq i \leq n\}$$

(An ordered n-tuple $< a_1, ..., a_n >$ is also denoted by $(a_1, ..., a_n)$.)

A *binary relation between A and B* is any subset R (possibly empty) of $A \times B$.

Given a relation R between A and B, the set

$$\{x \in A \ | \ \exists y \in B \ < x, y > \in R\},$$

is called the *domain* of R and denoted by $dom(R)$. The set

$$\{y \in B \ | \ \exists x \in A \ < x, y > \in R\}$$

is called the *range* of R and is denoted by $range(R)$.

When $A = B$, a relation R beween A and A is also called a relation *on* (or *over*) A. We will also use the notation xRy as an alternate to $(x, y) \in R$.

2.1.2 Partial Functions, Total Functions

A relation R between two sets A and B is *functional* iff, for all $x \in A$, and $y, z \in B$, $(x, y) \in R$ and $(x, z) \in R$ implies that $y = z$.

A *partial function* is a triple $f =< A, G, B >$, where A and B are arbitrary sets (possibly empty) and G is a functional relation (possibly empty) between A and B, called the *graph* of f.

Hence, a partial function is a functional relation such that every argument has at most one image under f. The graph of a function f is denoted as $graph(f)$. When no confusion can arise, a function f and its graph are usually identified.

A partial function $f =< A, G, B >$ is often denoted as $f : A \to B$. For every element x in the domain of a partial function f, the unique element y in the range of f such that $(x, y) \in graph(f)$ is denoted by $f(x)$. A partial function $f : A \to B$ is a *total function* iff $dom(f) = A$. It is customary to call a total function simply a function.

2.1.3 Composition of Relations and Functions

Given two binary relations R between A and B, and S between B and C, their *composition* denoted by $R \circ S$ is a relation between A and C defined by the following set of ordered pairs:

$$\{(a, c) \ | \ \exists b \in B, (a, b) \in R \text{ and } (b, c) \in S\}.$$

Given a set A, the *identity relation* of A is denoted by I_A and is the relation $\{(x,x) \mid x \in A\}$. Note that I_A is also a total function.

Given a relation R between A and B, its *converse* is the relation between B and A denoted by R^{-1} defined by the set

$$\{(b,a) \in B \times A \mid (a,b) \in R\}.$$

Given two partial or total functions $f : A \to B$ and $g : B \to C$, with $f = <A, G_1, B>$ and $g = <B, G_2, C>$, their composition denoted by $f \circ g$ (or $f.g$, or fg), is the partial or total function defined by $<A, G_1 \circ G_2, C>$. Notice that according to our notation, $f \circ g(x) = g(f(x))$, that is, f is applied first. Note also that composition is associative.

2.1.4 Injections, Surjections, Bijections

A function $f : A \to B$ is *injective* (or *one to one*) iff, for all $x, y \in A$, $f(x) = f(y)$ implies that $x = y$.

A function $f : A \to B$ is *surjective* (or *onto*) iff, for all $y \in B$, there is some $x \in A$ such that $f(x) = y$. Equivalently, the range of f is the set B.

A function is *bijective* iff it is both injective and surjective.

It can be shown that a function f is surjective if and only if there exists a function $g : B \to A$ such that $g \circ f = I_B$. If there exists a function $g : B \to A$ such that $f \circ g = I_A$, then $f : A \to B$ is injective. If $f : A \to B$ is injective and $A \neq \emptyset$, then there exists a function $g : B \to A$ such that $f \circ g = I_A$. As a consequence, it can be shown that a function $f : A \to B$ is bijective if and only if there is a unique function f^{-1} called its *inverse* such that $f \circ f^{-1} = I_A$ and $f^{-1} \circ f = I_B$.

2.1.5 Direct Image, Inverse Image

Given a (partial) function $f : A \to B$, for every subset X of A, the *direct image* (or for short, *image*) of X under f is the set

$$\{y \in B \mid \exists x \in X , f(x) = y\}$$

and is denoted by $f(X)$. For every subset Y of B, the *inverse image* of Y under f is the set

$$\{x \in A \mid \exists y \in Y, f(x) = y\}$$

and is denoted by $f^{-1}(Y)$.

Warning: The function f may not have an inverse. Hence, $f^{-1}(Y)$ should not be confused with $f^{-1}(y)$ for $y \in B$, which is only defined when f is a bijection.

2.1.6 Sequences

Given two sets I and X, an *I-indexed sequence* (or *sequence*) is any function $A : I \to X$, usually denoted by $(A_i)_{i \in I}$. The set I is called the *index set*. If X is a set of sets, $(A_i)_{i \in I}$ is called a *family* of sets.

2.1.7 Natural Numbers and Countability

The set of *natural numbers* (or nonnegative integers) is denoted by $\mathbf{N}$ and is the set $\{0, 1, 2, 3, ...\}$. A set A is *countable* (or *denumerable*) iff there is a surjection $h : \mathbf{N} \to A$ from $\mathbf{N}$ onto A. Otherwise, A is said to be *uncountable*. A set A is *countably infinite* iff there is a bijection $h : \mathbf{N} \to A$. The *cardinality* of a countably infinite set is denoted by ω. The set of positive integers is denoted by $\mathbf{N}_+$. For every positive integer $n \in \mathbf{N}_+$, the set $\{1, ..., n\}$ is denoted as $[n]$, and $[0]$ denotes the empty set. A set A is *finite* iff there is a bijection $h : [n] \to A$ for some natural number $n \in \mathbf{N}$. The natural number n is called the *cardinality* of the set A, which is also denoted by $|A|$. When I is the set $\mathbf{N}$ of natural numbers, a sequence $(A_i)_{i \in I}$ is called a *countable sequence*, and when I is some set $[n]$ with $n \in \mathbf{N}$, $(A_i)_{i \in I}$ is a *finite sequence*.

2.1.8 Equivalence Relations

A binary relation $R \subset A \times A$ is *reflexive* iff for all $x \in A$, $(x, x) \in R$.

The relation R is *symmetric* iff for all $x, y \in A$, $(x, y) \in R$ implies that $(y, x) \in R$.

The relation R is *transitive* iff for all $x, y, z \in A$, $(x, y) \in R$ and $(y, z) \in R$ implies that $(x, z) \in R$.

The relation R is an *equivalence relation* if it is reflexive, symmetric and transitive. Given an equivalence relation R on a set A, for every $x \in A$, the set $\{y \in A \mid (x, y) \in R\}$ is the *equivalence class of x modulo R* and is denoted by $[x]_R$, or $\overline{x}_R$, or simply $[x]$ or $\overline{x}$. The set of equivalence classes modulo R is the *quotient of A by R* and is denoted by A/R. The set A/R is also called a *partition* of A, since any two distinct equivalence classes are nonempty and disjoint, and their union is A itself. The surjective function $h_R : A \to A/R$ such that $h_R(x) = [x]_R$ is called the *canonical function* associated with R.

Given any relation R on a set A, we define the *powers* of R as follows: For every integer $n \geq 0$,

$$R^0 = I_A, \ R^1 = R, \text{ and } R^{n+1} = R^n \circ R.$$

The union

$$R^+ = \bigcup_{n \geq 1} R^n$$

called the *transitive closure* of R is the smallest transitive relation on A containing R, and

$$R^* = \bigcup_{n \geq 0} R^n$$

is called the *reflexive and transitive closure* of R and is the smallest reflexive and transitive relation on A containing R. It is obvious that $R^+ = R \circ R^* = R^* \circ R$, and that $R^* = I_A \cup R^+$. Thus, it can also be shown that for any relation R on a set A, $(R \cup R^{-1})^*$ is the least equivalence relation containing R.

2.1.9 Partial and Total Orders

A relation R on a set A is *antisymmetric* iff for all $x, y \in A$, $(x, y) \in R$ and $(y, x) \in R$ implies that $x = y$.

A relation R on a set A is a *partial order* iff R is reflexive, transitive and antisymmetric.

Given a partial order R on a set A, the pair $< A, R >$ is called a *partially ordered set* (or *poset*). A partial order is often denoted by the symbol $\leq$.

Given a partial order $\leq$ on a set A, given any subset X of A, X is a *chain* iff for all $x, y \in X$, either $x \leq y$, or $y \leq x$.

A partial order $\leq$ on a set A is a *total order* (or a *linear* order) iff A is a chain.

Given a partial order $\leq$ on a set A, given any subset X of A, an element $b \in A$ is a *lower bound* of X iff for all $x \in X$, $b \leq x$. An element $m \in A$ is an *upper bound* of X iff for all $x \in X$, $x \leq m$. Note that b or m may or may not belong to X. It can be easily shown that a lower bound (resp. upper bound) of X in X is unique. Hence the following definition is legitimate.

An element $b \in X$ is *the least element of* X iff for all $x \in X$, $b \leq x$. An element $m \in X$ is *the greatest element of* X iff for all $x \in X$, $x \leq m$. In view of the above remark, least and greatest elements are unique (when they exist).

Given a subset X of A, an element $b \in X$ is *minimal in* X iff for all $x \in X$, $x \leq b$ implies that $x = b$. An element $m \in X$ is *maximal in* X if for all $x \in X$, $m \leq x$ implies that $m = x$. Contrary to least and greatest elements, minimal or maximal elements are not necessarily unique.

An element $m \in A$ is *the least upper bound* of a subset X, iff the set of upper bounds of X is nonempty, and m is the least element of this set. An element $b \in A$ is *the greatest lower bound* of X if the set of lower bounds of X is nonempty, and b is the greatest element of this set.

Although the following fundamental result known as Zorn's lemma will not be used in the main text, it will be used in some of the problems. Hence, this result is stated without proof. For details and the proof, the reader is referred to Suppes, 1972; Levy, 1979; or Kuratowski and Mostowski, 1976.

Theorem 2.1.1 (Zorn's lemma) Given a partially ordered set $< A, \leq >$, if every (nonempty) chain in A has an upper bound, then A has some maximal element.

2.1.10 Well-Founded Sets and Complete Induction

A very general induction principle holds for the class of partially ordered sets having a well-founded ordering. Given a partial order $\leq$ on a set A, the *strict order* $<$ associated with $\leq$ is defined as follows:

$x < y$ if and only if $x \leq y$ and $x \neq y$.

A partially ordered set $< A, \leq >$ is *well-founded* iff it has no infinite decreasing sequence $(x_i)_{i \in \mathbf{N}}$, that is, sequence such that $x_{i+1} < x_i$ for all $i \geq 0$.

The following property of well-founded sets is fundamental.

Lemma 2.1.1 Given a partially ordered set $< A, \leq >$, $< A, \leq >$ is a well-founded set if and only if every nonempty subset of A has a minimal element.

Proof: First, assume that $< A, \leq >$ is well-founded. We proceed by contradiction. Let X be any nonempty subset of A, and assume that X does not have a minimal element. This means that for any $x \in X$, there is some $y \in X$ such that $y < x$, since otherwise there would be some minimal $x \in X$. Since X is nonempty, there is some x_0 in X. By the above remark, there is some $x_1 \in X$ such that $x_1 < x_0$. By repeating this argument (using induction on $\mathbf{N}$), an infinite decreasing sequence (x_i) can be defined in X, contradicting the fact that A is well-founded. Hence, X must have some minimal element.

Conversely, assume that every nonempty subset has a minimal element. If an infinite decreasing sequence (x_i) exists in A, (x_i) has some minimal element x_k. But this contradicts the fact that $x_{k+1} < x_k$. $\square$

The principle of *complete induction* (or *structural induction*) is now defined. Let $(A, \leq)$ be a well-founded poset, and let P be a property of the set A, that is, a function $P : A \rightarrow \{\mathbf{false}, \mathbf{true}\}$. We say that $P(x)$ *holds* if $P(x) = \mathbf{true}$.

Principle of Complete Induction

To prove that a property P holds for all $z \in A$, it suffices to show that, for every $x \in A$,

($*$) if $P(y)$ holds for all $y < x$,

($**$) then $P(x)$ holds.

The statement ($*$) is called the *induction hypothesis*, and the implication for all x, $(*)$ implies $(**)$

is called the *induction step*. Formally, the induction principle can be stated as:

(CI) $\qquad (\forall x \in A)[(\forall y \in A)(y < x \supset P(y)) \supset P(x)] \supset (\forall z \in A)P(z)$

Note that $P(x)$ has to be shown to be **true** for every minimal element x. These cases are called the *base cases*.

Complete induction is not valid for arbitrary posets (see the problems) but holds for well-founded sets as shown in the following lemma.

Lemma 2.1.2 The principle of complete induction holds for every well-founded set.

Proof: We proceed by contradiction. Assume that (CI) is false. Then,

(1) $\qquad (\forall x \in A)[(\forall y \in A)(y < x \supset P(y)) \supset P(x)]$

is **true** and

(2) $\qquad\qquad\qquad (\forall z \in A)P(z)$

is false, that is,

$$(\exists z \in A)(P(z) = \textbf{false})$$

is **true**.

Hence, the subset X of A defined by

$$X = \{x \in A \mid P(x) = \textbf{false}\}$$

is nonempty. Since A is well founded, by lemma 2.1.1, X has some minimal element b. Since (1) is **true** for all $x \in A$, letting $x = b$,

(3) $\qquad\qquad [(\forall y \in A)(y < b \supset P(y)) \supset P(b)]$

is **true**. If b is also minimal in A, there is no $y \in A$ such that $y < b$ and so,

$$(\forall y \in A)(y < b \supset P(y))$$

holds trivially and (3) implies that $P(b) = \textbf{true}$, which contradicts the fact that $b \in X$. Otherwise, for every $y \in A$ such that $y < b$, $P(y) = \textbf{true}$, since otherwise y would belong to X and b would not be minimal. But then,

$$(\forall y \in A)(y < b \supset P(y))$$

also holds and (3) implies that $P(b) = \textbf{true}$, contradicting the fact that $b \in X$. Hence, complete induction is valid for well-founded sets. $\square$

As an illustration of well-founded sets, we define the *lexicographic ordering*. Given a partially ordered set $(A, \leq)$, the *lexicographic ordering* $<<$ on $A \times A$ induced by $\leq$ is defined a follows: For all $x, y, x', y' \in A$,

$$(x, y) << (x', y') \text{ if and only if either}$$
$$x = x' \text{ and } y = y', \quad \text{or}$$
$$x < x' \quad \text{or}$$
$$x = x' \text{ and } y < y'.$$

We leave as an exercise the check that $<<$ is indeed a partial order on $A \times A$. The following lemma will be useful.

Lemma 2.1.3 If $< A, \leq>$ is a well-founded partially ordered set, the lexicographic ordering $<<$ on $A \times A$ is also well founded.

Proof: We proceed by contradiction. Assume that there is an infinite decreasing sequence $(< x_i, y_i >)_{i \in \mathbf{N}}$ in $A \times A$. Then, either,

(1) There is an infinite number of distinct x_i, or

(2) There is only a finite number of distinct x_i.

In case (1), the subsequence consisting of these distinct elements forms a decreasing sequence in A, contradicting the fact that $\leq$ is well founded. In case (2), there is some k such that for all $i \geq k$, $x_i = x_{i+1}$. By definition of $<<$, the sequence $(y_i)_{i \geq k}$ is a decreasing sequence in A, contradicting the fact that $\leq$ is well founded. Hence, $<<$ is well founded on $A \times A$. $\square$

As an illustration of the principle of complete induction, consider the following example in which it is shown that a function defined recursively is a total function.

EXAMPLE 2.1.1

(Ackermann's function) The following function over $\mathbf{N} \times \mathbf{N}$ known as *Ackermann's function* is well known in recursive function theory for its extraordinary rate of growth. It is defined recursively as follows:

$$A(x, y) = if \ x = 0 \ then \ y + 1$$
$$else \ if \ y = 0 \ then \ A(x - 1, 1)$$
$$else \ A(x - 1, A(x, y - 1))$$

It is actually not obvious that such a recursive definition defines a partial function, but this can be shown. The reader is referred to Machtey and Young, 1978; or Rogers, 1967, for more details.

We wish to prove that A is a total function. We proceed by complete induction over the lexicographic ordering on $\mathbf{N} \times \mathbf{N}$.

The base case is $x = 0$, $y = 0$. In this case, since $A(0, y) = y + 1$, $A(0, 0)$ is defined and equal to 1.

The induction hypothesis is that for any (m, n), $A(m', n')$ is defined for all $(m', n') << (m, n)$, with $(m, n) \neq (m', n')$.

For the induction step, we have three cases:

(1) If $m = 0$, since $A(0, y) = y + 1$, $A(0, n)$ is defined and equal to $n + 1$.

(2) If $m \neq 0$ and $n = 0$, since $(m - 1, 1) << (m, 0)$ and $(m - 1, 1) \neq (m, 0)$, by the induction hypothesis, $A(m - 1, 1)$ is defined, and so $A(m, 0)$ is defined since it is equal to $A(m - 1, 1)$.

(3) If $m \neq 0$ and $n \neq 0$, since $(m, n - 1) << (m, n)$ and $(m, n - 1) \neq (m, n)$, by the induction hypothesis, $A(m, n - 1)$ is defined. Since $(m - 1, y) << (m, z)$ and $(m - 1, y) \neq (m, z)$ no matter what y and z are, $(m - 1, A(m, n - 1)) << (m, n)$ and $(m - 1, A(m, n - 1)) \neq (m, n)$, and by the induction hypothesis, $A(m - 1, A(m, n - 1))$ is defined. But this is precisely $A(m, n)$, and so $A(m, n)$ is defined. This concludes the induction step. Hence, $A(x, y)$ is defined for all $x, y \geq 0$. $\square$

2.1.11 Restrictions and Extensions

We define a partial ordering $\subseteq$ on partial functions as follows: $f \subseteq g$ if and only $graph(f)$ is a subset of $graph(g)$. We say that g is an *extension* of f and that f is a *restriction* of g. The following lemma will be needed later.

Lemma 2.1.4 Let $(f_n)_{n \geq 0}$ be a sequence of partial functions $f_n : A \to B$ such that $f_n \subseteq f_{n+1}$ for all $n \geq 0$. Then, $g = (A, \bigcup graph(f_n), B)$ is a partial function. Furthermore, g is the least upper bound of the sequence (f_n).

Proof: First, we show that $G = \bigcup graph(f_n)$ is functional. Note that for every $(x, y) \in G$, there is some n such that $(x, y) \in graph(f_n)$. If $(x, y) \in G$ and $(x, z) \in G$, then there is some m such that $(x, y) \in graph(f_m)$ and some n such that $(x, z) \in graph(f_n)$. Letting $k = max(m, n)$, since (f_n) is a chain, we have $(x, y) \in graph(f_k)$ and $(x, z) \in graph(f_k)$. But since $graph(f_k)$ is functional, we must have $y = z$. Next, the fact that each relation $graph(f_n)$ is contained in G is obvious since $G = \bigcup graph(f_n)$. If h is any partial function such that $graph(f_n)$ is a subset of $graph(h)$ for all $n \geq 0$, by definition of a union, $G = \bigcup graph(f_n)$ is a subset of h. Hence, g is indeed the least upper bound of the chain (f_n). $\square$

2.1.12 Strings

Given any set A (even infinite), a *string* over A is any finite sequence $u : [n] \to A$, where n is a natural number. It is customary to call the set A an *alphabet*.

Given a string $u : [n] \to A$, the natural number n is called the *length* of u and is denoted by $|u|$. For $n = 0$, we have the string corresponding to the unique function from the empty set to A, called the *null string* (or *empty string*), and denoted by e_A, or for simplicity by e when the set A is understood. Given any set A (even infinite), the set of all strings over A is denoted by A^*. If $u : [n] \to A$ is a string and $n > 0$, for every $i \in [n]$, $u(i)$ is some element of A also denoted by u_i, and the string u is also denoted by $u_1...u_n$.

Strings can be *concatenated* as follows. Given any two strings $u : [m] \to A$ and $v : [n] \to A$, $(m, n \geq 0)$, their *concatenation* denoted by $u.v$ or uv is the string $w : [m + n] \to A$ such that:

$$w(i) = \begin{cases} u(i) & \text{if } 1 \leq i \leq m; \\ v(i - m) & \text{if } m + 1 \leq i \leq m + n. \end{cases}$$

One verifies immediately that for every string u, $u.e = e.u = u$. In other words, viewing concatenation as an algebraic operation on the set A^* of all strings, e is an identity element. It is also obvious that concatenation is associative, but not commutative in general.

Given a string u, a string v is a *prefix* (or *head*) of u if there is a string w such that $u = vw$. A string v is a *suffix* (or *tail*) of u if there is a string w such that $u = wv$. A string v is a *substring* of u if there are strings x and y such that $u = xvy$. A prefix v (suffix, substring) of a string u is *proper* if $v \neq u$.

2.2 Tree Domains and Trees

In order to define finite or infinite trees, we use the concept of a *tree domain* due to Gorn (Gorn, 1965).

2.2.1 Tree Domains

A *tree domain* D is a nonempty subset of strings in $\mathbf{N}_+^*$ satisfying the conditions:

(1) For each $u \in D$, every prefix of u is also in D.

(2) For each $u \in D$, for every $i \in \mathbf{N}_+$, if $ui \in D$ then, for every j, $1 \leq j \leq i$, uj is also in D.

EXAMPLE 2.2.1

The tree domain

$$D = \{e, 1, 2, 11, 21, 22, 221, 222, 2211\}$$

is represented as follows:

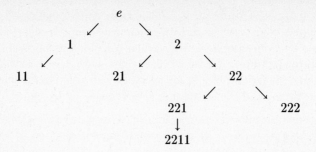

2.2.2 Trees

Given a set Σ of labels, a Σ-tree (for short, a *tree*) is a total function $t : D \to \Sigma$, where D is a tree domain.

The domain of a tree t is denoted by $dom(t)$. Every string u in $dom(t)$ is called a *tree address* or a *node*.

EXAMPLE 2.2.2

Let $\Sigma = \{f, g, h, a, b\}$. The tree $t : D \to \Sigma$, where D is the tree domain of example 2.2.1 and t is the function whose graph is

$$\{(e, f), (1, h), (2, g), (11, a), (21, a), (22, f), (221, h), (222, b), (2211, a)\}$$

is represented as follows:

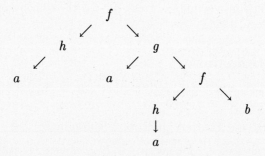

The *outdegree* (sometimes called *ramification*) $d(u)$ of a node u is the cardinality of the set $\{i \mid ui \in dom(t)\}$. Note that the outdegree of a node can be infinite. Most of the trees that we shall consider will be *finite-branching*, that is, for every node u, $d(u)$ will be an integer, and hence finite. A node of outdegree 0 is called a *leaf*. The node whose address is e is called the *root* of the tree. A tree is *finite* if its domain dom(t) is finite. Given a node u in $dom(t)$, every node of the form ui in $dom(t)$ with $i \in \mathbf{N}_+$ is called a *son* (or *immediate successor*) of u.

Tree addresses are totally ordered *lexicographically* as follows: $u \leq v$ if either u is a prefix of v or, there exist strings $x, y, z \in \mathbf{N}_+^*$ and $i, j \in \mathbf{N}_+$, with $i < j$, such that $u = xiy$ and $v = xjz$. In the first case, we say that u is an *ancestor* (or *predecessor*) of v (or u *dominates* v) and in the second case, that u is *to the left* of v. If $y = e$ and $z = e$, we say that xi is a *left brother* (or *left sibling*) of xj, $(i < j)$. Two tree addresses u and v are *independent* if u is not a prefix of v and v is not a prefix of u.

2.2.3 Paths

A *finite path* with *source* u and *target* v is a finite sequence of nodes $u_0, u_1, ..., u_n$ such that $u_0 = u$, $u_n = v$, and for all j, $1 \leq j \leq n$, $u_j = u_{j-1}i_j$ for some $i_j \in \mathbf{N}_+$. The *length of a path* $u_0, u_1, ..., u_n$ is n $(n \geq 0)$. When $n = 0$, we have the *null path* from u to u (of length 0). A *branch* (or *chain*) is a path from the root to a leaf. An *infinite path* with *source* u is an infinite sequence of nodes $u_0, u_1, ..., u_n, ...$, such that $u_0 = u$ and, for all $j \geq 1$, $u_j = u_{j-1}i_j$ for some $i_j \in \mathbf{N}_+$.

Given a finite tree t, the *height* of a node u in $dom(t)$ is equal to $max(\{length(p) \mid p \text{ is a path from } u \text{ to a leaf}\})$. The *depth* of a finite tree is the height of its root (the length of a longest path from the root to a leaf).

2.2.4 Subtrees

Given a tree t and a node u in $dom(t)$, the *subtree rooted at* u (also called *scope*) is the tree t/u whose domain is the set $\{v \mid uv \in dom(t)\}$ and such that $t/u(v) = t(uv)$ for all v in $dom(t/u)$.

Another important operation is the operation of tree replacement (or tree substitution).

2.2.5 Tree Replacement

Given two trees t_1 and t_2 and a tree address u in t_1, the *result of replacing* t_2 *at* u *in* t_1, denoted by $t_1[u \leftarrow t_2]$, is the function whose graph is the set of pairs

$$\{(v, t_1(v)) \mid u \text{ is not a prefix of } v\} \cup \{(uv, t_2(v)\}.$$

EXAMPLE 2.2.3

Let t_1 and t_2 be the trees defined by the following diagrams:

Tree t_1

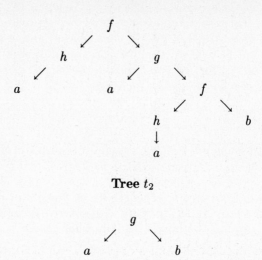

Tree t_2

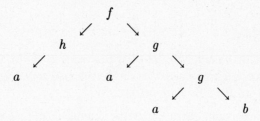

The tree $t_1[22 \leftarrow t_2]$ is defined by the following diagram:

2.2.6 Ranked Alphabets and Σ-Trees

In many situations, it is desirable to have a standard set of symbols to name operations taking a specified number of arguments. Such a set is called a *ranked alphabet* (or *simply stratified alphabet*, or *signature*).

A *ranked alphabet* is a set Σ together with a *rank function* $r : \Sigma \rightarrow \mathbf{N}$. Every symbol $f \in \Sigma$ has a *rank* (or *arity*) $r(f)$ indicating the fixed number of arguments of f. Symbols of arity 0 are also called *constants*. For every $n \geq 0$, Σ_n denotes the subset of Σ consisting of the function symbols of rank n.

If the set Σ of labels is a ranked alphabet, a Σ-*tree* is a function $t : dom(t) \rightarrow \Sigma$ as before, with the additional requirement that for every node u in $dom(t)$, $d(u) = r(t(u))$. In other words, the outdegree of a node is equal to the rank of its label.

EXAMPLE 2.2.4

Let $\Sigma = \{a, b, +, *\}$, where a, b have rank 0, and $+, *$ have rank 2. The following is a Σ-tree:

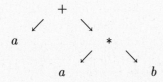

The set of all Σ-trees is denoted by CT_Σ and the set of all finite trees by T_Σ. Every one-node tree labeled with a constant a is also denoted by a.

2.3 Inductive Definitions

Most objects used in logic or computer science are *defined inductively*. By this we mean that we frequently define a set S of objects as:

The smallest set of objects containing a given set X of atoms, and closed under a given set F of constructors.

The purpose of this section is to define rigorously what the above sentence means.

2.3.1 Inductive Closures

Let us begin with an example.

EXAMPLE 2.3.1

Let $V = \{x_0, x_1, ...\}$ be a countable set of variables, let $X = V \cup \{0, 1\}$, let $+$ and $*$ two binary function symbols, let "(" denote the left parenthesis and ")" the right parenthesis. We wish to define the set $EXPR$ of arithmetic expressions defined using the variables in V, the constants 0,1, and the operators $+$ and $*$. The following definition is often given:

An arithmetic expression E is one of the following expressions:

(1) A variable in V, or 0, or 1;

(2) If E_1 and E_2 are arithmetic expressions, then so are $(E_1 + E_2)$ and $(E_1 * E_2)$;

(3) An expression is an arithmetic expression only if it is obtained by applications of clauses (1) and (2).

In such a definition called an *inductive definition*, clause (1) defines the atoms, clause (2) asserts some closure conditions, and clause (3) is supposed to assert that the set $EXPR$ of arithmetic expressions is the smallest set of expressions containing the atoms and closed under the operations described in (2). However, it is by no means clear that (1),(2),(3) really define a set, and that this set is the smallest set having properties defined by clauses (1) and (2).

The problem with the above definition is that the universe of all possible expressions is not defined, and that the operations defined by (2) are not clearly defined either. This can be remedied as follows. Let Σ be the alphabet $V \cup \{0, 1, (,), +, *\}$, and $A = \Sigma^*$ be the set of all strings over Σ. The set A is the universe of all possible expressions. Note that A contains a lot of expressions that are not arithmetic expressions, and the purpose of the above inductive definition is to define the subset $EXPR$ of A^* describing exactly all arithmetic expressions. We define the functions H_+ and H_* from $A \times A$ to A as follows: For all strings $u, v \in A$,

$$H_+(u, v) = (u + v)$$

$$H_*(u, v) = (u * v)$$

The string $(u + v)$ is the string obtained by concatenating the symbol "(", the string u, the symbol $+$, the string v, and the symbol ")", and similarly for the string $(u * v)$. Also, note that H_+ and H_* are defined for all strings in A, and not just legal arithmetic expressions. For example, if $u = 0$ and $v = *)$, $H_+(u, v) = (0 + *))$, which is not a legal arithmetic expression.

We say that a subset Y of A is *closed under H_+ and H_** if for all $u, v \in Y$, $H_+(u, v) \in Y$ and $H_*(u, v) \in Y$. We are now in a position to give a precise definition of the set $EXPR$ of arithmetic expressions. We define $EXPR$ as the least subset of A containing X and closed under H_+ and H_*. The only remaining problem is that we have not shown that such a set actually exists. This can be shown in two ways that turn out to be equivalent as we will prove shortly. The first method which might be called a top-down method, is to observe that:

(1) The family C of all subsets Y of A that contain X and are closed under H_+ and H_* is nonempty, since A satisfies these properties;

(2) Given any family of subsets of A containing X and closed under H_+ and H_*, the intersection of this family also contains X and is closed under H_+ and H_*.

Hence, the least subset X^+ of A containing X and closed under H_+ and H_* is the intersection of the family C.

The bottom-up method is to define a sequence $EXPR_i$ of subsets of A by induction as follows:

$$EXPR_0 = V \cup \{0, 1\};$$

$$EXPR_{i+1} = EXPR_i \cup \{H_+(u, v), H_*(u, v) | u, v \in EXPR_i\}, \text{ for } i \geq 0.$$

We let $X_+ = \bigcup EXPR_i$. We shall show below that $X_+ = X^+$ and therefore, $EXPR$ is equal to X_+.

Generalizing the method described in example 2.3.1, we give the following general definition.

Let A be a set, $X \subset A$ a subset of A, and F a set of functions $f : A^n \to A$, each having some arity $n > 0$. We say that a set Y is *inductive on X*, iff X is a subset of Y and Y is *closed under the functions in F*, that is: For every function $f : A^n \to A$ in F, for every $y_1, ..., y_n \in Y$, $f(y_1, ..., y_n)$ is also in Y. Clearly, A itself is inductive on X. The intersection of all inductive sets on X is also closed under F and it is called the *inductive closure of X under F*. Let us denote the inductive closure of X by X^+.

If X is nonempty, since every inductive set on X contains X and there is at least one inductive set on X (namely A), X^+ is nonempty. Note that X^+ is the *least inductive set* containing X. The above definition is what we might call a top-down definition. Frequently, X^+ is called the least set containing X and closed under F. There is also a bottom-up and more constructive way of charaterizing X^+. The sequence of sets $(X_i)_{i \geq 0}$ is defined by induction as follows:

$X_0 = X$ and

$$X_{i+1} = X_i \cup \{f(x_1, ..., x_n) \mid f \in F, x_1, ..., x_n \in X_i, n = r(f)\}.$$

It is clear that $X_i \subseteq X_{i+1}$ for all $i \geq 0$. Let

$$X_+ = \bigcup_{i \geq 0} X_i.$$

Lemma 2.3.1 $X^+ = X_+$.

Proof: First we show that X_+ is inductive on X. Since $X_0 = X$, X_+ contains X. Next, we show that X_+ is closed under F. For every f in F of arity $n > 0$ and for all $x_1, ..., x_n \in X_+$, by definition of X_+ there is some i such that $x_1, ..., x_n$ are all in X_i, and since $f(x_1, ..., x_n) \in X_{i+1}$ (by definition), $f(x_1, ..., x_n) \in X_+$. Since X_+ is inductive on X and X^+ is the least inductive set containing X, X^+ is a subset of X_+.

To prove that X_+ is a subset of X^+, we prove by induction that X_i is a subset of X^+ for every $i \geq 0$. But this is obvious since X^+ is closed under F. Hence, we have shown that $X^+ = X_+$. $\square$

The following *induction principle* for inductive sets is very useful:

Induction Principle for Inductive Sets

If X_+ is the inductive closure of X under F, for every subset Y of X_+, if Y contains X and is closed under F, then $Y = X_+$.

Lemma 2.3.2 The induction principle for inductive sets holds.

Proof: By hypothesis, Y is inductive on X. By lemma 2.3.1, $X_+ = X^+$ which is the least inductive set containing X. Hence, X_+ is a subset of Y. But Y is contained in X_+, so $Y = X_+$. $\square$

As an illustration of the induction principle, we prove that every arithmetic expression in $EXPR$ has the same number of left and right parentheses. Let Y be the subset of $EXPR$ consisting of all expressions having an equal number of left and right parentheses. Note that Y contains X since neither the variables nor 0 nor 1 contain parentheses. Y is closed under H_+ and H_* since these function introduce matching parentheses. Hence, by the induction principle, $Y = EXPR$.

2.3.2 Freely Generated Sets

One frequently needs to define functions *recursively* over an inductive closure. For example, one may want to define the process of evaluating arithmetic expressions.

EXAMPLE 2.3.2

Let E be the arithmetic expression $((x_0 + 1) * x_1)$. Assume that we want to evaluate the value of the expression E for the assignment to the variables given by $x_0 = 2$, $x_1 = 3$. Naturally, one will first compute the value of (x_0+1), which is $(2+1) = 3$, and then the value of $((x_0+1)*x_1)$ which is $(3 * 3) = 9$. Suppose that we now make the problem slightly more complicated. We want a method which, given any assignment $v : V \cup \{0,1\} \to \mathbf{N}$ of natural numbers to the variables such that $v(0) = 0$ and $v(1) = 1$, allows us to evaluate any expression E. The method is to evaluate expressions *recursively*. This means that we define the function $\widehat{v} : EXPR \to \mathbf{N}$ such that:

(0) $\widehat{v}(E) = v(x_i)$, if E is the variable x_i; $\widehat{v}(0) = 0$, $\widehat{v}(1) = 1$;

(1) $\widehat{v}(E) = \widehat{v}(E_1) + \widehat{v}(E_2)$, if E is $(E_1 + E_2)$;

(2) $\widehat{v}(E) = \widehat{v}(E_1) * \widehat{v}(E_2)$, if E is $(E_1 * E_2)$.

Note that $\widehat{v}$ is an extension of v, and in fact, it can be shown that it is the unique extension of v satisfying (1) and (2). However, it is not obvious that there is a function $\widehat{v}$ satisfying (0),(1),(2), and if such a function exists, it is not clear that it is unique. The existence and uniqueness of the function $\widehat{v}$ is a consequence of special properties of the inductive closure $EXPR$. In fact, given an inductive closure X_+ defined by a set X and a set F of functions, it is not always possible to define recursively a function extending a given function $v : X \to B$ (for some set B). We refer the reader to the problems of this chapter for a counter example. It turns out that functions are properly defined by recursion on an inductive closure exactly when this inductive closure is freely generated. The set $EXPR$ of expressions happens to be freely generated, and this is the reason functions are well defined by recursion.

To give an intuitive explanation of what freely generated means, observe that the bottom-up definition of X_+ suggests that each element of X_+ can be represented by a set of trees. Indeed, each atom, that is, each element x of X, is represented by the one-node tree labeled with that element, and each element $a = f(x_1, ..., x_n) \in X_{k+1}$ is represented by all trees of the form $f(t_1, ..., t_n)$, where each subtree t_i is any of the trees representing x_i. Each element of X_+ is usually represented by many different trees.

Roughly speaking, an inductive closure X_+ is freely generated by X and F if every element a of X_+ is represented by a *unique tree*.

EXAMPLE 2.3.3

Let $A = \{a, b, c\}$ and $* : A \times A \to A$ be the function defined by the following multiplication table:

$*$	a	b	c
a	a	b	c
b	b	c	a
c	c	a	b

Since $c = *(b, b)$ and $a = *(*(b, b), b)$, the inductive closure of $X = \{b\}$ is A. The element a is represented by the trees $*(b, c)$, $*(*(b, b), b)$, $*(*(a, b), c)$, $*(*(a, b), *(b, b))$, and in fact by infinitely many trees. As a consequence, A is not freely generated by X.

Technically, the definition of free generation is as follows.

Let A be a set, X a subset of A, F a set of functions on A, and X_+ the inductive closure of X under F. We say that X_+ is *freely generated by X and F* if the following conditions hold:

(1) The restriction of every function $f : A^m \to A$ in F to X_+^m is injective.

(2) For every $f : A^m \to A$, $g : A^n \to A$ in F, $f(X_+^m)$ is disjoint from $g(X_+^n)$ whenever $f \neq g$.

(3) For every $f : A^m \to A$ in F and every $(x_1, ..., x_m) \in X_+^m$, $f(x_1, ..., x_m) \notin X$.

Let $X_{-1} = \emptyset$. We now show the following lemma.

Lemma 2.3.3 If X_+ is freely generated by X and F, then for every $i \geq 0$, $X_{i-1} \neq X_i$ and $f(x_1, ..., x_n) \notin X_i$, for every f in F of arity n and every $(x_1, ..., x_n) \in X_i^n - X_{i-1}^n$.

Proof: We proceed by induction on $i \geq 0$. This is obvious for $i = 0$ since $X_{-1} = \emptyset$, $X_0 = X$ and by condition (3). For $i > 0$, we prove by induction on k, $0 \leq k \leq i$, that if $(x_1, ..., x_n) \in X_i^n - X_{i-1}^n$, then $f(x_1, ..., x_n) \notin X_k$. For $k = 0$, this follows from condition (3). Now, assume that if $(x_1, ..., x_n) \in X_i^n - X_{i-1}^n$, then $f(x_1, ..., x_n) \notin X_k$, for $0 \leq k \leq i - 1$. If $f(x_1, ..., x_n) \in X_{k+1}$, then

$f(x_1, ..., x_n) \in X_{k+1} - X_k$. By condition (2) and the definition of X_{k+1}, there is some $(y_1, ..., y_n)$ in X_k^n such that $f(x_1, ..., x_n) = f(y_1, ..., y_n)$. Since f is injective on X_+^n, we have $x_m = y_m$ for $1 \le m \le n$. Hence, we have $(x_1, ..., x_n) \in X_k^n$ for $k < i$, contradicting the hypothesis that $(x_1, ..., x_n) \in X_i^n - X_{i-1}^n$. Therefore, $f(x_1, ..., x_n) \notin X_{k+1}$, establishing the induction step on k. But this also shows that $X_i \ne X_{i+1}$, concluding the induction step on i. $\square$

It should be noted that conditions (1),(2),(3) apply to the *restrictions* of the functions in F to X_+. Indeed, there are cases in which the functions in F are not injective on A, and $f(A^m) \cap g(A^n) \ne \emptyset$ for distinct functions f, g, but conditions (1),(2),(3) hold and X_+ is freely generated. See problem 3.2.5. Lemma 2.3.3 can be used to formalize the statement that X_+ is freely generated by X and F iff every element has a unique tree representation. However, in order to define precisely what *representation by trees* means, it is necessary to show that trees are freely generated, and to define a function from trees to X_+ using theorem 2.3.1 proved next. For details of this representation, the reader is referred to the problems.

In logic, terms, formulae, and proofs are given by inductive definitions. Another important concept is that of a function defined recursively over an inductive set freely generated.

2.3.3 Functions Defined Recursively over Freely Generated Inductive Sets

Let A be a nonempty set, X a subset of A, F a set of functions on A, and X_+ the inductive closure of X under F. Let B be any nonempty set, and let G be a set of functions over the set B, such that there is a function $d : F \to G$ that associates with each function f of arity n in F, the function $d(f) : B^n \to B$ in G (d need not be a bijection).

Theorem 2.3.1 (Unique homomorphic extension theorem) If X_+ is freely generated by X and F, for every function $h : X \to B$, there is a unique function $\widehat{h} : X_+ \to B$ such that:

(1) For all $x \in X$, $\widehat{h}(x) = h(x)$;

For every function f of arity $n > 0$ in F, for every $x_1, ..., x_n \in X_+^n$,

(2) $\widehat{h}(f(x_1, ..., x_n)) = g(\widehat{h}(x_1), ..., \widehat{h}(x_n))$, where $g = d(f)$.

The diagram below illustrates the fact that $\widehat{h}$ extends h. The function η is the inclusion function of X into X_+.

$$
\begin{array}{ccc}
X & \xrightarrow{\;\eta\;} & X_+ \\
& {\scriptstyle h} \searrow & \downarrow {\scriptstyle \widehat{h}} \\
& & B
\end{array}
$$

The identities (1) and (2) mean that $\widehat{h}$ is a *homomorphism*, which is often called the *unique homomorphic extension* of h. Clause (2) can be described by the following commutative diagram:

$$
\begin{array}{ccc}
X_+^n & \xrightarrow{f} & X_+ \\
\widehat{h}^n \downarrow & & \downarrow \widehat{h} \\
B^n & \xrightarrow[d(f)]{} & B
\end{array}
$$

In the above diagram, the function $\widehat{h}^n$ is defined by $\widehat{h}^n(x_1, ..., x_n) = (\widehat{h}(x_1), ..., \widehat{h}(x_n))$. We say that this diagram is *commutative* if the composition $f \circ \widehat{h}$ is equal to the composition $\widehat{h}^n \circ g$.

Proof: We define by induction a sequence of functions $h_i : X_i \to B$ satisfying conditions (1) and (2) restricted to X_i. We set $h_0 = h$. Given h_i, let h_{i+1} have the graph:

$$\{(f(x_1, ..., x_n), g(h_i(x_1), ..., h_i(x_n))) \mid (x_1, ..., x_n) \in X_i^n - X_{i-1}^n, f \in F\} \cup$$
$$graph(h_i)$$

(with g=d(f).)

We have to check that this graph is indeed functional. Since X_+ is freely generated, by lemma 2.3.3, $f(x_1, ..., x_n) \in X_{i+1} - X_i$ whenever $(x_1, ..., x_n) \in X_i^n - X_{i-1}^n$, $(i \geq 0)$, and we only have to check functionality for the first part of the union. Since the elements of G are functions, by lemma 2.3.3, the only possibility for having $(x, y) \in graph(h_i)$ and $(x, z) \in graph(h_i)$ for some $x \in X_{i+1} - X_i$, is to have $x = f(x_1, ..., x_m) = f'(y_1, ..., y_n)$ for some $(x_1, ..., x_m) \in X_i^m - X_{i-1}^m$, $(y_1, ..., y_n) \in X_i^n - X_{i-1}^n$ and for some constructors f and f' in F. Since $f(X_+^m)$ and $f'(X_+^n)$ are disjoint whenever $f \neq f'$, $f(x_1, ..., x_m) = f'(y_1, ..., y_n)$ implies that $f = f'$ and $m = n$. Since every $f \in F$ is injective on X_+^n, we must also have $x_j = y_j$ for every j, $1 \leq j \leq n$. But then, $y = z = g(x_1, ..., x_n)$, with $g = d(f)$, showing functionality. Using lemma 2.1.4, $\widehat{h} = \bigcup_{i \geq 0} h_i$ is a partial function. Since $dom(\widehat{h}) = \bigcup dom(h_i) = \bigcup X_i = X_+$, $\widehat{h}$ is total on X_+. Furthermore, it is clear by definition of the h_i that $\widehat{h}$ satisfies (1) and (2). To prove that $\widehat{h}$ is unique, for any other function h' satisfying (1) and (2), it can be easily shown by induction that $\widehat{h}$ and h' agree on X_i for all $i \geq 0$. This proves the theorem. $\square$

EXAMPLE 2.3.4

Going back to example 2.3.2, the set A is Σ^*, the set F of functions is $\{H_+, H_*\}$, the set B is $\mathbf{N}$, the set G consists of *addition* and *multiplication* on the natural numbers, and the function $d : F \to G$ is given by $d(H_+) = addition$ and $d(H_*) = multiplication$. It can be shown that $EXPR$ is freely generated by $V \cup \{0, 1\}$ and $\{H_+, H_*\}$, but this is not obvious. Indeed, one has to prove rigorously conditions (1),(2),(3), for the functions H_+ and H_*, and this requires some work. A proof can be given by adapting the method used in theorem 3.2.1, and we leave it as an exercise. Since $EXPR$ is freely generated, for any function $v : V \cup \{0,1\} \to \mathbf{N}$ such that $v(0) = 0$ and $v(1) = 1$, by theorem 2.3.1, there is a unique function $\widehat{v}$ extending v which is a homomorphism.

Later on when we define satisfaction in first-order logic, we will need to define the concept of a *structure*, and we will have to reformulate slightly the notion of an inductive closure. This can be done conveniently by introducing the concept of an *algebra*. Since this material is only used in Chapter 5 and Chapter 10, it has been included in an appendix.

PROBLEMS

2.1.1. Show the following properties:

(a) If there exists a function $g : B \to A$ such that $f \circ g = I_A$, then $f : A \to B$ is injective. If $f : A \to B$ is injective and $A \neq \emptyset$, then there exists a function $g : B \to A$ such that $f \circ g = I_A$.

(b) A function f is surjective if and only if there exists a function $g : B \to A$ such that $g \circ f = I_B$.

(c) A function $f : A \to B$ is bijective if and only if there is a function f^{-1} called its *inverse* such that $f \circ f^{-1} = I_A$ and $f^{-1} \circ f = I_B$.

2.1.2. Prove that a function $f : A \to B$ is injective if and only if, for all functions $g, h : C \to A$, $g \circ f = h \circ f$ implies that $g = h$. A function $f : A \to B$ is surjective if and only if for all functions $g, h : B \to C$, $f \circ g = f \circ h$ implies that $g = h$.

2.1.3. Given a relation R on a set A, prove that R is transitive if and only if $R \circ R$ is a subset of R.

2.1.4. Given two equivalence relations R and S on a set A, prove that if $R \circ S = S \circ R$, then $R \circ S$ is the least equivalence relation containing R and S.

2.1.5. Prove that $R^+ = \bigcup_{n \geq 1} R^n$ is the smallest transitive relation on A containing R, and $R^* = \bigcup_{n \geq 0} R^n$ is the smallest reflexive and transitive relation on A containing R. Prove that for any relation R on a set A, $(R \cup R^{-1})^*$ is the least equivalence relation containing R.

2.1.6. Show that complete induction is not valid for posets that are not well-founded by giving a counter example.

2.1.7. Let $(A, \leq)$ and $(B, \leq')$ be two partially ordered sets. A function $f : A \to B$ is *monotonic* if, for all $a, b \in A$, $a \leq b$ implies $f(a) \leq' f(b)$.

(a) Show that the composition of monotonic functions is monotonic.

(b) Show that if f is monotonic and m is the least element of a subset S of A, then $f(m)$ is the least element of $f(S)$.

(c) Give a counter example showing that if f is monotonic and m is the least element of A, then $f(m)$ is not necessarily the least element of B. Give a counter example showing that if m is a minimal element of S, then $f(m)$ is not necessarily a minimal element of $f(S)$, even if f is surjective.

* **2.1.8.** Given a set A, a *multiset* over A is an unordered collection of elements of A that may have multiple occurrences of identical elements. Formally, a multiset over A may be defined as a function $M : A \to \mathbf{N}$, where $\mathbf{N}$ denotes the set of nonnegative integers. An element a in A has exactly n occurrences in M iff $M(a) = n$. In particular, a does not belong to M iff $M(a) = 0$. Let $M(A)$ denote the set of *finite* multisets over A, that is, the set of functions $M : A \to \mathbf{N}$ such that $M(a) \neq 0$ only for finitely many $a \in A$. Two (finite) multisets M and M' are equal iff every element occurring exactly n times in A also occurs exactly n times in B.

Multiset union and multiset difference is defined as follows: Given two multisets $M_1 : A \to \mathbf{N}$ and $M_2 : A \to \mathbf{N}$, their union is the multiset $M : A \to \mathbf{N}$, such that for all $x \in A$, $M(x) = M_1(x) + M_2(x)$. The *union* of M_1 and M_2 is also denoted as $M_1 \cup M_2$. The *difference* of M_1 and $\mathbf{M}_2$ is the multiset $M : A \to \mathbf{N}$ such that for all $x \in A$, $M(x) = M_1(x) - M_2(x)$ if $M_1(x) \geq M_2(x)$, $M(x) = 0$ otherwise. The difference of M_1 and M_2 is also denoted as $M_1 - M_2$. A multiset M_1 is a *submultiset* of a multiset M_2 if for all $x \in A$, $M_1(x) \leq M_2(x)$.

If A is partially ordered by $\leq$, the relation $<<$ on the set $M(A)$ of finite multisets is defined as follows:

$M << M'$ iff M is obtained from M' by removing zero or more elements from M', and replacing each such element x by zero or any finite number of elements from M', each strictly less than x (in the ordering $\leq$).

Formally, $M << M'$ iff $M = M'$, or there exist some finite multisets X, Y with X a nonempty submultiset of M' such that, $M = (M' - X) \cup Y$, and for all $y \in Y$, there is some $x \in X$ such that $y < x$.

(a) If $A = \mathbf{N} = \{0, 1, 2, ...\}$ and $\leq$ is the natural ordering on $\mathbf{N}$, give examples of pairs of multisets related by the relation $<<$.

(b) Prove that $<<$ is a partial ordering.

(c) Assume that $\leq$ is well founded. To prove that $<<$ is also well founded, we shall proceed by contradiction as follows. Assume that there is an infinite decreasing sequence $M_0 >> M_1 >> \ldots >>$ $M_n >> M_{n+1} >> \ldots$, and $M_n \neq M_{n+1}$ for all $n \geq 0$ ($M >> N$ holds iff $N << M$ holds). We build a tree in the following fashion:

Begin with a root node whose immediate descendants are labeled with the elements of M_0. Since $M_0 >> M_1$ (and $M_0 \neq M_1$), there exist multisets X and Y with X a nonempty multiset of M_0, such that $M_1 = (M_0 - X) \cup Y$, and for every $y \in Y$, there is some $x \in X$ and $y < x$. For each y in Y, choose some x in X such that $y < x$, and add a successor node labeled y to the node corresponding to that x. For every remaining x in X (element that is dropped and replaced by no elements at all), add a successor labeled with the special symbol $\bot$. This last step guarantees that at least one new node is added to the tree for every multiset M_n in the sequence. This is necessary in case Y is empty. Repeat the process for $M_1 >> M_2, M_2 >> M_3$, and so on. Let T be the resulting tree.

Note that by construction, the elements on any path form a strictly decreasing sequence in A (we can assume that $\bot$ is less than any element in A).

(i) Prove that the tree T is infinite and that each node has a finite number of successors. Then, by König's lemma (if a tree is finite branching and infinite then it contains an infinite path), there must be an infinite path in T.

(ii) Prove that there is a path in T corresponding to an infinite decreasing sequence of elements in A. Conclude that the partial ordering $<<$ is well founded.

2.2.1. Let t be a tree and let u and v be two independent tree addresses in $dom(t)$ (that is, u is not a prefix of v and v is not a prefix of u). Prove that for any trees t_1 and t_2,

$$t[u \leftarrow t_1][v \leftarrow t_2] = t[v \leftarrow t_2][u \leftarrow t_1].$$

2.3.1. Let $A = \{a, b, c\}$ and $* : A \times A \to A$ be the function defined by the following table:

$*$	a	b	c
a	a	b	c
b	b	c	a
c	c	a	b

(a) Show that the inductive closure of $X = \{b\}$ is A.

(b) If **N** denotes the set of nonnegative integers and $+$ is addition (of integers), show that there is some function $h : X \to \mathbf{N}$ which does not have any homomorphic extension to A (a function g is a homomorphic extension of h if, $g(b) = h(b)$ and $g(x*y) = g(x)+g(y)$, for all $x, y \in A$).

Find an infinite set that is the inductive closure of a finite set, but is not freely generated.

2.3.2. Show that if X_+ is freely generated by X and F, then $X_i^n - X_{i-1}^n \neq (X_i - X_{i-1})^n$. Show that if X_+ is not freely generated, then $X_i^n - X_{i-1}^n = (X_i - X_{i-1})^n$ is possible.

$*$ **2.3.3.** Recall from Subsection 2.2.6 that T_Σ denotes the set of all finite Σ-trees over the ranked alphabet Σ. Every function symbol f of rank $n > 0$ defines the function $\overline{f} : T_\Sigma^n \to T_\Sigma$ as follows: For every $t_1, t_2, ..., t_n \in T_\Sigma$, $\overline{f}(t_1, t_2, ..., t_n)$ is the tree denoted by $ft_1t_2...t_n$ and whose graph is the set of pairs

$$\{(e, f)\} \cup \bigcup_{i=1}^{i=n} \{(iu, t_i(u)) \mid u \in dom(t_i)\}.$$

The tree $ft_1...t_n$ is the tree with f at the root and t_i as the subtree at address i.

(a) Prove that T_Σ is freely generated by the set Σ_0 of constant symbols in Σ and the functions $\overline{f}$ defined above.

Hint: See the proof of lemma 2.4.2.

Let A be a set, X a subset of A, F a set of functions on A, and X_+ the inductive closure of X under F. We define the ranked alphabet Σ as follows:

$$\Sigma_0 = X,$$
$$\Sigma_n = \{f \mid f \in F \text{ of rank } n\}.$$

(b) Prove that the unique homomorphic extension $h : T_\Sigma \to X_+$ of the inclusion function $J : X \to X_+$ is surjective. We say that a tree $t \in T_\Sigma$ *represents* an element $x \in X_+$ iff $h(t) = x$.

(c) Prove that X_+ is freely generated by X and F iff h is a bijection.

$*$ **2.3.4.** Prove that $EXPR$ is freely generated by $V \cup \{0, 1\}$ and $\{H_+, H_*\}$.

Hint: Use the proof technique of theorem 3.2.1.

Chapter 3

Propositional Logic

3.1 Introduction

Every logic comprises a (formal) language for making statements about objects and reasoning about properties of these objects. This view of logic is very general and actually we will restrict our attention to mathematical objects, programs, and data structures in particular. Statements in a logical language are constructed according to a predefined set of formation rules (depending on the language) called *syntax rules*.

One might ask why a special language is needed at all, and why English (or any other natural language) is not adequate for carrying out logical reasoning. The first reason is that English (and any natural language in general) is such a rich language that it cannot be formally described. The second reason, which is even more serious, is that the meaning of an English sentence can be ambiguous, subject to different interpretations depending on the context and implicit assumptions. If the object of our study is to carry out precise rigorous arguments about assertions and proofs, a precise language whose syntax can be completely described in a few simple rules and whose semantics can be defined unambiguously is required.

Another important factor is conciseness. Natural languages tend to be verbose, and even fairly simple mathematical statements become exceedingly long (and unclear) when expressed in them. The logical languages that we shall define contain special symbols used for abbreviating syntactical constructs.

A logical language can be used in different ways. For instance, a language can be used as a *deduction system* (or proof system); that is, to construct proofs or refutations. This use of a logical language is called *proof theory*. In this case, a set of facts called axioms and a set of deduction rules (inference rules) are given, and the object is to determine which facts follow from the axioms and the rules of inference. When using logic as a proof system, one is not concerned with the meaning of the statements that are manipulated, but with the arrangement of these statements, and specifically, whether proofs or refutations can be constructed. In this sense, statements in the language are viewed as cold facts, and the manipulations involved are purely mechanical, to the point that they could be carried out by a computer. This does not mean that finding a proof for a statement does not require creativity, but that the interpetation of the statements is irrelevant. This use of logic is similar to game playing. Certain facts and rules are given, and it is assumed that the players are perfect, in the sense that they always obey the rules. Occasionally, it may happen that following the rules leads to inconsistencies, in which case it may be necessary to revise the rules.

However, the statements expressed in a logical language often have an intended meaning. The second use of a formal language is for expressing statements that receive a meaning when they are given what is called an interpretation. In this case, the language of logic is used to formalize properties of structures, and determine when a statement is true of a structure. This use of a logical language is called *model theory*.

One of the interesting aspects of model theory is that it forces us to have a precise and rigorous definition of the concept of truth in a structure. Depending on the interpretation that one has in mind, truth may have quite a different meaning. For instance, whether a statement is true or false may depend on parameters. A statement true under all interpretations of the parameters is said to be valid. A useful (and quite reasonable) mathematical assumption is that the truth of a statement can be obtained from the truth (or falsity) of its parts (substatements). From a technical point of view, this means that the truth of a statement is defined by recursion on the syntactical structure of the statement. The notion of truth that we shall describe (due to Tarski) formalizes the above intuition, and is firmly justified in terms of the concept of an algebra presented in Section 2.4 and the unique homomorphic extension theorem (theorem 2.4.1).

The two aspects of logic described above are actually not independent, and it is the interaction between model theory and proof theory that makes logic an interesting and effective tool. One might say that model theory and proof theory form a couple in which the individuals complement each other. To illustrate this point, consider the problem of finding a procedure for listing all statements true in a certain class of stuctures. It may be that checking the truth of a statement requires an infinite computation. Yet, if the class of structures can be axiomatized by a finite set of axioms, we might be able to find a proof procedure that will give us the answer.

Conversely, suppose that we have a set of axioms and we wish to know whether the resulting theory (the set of consequences) is consistent, in the sense that no statement and its negation follow from the axioms. If one discovers a structure in which it can be shown that the axioms and their consequences are true, one will know that the theory is consistent, since otherwise some statement and its negation would be true (in this structure).

To summarize, a logical language has a certain *syntax*, and the meaning, or *semantics*, of statements expressed in this language is given by an interpretation in a structure. Given a logical language and its semantics, one usually has one or more *proof systems* for this logical system.

A proof system is acceptable only if every provable formula is indeed valid. In this case, we say that the proof system is *sound*. Then, one tries to prove that the proof system is *complete*. A proof system is complete if every valid formula is provable. Depending on the complexity of the semantics of a given logic, it is not always possible to find a complete proof system for that logic. This is the case, for instance, for second-order logic. However, there are complete proof systems for propositional logic and first-order logic. In the first-order case, this only means that a procedure can be found such that, if the input formula is valid, the procedure will halt and produce a proof. But this does not provide a decision procedure for validity. Indeed, as a consequence of a theorem of Church, there is no procedure that will halt for every input formula and decide whether or not a formula is valid.

There are many ways of proving the completeness of a proof system. Oddly, most proofs establishing completeness only show that if a formula A is valid, then there *exists* a proof of A. However, such arguments do not actually yield a method for *constructing* a proof of A (in the formal system). Only the existence of a proof is shown. This is the case in particular for so-called *Henkin proofs*. To illustrate this point in a more colorful fashion, the above situation is comparable to going to a restaurant where you are told that excellent dinners exist on the menu, but that the inexperienced chef does not know how to prepare these dinners. This may be satisfactory for a philosopher, but not for a hungry computer scientist! However, there is an approach that does yield a procedure for constructing a formal proof of a formula if it is valid. This is the approach using Gentzen systems (or tableaux systems). Furthermore, it turns out that all of the basic theorems of first-order logic can be obtained using this approach. Hence, this author feels that a student (especially a computer scientist) has nothing to lose, and in fact will reap extra benefits by learning Gentzen systems first.

Propositional logic is the system of logic with the simplest semantics. Yet, many of the concepts and techniques used for studying propositional logic generalize to first-order logic. Therefore, it is pedagogically sound to begin by studying propositional logic as a "gentle" introduction to the methods used in first-order logic.

In propositional logic, there are atomic assertions (or atoms, or propositional letters) and compound assertions built up from the atoms and the logical connectives, *and, or, not, implication* and *equivalence*. The atomic facts are interpreted as being either true or false. In propositional logic, once the atoms in a proposition have received an interpretation, the truth value of the proposition can be computed. Technically, this is a consequence of the fact that the set of propositions is a freely generated inductive closure. Certain propositions are true for all possible interpretations. They are called *tautologies*. Intuitively speaking, a tautology is a *universal truth*. Hence, tautologies play an important role.

For example, let "John is a teacher," "John is rich," and "John is a rock singer" be three atomic propositions. Let us abbreviate them as A,B,C. Consider the following statements:

"John is a teacher";

It is false that "John is a teacher" and "John is rich";

If "John is a rock singer" then "John is rich."

We wish to show that the above assumptions imply that

It is false that "John is a rock singer."

This amounts to showing that the (formal) proposition

(∗) (A and not(A and B) and (C implies B)) implies (not C)

is a tautology. Informally, this can be shown by contradiction. The statement (∗) is false if the premise (A and not(A and B) and (C implies B)) is true and the conclusion (not C) is false. This implies that C is true. Since C is true, then, since (C implies B) is assumed to be true, B is true, and since A is assumed to be true, (A and B) is true, which is a contradiction, since not(A and B) is assumed to be true.

Of course, we have assumed that the reader is familiar with the semantics and the rules of propositional logic, which is probably not the case. In this chapter, such matters will be explained in detail.

3.2 Syntax of Propositional Logic

The syntax of propositional logic is described in this section. This presentation will use the concept of an *inductive closure* explained in Section 2.3, and the reader is encouraged to review it.

3.2.1 The Language of Propositional Logic

Propositional formulae (or propositions) are strings of symbols from a countable alphabet defined below, and formed according to certain rules stated in definition 3.2.2.

Definition 3.2.1 (The alphabet for propositional formulae) This alphabet consists of:

(1) A countable set **PS** of *proposition symbols*: $P_0, P_1, P_2...$;

(2) The *logical connectives*: $\wedge$ (and), $\vee$ (or), $\supset$ (implication), $\neg$ (not), and sometimes $\equiv$ (equivalence) and the constant $\perp$ (false);

(3) *Auxiliary symbols*: "(" (left parenthesis), ")" (right parenthesis).

The set $PROP$ of propositional formulae (or propositions) is defined as the inductive closure (as in Section 2.3) of a certain subset of the alphabet of definition 3.2.1 under certain operations defined below.

Definition 3.2.2 Propositional formulae. The set $PROP$ of *propositional formulae* (or *propositions*) is the inductive closure of the set $\mathbf{PS} \cup \{\perp\}$ under the functions $C_\neg$, $C_\wedge$, $C_\vee$, $C_\supset$ and $C_\equiv$, defined as follows: For any two strings A, B over the alphabet of definition 3.2.1,

$$C_\neg(A) = \neg A,$$
$$C_\wedge(A, B) = (A \wedge B),$$
$$C_\vee(A, B) = (A \vee B),$$
$$C_\supset(A, B) = (A \supset B) \; and$$
$$C_\equiv(A, B) = (A \equiv B).$$

The above definition is the official definition of $PROP$ as an inductive closure, but is a bit formal. For that reason, it is often stated less formally as follows:

The set $PROP$ of propositions is the smallest set of strings over the alphabet of definition 3.2.1, such that:

(1) Every proposition symbol P_i is in $PROP$ and $\perp$ is in $PROP$;

(2) Whenever A is in $PROP$, $\neg A$ is also in $PROP$;

(3) Whenever A, B are in $PROP$, $(A \vee B)$, $(A \wedge B)$, $(A \supset B)$ and $(A \equiv B)$ are also in $PROP$.

(4) A string is in $PROP$ only if it is formed by applying the rules (1),(2),(3).

The official inductive definition of $PROP$ will be the one used in proofs.

3.2.2 Free Generation of PROP

The purpose of the parentheses is to ensure unique readability; that is, to ensure that $PROP$ is freely generated on **PS**. This is crucial in order to give a proper definition of the semantics of propositions. Indeed, the meaning of a proposition will be given by a function defined recursively over the set $PROP$, and from theorem 2.4.1, we know that such a function exists and is unique when an inductive closure is freely generated.

There are other ways of defining the syntax in which parentheses are unnecessary, for example the prefix (or postfix) notation, which will be discussed later.

It is necessary for clarity and to avoid contradictions to distinguish between the formal language that is the object of our study (the set $PROP$ of propositions), and the (informal) language used to talk about the object of study. The first language is usually called the *object language* and the second, the *meta-language*. It is often tedious to maintain a clear notational distinction between the two languages, since this requires the use of a formidable number of symbols. However, one should always keep in mind this distinction to avoid confusion (and mistakes !).

For example, the symbols P, Q, R, ... will usually range over propositional symbols, and the symbols A, B, C, ... over propositions. Such symbols are called *meta-variables*.

Let us give a few examples of propositions:

EXAMPLE 3.2.1

The following strings are propositions.

$$P_1, \qquad P_2, \qquad (P_1 \lor P_2),$$

$$((P_1 \supset P_2) \equiv (\neg P_1 \lor P_2)), \qquad (\neg P_1 \equiv (P_1 \supset \bot)),$$

$$(((P_1 \supset P_2) \land \neg P_2) \supset \neg P_1), \qquad (P_1 \lor \neg P_1).$$

On the other hand, strings such as

$$(()), \quad \text{or} \qquad (P_1 \lor P_2)\land$$

are not propositions, because they cannot be constructed from **PS** and $\bot$ and the logical connectives.

Since $PROP$ is inductively defined on **PS**, the induction principle (of Section 2.3) applies. We are now going to use this induction principle to show that $PROP$ is freely generated by the propositional symbols (and $\bot$) and the logical connectives.

Lemma 3.2.1 (i) Every proposition in $PROP$ has the same number of left and right parentheses.

(ii) Any proper prefix of a proposition is either the empty string, a (nonempty) string of negation symbols, or it contains an excess of left parentheses.

(iii) No proper prefix of a proposition can be a proposition.

Proof: (i) Let S be the set of propositions in $PROP$ having an equal number of left and right parentheses. We show that S is inductive on the set of propositional symbols and $\perp$. By the induction principle, this will show that $S = PROP$, as desired. It is obvious that S contains the propositional symbols (no parentheses) and $\perp$. It is also obvious that the rules in definition 3.2.2 introduce matching parentheses and so, preserve the above property. This concludes the first part of the proof.

(ii) Let S be the set of propositions in $PROP$ such that any proper prefix is either the empty string, a string of negations, or contains an excess of left parentheses. We also prove that S is inductive on the set of propositional symbols and $\perp$. First, it is obvious that every propositional symbol is in S, as well as $\perp$. Let us verify that S is closed under $C_\wedge$, leaving the other cases as an exercise. Let A and B be in S. The nonempty proper prefixes of $C_\wedge(A, B) = (A \wedge B)$ are:

$($
$(C$ where C is a proper prefix of A
$(A$
$(A\wedge$
$(A \wedge C$ where C is a proper prefix of B
$(A \wedge B$

Applying the induction hypothesis that A and B are in S, we obtain the desired conclusion.

Clause (iii) of the lemma follows from the two previous properties. If a proper prefix of a proposition is a proposition, then by (i), it has the same number of left and right parentheses. If a proper prefix has no parentheses, it is either the empty string or a string of negations, but neither is a proposition. If it has parentheses, by property (ii), it has an excess of left parentheses, a contradiction. $\square$

The above lemma allows us to show the theorem:

Theorem 3.2.1 The set $PROP$ of propositions is freely generated by the propositional symbols in **PS**, $\perp$, and the logical connectives.

Proof: First, we show that the restrictions of the functions $C_\neg$, $C_\wedge$, $C_\vee$, $C_\supset$ and $C_\equiv$ to $PROP$ are injective. This is obvious for $C_\neg$ and we only check this for $C_\wedge$, leaving the other cases as an exercise. If $(A \wedge B) = (C \wedge D)$, then $A \wedge B) = C \wedge D)$. Either $A = C$, or A is a proper prefix of C, or C is a proper prefix of A. But the last two cases are impossible by lemma 3.2.1. Then $\wedge B) = \wedge D)$, which implies $B = D$.

Next, we have to show that the ranges of the restrictions of the above functions to $PROP$ are disjoint. We only discuss one case, leaving the others as an exercise. For example, if $(A \wedge B) = (C \supset D)$, then $A \wedge B) = C \supset D)$. By the same reasoning as above, we must have $A = C$. But then, we must have $\wedge = \supset$, which is impossible. Finally, since all the functions yield a string of length greater than that of its arguments, all the conditions for being freely generated are met. $\square$

The above result allows us to define functions over $PROP$ recursively. Every function with domain $PROP$ is uniquely determined by its restriction to the set **PS** of propositional symbols and to $\bot$. We are going to use this fact in defining the semantics of propositional logic. As an illustration of theorem 3.2.1, we give a recursive definition of the set of propositional letters occurring in a proposition.

EXAMPLE 3.2.2

The function $symbols : PROP \to 2^{\mathbf{PS}}$ is defined recursively as follows:

$$symbols(\bot) = \emptyset,$$
$$symbols(P_i) = \{P_i\},$$
$$symbols((B * C)) = symbols(B) \cup symbols(C), \text{ for } * \in \{\wedge, \vee, \supset, \equiv\},$$
$$symbols(\neg A) = symbols(A).$$

For example,

$$symbols(((P_1 \supset P_2) \vee \neg P_3) \wedge P_1)) = \{P_1, P_2, P_3\}.$$

In order to minimize the number of parentheses, a *precedence* is assigned to the logical connectives and it is assumed that they are left associative. Starting from highest to lowest precedence we have:

$$\neg$$

$$\wedge$$

$$\vee$$

$$\supset, \equiv.$$

EXAMPLE 3.2.3

$A \wedge B \supset C$ is an abbreviation for $((A \wedge B) \supset C)$,

$A \vee B \wedge C$ an abbreviation for $(A \vee (B \wedge C))$, and

$A \vee B \vee C$ is an abbreviation for $((A \vee B) \vee C)$.

Parentheses can be used to disambiguate expressions. These conventions are consistent with the semantics of the propositional calculus, as we shall see in Section 3.3.

Another way of avoiding parentheses is to use the *prefix notation*. In prefix notation, $(A \vee B)$ becomes $\vee AB$, $(A \wedge B)$ becomes $\wedge AB$, $(A \supset B)$ becomes $\supset AB$ and $(A \equiv B)$ becomes $\equiv AB$.

In order to justify the legitimacy of the prefix notation, that is, to show that the set of propositions in prefix notation is freely generated, we have to show that every proposition can be written in a unique way. We shall come back to this when we consider terms in first-order logic.

PROBLEMS

3.2.1. Let *PROP* be the set of all propositions over the set **PS** of propositional symbols. The *depth* $d(A)$ of a proposition A is defined recursively as follows:

$$d(\bot) = 0,$$
$$d(P) = 0, \text{ for each symbol } P \in \mathbf{PS},$$
$$d(\neg A) = 1 + d(A),$$
$$d(A \vee B) = 1 + max(d(A), d(B)),$$
$$d(A \wedge B) = 1 + max(d(A), d(B)),$$
$$d(A \supset B) = 1 + max(d(A), d(B)),$$
$$d(A \equiv B) = 1 + max(d(A), d(B)).$$

If PS_i is the i-th stage of the inductive definition of $PROP = (\mathbf{PS} \cup \{\bot\})_+$ (as in Section 2.3), show that PS_i consists exactly of all propositions of depth less than or equal to i.

3.2.2. Which of the following are propositions? Justify your answer.

$\neg\neg\neg P_1$

$\neg P_1 \vee \neg P_2$

$\neg(P_1 \vee P_2)$

$(\neg P_1 \supset \neg P_2)$

$\neg(P_1 \vee (P_2 \wedge (P_3 \vee P_4) \wedge (P_1 \wedge (P_3 \wedge \neg P_1) \vee (P_4 \vee P_1)))$

(*Hint:* Use problem 3.2.1, lemma 3.2.1.)

3.2.3. Finish the proof of the cases in lemma 3.2.1.

3.2.4. The function $sub : PROP \rightarrow 2^{PROP}$ which assigns to any proposition
A the set $sub(A)$ of all its subpropositions is defined recursively as
follows:

$$sub(\bot) = \{\bot\},$$
$$sub(P_i) = \{P_i\}, \text{ for a propositional symbol } P_i,$$
$$sub(\neg A) = sub(A) \cup \{\neg A\},$$
$$sub((A \vee B)) = sub(A) \cup sub(B) \cup \{(A \vee B)\},$$
$$sub((A \wedge B)) = sub(A) \cup sub(B) \cup \{(A \wedge B)\},$$
$$sub((A \supset B)) = sub(A) \cup sub(B) \cup \{(A \supset B)\},$$
$$sub((A \equiv B)) = sub(A) \cup sub(B) \cup \{(A \equiv B)\}.$$

Prove that if a proposition A has n connectives, then $sub(A)$ contains
at most $2n + 1$ propositions.

3.2.5. Give an example of propositions A and B and of strings u and v such
that $(A \vee B) = (u \vee v)$, but $u \neq A$ and $v \neq B$. Similarly give an
example such that $(A \vee B) = (u \wedge v)$, but $u \neq A$ and $v \neq B$.

∗ 3.2.6. The set of propositions can be defined by a context-free grammar,
provided that the propositional symbols are encoded as strings over a
finite alphabet. Following Lewis and Papadimitriou, 1981, the symbol
P_i, $(i \geq 0)$ will be encoded as $PI...I\$$, with a number of I's equal
to i. Then, $PROP$ is the language $L(G)$ defined by the following
context-free grammar $G = (V, \Sigma, R, S)$:

$$\Sigma = \{P, I, \$, \wedge, \vee, \supset, \equiv, \neg, \bot\}, V = \Sigma \cup \{S, N\},$$

$$R = \{N \rightarrow e,$$
$$N \rightarrow NI,$$
$$S \rightarrow PN\$,$$
$$S \rightarrow \bot,$$
$$S \rightarrow (S \vee S),$$
$$S \rightarrow (S \wedge S),$$
$$S \rightarrow (S \supset S),$$
$$S \rightarrow (S \equiv S),$$
$$S \rightarrow \neg S\}.$$

Prove that the grammar G is unambiguous.

Note: The above language is actually SLR(1). For details on parsing
techniques, consult Aho and Ullman, 1977.

3.2.7. The set of propositions in prefix notation is the inductive closure of
PS $\cup \{\bot\}$ under the following functions:

For all strings A, B over the alphabet of definition 3.2.1, excluding parentheses,

$$C_\wedge(A, B) = \wedge AB,$$
$$C_\vee(A, B) = \vee AB,$$
$$C_\supset(A, B) = \supset AB,$$
$$C_\equiv(A, B) = \equiv AB,$$
$$C_\neg(A) = \neg A.$$

In order to prove that the set of propositions in prefix notation is freely generated, we define the function K as follows:

$K(\wedge) = -1$; $K(\vee) = -1$; $K(\supset) = -1$; $K(\equiv) = -1$; $K(\neg) = 0$; $K(\bot) = 1$; $K(P_i) = 1$, for every propositional symbol P_i.

The function K is extended to strings as follows: For every string $w_1...w_k$ (over the alphabet of definition 3.2.1, excluding parentheses), $K(w_1...w_k) = K(w_1) + ... + K(w_k)$.

(i) Prove that for any proposition A, $K(A) = 1$.

(ii) Prove that for any proper prefix w of a proposition, $K(w) \leq 0$.

(iii) Prove that no proper prefix of a proposition is a proposition.

(iv) Prove that the set of propositions in prefix notation is freely generated.

3.2.8. Suppose that we modify definition 3.2.2 by omitting all right parentheses. Thus, instead of

$$((P \wedge \neg Q) \supset (R \vee S)),$$

we have

$$((P \wedge \neg Q \supset (R \vee S.$$

Formally, we define the functions:

$$C_\wedge(A, B) = (A \wedge B,$$
$$C_\vee(A, B) = (A \vee B,$$
$$C_\supset(A, B) = (A \supset A,$$
$$C_\equiv(A, B) = (A \equiv A,$$
$$C_\neg(A) = \neg A.$$

Prove that the set of propositions defined in this fashion is still freely generated.

3.3 Semantics of Propositional Logic

In this section, we present the semantics of propositional logic and define the concepts of satisfiability and tautology.

3.3.1 The Semantics of Propositions

The semantics of the propositional calculus assigns a truth function to each proposition in $PROP$. First, it is necessary to define the meaning of the logical connectives. We first define the domain $BOOL$ of truth values.

Definition 3.3.1 The set of *truth values* is the set $BOOL = \{\mathbf{T}, \mathbf{F}\}$. It is assumed that $BOOL$ is (totally) ordered with $\mathbf{F} < \mathbf{T}$.

Each logical connective X is interpreted as a function H_X with range $BOOL$. The logical connectives are interpreted as follows.

Definition 3.3.2 The graphs of the logical connectives are represented by the following table:

P	Q	$H_\neg(P)$	$H_\wedge(P,Q)$	$H_\vee(P,Q)$	$H_\supset(P,Q)$	$H_\equiv(P,Q)$
T	**T**	**F**	**T**	**T**	**T**	**T**
T	**F**	**F**	**F**	**T**	**F**	**F**
F	**T**	**T**	**F**	**T**	**T**	**F**
F	**F**	**T**	**F**	**F**	**T**	**T**

The logical constant $\bot$ is interpreted as $\mathbf{F}$.

The above table is what is called a *truth table*. We have introduced the function H_X to distinguish between the symbol X and its meaning H_X. This is a heavy notational burden, but it is essential to distinguish between syntax and semantics, until the reader is familiar enough with these concepts. Later on, when the reader has assimilated these concepts, we will often use X for H_X to simplify the notation.

We now define the semantics of formulae in $PROP$.

Definition 3.3.3 A *truth assignment* or *valuation* is a function $v : \mathbf{PS} \to BOOL$ assigning a truth value to all the propositional symbols. From theorem 2.4.1, since $PROP$ is freely generated by $\mathbf{PS}$, every valuation v extends to a unique function $\hat{v} : PROP \to BOOL$ satisfying the following clauses for all $A, B \in PROP$:

$$\widehat{v}(\bot) = \mathbf{F},$$
$$\widehat{v}(P) = v(P), \text{ for all } P \in \mathbf{PS},$$
$$\widehat{v}(\neg A) = H_\neg(\widehat{v}(A)),$$
$$\widehat{v}((A \wedge B)) = H_\wedge(\widehat{v}(A), \widehat{v}(B)),$$
$$\widehat{v}((A \vee B)) = H_\vee(\widehat{v}(A), \widehat{v}(B)),$$
$$\widehat{v}((A \supset B)) = H_\supset(\widehat{v}(A), \widehat{v}(B)),$$
$$\widehat{v}((A \equiv B)) = H_\equiv(\widehat{v}(A), \widehat{v}(B)).$$

In the above definition, the *truth value* $\widehat{v}(A)$ of a proposition A is defined for a truth assignment v assigning truth values to all propositional symbols, including infinitely many symbols not occurring in A. However, for any formula A and any valuation v, the value $\widehat{v}(A)$ only depends on the propositional symbols actually occurring in A. This is justified by the following lemma.

Lemma 3.3.1 For any proposition A, for any two valuations v and v' such that $v(P) = v'(P)$ for all proposition symbols occurring in A, $\widehat{v}(A) = \widehat{v'}(A)$.

Proof: We proceed by induction. The lemma is obvious for $\bot$, since $\bot$ does not contain any propositional symbols. If A is the propositional symbol P_i, since $v(P_i) = v'(P_i)$, $\widehat{v}(P_i) = v(P_i)$, and $\widehat{v'}(P_i) = v'(P_i)$, the Lemma holds for propositional symbols.

If A is of the form $\neg B$, since the propositional symbols occurring in B are the propositional symbols occurring in A, by the induction hypothesis, $\widehat{v}(B) = \widehat{v'}(B)$. Since

$$\widehat{v}(A) = H_\neg(\widehat{v}(B)) \quad \text{and} \quad \widehat{v'}(A) = H_\neg(\widehat{v'}(B)),$$

we have

$$\widehat{v}(A) = \widehat{v'}(A).$$

If A is of the form $(B * C)$, for a connective $* \in \{\vee, \wedge, \supset, \equiv\}$, since the sets of propositional letters occurring in B and C are subsets of the set of propositional letters occurring in A, the induction hypothesis applies to B and C. Hence,

$$\widehat{v}(B) = \widehat{v'}(B) \quad \text{and} \quad \widehat{v}(C) = \widehat{v'}(C).$$

But

$$\widehat{v}(A) = H_*(\widehat{v}(B), \widehat{v}(C)) = H_*(\widehat{v'}(B), \widehat{v'}(C)) = \widehat{v'}(A),$$

showing that

$$\widehat{v}(A) = \widehat{v'}(A).$$

$\square$

Using lemma 3.3.1, we observe that given a proposition A containing the set of propositional symbols $\{P_1, ..., P_n\}$, its truth value for any assignment v can be computed recursively and only depends on the values $v(P_1), ..., v(P_n)$.

EXAMPLE 3.3.1

Let
$$A = ((P \supset Q) \equiv (\neg Q \supset \neg P)).$$

Let v be a truth assignment whose restriction to $\{P, Q\}$ is $v(P) = \mathbf{T}$, $v(Q) = \mathbf{F}$. According to definition 3.3.3,

$$\widehat{v}(A) = H_\equiv(\widehat{v}((P \supset Q)), \widehat{v}((\neg Q \supset \neg P))).$$

In turn,
$$\widehat{v}((P \supset Q)) = H_\supset(\widehat{v}(P), \widehat{v}(Q)) \quad \text{and}$$
$$\widehat{v}((\neg Q \supset \neg P)) = H_\supset(\widehat{v}(\neg Q), \widehat{v}(\neg P)).$$

Since
$$\widehat{v}(P) = v(P) \quad \text{and} \quad \widehat{v}(Q) = v(Q),$$

we have
$$\widehat{v}((P \supset Q)) = H_\supset(\mathbf{T}, \mathbf{F}) = \mathbf{F}.$$

We also have

$$\widehat{v}(\neg Q) = H_\neg(\widehat{v}(Q)) = H_\neg(v(Q)) \quad \text{and}$$

$$\widehat{v}(\neg P) = H_\neg(\widehat{v}(P)) = H_\neg(v(P)).$$

Hence,
$$\widehat{v}(\neg Q) = H_\neg(\mathbf{F}) = \mathbf{T},$$
$$\widehat{v}(\neg P) = H_\neg(\mathbf{T}) = \mathbf{F} \quad \text{and}$$
$$\widehat{v}((\neg Q \supset \neg P)) = H_\supset(\mathbf{T}, \mathbf{F}) = \mathbf{F}.$$

Finally,
$$\widehat{v}(A) = H_\equiv(\mathbf{F}, \mathbf{F}) = \mathbf{T}.$$

The above recursive computation can be conveniently described by a truth table as follows:

P	Q	$\neg P$	$\neg Q$	$(P \supset Q)$	$(\neg Q \supset \neg P)$	$((P \supset Q) \equiv (\neg Q \supset \neg P))$
$\mathbf{T}$	$\mathbf{F}$	$\mathbf{F}$	$\mathbf{T}$	$\mathbf{F}$	$\mathbf{F}$	$\mathbf{T}$

If $\widehat{v}(A) = \mathbf{T}$ for a valuation v and a proposition A, we say that v *satisfies* A, and this is denoted by $v \models A$. If v does not satisfy A, we say that v *falsifies* A, and this is denoted by, *not* $v \models A$, or $v \not\models A$.

An expression such as $v \models A$ (or $v \not\models A$) is merely a notation used in the meta-language to express concisely statements about satisfaction. The reader should be well aware that such a notation is not a proposition in the object language, and should be used with care. An illustation of the danger of mixing the meta-language and the object language is given by the definition (often found in texts) of the notion of satisfaction in terms of the notation $v \models A$. Using this notation, the recursive clauses of the definition of $\widehat{v}$ can be stated informally as follows:

$$v \not\models \bot,$$
$$v \models P_i \text{ iff } v(P_i) = \mathbf{T},$$
$$v \models \neg A \text{ iff } v \not\models A,$$
$$v \models A \wedge B \text{ iff } v \models A \text{ and } v \models B,$$
$$v \models A \vee B \text{ iff } v \models A \text{ or } v \models B,$$
$$v \models A \supset B \text{ iff } v \not\models A \text{ or } v \models B,$$
$$v \models A \equiv B \text{ iff } (v \models A \text{ iff } v \models B).$$

The above definition is not really satisfactory because it mixes the object language and the meta-language too much. In particular, the meaning of the words *not*, *or*, *and* and *iff* is ambiguous. What is worse is that the legitimacy of the recursive definition of "$\models$" is far from being clear. However, the definition of $\models$ using the recursive definition of $\widehat{v}$ is rigorously justified by theorem 2.4.1.

An important subset of *PROP* is the set of all propositions that are true in *all* valuations. These propositions are called *tautologies*.

3.3.2 Satisfiability, Unsatisfiability, Tautologies

First, we define the concept of a tautology.

Definition 3.3.4 A proposition A is a *tautology* iff $\widehat{v}(A) = \mathbf{T}$ for all valuations v. This is abbreviated as $\models A$.

A proposition is *satisfiable* if there is a valuation (or truth assignment) v such that $\widehat{v}(A) = \mathbf{T}$. A proposition is *unsatisfiable* if it is not satisfied by any valuation.

Given a set of propositions Γ, we say that A is a *semantic consequence* of Γ, denoted by $\Gamma \models A$, if for all valuations v, $\widehat{v}(B) = \mathbf{T}$ for all B in Γ implies that $\widehat{v}(A) = \mathbf{T}$.

The problem of determining whether any arbitrary proposition is satisfiable is called the *satisfiability problem*. The problem of determining whether any arbitrary proposition is a tautology is called the *tautology problem*.

EXAMPLE 3.3.2

The following propositions are tautologies:

$$A \supset A,$$

$$\neg\neg A \supset A,$$
$$(P \supset Q) \equiv (\neg Q \supset \neg P).$$

The proposition
$$(P \vee Q) \wedge (\neg P \vee \neg Q)$$
is satisfied by the assignment $v(P) = \mathbf{F}$, $v(Q) = \mathbf{T}$.

The proposition
$$(\neg P \vee Q) \wedge (\neg P \vee \neg Q) \wedge P$$
is unsatisfiable. The following are valid consequences.

$$A, (A \supset B) \models B,$$
$$A, B \models (A \wedge B),$$
$$(A \supset B), \neg B \models \neg A.$$

Note that $P \supset Q$ is false if and only if both P is true and Q is false. In particular, observe that $P \supset Q$ is true when P is false.

The relationship between satisfiability and being a tautology is recorded in the following useful lemma.

Lemma 3.3.2 A proposition A is a tautology if and only if $\neg A$ is unsatisfiable.

Proof: Assume that A is a tautology. Hence, for all valuations v,

$$\widehat{v}(A) = \mathbf{T}.$$

Since

$$\widehat{v}(\neg A) = H_\neg(\widehat{v}(A)),$$
$$\widehat{v}(A) = \mathbf{T} \quad \text{if and only if} \quad \widehat{v}(\neg A) = \mathbf{F}.$$

This shows that for all valuations v,

$$\widehat{v}(\neg A) = \mathbf{F},$$

which is the definition of unsatisfiability. Conversely, if $\neg A$ is unsatisfiable, for all valuations v,
$$\widehat{v}(\neg A) = \mathbf{F}.$$

By the above reasoning, for all v,

$$\widehat{v}(A) = \mathbf{T},$$

which is the definition of being a tautology. $\square$

The above lemma suggests two different approaches for proving that a proposition is a tautology. In the first approach, one attempts to show *directly* that A is a tautology. The method in Section 3.4 using Gentzen systems illustrates the first approach (although A is proved to be a tautology if the attempt to falsify A fails). In the second approach, one attempts to show *indirectly* that A is a tautology, by showing that $\neg A$ is unsatisfiable. The method in Chapter 4 using resolution illustrates the second approach.

As we saw in example 3.3.1, the recursive definition of the unique extension $\hat{v}$ of a valuation v suggests an algorithm for computing the truth value $\hat{v}(A)$ of a proposition A. The algorithm consists in computing recursively the truth tables of the parts of A. This is known as the *truth table method*.

The truth table method clearly provides an algorithm for testing whether a formula is a tautology: If A contains n propositional letters, one constructs a truth table in which the truth value of A is computed for all valuations depending on n arguments. Since there are 2^n such valuations, the size of this truth table is at least 2^n. It is also clear that there is an algorithm for deciding whether a proposition is satisfiable: Try out all possible valuations (2^n) and compute the corresponding truth table.

EXAMPLE 3.3.3

Let us compute the truth table for the proposition $A = ((P \supset Q) \equiv (\neg Q \supset \neg P))$.

P	Q	$\neg P$	$\neg Q$	$(P \supset Q)$	$(\neg Q \supset \neg P)$	$((P \supset Q) \equiv (\neg Q \supset \neg P))$
F	F	T	T	T	T	T
F	T	T	F	T	T	T
T	F	F	T	F	F	T
T	T	F	F	T	T	T

Since the last column contains only the truth value **T**, the proposition A is a tautology.

The above method for testing whether a proposition is satisfiable or a tautology is computationally expensive, in the sense that it takes an exponential number of steps. One might ask if it is possible to find a more efficient procedure. Unfortunately, the satisfiability problem happens to be what is called an NP-complete problem, which implies that there is probably no fast algorithm for deciding satisfiability. By a fast algorithm, we mean an algorithm that runs in a number of steps bounded by $p(n)$, where n is the length of the input, and p is a (fixed) polynomial. NP-complete problems and their significance will be discussed at the end of this section.

Is there a better way of testing whether a proposition A is a tautology than computing its truth table (which requires computing at least 2^n entries, where n is the number of proposition symbols occurring in A)? One possibility is to work backwards, trying to find a truth assignment which makes the proposition false. In this way, one may detect failure much earlier. This is the essence of Gentzen systems to be discussed shortly.

As we said at the beginning of Section 3.3, every proposition defines a function taking truth values as arguments, and yielding truth values as results. The truth function associated with a proposition is defined as follows.

3.3.3 Truth Functions and Functionally Complete Sets of Connectives

We now show that the logical connectives are not independent. For this, we need to define what it means for a proposition to define a truth function.

Definition 3.3.5 Let A be a formula containing exactly n distinct propositional symbols. The function $H_A : BOOL^n \to BOOL$ is defined such that, for every $(a_1, ..., a_n) \in BOOL^n$,

$$H_A(a_1, ..., a_n) = \widehat{v}(A),$$

with v any valuation such that $v(P_i) = a_i$ for every propositional symbol P_i occurring in A.

For simplicity of notation we will often name the function H_A as A. H_A is a *truth function*. In general, every function $f : BOOL^n \to BOOL$ is called an n-ary truth function.

EXAMPLE 3.3.4

The proposition

$$A = (P \wedge \neg Q) \vee (\neg P \wedge Q)$$

defines the truth function $\oplus$ given by the following truth table:

P	Q	$\neg P$	$\neg Q$	$(P \wedge \neg Q)$	$(\neg P \wedge Q)$	$(P \wedge \neg Q) \vee (\neg P \wedge Q))$
F	F	T	T	F	F	F
F	T	T	F	F	T	T
T	F	F	T	T	F	T
T	T	F	F	F	F	F

Observe that the function $\oplus$ takes the value **T** if and only if its arguments have different truth values. For this reason, it is called the *exclusive OR* function.

It is natural to ask whether every truth function f can be realized by some proposition A, in the sense that $f = H_A$. This is indeed the case. We say that the boolean connectives form a *functionally complete set* of connectives.

The significance of this result is that for any truth function f of n arguments, we do not enrich the collection of assertions that we can make by adding a new symbol to the syntax, say F, and interpreting F as f. Indeed, since f is already definable in terms of $\lor$, $\land$, $\neg$, $\supset$, $\equiv$, every extended proposition A containing F can be converted to a proposition A' in which F does not occur, such that for every valuation v, $\widehat{v}(A) = \widehat{v}(A')$. Hence, there is no loss of generality in restricting our attention to the connectives that we have introduced. In fact, we shall prove that each of the sets $\{\lor, \neg\}$, $\{\land, \neg\}$, $\{\supset, \neg\}$ and $\{\supset, \bot\}$ is functionally complete.

First, we prove the following two lemmas.

Lemma 3.3.3 Let A and B be any two propositions, let $\{P_1, ..., P_n\}$ be the set of propositional symbols occurring in $(A \equiv B)$, and let H_A and H_B be the functions associated with A and B, considered as functions of the arguments in $\{P_1, ..., P_n\}$. The proposition $(A \equiv B)$ is a tautology if and only if for all valuations v, $\widehat{v}(A) = \widehat{v}(B)$, if and only if $H_A = H_B$.

Proof: For any valuation v,

$$\widehat{v}((A \equiv B)) = H_\equiv(\widehat{v}(A), \widehat{v}(B)).$$

Consulting the truth table for $H_\equiv$, we see that

$$\widehat{v}((A \equiv B)) = \mathbf{T} \quad \text{if and only if} \quad \widehat{v}(A) = \widehat{v}(B).$$

By lemma 3.3.1 and definition 3.3.5, this implies that $H_A = H_B$. $\square$

By constructing truth tables, the propositions listed in lemma 3.3.4 below can be shown to be tautologies.

Lemma 3.3.4 The following properties hold:

$$\models (A \equiv B) \equiv ((A \supset B) \land (B \supset A)); \tag{1}$$
$$\models (A \supset B) \equiv (\neg A \lor B); \tag{2}$$
$$\models (A \lor B) \equiv (\neg A \supset B); \tag{3}$$
$$\models (A \lor B) \equiv \neg(\neg A \land \neg B); \tag{4}$$
$$\models (A \land B) \equiv \neg(\neg A \lor \neg B); \tag{5}$$
$$\models \neg A \equiv (A \supset \bot); \tag{6}$$
$$\models \bot \equiv (A \land \neg A). \tag{7}$$

Proof: We prove (1), leaving the other cases as an exercise. By lemma 3.3.3, it is sufficient to verify that the truth tables for $(A \equiv B)$ and $((A \supset B) \land (B \supset A))$ are identical.

A	B	$A \supset B$	$B \supset A$	$(A \equiv B)$	$((A \supset B) \wedge (B \supset A))$
F	F	T	T	T	T
F	T	T	F	F	F
T	F	F	T	F	F
T	T	T	T	T	T

Since the columns for $(A \equiv B)$ and $((A \supset B) \wedge (B \supset A))$ are identical, (1) is a tautology. $\square$

We now show that $\{\vee, \wedge, \neg\}$ is a functionally complete set of connectives.

Theorem 3.3.1 For every n-ary truth function f, there is a proposition A only using the connectives $\wedge$, $\vee$ and $\neg$ such that $f = H_A$.

Proof: We proceed by induction on the arity n of f. For $n = 1$, there are four truth functions whose truth tables are:

P	1	2	3	4
F	T	F	F	T
T	T	F	T	F

Clearly, the propositions $P \vee \neg P$, $P \wedge \neg P$, P and $\neg P$ do the job. Let f be of arity $n + 1$ and assume the induction hypothesis for n. Let

$$f_1(x_1, ..., x_n) = f(x_1, ..., x_n, \mathbf{T}) \quad \text{and}$$
$$f_2(x_1, ..., x_n) = f(x_1, ..., x_n, \mathbf{F}).$$

Both f_1 and f_2 are n-ary. By the induction hypothesis, there are propositions B and C such that

$$f_1 = H_B \quad \text{and} \quad f_2 = H_C.$$

But then, letting A be the formula

$$(P_{n+1} \wedge B) \vee (\neg P_{n+1} \wedge C),$$

where P_{n+1} occurs neither in B nor C, it is easy to see that $f = H_A$. $\square$

Using lemma 3.3.4, it follows that $\{\vee, \neg\}$, $\{\wedge, \neg\}$, $\{\supset, \neg\}$ and $\{\supset, \perp\}$ are functionally complete. Indeed, using induction on propositions, $\wedge$ can be expressed in terms of $\vee$ and $\neg$ by (5), $\supset$ can be expressed in terms of $\neg$ and $\vee$ by (2), $\equiv$ can be expressed in terms of $\supset$ and $\wedge$ by (1), and $\perp$ can be expressed in terms of $\wedge$ and $\neg$ by (7). Hence, $\{\vee, \neg\}$ is functionally complete. Since $\vee$ can be expressed in terms of $\wedge$ and $\neg$ by (4), the set $\{\wedge, \neg\}$ is functionally complete, since $\{\vee, \neg\}$ is. Since $\vee$ can be expressed in terms of $\supset$ and $\neg$ by (3), the set $\{\supset, \neg\}$ is functionally complete, since $\{\vee, \neg\}$ is. Finally, since $\neg$

can be expressed in terms of $\supset$ and $\perp$ by (6), the set $\{\supset, \perp\}$ is functionally complete since $\{\supset, \neg\}$ is.

In view of the above theorem, we may without loss of generality restrict our attention to propositions expressed in terms of the connectives in some functionally complete set of our choice. The choice of a suitable functionally complete set of connectives is essentially a matter of convenience and taste. The advantage of using a small set of connectives is that fewer cases have to be considered in proving properties of propositions. The disadvantage is that the meaning of a proposition may not be as clear for a proposition written in terms of a smaller complete set, as it is for the same proposition expressed in terms of the full set of connectives used in definition 3.2.1. Furthermore, depending on the set of connectives chosen, the representations can have very different lengths. For example, using the set $\{\supset, \perp\}$, the proposition $(A \wedge B)$ takes the form

$$(((A \supset \perp) \supset \perp) \supset (B \supset \perp)) \supset \perp .$$

I doubt that many readers think that the second representation is more perspicuous than the first!

In this book, we will adopt the following compromise between mathematical conciseness and intuitive clarity. The set $\{\wedge, \vee, \neg, \supset\}$ will be used. Then, $(A \equiv B)$ will be considered as an abbreviation for $((A \supset B) \wedge (B \supset A))$, and $\perp$ as an abbreviation for $(P \wedge \neg P)$.

We close this section with some results showing that the set of propositions has a remarkable structure called a boolean algebra. (See Subsection 2.4.1 for the definition of an algebra.)

3.3.4 Logical Equivalence and Boolean Algebras

First, we show that lemma 3.3.3 implies that a certain relation on $PROP$ is an equivalence relation.

Definition 3.3.6 The relation $\simeq$ on $PROP$ is defined so that for any two propositions A and B, $A \simeq B$ if and only if $(A \equiv B)$ is a tautology. We say that A and B are *logically equivalent*, or for short, equivalent.

From lemma 3.3.3, $A \simeq B$ if and only if $H_A = H_B$. This implies that the relation $\simeq$ is reflexive, symmetric, and transitive, and therefore it is an equivalence relation. The following additional properties show that it is a congruence in the sense of Subsection 2.4.6.

Lemma 3.3.5 For all propositions A, A', B, B', the following properties hold:

If $A \simeq A'$ and $B \simeq B'$,   then for $* \in \{\wedge, \vee, \supset, \equiv\}$,

$$(A * B) \simeq (A' * B') \quad \text{and}$$
$$\neg A \simeq \neg A'.$$

Proof: By definition 3.3.6,

$$(A * B) \simeq (A' * B') \quad \text{if and only if} \quad \models (A * B) \equiv (A' * B').$$

By lemma 3.3.3, it is sufficient to show that for all valuations v,

$$\widehat{v}(A * B) = \widehat{v}(A' * B').$$

Since

$$A \simeq A' \quad \text{and} \quad B \simeq B' \quad \text{implies that}$$
$$\widehat{v}(A) = \widehat{v}(A') \quad \text{and} \quad \widehat{v}(B) = \widehat{v}(B'),$$

we have

$$\widehat{v}(A * B) = H_*(\widehat{v}(A), \widehat{v}(B)) = H_*(\widehat{v}(A'), \widehat{v}(B')) = \widehat{v}(A' * B').$$

Similarly,

$$\widehat{v}(\neg A) = H_\neg(\widehat{v}(A)) = H_\neg(\widehat{v}(A')) = \widehat{v}(\neg A').$$

$\square$

In the rest of this section, it is assumed that the constant symbol $\top$ is added to the alphabet of definition 3.2.1, yielding the set of propositions $PROP'$, and that $\top$ is interpreted as **T**. The proof of the following properties is left as an exercise.

Lemma 3.3.6 The following properties hold for all propositions in $PROP'$.

Associativity rules:
$$((A \vee B) \vee C) \simeq (A \vee (B \vee C)) \quad ((A \wedge B) \wedge C) \simeq (A \wedge (B \wedge C))$$

Commutativity rules:
$$(A \vee B) \simeq (B \vee A) \quad (A \wedge B) \simeq (B \wedge A)$$

Distributivity rules:
$$(A \vee (B \wedge C)) \simeq ((A \vee B) \wedge (A \vee C))$$
$$(A \wedge (B \vee C)) \simeq ((A \wedge B) \vee (A \wedge C))$$

De Morgan's rules:
$$\neg(A \vee B) \simeq (\neg A \wedge \neg B) \quad \neg(A \wedge B) \simeq (\neg A \vee \neg B)$$

Idempotency rules:
$$(A \vee A) \simeq A \quad (A \wedge A) \simeq A$$

Double negation rule:
$$\neg\neg A \simeq A$$

Absorption rules:
$$(A \vee (A \wedge B)) \simeq A \quad (A \wedge (A \vee B)) \simeq A$$

Laws of zero and one:
$$(A \vee \bot) \simeq A \quad (A \wedge \bot) \simeq \bot$$
$$(A \vee \top) \simeq \top \quad (A \wedge \top) \simeq A$$
$$(A \vee \neg A) \simeq \top \quad (A \wedge \neg A) \simeq \bot$$

Let us denote the equivalence class of a proposition A modulo $\simeq$ as $[A]$, and the set of all such equivalence classes as $\mathbf{B}_{PROP}$. We define the operations $+$, $*$ and $\quad$ on $\mathbf{B}_{PROP}$ as follows:

$$[A] + [B] = [A \vee B],$$
$$[A] * [B] = [A \wedge B],$$
$$\overline{[A]} = [\neg A].$$

Also, let $0 = [\bot]$ and $1 = [\top]$.

By lemma 3.3.5, the above functions (and constants) are independent of the choice of representatives in the equivalence classes. But then, the properties of lemma 3.3.6 are identities valid on the set $\mathbf{B}_{PROP}$ of equivalence classes modulo $\simeq$. The structure $\mathbf{B}_{PROP}$ is an algebra in the sense of Subsection 2.4.1. Because it satisfies the identities of lemma 3.3.6, it is a very rich structure called a *boolean algebra*. $\mathbf{B}_{PROP}$ is called the *Lindenbaum algebra* of *PROP*. In this book, we will only make simple uses of the fact that $\mathbf{B}_{PROP}$ is a boolean algebra (the properties in lemma 3.3.6, associativity, commutativity, distributivity, and idempotence in particular) but the reader should be aware that there are important and interesting consequences of this fact. However, these considerations are beyond the scope of this text. We refer the reader to Halmos, 1974, or Birkhoff, 1973 for a comprehensive study of these algebras.

* 3.3.5 NP-Complete Problems

It has been remarked earlier that both the satisfiability problem (SAT) and the tautology problem $(TAUT)$ are computationally hard problems, in the sense that known algorithms to solve them require an exponential number of steps in the length of the input. Even though modern computers are capable of performing very complex computations much faster than humans can, there are problems whose computational complexity is such that it would take too much time or memory space to solve them with a computer. Such problems are called *intractable*. It should be noted that this does not mean that we do not have algorithms to solve such problems. This means that all known algorithms solving these problems in theory either require too much time or too much memory space to solve them in practice, except perhaps in rather trivial cases. An algorithm is considered intractable if either it requires an exponential number of steps, or an exponential amount of space, in the length of the input. This is because exponential functions grow very fast. For example, $2^{10} = 1024$, but 2^{1000} is equal to $10^{1000 log_{10} 2}$, which has over 300 digits! A problem that can be solved in polynomial time and polynomial space is considered to be tractable.

It is not known whether SAT or $TAUT$ are tractable, and in fact, it is conjectured that they are not. But SAT and $TAUT$ play a special role for another reason. There is a class of problems (NP) which contains problems for which no polynomial-time algorithms are known, but for which polynomial-time solutions exist, if we are allowed to make guesses, and if we are not charged for checking wrong guesses, but only for successful guesses leading to an answer. SAT is such a problem. Indeed, given a proposition A, if one is allowed to guess valuations, it is not difficult to design a polynomial-time algorithm to check that a valuation v satisfies A. The satisfiability problem can be solved nondeterministically by guessing valuations and checking that they satisfy the given proposition. Since we are not "charged" for checking wrong guesses, such a procedure works in polynomial-time.

A more accurate way of describing such algorithms is to say that free backtracking is allowed. If the algorithm reaches a point where several choices are possible, any choice can be taken, but if the path chosen leads to a dead end, the algorithm can jump back (backtrack) to the latest choice point, with no cost of computation time (and space) consumed on a wrong path involved. Technically speaking, such algorithms are called *nondeterministic*. A nondeterministic algorithm can be simulated by a deterministic algorithm, but the deterministic algorithm needs to keep track of the nondeterministic choices explicitly (using a stack), and to use a backtracking technique to handle unsuccessful computations. Unfortunately, all known backtracking techniques yield exponential-time algorithms.

In order to discuss complexity issues rigorously, it is necessary to define a model of computation. Such a model is the *Turing machine* (invented by the mathematician Turing, circa 1935). We will not present here the theory of Turing Machines and complexity classes, but refer the interested reader to Lewis and Papadimitriou, 1981, or Davis and Weyuker, 1983. We will instead conclude with an informal discussion of the classes P and NP.

In dealing with algorithms for solving classes of problems, it is convenient to assume that problems are encoded as sets of strings over a finite alphabet Σ. Then, an algorithm for solving a problem A is an algorithm for deciding whether for any string $u \in \Sigma^*$, u is a member of A. For example, the satisfiability problem is encoded as the set of strings representing satisfiable propositions, and the tautology problem as the set of strings representing tautologies.

A Turing machine is an abstract computing device used to accept sets of strings. Roughly speaking, a Turing machine M consists of a finite set of states and of a finite set of instructions. The set of states is partitioned into two subsets of *accepting* and *rejecting* states.

To explain how a Turing machine operates, we define the notion of an instantaneous description (ID) and of a computation. An instantaneous description is a sort of snapshot of the configuration of the machine during a computation that, among other things, contains a state component. A Turing machine operates in discrete steps. Every time an instruction is executed, the ID describing the current configuration is updated. The intuitive idea is that executing an instruction I when the machine is in a configuration described by an instantaneous description C_1 yields a new configuration described by C_2. A computation is a finite or infinite sequence of instantaneous descriptions $C_0, ..., C_n$ (or $C_0, ..., C_n, C_{n+1}, ...$, if it is infinite), where C_0 is an initial instantaneous description containing the input, and each C_{i+1} is obtained from C_i by application of some instruction of M. A finite computation is called a halting computation, and the last instantaneous description C_n is called a final instantaneous description.

A Turing machine is deterministic, if for every instantaneous description C_1, at most one instantaneous description C_2 follows from C_1 by execution of some instruction of M. It is nondeterministic if for any ID C_1, there may be several successors C_2 obtained by executing (different) instructions of M.

Given a halting computation $C_0, ..., C_n$, if the final ID C_n contains an accepting state, we say that the computation is an *accepting computation*, otherwise it is a *rejecting computation*.

A set A (over Σ) is accepted deterministically in polynomial time if there is a deterministic Turing machine M and a polynomial p such that, for every input u,

(i) $u \in A$ iff the computation $C_0, ..., C_n$ on input u is an accepting computation such that $n \leq p(|u|)$ and,

(ii) $u \notin A$ iff the computation $C_0, ..., C_n$ on input u is a rejecting computation such that $n \leq p(|u|)$.

A set A (over Σ) is accepted nondeterministically in polynomial time if there is a nondeterministic Turing machine M and a polynomial p, such that, for every input u, there is *some* accepting computation $C_0, ..., C_n$ such that $n \leq p(|u|)$.

It should be noted that in the nondeterministic case, a string u is rejected by a Turing machine M (that is, $u \notin A$) iff every computation of M is either a rejecting computation $C_0, ..., C_n$ such that $n \leq p(|u|)$, or a computation that takes more than $p(|u|)$ steps on input u.

The class of sets accepted deterministically in polynomial time is denoted by P, and the class of sets accepted nondeterministically in polynomial time is denoted by NP. It is obvious from the definitions that P is a subset of NP. However, whether $P = NP$ is unknown, and in fact, is a famous open problem.

The importance of the class P resides in the widely accepted (although somewhat controversial) opinion that P consists of the problems that can be realistically solved by computers. The importance of NP lies in the fact that many problems for which efficient algorithms would be highly desirable, but are yet unknown, belong to NP. The traveling salesman problem and the integer programming problem are two such problems among many others.

For an extensive list of problems in NP, the reader should consult Garey and Johnson, 1979.

The importance of the satisfiability problem SAT is that it is *NP-complete*. This implies the remarkable fact that if SAT is in P, then $P = NP$. In other words, the existence of a polynomial-time algorithm for SAT implies that all problems in NP have polynomial-time algorithms, which includes many interesting problems apparently untractable at the moment.

In order to explain the notion of NP-completeness, we need the concept of polynomial-time reducibility. Deterministic Turing machines can also be used to compute functions. Given a function $f : \Sigma^* \to \Sigma^*$, a deterministic Turing machine M computes f if for every input u, there is a halting computation $C_0, ..., C_n$ such that C_0 contains u as input and C_n contains $f(u)$ as output. The machine M computes f in polynomial time iff there is a polynomial p such that, for every input u, $n \le p(|u|)$, where n is the number of steps in the computation on input u. Then, we say that a set A is polynomially reducible to a set B if there is a function $f : \Sigma^* \to \Sigma^*$ computable in polynomial time such that, for every input u,

$$u \in A \quad \text{if and only if} \quad f(u) \in B.$$

A set B is *NP-hard* if every set A in NP is reducible to B. A set B is *NP-complete* if it is in NP, and it is *NP-hard*.

The significance of NP-complete problems lies in the fact that if one finds a polynomial time algorithm for any NP-complete problem B, then $P = NP$. Indeed, given any problem $A \in NP$, assuming that the deterministic Turing machine M solves B in polynomial time, we could construct a deterministic Turing machine M' solving A as follows. Let M_f be the deterministic Turing machine computing the reduction function f. Then, to decide whether any arbitrary input u is in A, run M_f on input u, producing $f(u)$, and then run M on input $f(u)$. Since $u \in A$ if and only if $f(u) \in B$, the above procedure solves A. Furthermore, it is easily shown that a deterministic Turing machine M' simulating the composition of M_f and M can be constructed, and that it runs in polynomial time (because the functional composition of polynomials is a polynomial).

The importance of SAT lies in the fact that it was shown by S. A. Cook (Cook, 1971) that SAT is an NP-complete problem. In contrast, whether $TAUT$ is in NP is an open problem. But $TAUT$ is interesting for two other reasons. First, it can be shown that if $TAUT$ is in P, then $P = NP$. This is unlikely since we do not even know whether $TAUT$ is in NP. The second reason is related to the closure of NP under complementation. NP is said to be *closed under complementation* iff for every set A in NP, its complement $\Sigma^* - A$ is also in NP.

The class P is closed under complementation, but this is an open problem for the class NP. Given a deterministic Turing machine M, in order to accept the complement of the set A accepted by M, one simply has to create the machine $\overline{M}$ obtained by swapping the accepting and the rejecting states of M. Since for every input u, the computation $C_0, ..., C_n$ of M on input u halts in $n \leq p(|u|)$ steps, the modified machine $\overline{M}$ accepts $\Sigma^* - A$ is polynomial time. However, if M is nondeterministic, M may reject some input u because *all* computations on input u exceed the polynomial time bound $p(|u|)$. Thus, for this input u, there is no computation of the modified machine $\overline{M}$ which accepts u within $p(|u|)$ steps. The trouble is not that $\overline{M}$ cannot tell that u is rejected by M, but that $\overline{M}$ cannot report this fact in fewer than $p(|u|)$ steps. This shows that in the nondeterministic case, a different construction is required. Until now, no such construction has been discovered, and is rather unlikely that it will. Indeed, it can be shown that $TAUT$ is in NP if and only if NP is closed under complementation. Furthermore, since P is closed under complementation if NP is not closed under complementation, then $NP \neq P$.

Hence, one approach for showing that $NP \neq P$ would be to show that $TAUT$ is not in NP. This explains why a lot of efforts have been spent on the complexity of the tautology problem.

To summarize, the satisfiability problem SAT and the tautology problem $TAUT$ are important because of the following facts:

$$SAT \in P \quad \text{if and only if} \quad P = NP;$$
$$TAUT \in NP \quad \text{if and only if} \quad NP \text{ is closed under complementation;}$$
$$\text{If } TAUT \in P, \text{ then } P = NP;$$
$$\text{If } TAUT \notin NP, \text{ then } NP \neq P.$$

Since the two questions $P = NP$ and the closure of NP under complementation appear to be very hard to solve, and it is usually believed that their answer is negative, this gives some insight to the difficulty of finding efficient algorithms for SAT and $TAUT$. Also, the tautology problem appears to be harder than the satisfiability problem. For more details on these questions, we refer the reader to the article by S. A. Cook and R. A. Reckhow (Cook and Reckhow, 1971).

PROBLEMS

3.3.1. In this problem, it is assumed that the language of propositional logic is extended by adding the constant symbol $\top$, which is interpreted as **T**. Prove that the following propositions are tautologies by constructing truth tables.

Associativity rules:
$$((A \vee B) \vee C) \equiv (A \vee (B \vee C)) \quad ((A \wedge B) \wedge C) \equiv (A \wedge (B \wedge C))$$

Commutativity rules:

$$(A \vee B) \equiv (B \vee A) \quad (A \wedge B) \equiv (B \wedge A)$$

Distributivity rules:

$$(A \vee (B \wedge C)) \equiv ((A \vee B) \wedge (A \vee C))$$
$$(A \wedge (B \vee C)) \equiv ((A \wedge B) \vee (A \wedge C))$$

De Morgan's rules:

$$\neg(A \vee B) \equiv (\neg A \wedge \neg B) \quad \neg(A \wedge B) \equiv (\neg A \vee \neg B)$$

Idempotency rules:

$$(A \vee A) \equiv A \quad (A \wedge A) \equiv A$$

Double negation rule:

$$\neg\neg A \equiv A$$

Absorption rules:

$$(A \vee (A \wedge B)) \equiv A \quad (A \wedge (A \vee B)) \equiv A$$

Laws of zero and one:

$$(A \vee \bot) \equiv A \quad (A \wedge \bot) \equiv \bot$$
$$(A \vee \top) \equiv \top \quad (A \wedge \top) \equiv A$$
$$(A \vee \neg A) \equiv \top \quad (A \wedge \neg A) \equiv \bot$$

3.3.2. Show that the following propositions are tautologies.

$$A \supset (B \supset A)$$
$$(A \supset B) \supset ((A \supset (B \supset C)) \supset (A \supset C))$$
$$A \supset (B \supset (A \wedge B))$$
$$A \supset (A \vee B) \qquad B \supset (A \vee B)$$
$$(A \supset B) \supset ((A \supset \neg B) \supset \neg A)$$
$$(A \wedge B) \supset A \qquad (A \wedge B) \supset B$$
$$(A \supset C) \supset ((B \supset C) \supset ((A \vee B) \supset C))$$
$$\neg\neg A \supset A$$

3.3.3. Show that the following propositions are tautologies.

$$(A \supset B) \supset ((B \supset A) \supset (A \equiv B))$$
$$(A \equiv B) \supset (A \supset B)$$
$$(A \equiv B) \supset (B \supset A)$$

3.3.4. Show that the following propositions are not tautologies. Are they satisfiable? If so, give a satisfying valuation.

$$(A \supset C) \supset ((B \supset D) \supset ((A \vee B) \supset C))$$
$$(A \supset B) \supset ((B \supset \neg C) \supset \neg A)$$

3.3.5. Prove that the propositions of lemma 3.3.4 are tautologies.

* **3.3.6.** Given a function f of m arguments and m functions $g_1, ..., g_m$ each of n arguments, the composition of f and $g_1, ..., g_m$ is the function h of n arguments such that for all $x_1, ..., x_n$,

$$h(x_1, ..., x_n) = f(g_1(x_1, ..., x_n), ..., g_m(x_1, ..., x_n)).$$

For every integer $n \geq 1$, we let P_i^n, $(1 \leq i \leq n)$, denote the projection function such that, for all $x_1, .., x_n$,

$$P_i^n(x_1, ..., x_n) = x_i.$$

Then, given any k truth functions $H_1, ..., H_k$, let $\mathbf{TF}_n$ be the inductive closure of the set of functions $\{H_1, ..., H_k, P_1^n, ..., P_n^n\}$ under composition. We say that a truth function H of n arguments is *definable* from $H_1, ..., H_k$ if H belongs to $\mathbf{TF}_n$.

Let $H_{d,n}$ be the n-ary truth function such that

$$H_{d,n}(x_1, ..., x_n) = \mathbf{F} \quad \text{if and only if} \quad x_1 = ... = x_n = \mathbf{F},$$

and $H_{c,n}$ the n-ary truth function such that

$$H_{c,n}(x_1, ..., x_n) = \mathbf{T} \quad \text{if and only if} \quad x_1 = ... = x_n = \mathbf{T}.$$

(i) Prove that every n-ary truth function is definable in terms of $H_\neg$ and some of the functions $H_{d,n}$, $H_{c,n}$.

(ii) Prove that $H_\neg$ is not definable in terms of $H_\vee$, $H_\wedge$, $H_\supset$, and $H_\equiv$.

* **3.3.7.** Let H_{nor} be the binary truth function such that

$$H_{nor}(x, y) = \mathbf{T} \quad \text{if and only if} \quad x = y = \mathbf{F}.$$

Show that $H_{nor} = H_A$, where A is the proposition $(\neg P \wedge \neg Q)$. Show that $\{H_{nor}\}$ is functionally complete.

* **3.3.8.** Let H_{nand} be the binary truth function such that $H_{nand}(x, y) = \mathbf{F}$ if and only if $x = y = \mathbf{T}$. Show that $H_{nand} = H_B$, where B is the proposition $(\neg P \vee \neg Q)$. Show that $\{H_{nand}\}$ is functionally complete.

* **3.3.9.** An n-ary truth function H is *singulary* if there is a unary truth function H' and some i, $1 \leq i \leq n$, such that for all $x_1, ..., x_n$, $H(x_1, ..., x_n) = H'(x_i)$.

(i) Prove that if H is singulary, then every n-ary function definable in terms of H is also singulary. (See problem 3.3.6.)

(ii) Prove that if H is a binary truth function and $\{H\}$ is functionally complete, then either $H = H_{nor}$ or $H = H_{nand}$.

Hint: Show that $H(\mathbf{T}, \mathbf{T}) = \mathbf{F}$ and $H(\mathbf{F}, \mathbf{F}) = \mathbf{T}$, that only four binary truth functions have that property, and use (i).

3.3.10. A *substitution* is a function $s : \mathbf{PS} \to PROP$. Since $PROP$ is freely generated by $\mathbf{PS}$ and $\perp$, every substitution s extends to a unique function $\hat{s} : PROP \to PROP$ defined by recursion. Let A be any proposition containing the propositional symbols $\{P_1, ..., P_n\}$, and s_1 and s_2 be any two substitutions such that for every $P_i \in \{P_1, ..., P_n\}$, the propositions $s_1(P_i)$ and $s_2(P_i)$ are equivalent (that is $s_1(P_i) \equiv s_2(P_i)$ is a tautology).

Prove that the propositions $\hat{s}_1(A)$ and $\hat{s}_2(A)$ are equivalent.

3.3.11. Show that for every set Γ of propositions,

$$(i)\ \Gamma, A \models B \quad \text{if and only if} \quad \Gamma \models (A \supset B);$$
$$(ii)\ \text{If}\ \ \Gamma, A \models B \quad \text{and} \quad \Gamma, A \models \neg B, \quad \text{then} \quad \Gamma \models \neg A;$$
$$(iii)\ \text{If}\ \ \Gamma, A \models C \quad \text{and} \quad \Gamma, B \models C, \quad \text{then} \quad \Gamma, (A \vee B) \models C.$$

3.3.12. Assume that we consider propositions expressed only in terms of the set of connectives $\{\vee, \wedge, \neg\}$. The *dual* of a proposition A, denoted by A^*, is defined recursively as follows:

$$P^* = P, \quad \text{for every propositional symbol } P;$$
$$(A \vee B)^* = (A^* \wedge B^*);$$
$$(A \wedge B)^* = (A^* \vee B^*);$$
$$(\neg A)^* = \neg A^*.$$

(a) Prove that, for any two propositions A and B,

$$\models A \equiv B \quad \text{if and only if} \quad \models A^* \equiv B^*.$$

(b) If we change the definition of A^* so that for every propositional letter P, $P^* = \neg P$, prove that A^* and $\neg A$ are logically equivalent (that is, $(A^* \equiv \neg A)$ is a tautology).

3.3.13. A *literal* is either a propositional symbol P, or the negation $\neg P$ of a propositional symbol. A proposition A is in *disjunctive normal form* (DNF) if it is of the form $C_1 \vee ... \vee C_n$, where each C_i is a conjunction of literals.

(i) Show that the satisfiability problem for propositions in DNF can be decided in linear time. What does this say about the complexity of any algorithm for converting a proposition to disjunctive normal form?

(ii) Using the proof of theorem 3.3.1 and some of the identities of lemma 3.3.6, prove that every proposition A containing n propositional symbols is equivalent to a proposition A' is in disjunctive normal form, such that each disjunct C_i contains exactly n literals.

* **3.3.14.** Let $\oplus$ (*exclusive OR*) be the truth function defined by the proposition

$$(P \wedge \neg Q) \vee (\neg P \wedge Q).$$

(i) Prove that $\oplus$ is commutative and associative.

(ii) In this question, assume that the constant for false is denoted by 0 and that the constant for true is denoted by 1. Prove that the following are tautologies.

$$A \vee B \equiv A \wedge B \oplus A \oplus B$$
$$\neg A \equiv A \oplus 1$$
$$A \oplus 0 \equiv A$$
$$A \oplus A \equiv 0$$
$$A \wedge 1 \equiv A$$
$$A \wedge A \equiv A$$
$$A \wedge (B \oplus C) \equiv A \wedge B \oplus A \wedge C$$
$$A \wedge 0 \equiv 0$$

(iii) Prove that $\{\oplus, \wedge, 1\}$ is functionally complete.

* **3.3.15.** Using problems 3.3.13 and 3.3.14, prove that every proposition A is equivalent to a proposition A' which is either of the form 0, 1, or $C_1 \oplus \ldots \oplus C_n$, where each C_i is either 1 or a conjunction of positive literals. Furthermore, show that A' can be chosen so that the C_i are distinct, and that the positive literals in each C_i are all distinct (such a proposition A' is called a *reduced exclusive-OR normal form*).

* **3.3.16.** (i) Prove that if A' and A'' are reduced exclusive-OR normal forms of a same proposition A, then they are equal up to commutativity, that is:

$$\text{Either} \quad A' = A'' = 0, \text{ or } A' = A'' = 1, \text{ or}$$
$$A' = C_1' \wedge \ldots \wedge C_n', \quad A'' = C_1'' \wedge \ldots \wedge C_n'',$$

where each C_i' is a permutation of some C_j'' (and conversely).

Hint: There are 2^{2^n} truth functions of n arguments.

(ii) Prove that a proposition is a tautology if and only if its reduced exclusive-OR normal form is 1. What does this say about the complexity of any algorithm for converting a proposition to reduced exclusive-OR normal form?

* **3.3.17.** A set Γ of propositions is *independent* if, for every $A \in \Gamma$,

$$\Gamma - \{A\} \not\models A.$$

(a) Prove that every finite set Γ has a finite independent subset Δ such that, for every $A \in \Gamma$, $\Delta \models A$.

(b) Let Γ be ordered as the sequence $< A_1, A_2, ... >$. Find a sequence $\Gamma' =< B_1, B_2, ... >$ equivalent to Γ (that is, for every $i \geq 1$, $\Gamma \models B_i$ and $\Gamma' \models A_i$), such that, for every $i \geq 1$, $\models (B_{i+1} \supset B_i)$, but $\not\models (B_i \supset B_{i+1})$. Note that Γ' may be finite.

(c) Consider a countable sequence Γ' as in (b). Define $C_1 = B_1$, and for every $i \geq 1$, $C_{n+1} = (B_n \supset B_{n+1})$. Prove that $\Delta =< C_1, C_2, ... >$ is equivalent to Γ' and independent.

(d) Prove that every countable set Γ is equivalent to an independent set.

(e) Show that Δ need not be a subset of Γ.

Hint: Consider

$$\{P_0, P_0 \wedge P_1, P_0 \wedge P_1 \wedge P_1, ...\}.$$

* **3.3.18.** See problem 3.3.13 for the definition of a *literal*. A proposition A is a *basic Horn formula* iff it is a disjunction of literals, with at most one positive literal (literal of the form P). A proposition is a *Horn formula* iff it is a conjunction of basic Horn formulae.

(a) Show that every Horn formula A is equivalent to a conjunction of distinct formulae of the form,

$$P_i, \quad \text{or}$$
$$\neg P_1 \vee .... \vee \neg P_n, (n \geq 1), \quad \text{or}$$
$$\neg P_1 \vee .... \vee \neg P_n \vee P_{n+1}, (n \geq 1),$$

where all the P_i are distinct. We say that A is *reduced*.

(b) Let A be a reduced Horn formula $A = C_1 \wedge C_2 \wedge ... \wedge C_n$, where the C_i are distinct and each C_i is reduced as in (a). Since $\vee$ is commutative and associative (see problem 3.3.1), we can view each conjunct C_i as a set.

(i) Show that if no conjunct C_i is a positive literal, or every conjunct containing a negative literal also contains a positive literal, then A is satisfiable.

(ii) Assume that A contains some conjunct C_i having a single positive literal P, and some conjunct C_j distinct from C_i, such that C_j contains $\neg P$. Let $D_{i,j}$ be obtained by deleting $\neg P$ from C_j. Let A' be the conjunction obtained from A by replacing C_j by the conjunct $D_{i,j}$, provided that $D_{i,j}$ is not empty.

Show that A is satisfiable if and only if A' is satisfiable and $D_{i,j}$ is not empty.

(iii) Using the above, prove that the satisfiability of a Horn formula A can be decided in time polynomial in the length of A.

Note: Linear-time algorithms are given in Dowling and Gallier, 1984.

3.4 Proof Theory of Propositional Logic: The Gentzen System G'

In this section, we present a proof system for propositional logic and prove some of its main properties: soundness and completeness.

3.4.1 Basic Idea: Searching for a Counter Example

As we have suggested in Section 3.3, another perhaps more effective way of testing whether a proposition A is a tautology is to search for a valuation that falsifies A. In this section, we elaborate on this idea. As we progress according to this plan, we will be dealing with a tree whose nodes are labeled with pairs of finite lists of propositions. In our attempt to falsify A, the tree is constructed in such a way that we are trying to find a valuation that makes every proposition occurring in the first component of a pair labeling a node true, and all propositions occurring in the second component of that pair false. Hence, we are naturally led to deal with pairs of finite sequences of propositions called *sequents*. The idea of using sequents originates with Gentzen, although Gentzen's motivations were quite different. A proof system using sequents is very natural because the rules reflect very clearly the semantics of the connectives. The idea of searching for a valuation falsifying the given proposition is simple, and the tree-building algorithm implementing this search is also simple. Let us first illustrate the falsification procedure by means of an example.

EXAMPLE 3.4.1

Let

$$A = (P \supset Q) \supset (\neg Q \supset \neg P).$$

Initially, we start with a one-node tree labeled with the pair

$$(<>, < (P \supset Q) \supset (\neg Q \supset \neg P) >)$$

whose first component is the empty sequence and whose second component is the sequence containing the proposition A that we are attempting to falsify. In order to make A false, we must make $P \supset Q$ true and $\neg Q \supset \neg P$ false. Hence, we build the following tree:

$$\frac{(< P \supset Q >, < \neg Q \supset \neg P >)}{(<>, < (P \supset Q) \supset (\neg Q \supset \neg P) >)}$$

Now, in order to make $P \supset Q$ true, we must *either* make P false *or* Q true. The tree must therefore split as shown:

$$\frac{\dfrac{(<>, < P, \neg Q \supset \neg P >) \qquad\qquad (< Q >, < \neg Q \supset \neg P >)}{(< P \supset Q >, < \neg Q \supset \neg P >)}}{(<>, < (P \supset Q) \supset (\neg Q \supset \neg P) >)}$$

We continue the same procedure with each leaf. Let us consider the leftmost leaf first. In order to make $\neg Q \supset \neg P$ false, we must make $\neg Q$ true and $\neg P$ false. We obtain the tree:

$$\frac{\dfrac{\dfrac{(< \neg Q >, < P, \neg P >)}{(<>, < P, \neg Q \supset \neg P >)} \qquad\qquad (< Q >, < \neg Q \supset \neg P >)}{(< P \supset Q >, < \neg Q \supset \neg P >)}}{(<>, < (P \supset Q) \supset (\neg Q \supset \neg P) >)}$$

But now, in order to falsify the leftmost leaf, we must make both P and $\neg P$ false and $\neg Q$ true. This is impossible. We say that this leaf of the tree is *closed*. We still have to continue the procedure with the rightmost leaf, since there may be a way of obtaining a falsifying valuation this way. To make $\neg Q \supset \neg P$ false, we must make $\neg Q$ true and $\neg P$ false, obtaining the tree:

$$\frac{\dfrac{\dfrac{(< \neg Q >, < P, \neg P >)}{(<>, < P, \neg Q \supset \neg P >)} \qquad\qquad \dfrac{(< Q, \neg Q >, < \neg P >)}{(< Q >, < \neg Q \supset \neg P >)}}{(< P \supset Q >, < \neg Q \supset \neg P >)}}{(<>, < (P \supset Q) \supset (\neg Q \supset \neg P) >)}$$

This time, we must try to make $\neg P$ false and both Q and $\neg Q$ false, which is impossible. Hence, this branch of the tree is also closed, and our attempt to falsify A has failed. However, this failure to falsify A is really a success, since, as we shall prove shortly, this demonstrates that A is valid!

Trees as above are called *deduction trees*. In order to describe precisely the algorithm we have used in our attempt to falsify the proposition A, we need to state clearly the rules that we have used in constructing the tree.

3.4.2 Sequents and the Gentzen System G'

First, we define the notion of a sequent.

Definition 3.4.1 A *sequent* is a pair (Γ, Δ) of finite (possibly empty) sequences $\Gamma =< A_1, ..., A_m >$, $\Delta =< B_1, ..., B_n >$ of propositions.

Instead of using the notation (Γ, Δ), a sequent is usually denoted as $\Gamma \rightarrow \Delta$. For simplicity, a sequence $< A_1, ..., A_m >$ is denoted as $A_1, ..., A_m$. If Γ is the empty sequence, the corresponding sequent is denoted as $\rightarrow \Delta$; if Δ is empty, the sequent is denoted as $\Gamma \rightarrow$. and if both Γ and Δ are empty, we have the special sequent $\rightarrow$ (the *inconsistent sequent*). Γ is called the *antecedent* and Δ the *succedent*.

The intuitive meaning of a sequent is that a valuation v makes a sequent $A_1, ..., A_m \rightarrow B_1, ..., B_n$ true iff

$$v \models (A_1 \wedge ... \wedge A_m) \supset (B_1 \vee ... \vee B_n).$$

Equivalently, v makes the sequent false if v makes $A_1, ..., A_m$ all true and $B_1, ..., B_n$ all false.

It should be noted that the semantics of sequents suggests that instead of using sequences, we could have used sets. We could indeed define sequents as pairs (Γ, Δ) of finite *sets* of propositions, and all the results in this section would hold. The results of Section 3.5 would also hold, but in order to present the generalization of the tree construction procedure, we would have to order the sets present in the sequents anyway. Rather than switching back and forth between sets and sequences, we think that it is preferable to stick to a single formalism. Using sets instead of sequences can be viewed as an optimization.

The rules operating on sequents fall naturally into two categories: those operating on a proposition occurring in the antecedent, and those on a proposition occurring in the succedent. Both kinds of rules break the proposition on which the rule operates into subpropositions that may also be moved from the antecedent to the succedent, or vice versa. Also, the application of a rule may cause a sequent to be split into two sequents. This causes branching in the trees. Before stating the rules, let us mention that it is traditional in logic to represent trees with their root at the bottom instead of the root at the top as it is customary in computer science. The main reason is that a tree obtained in failing to falsify a given proposition can be viewed as a formal proof of the proposition. The proposition at the root of the tree is the logical conclusion of a set of inferences, and it is more natural to draw a proof tree in such a way that each premise in a rule occurs above its conclusion. However, this may be a matter of taste (and perhaps, aesthetics).

In the rest of this section, it will be assumed that the set of connectives used is $\{\wedge, \vee, \supset, \neg\}$, and that $(A \equiv B)$ is an abbreviation for $(A \supset B) \wedge (B \supset A)$, and $\perp$ an abbreviation for $(P \wedge \neg P)$.

Definition 3.4.2 The Gentzen system G'. The symbols Γ, Δ, Λ will be used to denote arbitrary sequences of propositions and A, B to denote propositions. The *inference rules* of the sequent calculus G' are the following:

$$\frac{\Gamma, A, B, \Delta \rightarrow \Lambda}{\Gamma, A \wedge B, \Delta \rightarrow \Lambda} \; (\wedge : left) \qquad \frac{\Gamma \rightarrow \Delta, A, \Lambda \quad \Gamma \rightarrow \Delta, B, \Lambda}{\Gamma \rightarrow \Delta, A \wedge B, \Lambda} \; (\wedge : right)$$

$$\frac{\Gamma, A, \Delta \rightarrow \Lambda \quad \Gamma, B, \Delta \rightarrow \Lambda}{\Gamma, A \vee B, \Delta \rightarrow \Lambda} \; (\vee : left) \qquad \frac{\Gamma \rightarrow \Delta, A, B, \Lambda}{\Gamma \rightarrow \Delta, A \vee B, \Lambda} \; (\vee : right)$$

$$\frac{\Gamma, \Delta \rightarrow A, \Lambda \quad B, \Gamma, \Delta \rightarrow \Lambda}{\Gamma, A \supset B, \Delta \rightarrow \Lambda} \; (\supset : left) \qquad \frac{A, \Gamma \rightarrow B, \Delta, \Lambda}{\Gamma \rightarrow \Delta, A \supset B, \Lambda} \; (\supset : right)$$

$$\frac{\Gamma, \Delta \rightarrow A, \Lambda}{\Gamma, \neg A, \Delta \rightarrow \Lambda} \; (\neg : left) \qquad \frac{A, \Gamma \rightarrow \Delta, \Lambda}{\Gamma \rightarrow \Delta, \neg A, \Lambda} \; (\neg : right)$$

The name of every rule is stated immediately to its right. Every rule consists of one or two upper sequents called *premises* and of a lower sequent called the *conclusion*. The above rules are called *inference rules*. For every rule, the proposition to which the rule is applied is called the *principal formula*, the propositions introduced in the premises are called the *side formulae*, and the other propositions that are copied unchanged are called the *extra formulae*.

Note that every inference rule can be represented as a tree with two nodes if the rule has a single premise, or three nodes if the rule has two premises. In both cases, the root of the tree is labeled with the conclusion of the rule and the leaves are labeled with the premises. If the rule has a single premise, it is a tree of the form

$$(1) \quad S_1$$
$$|$$
$$(e) \quad S_2$$

where the premise labels the node with tree address 1, and the conclusion labels the node with tree address e. If it has two premises, it is a tree of the form

$$(1) \quad S_1 \diagdown \quad \diagup S_2 \quad (2)$$
$$S_3$$
$$(e)$$

where the first premise labels the node with tree address 1, the second premise labels the node with tree address 2, and the conclusion labels the node with tree address e.

EXAMPLE 3.4.2

Consider the following instance of the $\supset$:left rule:

$$\frac{A, B \rightarrow P, D \qquad Q, A, B \rightarrow D}{A, (P \supset Q), B \rightarrow D}$$

In the above inference, $(P \supset Q)$ is the principal formula, P and Q are side formulae, and A, B, D are extra formulae.

A careful reader might have observed that the rules ($\supset$:left), ($\supset$:right), ($\neg$:left), and ($\neg$:right) have been designed in a special way. Notice that the side proposition added to the antecedent of an upper sequent is added at the front, and similarly for the side proposition added to the succedent of an upper sequent. We have done so to facilitate the generalization of the *search* procedure presented below to infinite sequents.

We will now prove that the above rules achieve the falsification procedure sketched in example 3.4.1.

3.4.3 Falsifiable and Valid Sequents

First, we extend the concepts of falsifiability and validity to sequents.

Definition 3.4.3 A sequent $A_1, ..., A_m \rightarrow B_1, ..., B_n$ is *falsifiable* iff there exists a valuation v such that

$$v \models (A_1 \wedge ... \wedge A_m) \wedge (\neg B_1 \wedge ... \wedge \neg B_n).$$

A sequent as above is *valid* iff for every valuation v,

$$v \models (A_1 \wedge ... \wedge A_m) \supset (B_1 \vee ... \vee B_n).$$

This is also denoted by

$$\models A_1, ..., A_m \rightarrow B_1, ..., B_n.$$

If $m = 0$, the sequent $\rightarrow B_1, ..., B_n$ is falsifiable iff the proposition $(\neg B_1 \wedge ... \wedge \neg B_n)$ is satisfiable, valid iff the proposition $(B_1 \vee ... \vee B_n)$ is valid. If $n = 0$, the sequent $A_1, ..., A_m \rightarrow$ is falsifiable iff the proposition $(A_1 \wedge ... \wedge A_m)$ is satisfiable, valid iff the proposition $(A_1 \wedge ... \wedge A_m)$ is not satisfiable. Note that a sequent $\Gamma \rightarrow \Delta$ is valid if and only if it is not falsifiable.

Lemma 3.4.1 For each of the rules given in definition 3.4.2, a valuation v falsifies the sequent occurring as the conclusion of the rule if and only if v falsifies at least one of the sequents occurring as premises. Equivalently, v makes the conclusion of a rule true if and only if v makes all premises of that rule true.

Proof: The proof consists in checking the truth tables of the logical connectives. We treat one case, leaving the others as an exercise. Consider the ($\supset$:left) rule:

$$\frac{\Gamma, \Delta \rightarrow A, \Lambda \qquad B, \Gamma, \Delta \rightarrow \Lambda}{\Gamma, (A \supset B), \Delta \rightarrow \Lambda}$$

For every valuation v, v falsifies the conclusion if and only if v satisfies all propositions in Γ and Δ, and satisfies $(A \supset B)$, and falsifies all propositions in Λ. From the truth table of $(A \supset B)$, v satisfies $(A \supset B)$ if either v falsifies A, or v satisfies B. Hence, v falsifies the conclusion if and only if, either

(1) v satisfies Γ and Δ, and falsifies A and Λ, or

(2) v satisfies B, Γ and Δ, and falsifies Λ. $\square$

3.4.4 Axioms, Deduction Trees, Proof Trees, Counter Example Trees

The central concept in any proof system is the notion of *proof*. First, we define the axioms of the system G'.

Definition 3.4.4 An *axiom* is any sequent $\Gamma \rightarrow \Delta$ such that Γ and Δ contain some common proposition.

Lemma 3.4.2 No axiom is falsifiable. Equivalently, every axiom is valid.

Proof: The lemma follows from the fact that in order to falsify an axiom, a valuation would have to make some proposition true on the left hand side, and that same proposition false on the right hand side, which is impossible. $\square$

Proof trees are given by the following inductive definition.

Definition 3.4.5 The set of *proof trees* is the least set of trees containing all one-node trees labeled with an axiom, and closed under the rules of definition 3.4.2 in the following sense:

(1) For any proof tree T_1 whose root is labeled with a sequent $\Gamma \rightarrow \Delta$, for any instance of a one-premise inference rule with premise $\Gamma \rightarrow \Delta$ and conclusion $\Lambda \rightarrow \Theta$, the tree T whose root is labeled with $\Lambda \rightarrow \Theta$ and whose subtree $T/1$ is equal to T_1 is a proof tree.

(2) For any two proof trees T_1 and T_2 whose roots are labeled with sequents $\Gamma \rightarrow \Delta$ and $\Gamma' \rightarrow \Delta'$ respectively, for every instance of a two-premise inference rule with premises $\Gamma \rightarrow \Delta$ and $\Gamma' \rightarrow \Delta'$ and conclusion $\Lambda \rightarrow \Theta$, the tree T whose root is labeled with $\Lambda \rightarrow \Theta$ and whose subtrees $T/1$ and $T/2$ are equal to T_1 and T_2 respectively is a proof tree.

The set of *deduction trees* is defined inductively as the least set of trees containing *all* one-node trees (not necessarily labeled with an axiom), and closed under (1) and (2) as above.

A deduction tree such that some leaf is labeled with a sequent $\Gamma \rightarrow \Delta$ where Γ, Δ consist of propositional letters and are disjoint is called a *counterexample tree*. The sequent labeling the root of a proof tree (deduction tree) is called the *conclusion* of the proof tree (deduction tree). A sequent is *provable* iff there exists a proof tree of which it is the conclusion. If a sequent $\Gamma \rightarrow \Delta$ is provable, this is denoted by

$$\vdash \Gamma \rightarrow \Delta.$$

EXAMPLE 3.4.3

The deduction tree below is a proof tree.

$$\dfrac{\dfrac{\dfrac{P, \neg Q \rightarrow P}{\neg Q \rightarrow \neg P, P}}{\rightarrow P, (\neg Q \supset \neg P)} \qquad \dfrac{\dfrac{Q \rightarrow Q, \neg P}{\neg Q, Q \rightarrow \neg P}}{Q \rightarrow (\neg Q \supset \neg P)}}{\dfrac{(P \supset Q) \rightarrow (\neg Q \supset \neg P)}{\rightarrow (P \supset Q) \supset (\neg Q \supset \neg P)}}$$

The above tree is a proof tree obtained from the proof tree

$$\dfrac{\dfrac{\dfrac{P, \neg Q \rightarrow P}{\neg Q \rightarrow \neg P, P}}{\rightarrow P, (\neg Q \supset \neg P)} \qquad \dfrac{\dfrac{Q \rightarrow Q, \neg P}{\neg Q, Q \rightarrow \neg P}}{Q \rightarrow (\neg Q \supset \neg P)}}{(P \supset Q) \rightarrow (\neg Q \supset \neg P)}$$

and the rule

$$\frac{(P \supset Q) \rightarrow (\neg Q \supset \neg P)}{\rightarrow (P \supset Q) \supset (\neg Q \supset \neg P)}$$

In contrast, the deduction tree below is a counter-example tree.

$$\frac{\dfrac{\dfrac{Q \rightarrow P}{Q, \neg P \rightarrow}}{\dfrac{\neg P \rightarrow \neg Q}{\rightarrow (\neg P \supset \neg Q)}}}{\rightarrow (P \supset Q) \wedge (\neg P \supset \neg Q)}$$

$$\frac{P \rightarrow Q}{\rightarrow (P \supset Q)}$$

The above tree is obtained from the two counter-example trees

$$\frac{P \rightarrow Q}{\rightarrow (P \supset Q)} \qquad \frac{\dfrac{\dfrac{Q \rightarrow P}{Q, \neg P \rightarrow}}{\neg P \rightarrow \neg Q}}{\rightarrow (\neg P \supset \neg Q)}$$

and the rule

$$\frac{\rightarrow (P \supset Q) \qquad \rightarrow (\neg P \supset \neg Q)}{\rightarrow (P \supset Q) \wedge (\neg P \supset \neg Q)}$$

It is easily shown that a deduction tree T is a proof tree if and only if every leaf sequent of T is an axiom.

Since proof trees (and deduction trees) are defined inductively, the induction principle applies. As an application, we now show that every provable sequent is valid.

3.4.5 Soundness of the Gentzen System G'

Lemma 3.4.3 Soundness of the system G'. If a sequent $\Gamma \rightarrow \Delta$ is provable, then it is valid.

Proof: We use the induction principle applied to proof trees. By lemma 3.4.2, every one-node proof tree (axiom) is valid. There are two cases in the induction step.

Case 1: The root of the proof tree T has a single descendant. In this case, T is obtained from some proof tree T_1 and some instance of a rule

$$\frac{S_1}{S_2}$$

By the induction hypothesis, S_1 is valid. Since by Lemma 3.4.1, S_1 is valid if and only if S_2 is valid, Lemma 3.4.3 holds.

Case 2: The root of the proof tree T has two descendants. In this case, T is obtained from two proof trees T_1 and T_2 and some instance of a rule

$$\frac{S_1 \qquad S_2}{S_3}$$

By the induction hypothesis, both S_1 and S_2 are valid. Since by lemma 3.4.1, S_3 is valid if and only if both S_1 and S_2 are, lemma 3.4.3 holds. $\square$

Next, we shall prove the fundamental theorem for the propositional sequent calculus G'. Roughly speaking, the fundamental theorem states that there exists a procedure for constructing a candidate counter-example tree, and that this procedure always terminates (is an algorithm). If the original sequent is valid, the algorithm terminates with a tree which is in fact a proof tree. Otherwise, the counter-example tree yields a falsifying valuation (in fact, all falsifying valuations). The fundamental theorem implies immediately the completeness of the sequent calculus G'.

3.4.6 The Search Procedure

The algorithm searching for a candidate counter-example tree builds this tree in a systematic fashion. We describe an algorithm that builds the tree in a breadth-first fashion. Note that other strategies for building such a tree could be used, (depth-first, in particular). A breadth-first expansion strategy was chosen because it is the strategy that works when we generalize the *search* procedure to infinite sequents. We will name this algorithm the *search procedure*.

Let us call a leaf of a tree *finished* iff the sequent labeling it is either an axiom, or all propositions in it are propositional symbols. We assume that a boolean function named *finished* testing whether a leaf is finished is available. A proposition that is a propositional symbol will be called *atomic*, and other propositions will be called *nonatomic*. A tree is finished when all its leaves are finished.

The procedure *search* traverses all leaves of the tree from left to right as long as not all of them are finished. For every unfinished leaf, the procedure *expand* is called. Procedure *expand* builds a subtree by applying the appropriate inference rule to every nonatomic proposition in the sequent labeling that leaf (proceeding from left to right). When the tree is finished, that is when all leaves are finished, either all leaves are labeled with axioms or some of the leaves are falsifiable. In the first case we have a proof tree, and in the second, all falsifying valuations can be found.

Definition 3.4.6 Procedure *search*. The input to *search* is a one-node tree labeled with a sequent $\Gamma \rightarrow \Delta$. The output is a finished tree T called a *systematic deduction tree*.

Procedure Search

procedure *search*$(\Gamma \to \Delta : sequent;$ **var** $T : tree);$
   begin
      let T be the one-node tree labeled with $\Gamma \to \Delta;$
      while not *all leaves of T are finished*   **do**
         $T_0 := T;$
         for each *leaf node of T_0*
         (in lexicographic order) **do**
            if not *finished(node)* **then**
               expand(node, T)
            endif
         endfor
      endwhile;
      if *all leaves are axioms*
      then
         write *('T is a proof of $\Gamma \to \Delta$')*
      else
         write *('$\Gamma \to \Delta$ is falsifiable')*
      endif
   end

Procedure Expand

procedure *expand(node : tree-address;* **var** $T : tree);$
   begin
      let $A_1, ..., A_m \to B_1, ..., B_n$ be the label of node;
      let S be the one-node tree labeled with
         $A_1, ..., A_m \to B_1, ..., B_n;$
      for $i := 1$ **to** m **do**
         if *nonatomic(A_i)* **then**
            $S :=$ *the new tree obtained from S by*
            applying to the descendant of A_i in
            every nonaxiom leaf of S the
            left rule applicable to A_i;
         endif
      endfor;
      for $i := 1$ **to** n **do**
         if *nonatomic(B_i)* **then**
            $S :=$ *the new tree obtained from S by*
            applying to the descendant of B_i in
            every nonaxiom leaf of S the
            right rule applicable to B_i;
         endif
      endfor;
      $T := dosubstitution(T, node, S)$
   end

The function *dosubstitution* yields the tree $T[node \leftarrow S]$ obtained by substituting the tree S at the address *node* in the tree T. Since a sequent $A_1, ..., A_m \rightarrow B_1, ..., B_n$ is processed from left to right, if the propositions $A_1, ..., A_{i-1}$ have been expanded so far, since the propositions A_i, ..., A_m, B_1, ..., B_n are copied unchanged, every leaf of the tree S obtained so far is of the form $\Gamma, A_i, ..., A_m \rightarrow \Delta, B_1, ..., B_n$. The occurrence of A_i following Γ is called the descendant of A_i in the sequent. Similarly, if the propositions $A_1, ..., A_m, B_1, ..., B_{i-1}$ have been expanded so far, every leaf of S is of the form $\Gamma \rightarrow \Delta, B_i, ..., B_n$, and the occurrence of B_i following Δ is called the descendant of B_i in the sequent. Note that the two *for* loops may yield a tree S of depth $m + n$.

A call to procedure *expand* is also called an *expansion step*. A *round* is the sequence of expansion calls performed during the *for* loop in procedure *search*, in which each unfinished leaf of the tree (T_0) is expanded. Note that if we change the function *finished* so that a leaf is finished if all propositions in the sequent labeling it are propositional symbols, the procedure *search* will produce trees in which leaves labeled with axioms also consist of sequents in which all propositions are atomic. Trees obtained in this fashion are called *atomically closed*.

EXAMPLE 3.4.4

Let us trace the construction of the systematic deduction tree (which is a proof tree) for the sequent

$$(P \wedge \neg Q), (P \supset Q), (T \supset R), (P \wedge S) \rightarrow T.$$

The following tree is obtained at the end of the first round:

$$
\cfrac{
\cfrac{
\cfrac{Q, P, \neg Q, P, S \rightarrow T, T}{Q, P, \neg Q, (P \wedge S) \rightarrow T, T} \qquad
\cfrac{R, Q, P, \neg Q, P, S \rightarrow T}{R, Q, P, \neg Q, (P \wedge S) \rightarrow T}
}{Q, P, \neg Q, (T \supset R), (P \wedge S) \rightarrow T}
}{
\cfrac{P, \neg Q, (P \supset Q), (T \supset R), (P \wedge S) \rightarrow T}{(P \wedge \neg Q), (P \supset Q), (T \supset R), (P \wedge S) \rightarrow T}
}
$$

$$\Gamma \rightarrow \Delta$$

where $\Gamma \rightarrow \Delta = P, \neg Q, (T \supset R), (P \wedge S) \rightarrow P, T$.

The leaves $Q, P, \neg Q, P, S \rightarrow T, T$ and $R, Q, P, \neg Q, P, S \rightarrow T$ are not axioms (yet). Note how the same rule was applied to $(P \wedge S)$ in the nodes labeled $Q, P, \neg Q, (P \wedge S) \rightarrow T, T$ and $R, Q, P, \neg Q, (P \wedge S) \rightarrow T$, because these occurrences are descendants of $(P \wedge S)$ in $(P \wedge \neg Q), (P \supset Q), (T \supset R), (P \wedge S) \rightarrow T$. After the end of the second round, we have the following tree, which is a proof tree.

$$\cfrac{\cfrac{\cfrac{Q,P,P,S \to Q,T,T}{Q,P,\neg Q,P,S \to T,T}}{Q,P,\neg Q,(P \wedge S) \to T,T} \qquad \cfrac{\cfrac{R,Q,P,P,S \to Q,T}{R,Q,P,\neg Q,P,S \to T}}{R,Q,P,\neg Q,(P \wedge S) \to T}}{\cfrac{\cfrac{Q,P,\neg Q,(T \supset R),(P \wedge S) \to T}{P,\neg Q,(P \supset Q),(T \supset R),(P \wedge S) \to T}}{(P \wedge \neg Q),(P \supset Q),(T \supset R),(P \wedge S) \to T}}$$

$$\Gamma \to \Delta$$

where $\Gamma \to \Delta = P, \neg Q, (T \supset R), (P \wedge S) \to P, T$.

It should also be noted that the algorithm of definition 3.4.6 could be made more efficient. For example, during a round through a sequent we could delay the application of two-premise rules. This can be achieved if a two-premise rule is applied only if a sequent does not contain propositions to which a one-premise rule applies. Otherwise the *expand* procedure is called only for those propositions to which a one-premise rule applies. In this fashion, smaller trees are usually obtained. For more details, see problem 3.4.13.

3.4.7 Completeness of the Gentzen System G'

We are now ready to prove the fundamental theorem of this section.

Theorem 3.4.1 The procedure *search* terminates for every finite input sequent. If the input sequent $\Gamma \to \Delta$ is valid, procedure *search* produces a proof tree for $\Gamma \to \Delta$; if $\Gamma \to \Delta$ is falsifiable, *search* produces a tree from which all falsifying valuations can be found.

Proof: Define the *complexity* of a proposition A as the number of logical connectives occurring in A (hence, propositional symbols have complexity 0). Given a sequent $A_1, ..., A_m \to B_1, ..., B_n$, define its *complexity* as the sum of the complexities of $A_1, ..., A_m, B_1, ..., B_n$. Then, observe that for every call to procedure *expand*, the complexity of every upper sequent involved in applying a rule is strictly smaller than the complexity of the lower sequent (to which the rule is applied). Hence, either all leaves will become axioms or their complexity will become 0, which means that the *while* loop will always terminate. This proves termination. We now prove the following claim.

Claim: Given any deduction tree T, a valuation v falsifies the sequent $\Gamma \to \Delta$ labeling the root of T if and only if v falsifies some sequent labeling a leaf of T.

Proof of claim: We use the induction principle for deduction trees. In case of a one-node tree, the claim is obvious. Otherwise, the deduction tree is either of the form

$$T_2$$
$$\frac{S_2}{S_1}$$

where the bottom part of the tree is a one-premise rule, or of the form

$$\frac{\begin{array}{cc} T_2 & T_3 \\ S_2 & S_3 \end{array}}{S_1}$$

where the bottom part of the tree is a two-premise rule. We consider the second case, the first one being similar. By the induction hypothesis, v falsifies S_2 if and only if v falsifies some leaf of T_2, and v falsifies S_3 if and only if v falsifies some leaf of T_3. By lemma 3.4.1, v falsifies S_1 if and only if v falsifies S_2 or S_3. Hence, v falsifies S_1 if and only if either v falsifies some leaf of T_2 or some leaf of T_3, that is, v falsifies some leaf of T. $\square$

As a consequence of the claim, the sequent $\Gamma \rightarrow \Delta$ labeling the root of the deduction tree is valid, if and only if all leaf sequents are valid. It is easy to check that *search* builds a deduction tree (in a breadth-first fashion). Now, either $\Gamma \rightarrow \Delta$ is falsifiable, or it is valid. In the first case, by the above claim, if v falsifies $\Gamma \rightarrow \Delta$, then v falsifies some leaf sequent of the deduction tree T. By the definition of *finished*, such a leaf sequent must be of the form $P_1, ..., P_m \rightarrow Q_1, ..., Q_n$ where the P_i and Q_j are propositional symbols, and the sets $\{P_1, ..., P_m\}$ and $\{Q_1, ..., Q_n\}$ are disjoint since the sequent is not an axiom. Hence, if $\Gamma \rightarrow \Delta$ is falsifiable, the deduction tree T is not a proof tree. Conversely, if T is not a proof tree, some leaf sequent of T is not an axiom. By the definition of *finished*, this sequent must be of the form $P_1, ..., P_m \rightarrow Q_1, ..., Q_n$ where the P_i and Q_j are propositional symbols, and the sets $\{P_1, ..., P_m\}$ and $\{Q_1, ..., Q_n\}$ are disjoint. The valuation v which makes every P_i true and every Q_j false falsifies the leaf sequent $P_1, ..., P_m \rightarrow Q_1, ..., Q_n$, and by the above claim, it also falsifies the sequent $\Gamma \rightarrow \Delta$. Therefore, we have shown that $\Gamma \rightarrow \Delta$ is falsifiable if and only if the deduction tree T is not a proof tree, or equivalently, that $\Gamma \rightarrow \Delta$ is valid if and only if the deduction tree T is a proof tree. Furthermore, the above proof also showed that if the deduction tree T is not a proof tree, all falsifying valuations for $\Gamma \rightarrow \Delta$ can be found by inspecting the nonaxiom leaves of T. $\square$

Corollary   Completeness of G'. Every valid sequent is provable. Furthermore, there is an algorithm for deciding whether a sequent is valid and if so, a proof tree is obtained.

As an application of the main theorem we obtain an algorithm to convert a proposition to conjunctive (or disjunctive) normal form.

3.4.8 Conjunctive and Disjunctive Normal Form

Definition 3.4.7 A proposition A is in *conjunctive normal form* (for short, CNF) if it is a conjunction $C_1 \wedge ... \wedge C_m$ of disjunctions $C_i = B_{i,1} \vee ... \vee B_{i,n_i}$, where each $B_{i,j}$ is either a propositional symbol P or the negation $\neg P$ of a propositional symbol. A proposition A is in *disjunctive normal form* (for short, DNF) if it is a disjunction $C_1 \vee ... \vee C_m$ of conjunctions $C_i = B_{i,1} \wedge ... \wedge B_{i,n_i}$, where each $B_{i,j}$ is either a propositional symbol P or the negation $\neg P$ of a propositional symbol.

Theorem 3.4.2 For every proposition A, a proposition A' in conjunctive normal form can be found such that $\models A \equiv A'$. Similarly, a proposition A'' in disjunctive normal form can be found such that $\models A \equiv A''$.

Proof: Starting with the input sequent $\rightarrow A$, let T be the tree given by the algorithm *search*. By theorem 3.4.1, either A is valid in which case all leaves are axioms, or A is falsifiable. In the first case, let $A' = P \vee \neg P$. Clearly, $\models A \equiv A'$. In the second case, the proof of theorem 3.4.1 shows that a valuation v makes A true if and only if it makes every leaf sequent true. For every nonaxiom leaf sequent $A_1, ..., A_m \rightarrow B_1, ..., B_n$, let

$$C = \neg A_1 \vee ... \vee \neg A_m \vee B_1 ... \vee B_n$$

and let A' be the conjunction of these propositions. Clearly, a valuation v makes A' true if and only if it makes every nonaxiom leaf sequent $A_1, ..., A_m \rightarrow B_1, ..., B_n$ true, if and only if it makes A true. Hence, $\models A \equiv A'$.

To get an equivalent proposition in disjunctive normal form, start with the sequent $A \rightarrow$. Then, a valuation v makes $A \rightarrow$ false if and only if v makes at least some of the sequent leaves false. Also, v makes $A \rightarrow$ false if and only if v makes A true. For every nonaxiom sequent leaf $A_1, ..., A_m \rightarrow B_1, ..., B_n$, let

$$C = A_1 \wedge ... \wedge A_m \wedge \neg B_1 \wedge ... \wedge \neg B_n$$

and let A'' be the disjunction of these propositions. We leave as an exercise to check that a valuation v makes some of the non-axiom leaves false if and only if it makes the disjunction A'' true. Hence $\models A \equiv A''$. $\square$

EXAMPLE 3.4.5

Counter-example tree for

$$\rightarrow (\neg P \supset Q) \supset (\neg R \supset S).$$

$$\frac{\dfrac{P \to R, S}{\to R, S, \neg P}}{\dfrac{\neg R \to S, \neg P}{\to \neg P, \neg R \supset S}} \qquad \frac{\dfrac{Q \to R, S}{\neg R, Q \to S}}{Q \to \neg R \supset S}$$

$$\frac{\neg P \supset Q \to \neg R \supset S}{\to (\neg P \supset Q) \supset (\neg R \supset S)}$$

An equivalent proposition in conjunctive normal form is:

$$(\neg Q \vee R \vee S) \wedge (\neg P \vee R \vee S).$$

EXAMPLE 3.4.6

Counter-example tree for

$$(\neg P \supset Q) \supset (\neg R \supset S) \to .$$

$$\frac{\dfrac{\to P, Q}{\neg P \to Q}}{\to \neg P \supset Q} \qquad \frac{\dfrac{R \to}{\to \neg R} \qquad S \to}{\neg R \supset S \to}$$

$$(\neg P \supset Q) \supset (\neg R \supset S) \to$$

An equivalent proposition in disjunctive normal form is:

$$S \vee R \vee (\neg P \wedge \neg Q).$$

We present below another method for converting a proposition to conjunctive normal form that does not rely on the construction of a deduction tree. This method is also useful in conjunction with the resolution method presented in Chapter 4. First, we define the negation normal form of a proposition.

3.4.9 Negation Normal Form

The set of propositions in negation normal form is given by the following inductive definition.

Definition 3.4.8 The set of propositions in *negation normal form* (for short, NNF) is the inductive closure of the set of propositions $\{P, \neg P \mid P \in \mathbf{PS}\}$ under the constructors $C_\vee$ and $C_\wedge$.

More informally, propositions in NNF are defined as follows:

(1) For every propositional letter P, P and $\neg P$ are in NNF;

(2) If A and B are in NNF, then $(A \vee B)$ and $(A \wedge B)$ are in NNF.

Lemma 3.4.4 Every proposition is equivalent to a proposition in NNF.

Proof: By theorem 3.3.1, we can assume that the proposition A is expressed only in terms of the connectives $\vee$ and $\wedge$ and $\neg$. The rest of the proof proceeds by induction on the number of connectives. By clause (1) of definition 3.4.8, every propositional letter is in NNF. Let A be of the form $\neg B$. If B is a propositional letter, by clause (1) of definition 3.4.8, the property holds. If B is of the form $\neg C$, by lemma 3.3.6, $\neg\neg C$ is equivalent to C, by the induction hypothesis, C is equivalent to a proposition C' in NNF, and by lemma 3.3.5, A is equivalent to C', a proposition in NNF. If B is of the form $(C \vee D)$, by lemma 3.3.6, $\neg(C \vee D)$ is equivalent to $(\neg C \wedge \neg D)$. Note that both $\neg C$ and $\neg D$ have fewer connectives than A. Hence, by the induction hypothesis, $\neg C$ and $\neg D$ are equivalent to propositions C' and D' in NNF. By lemma 3.3.5, A is equivalent to $(C' \wedge D')$, which is in NNF. If B is of the form $(C \wedge D)$, by lemma 3.3.6, $\neg(C \wedge D)$ is equivalent to $(\neg C \vee \neg D)$. As in the previous case, by the induction hypothesis, $\neg C$ and $\neg D$ are equivalent to propositions C' and D' in NNF. By lemma 3.3.5, A is equivalent to $(C' \vee D')$, which is in NNF. Finally, if A is of the form $(B * C)$ where $* \in \{\wedge, \vee\}$, by the induction hypothesis, C and D are equivalent to propositions C' and D' in NNF, and by lemma 3.3.5, A is equivalent to $(C' * D')$ which is in NNF. $\square$

Lemma 3.4.5 Every proposition A (containing only the connectives $\vee, \wedge, \neg$) can be transformed into an equivalent proposition in conjunctive normal form, by application of the following identities:

$$\neg\neg A \simeq A$$
$$\neg(A \wedge B) \simeq (\neg A \vee \neg B)$$
$$\neg(A \vee B) \simeq (\neg A \wedge \neg B)$$
$$A \vee (B \wedge C) \simeq (A \vee B) \wedge (A \vee C)$$
$$(B \wedge C) \vee A \simeq (B \vee A) \wedge (C \vee A)$$
$$(A \wedge B) \wedge C \simeq A \wedge (B \wedge C)$$
$$(A \vee B) \vee C \simeq A \vee (B \vee C)$$

Proof: The proof of lemma 3.4.4 only uses the first three tautologies. Hence, given a proposition A, we can assume that it is already in NNF. We prove by induction on propositions that a proposition in NNF can be converted to a proposition in CNF using the last four tautologies. If A is of the form P or $\neg P$, we are done. If A is of the form $(B \vee C)$ with B and C in NNF, by the induction hypothesis both B and C are equivalent to propositions B' and C' in CNF. If both B' and C' consist of a single conjunct, $(B' \vee C')$ is a disjunction of propositional letters or negations of propositional letters and by lemma 3.3.5, A is equivalent to $(B' \vee C')$ which is in CNF. Otherwise, let $B' = B'_1 \wedge ... \wedge B'_m$ and $C' = C'_1 \wedge ... \wedge C'_n$, with either $m > 1$ or $n > 1$. By repeated applications of the distributivity and associativity rules (to be rigorous, by induction on $m + n$),

$$(B' \vee C') = (B'_1 \wedge ... \wedge B'_m) \vee (C'_1 \wedge ... \wedge C'_n)$$
$$\simeq ((B'_1 \wedge ... \wedge B'_m) \vee C'_1) \wedge ... \wedge ((B'_1 \wedge ... \wedge B'_m) \vee C'_n)$$
$$\simeq \bigwedge \{(B'_i \vee C'_j) \mid 1 \leq i \leq m, \ 1 \leq j \leq n\}.$$

The resulting proposition is in CNF, and by lemma 3.3.5, A is equivalent to a proposition in CNF. If A is of the form $(B \wedge C)$ where B and C are in NNF, by the induction hypothesis, B and C are equivalent to propositions B' and C' in CNF. But then, $(B' \wedge C')$ is in CNF, and by lemma 3.4.5, A is equivalent to $(B' \wedge C')$. $\square$

The conjunctive normal form of a proposition may be simplified by using the commutativity rules and the idempotency rules given in lemma 3.3.6. A lemma similar to lemma 3.4.5 can be shown for the disjunctive normal form of a proposition.

EXAMPLE 3.4.7

Consider the proposition

$$A = (\neg P \supset Q) \supset (\neg R \supset S).$$

First, we eliminate $\supset$ using the fact that $(\neg B \vee C)$ is equivalent to $(B \supset C)$. We get

$$(\neg(\neg\neg P \vee Q)) \vee (\neg\neg R \vee S).$$

Then, we put this proposition in NNF. We obtain

$$(\neg P \wedge \neg Q) \vee (R \vee S).$$

Using distributivity we obtain

$$(\neg P \vee R \vee S) \wedge (\neg Q \vee R \vee S),$$

which is the proposition obtained in example 3.4.5 (up to commutativity). However, we could also have obtained the proposition

$$(\neg P \vee R) \wedge (\neg P \vee S) \wedge (\neg Q \vee R) \wedge (\neg Q \vee S),$$

by applying the distributivity rules in a different order. This illustrates the fact that the CNF (or DNF) of a proposition is generally not unique.

PROBLEMS

3.4.1. Give proof trees for the following tautologies:

$$A \supset (B \supset A)$$
$$(A \supset B) \supset ((A \supset (B \supset C)) \supset (A \supset C))$$
$$A \supset (B \supset (A \wedge B))$$
$$A \supset (A \vee B) \qquad B \supset (A \vee B)$$
$$(A \supset B) \supset ((A \supset \neg B) \supset \neg A)$$
$$(A \wedge B) \supset A \qquad (A \wedge B) \supset B$$
$$(A \supset C) \supset ((B \supset C) \supset ((A \vee B) \supset C))$$
$$\neg\neg A \supset A$$

3.4.2. Using counter-example trees, give propositions in conjunctive and disjunctive normal form equivalent to the following propositions:

$$(A \supset C) \supset ((B \supset D) \supset ((A \vee B) \supset C))$$
$$(A \supset B) \supset ((B \supset \neg C) \supset \neg A)$$

3.4.3. Recall that $\perp$ is a constant symbol always interpreted as **F**.

(i) Show that the following equivalences are valid.

$$\neg A \equiv (A \supset \perp)$$
$$(A \vee B) \equiv ((A \supset \perp) \supset B)$$
$$(A \equiv B) \equiv (A \supset B) \wedge (B \supset A)$$
$$(A \wedge B) \equiv \neg(\neg A \vee \neg B)$$

(ii) Show that every proposition A is equivalent to a proposition A' using only the connective $\supset$ and the constant symbol $\perp$.

(iii) Consider the following Gentzen-like rules for propositions over the language consisting of the propositional letters, $\supset$ and $\perp$.

The symbols Γ, Δ, Λ denote finite arbitrary sequences of propositions (possibly empty):

$$\frac{\Gamma, \Delta \to A, \Lambda \quad B, \Gamma, \Delta \to \Lambda}{\Gamma, (A \supset B), \Delta \to \Lambda} \qquad \frac{A, \Gamma \to B, \Delta, \Lambda}{\Gamma \to \Delta, (A \supset B), \Lambda}$$

$$\frac{\Gamma \to \Delta, \Lambda}{\Gamma \to \Delta, \perp, \Lambda}$$

The axioms of this systems are all sequents of the form $\Gamma \to \Delta$ where Γ and Δ contain a common proposition, and all sequents of the form $\Gamma, \perp, \Delta \to \Lambda$.

(a) Prove that for every valuation v, the conclusion of a rule is falsifiable if and only if one of the premises is falsifiable, and that the axioms are not falsifiable.

(b) Prove that the above Gentzen-like system is complete.

(c) Convert $(P \supset Q) \supset (\neg Q \supset \neg P)$ to a proposition involving only $\supset$ and $\perp$. Give a proof tree for this proposition in the above system.

3.4.4. Let C and D be propositions and P a propositional letter. Prove that the following equivalence (the resolution rule) holds by giving a proof tree:

$$(C \vee P) \wedge (D \vee \neg P) \equiv (C \vee P) \wedge (D \vee \neg P) \wedge (C \vee D)$$

Show that the above also holds if either C or D is missing.

3.4.5. Give Gentzen-like rules for equivalence ($\equiv$).

3.4.6. Give proof trees for the following tautologies:

Associativity rules:
$$((A \vee B) \vee C) \equiv (A \vee (B \vee C)) \quad ((A \wedge B) \wedge C) \equiv (A \wedge (B \wedge C))$$
Commutativity rules:
$$(A \vee B) \equiv (B \vee A) \quad (A \wedge B) \equiv (B \wedge A)$$
Distributivity rules:
$$(A \vee (B \wedge C)) \equiv ((A \vee B) \wedge (A \vee C))$$
$$(A \wedge (B \vee C)) \equiv ((A \wedge B) \vee (A \wedge C))$$
De Morgan's rules:
$$\neg(A \vee B) \equiv (\neg A \wedge \neg B) \quad \neg(A \wedge B) \equiv (\neg A \vee \neg B)$$
Idempotency rules:
$$(A \vee A) \equiv A \quad (A \wedge A) \equiv A$$
Double negation rule:
$$\neg\neg A \equiv A$$
Absorption rules:
$$(A \vee (A \wedge B)) \equiv A \quad (A \wedge (A \vee B)) \equiv A$$
Laws of zero and one:
$$(A \vee \perp) \equiv A \quad (A \wedge \perp) \equiv \perp$$
$$(A \vee \top) \equiv \top \quad (A \wedge \top) \equiv A$$
$$(A \vee \neg A) \equiv \top \quad (A \wedge \neg A) \equiv \perp$$

3.4.7. Instead of defining logical equivalence ($\simeq$) semantically as in definition 3.3.6, let us define $\simeq$ proof-theoretically so that, for all $A, B \in PROP$,

$$A \simeq B \text{ if and only if } \vdash (A \supset B) \wedge (B \supset A) \text{ in } G'.$$

Prove that $\simeq$ is an equivalence relation satisfying the properties of lemma 3.3.5 (Hence, $\simeq$ is a congruence).

3.4.8. Give Gentzen-like rules for the connective $\oplus$ (exclusive-or), $\oplus$ being the binary truth function defined by the proposition $(P \wedge \neg Q) \vee (\neg P \vee Q)$.

$*$ **3.4.9.** The *Hilbert system* H for propositional logic is defined below. For simplicity, it is assumed that only the connectives $\wedge$, $\vee$, $\supset$ and $\neg$ are used.

The *axioms* are all propositions given below, where A,B,C denote arbitrary propositions.

$$A \supset (B \supset A)$$
$$(A \supset B) \supset ((A \supset (B \supset C)) \supset (A \supset C))$$
$$A \supset (B \supset (A \wedge B))$$
$$A \supset (A \vee B), \qquad B \supset (A \vee B)$$
$$(A \supset B) \supset ((A \supset \neg B) \supset \neg A)$$
$$(A \wedge B) \supset A, \qquad (A \wedge B) \supset B$$
$$(A \supset C) \supset ((B \supset C) \supset ((A \vee B) \supset C))$$
$$\neg \neg A \supset A$$

There is a single inference rule, called *modus ponens* given by:

$$\frac{A \qquad (A \supset B)}{B}$$

Let $\{A_1, ..., A_m\}$ be any set of propositions. The concept of a *deduction tree* (in the system H) for a proposition B from the set $\{A_1, ..., A_m\}$ is defined inductively as follows:

(i) Every one-node tree labeled with an axiom B or a proposition B in $\{A_1, ..., A_m\}$ is a deduction tree of B from $\{A_1, ..., A_m\}$.

(ii) If T_1 is a deduction tree of A from $\{A_1, ..., A_m\}$ and T_2 is a deduction tree of $(A \supset B)$ from $\{A_1, ..., A_m\}$, then the following tree is a deduction tree of B from $\{A_1, ..., A_m\}$:

$$\frac{\dfrac{T_1}{A} \qquad \dfrac{T_2}{(A \supset B)}}{B}$$

A proof tree is a deduction tree whose leaves are labeled with axioms. Given a set $\{A_1, ..., A_m\}$ of propositions and a proposition B, we use the notation $A_1, ..., A_m \vdash B$ to denote that there is deduction of B from $\{A_1, ..., A_m\}$. In particular, if the set $\{A_1, ..., A_m\}$ is empty, the tree is a proof tree and we write $\vdash B$.

(i) Prove that *modus ponens* is a sound rule, in the sense that if both premises are valid, then the conclusion is valid. Prove that the system H is sound; that is, every proposition provable in H is valid.

(ii) Prove that for every proposition A, $\vdash (A \supset A)$.

Hint: Use the axioms $A \supset (B \supset A)$ and $(A \supset B) \supset ((A \supset (B \supset C)) \supset (A \supset C))$.

(iii) Prove the following:

(a) $A_1, ..., A_m \vdash A_i$, for every $i, 1 \leq i \leq m$.

(b)   If $A_1, ..., A_m \vdash B_i$ for every $i, 1 \leq i \leq m$ and
$B_1, ..., B_m \vdash C$, then $A_1, ..., A_m \vdash C$.

* **3.4.10.** In this problem, we are also considering the proof system H of problem 3.4.9. The *deduction theorem* states that, for arbitrary propositions $A_1, ..., A_m, A, B$,

$$\text{if } A_1, ..., A_m, A \vdash B, \text{ then}$$
$$A_1, ..., A_m \vdash (A \supset B).$$

Prove the deduction theorem.

Hint: Use induction on deduction trees. The base case is relatively easy. For the induction step, assume that the deduction tree is of the form

$$\frac{\begin{matrix} T_1 & T_2 \\ B_1 & (B_1 \supset B) \end{matrix}}{B}$$

where the leaves are either axioms or occurrences of the propositions $A_1, ..., A_m, A$. By the induction hypothesis, there are deduction trees T_1' for $(A \supset B_1)$ and T_2' for $(A \supset (B_1 \supset B))$, where the leaves of T_1' and T_2' are labeled with axioms or the propositions $A_1, ..., A_m$. Show how a deduction tree whose leaves are labeled with axioms or the propositions $A_1, ..., A_m$ can be obtained for $(A \supset B)$.

* **3.4.11.** In this problem, we are still considering the proof system H of problem 3.4.9. Prove that the following meta-rules hold about deductions in

the system H: For all propositions A,B,C and finite sequence Γ of propositions (possibly empty), we have:

Introduction	Elimination
$\supset$ If $\Gamma, A \vdash B$, then $\Gamma \vdash (A \supset B)$	$A, (A \supset B) \vdash B$
$\wedge$ $A, B \vdash (A \wedge B)$	$(A \wedge B) \vdash A$ $(A \wedge B) \vdash B$
$\vee$ $A \vdash (A \vee B)$	If $\Gamma, A \vdash C$ and $\Gamma, B \vdash C$ then $\Gamma, (A \vee B) \vdash C$
$\neg$ If $\Gamma, A \vdash B$ and $\Gamma, A \vdash \neg B$ then $\Gamma \vdash \neg A$ (reductio ad absurdum)	$\neg\neg A \vdash A$ (double negation elimination) $A, \neg A \vdash B$ (weak negation elimination)

Hint: Use problem 3.4.9(iii) and the deduction theorem.

* **3.4.12.** In this problem it is shown that the Hilbert system H is complete, by proving that for every Gentzen proof T of a sequent $\rightarrow A$, where A is any proposition, there is a proof in the system H.

(i) Prove that for arbitrary propositions $A_1, ..., A_m,\ B_1, ..., B_n$,

(a) in H, for $n > 0$,

$$A_1, ..., A_m, \neg B_1, ..., \neg B_n \vdash P \wedge \neg P \text{ if and only if}$$
$$A_1, ..., A_m, \neg B_1, ..., \neg B_{n-1} \vdash B_n, \text{ and}$$

(b) in H, for $m > 0$,

$$A_1, ..., A_m, \neg B_1, ..., \neg B_n \vdash P \wedge \neg P \text{ if and only if}$$
$$A_2, ..., A_m, \neg B_1, ..., \neg B_n \vdash \neg A_1.$$

(ii) Prove that for any sequent $A_1, ..., A_m \rightarrow B_1, ..., B_n$, if

$$A_1, ..., A_m \rightarrow B_1, ..., B_n$$

is provable in the Gentzen system G' then

$$A_1, ..., A_m, \neg B_1, ..., \neg B_n \vdash (P \wedge \neg P)$$

is a deduction in the Hilbert system H. Conclude that H is complete.

Hint: Use problem 3.4.11.

3.4.13. Consider the modification of the algorithm of definition 3.4.6 obtained by postponing applications of two-premise rules. During a round, a

two-premise rule is applied only if a sequent does not contain propositions to which a one-premise rule applies. Otherwise, during a round
the *expand* procedure is called only for those propositions to which a
one-premise rule applies.

Show that theorem 3.4.1 still holds for the resulting algorithm. Compare the size of the proof trees obtained from both versions of the
search algorithm, by trying a few examples.

3.4.14. Write a computer program (preferably in PASCAL or C) implementing the *search* procedure of definition 3.4.6.

3.5 Proof Theory for Infinite Sequents: Extended Completeness of G'

In this section, we obtain some important results for propositional logic (extended completeness, compactness, model existence) by generalizing the procedure *search* to infinite sequents.

3.5.1 Infinite Sequents

We extend the concept of a sequent $\Gamma \to \Delta$ by allowing Γ or Δ to be countably infinite sequences of propositions. The method of this section is very
important because it can be rather easily adapted to show the completeness
of a Gentzen system obtained by adding quantifier rules to G' for first-order
logic (see Chapter 5).

By suitably modifying the *search* procedure, we can generalize the main
theorem (theorem 3.4.1) and obtain both the extended completeness theorem
and the compactness theorem. The procedure *search* is no longer an algorithm
since it can go on forever in some cases. However, if the input sequent is
valid, a finite proof will be obtained. Also, if the input sequent is falsifiable,
a falsifying valuation will be (nonconstructively) obtained.

We will now allow sequents $\Gamma \to \Delta$ in which Γ or Δ can be countably
infinite sequences. It is convenient to assume that Γ is written as $A_1, ..., A_m, ...$
(possibly infinite to the right) and that Δ is written as $B_1, ..., B_n, ...$ (possibly
infinite to the right). Hence, a sequent will be denoted as

$$A_1, ..., A_m, ... \to B_1, ..., B_n, ...$$

where the lists on both sides of $\to$ are finite or (countably) infinite.

In order to generalize the *search* procedure, we need to define the functions *head* and *tail* operating on possibly infinite sequences. Let us denote
the empty sequence as $<>$.

$head(<>) = <>$; otherwise $head(A_1, ..., A_m, ...) = A_1$.

$tail(<>) = <>$; otherwise $tail(A_1, ..., A_m, ...) = A_2, ..., A_m, ...$. In particular, $tail(A_1) = <>$. The predicate *atomic* is defined such that $atomic(A)$ is true if and only if A is a propositional symbol.

3.5.2 The Search Procedure for Infinite Sequents

Every node of the systematic tree constructed by *search* is still labeled with a *finite* sequent $\Gamma \rightarrow \Delta$. We will also use two global variables L and R, which are possibly countably infinite sequences of propositions. The initial value of L is $tail(\Gamma_0)$ and the initial value of R is $tail(\Delta_0)$, where $\Gamma_0 \rightarrow \Delta_0$ is the initial sequent.

A leaf of the tree is an *axiom* (or is *closed*) iff $\Gamma \rightarrow \Delta$ is an axiom. A leaf is *finished* iff either

(1) it is closed, or

(2) the sequences L and R are empty and all propositions in Γ, and Δ are atomic. The new versions of procedures *search* and *expand* are given as follows.

Definition 3.5.1 Procedure *search*.

```
procedure search(Γ₀ → Δ₀ : sequent; var T : tree);
  begin
    L := tail(Γ₀); Γ := head(Γ₀);
    R := tail(Δ₀); Δ := head(Δ₀);
    let T be the one-node tree labeled with Γ → Δ;
    while not all leaves of T are finished do
      T₀ := T;
      for each leaf node of T₀
      (in lexicographic order) do
        if not finished(node) then
          expand(node, T)
        endif
      endfor;
      L := tail(L); R := tail(R)
    endwhile;
    if all leaves are closed
    then
      write ('T is a proof of Γ₀ → Δ₀')
    else
      write ('Γ₀ → Δ₀ is falsifiable')
    endif
  end
```

The input to *search* is a one-node tree labeled with a possibly infinite sequent $\Gamma \rightarrow \Delta$. Procedure *search* builds a possibly infinite systematic deduction tree using procedure *expand*.

Procedure *expand* is modified as follows: For every leaf u created during an expansion step, if $\Gamma \rightarrow \Delta$ is the label of u, the finite sequent $\Gamma \rightarrow \Delta$ is extended to $\Gamma, head(L) \rightarrow \Delta, head(R)$. At the end of a round, the heads of both L and R are deleted. Hence, every proposition will eventually be considered.

Procedure Expand

```
procedure expand(node : tree-address; var T : tree);
  begin
    let A₁, ..., Aₘ → B₁, ..., Bₙ  be the label of node;
    let S be the one-node tree labeled with
      A₁, ..., Aₘ → B₁, ..., Bₙ;
    for i := 1 to m do
      if nonatomic(Aᵢ) then
        S := the new tree obtained from S by
        applying to the descendant of Aᵢ in
        every nonaxiom leaf of S the
        left rule applicable to  Aᵢ;
        (only the sequent part is modified,
        L and R are unchanged)
      endif
    endfor;
    for i := 1 to n do
      if nonatomic(Bᵢ) then
        S := the new tree obtained from S by
        applying to the descendant of Bᵢ in
        every nonaxiom leaf of S the
        right rule applicable to  Bᵢ;
        (only the sequent part is modified,
        L and R are unchanged)
      endif
    endfor;
    for each nonaxiom leaf u of S do
      let Γ → Δ  be the label of u;
      Γ' := Γ, head(L);
      Δ' := Δ, head(R);
      create a new leaf u1, son of u,
      labeled with the sequent Γ' → Δ'
    endfor;
    T := dosubstitution(T, node, S)
  end
```

If *search* terminates with a systematic tree whose leaves are all closed, we say that *search* terminates with a *closed tree*. We will also need the following definitions.

Definition 3.5.2 Given a possibly infinite sequent $\Gamma \to \Delta$, a valuation v *falsifies* $\Gamma \to \Delta$ if

$$v \models A$$

for every proposition A in Γ, and

$$v \models \neg B$$

for every proposition B in Δ. We say that $\Gamma \to \Delta$ is *falsifiable*.

A valuation v *satisfies* $\Gamma \to \Delta$ if, whenever

$$v \models A$$

for every proposition A in Γ, then there is some proposition B in Δ such that

$$v \models B.$$

We say that $\Gamma \to \Delta$ is *satisfiable*.

The sequent $\Gamma \to \Delta$ is *valid* if it is satisfied by every valuation. This is denoted by

$$\models \Gamma \to \Delta.$$

The sequent $\Gamma \to \Delta$ is *provable* if there exist finite subsequences $C_1, ..., C_m$ and $D_1, ..., D_n$ of Γ and Δ respectively, such that the finite sequent

$$C_1, ..., C_m \to D_1, ..., D_n$$

is provable. This is denoted by

$$\vdash \Gamma \to \Delta.$$

Note that if an infinite sequent is provable, then it is valid. Indeed, if $\Gamma \to \Delta$ is provable, some subsequent $C_1, ..., C_m \to D_1, ..., D_n$ is provable, and therefore valid. But this implies that $\Gamma \to \Delta$ is valid, since $D_1, ..., D_n$ is a subsequence of Δ.

EXAMPLE 3.5.1

Consider the sequent $\Gamma_0 \to \Delta_0$ where

$$\Gamma_0 = <P_0, (P_0 \supset P_1), (P_1 \supset P_2), ..., (P_i \supset P_{i+1}), ... >,$$

and

$$\Delta_0 = <Q, P_3 > .$$

Initially,

$$\Gamma = <P_0>,$$
$$L = <(P_0 \supset P_1), (P_1 \supset P_2), ..., (P_i \supset P_{i+1}), ...>,$$
$$\Delta = <Q>, \quad \text{and}$$
$$R = <P_3>.$$

At the end of the first round, we have the following tree:

$$\frac{P_0, (P_0 \supset P_1) \to Q, P_3}{P_0 \to Q}$$

Note how $(P_0 \supset P_1)$, the head of L, was added in the premise of the top sequent, and P_3, the head of R, was added to the conclusion of the top sequent. At the end of this round,

$$L = <(P_1 \supset P_2), ..., (P_i \supset P_{i+1}), ...> \quad \text{and } R = <>.$$

At the end of the second round, we have:

$$\frac{P_0 \to P_0, Q, P_3 \qquad \dfrac{\dfrac{P_1, P_0, (P_1 \supset P_2) \to Q, P_3}{P_1, P_0 \to Q, P_3}}{P_0, (P_0 \supset P_1) \to Q, P_3}}{P_0 \to Q}$$

We have

$$L = <(P_2 \supset P_3), ..., (P_i \supset P_{i+1}), ...> \quad \text{and } R = <>.$$

At the end of the third round, we have the tree:

$$\frac{P_0 \to P_0, Q, P_3 \qquad \dfrac{P_1, P_0 \to P_1, Q, P_3 \qquad \dfrac{\dfrac{P_2, P_1, P_0, (P_2 \supset P_3) \to Q, P_3}{P_2, P_1, P_0 \to Q, P_3}}{P_1, P_0, (P_1 \supset P_2) \to Q, P_3}}{P_1, P_0 \to Q, P_3}}{\dfrac{P_0, (P_0 \supset P_1) \to Q, P_3}{P_0 \to Q}}$$

We have

$$L = <(P_3 \supset P_4), ..., (P_i \supset P_{i+1}), ...> \quad \text{and } R = <>.$$

At the end of the fourth round, we have the closed tree:

$$
\cfrac{
  \cfrac{
    \cfrac{
      \cfrac{
        \cfrac{
          P_2, P_1, P_0 \to P_2, Q, P_3 \quad P_3, P_2, P_1, P_0 \to Q, P_3
        }{
          P_2, P_1, P_0, (P_2 \supset P_3) \to Q, P_3
        }
      }{
        P_2, P_1, P_0 \to Q, P_3
      }
    }{
      P_1, P_0, (P_1 \supset P_2) \to Q, P_3
    } \quad P_1, P_0 \to P_1, Q, P_3
  }{
    P_1, P_0 \to Q, P_3
  }
}{
  P_0, (P_0 \supset P_1) \to Q, P_3
}
$$

$$
\Pi_1 \qquad \cfrac{P_0, (P_0 \supset P_1) \to Q, P_3}{P_0 \to Q}
$$

where

$$
\Pi_1 = P_0 \to P_0, Q, P_3.
$$

The above tree is not quite a proof tree, but a proof tree can be constructed from it as follows. Starting from the root and proceeding from bottom-up, for every sequent at depth k in which a proposition of the form $head(L)$ or $head(R)$ was introduced, add $head(L)$ after the rightmost proposition of the premise of every sequent at depth less than k, and add $head(R)$ after the rightmost proposition of the conclusion of every sequent at depth less than k:

$$
\cfrac{
  \cfrac{
    \cfrac{
      \cfrac{
        \cfrac{
          P_2, P_1, P_0 \to P_2, Q, P_3 \quad P_3, P_2, P_1, P_0 \to Q, P_3
        }{
          P_2, P_1, P_0, (P_2 \supset P_3) \to Q, P_3
        }
      }{
        P_2, P_1, P_0, (P_2 \supset P_3) \to Q, P_3
      }
    }{
      P_1, P_0, (P_1 \supset P_2), (P_2 \supset P_3) \to Q, P_3
    }
  }{
    P_1, P_0, (P_1 \supset P_2), (P_2 \supset P_3) \to Q, P_3
  }
}{
  P_0, (P_0 \supset P_1), (P_1 \supset P_2), (P_2 \supset P_3) \to Q, P_3
}
$$

$$
\Pi_3 \qquad \Pi_2
$$

$$
\cfrac{P_0, (P_0 \supset P_1), (P_1 \supset P_2), (P_2 \supset P_3) \to Q, P_3}{P_0, (P_0 \supset P_1), (P_1 \supset P_2), (P_2 \supset P_3) \to Q, P_3}
$$

with

$$
\Pi_2 = P_0, (P_1 \supset P_2), (P_2 \supset P_3) \to P_0, Q, P_3
$$
$$
\text{and} \quad \Pi_3 = P_1, P_0, (P_2 \supset P_3) \to P_1, Q, P_3.
$$

Then, delete duplicate nodes, obtaining a legal proof tree:

$$\dfrac{P_2, P_1, P_0 \to P_2, Q, P_3 \quad P_3, P_2, P_1, P_0 \to Q, P_3}{\text{(continued)}}$$

$$\Pi_3 \quad \dfrac{}{P_2, P_1, P_0, (P_2 \supset P_3) \to Q, P_3}$$

$$\Pi_2 \quad \dfrac{P_1, P_0, (P_1 \supset P_2), (P_2 \supset P_3) \to Q, P_3}{P_0, (P_0 \supset P_1), (P_1 \supset P_2), (P_2 \supset P_3) \to Q, P_3}$$

with

$$\Pi_2 = P_0, (P_1 \supset P_2), (P_2 \supset P_3) \to P_0, Q, P_3$$
$$\text{and} \quad \Pi_3 = P_1, P_0, (P_2 \supset P_3) \to P_1, Q, P_3.$$

EXAMPLE 3.5.2

Consider the sequent $\Gamma_0 \to \Delta_0$ where

$$\Gamma_0 = < P_0, (P_0 \supset P_1), (P_1 \supset P_2), ..., (P_i \supset P_{i+1}), ... >,$$

and

$$\Delta_0 = < Q > .$$

Note that the only difference is the absence of P_3 in the conclusion. This time, the *search* procedure does not stop. Indeed, the rightmost branch of the tree is infinite, since every sequent in it is of the form

$$P_n, P_{n-1}, ... P_1, P_0 \to Q.$$

Let U be the union of all the propositions occurring as premises in the sequents on the infinite branch of the tree, and V be the union of all the propositions occurring as conclusions in such sequents. We have

$$U = \{(P_0 \supset P_1), ..., (P_i \supset P_{i+1}), ..., P_0, P_1, ..., P_i, ...\},$$

and

$$V = \{Q\}.$$

The pair (U, V) can be encoded as a single set if we prefix every proposition in U with the letter "T" (standing for **true**), and every proposition in V with the letter "F" (standing for **false**). The resulting set

$$\{T(P_0 \supset P_1), ..., T(P_i \supset P_{i+1}), ..., TP_0, TP_1, ..., TP_i, ..., FQ\}$$

is a set having some remarkable properties, and called a *Hintikka set*. The crucial property of Hintikka sets is that they are always satisfiable. For instance, it is easy to see that the valuation such that $v(P_i) = \mathbf{T}$ for all $i \geq 0$ and $v(Q) = \mathbf{F}$ satisfies the above Hintikka set.

Roughly speaking, the new version of the *search* procedure is complete because:

(1) If the input sequent is valid, a proof tree can be constructed from the output tree (as in example 3.5.1);

(2) If the sequent is falsifiable, the output tree contains a path from which a Hintikka set can be constructed (as in example 3.5.2). Hence, a counter example exists.

In order to prove rigorously properties (1) and (2), we will need some auxiliary definitions and lemmas. First, we shall need the following result about infinite finite-branching trees known as König's lemma.

3.5.3 König's Lemma

Recall from Subsection 2.2.2 that a tree T is finite branching iff every node has finite outdegree (finite number of successors).

Lemma 3.5.1 (König's lemma) If T is a finite-branching tree with infinite domain, then there is some infinite path in T.

Proof: We show by induction on the distance from the root that an infinite path must exist. Let u_0 be the root of the tree. Since the domain of T is infinite and u_0 has a finite number of successors, one of the subtrees of u_0 must be infinite (otherwise, T would be finite). Let u_1 be the root of the leftmost infinite subtree of u_0. Now, assume by induction that a path $u_0, ..., u_n$ has been defined and that the subtree T/u_n is infinite. Since u_n has a finite number of successors and since T/u_n is infinite, using the same reasoning as above, u_n must have a successor which is the root of an infinite tree. Let u_{n+1} be the leftmost such node. It is clear that the above inductive construction yields an infinite path in T (in fact, the leftmost such path). □

Remark: The above proof only shows the *existence* of an infinite path. In particular, since there is in general no effective way of testing whether a tree is infinite, there is generally no algorithm to find the above nodes.

In example 3.5.2, the two sets U and V play a crucial role since they yield a falsifying valuation. The union of U and V is a set having certain remarkable properties first investigated by Hintikka and that we now describe. For this, it is convenient to introduce the concept of a *signed formula* as in Smullyan, 1968.

3.5.4 Signed Formulae

Following Smullyan, we will define the concept of an *a-formula* and of a *b-formula*, and describe their components. Using this device greatly reduces the number of cases in the definition of a Hintikka set, as well as in some proofs.

Definition 3.5.3 A *signed formula* is any expression of the form TA or FA, where A is an arbitrary proposition. Given any sequent (even infinite) $\Gamma \rightarrow \Delta$, we define the *signed set of formulae*

$$\{TA \mid A \in \Gamma\} \cup \{FB \mid B \in \Delta\}.$$

Definition 3.5.4 *type-a and type-b signed formulae* and their *components* are defined in the following tables. If A is a signed formula of type a, it has two components denoted by A_1 and A_2. Similarly, if B is a formula of type b, it has two components denoted by B_1 and B_2.

Type-a formulae

A	A_1	A_2
$T(X \wedge Y)$	TX	TY
$F(X \vee Y)$	FX	FY
$F(X \supset Y)$	TX	FY
$T(\neg X)$	FX	FX
$F(\neg X)$	TX	TX

Type-b formulae

B	B_1	B_2
$F(X \wedge Y)$	FX	FY
$T(X \vee Y)$	TX	TY
$T(X \supset Y)$	FX	TY

Definition 3.5.5 A valuation v makes the signed formula TA **true** iff v makes A **true** and v makes FA **true** iff v makes A **false**. A valuation v *satisfies a signed set* S iff v makes every signed formula in S **true**.

Note that for any valuation, a signed formula A of type a is **true** if and only if *both* A_1 *and* A_2 are **true**. Accordingly, we also refer to an *a-formula* as a formula of *conjunctive* type. On the other hand, for any valuation, a signed formula B of type b is **true** if and only if *at least* one of B_1, B_2 is **true**. Accordingly, a *b-formula* is also called a formula of *disjunctive* type.

Definition 3.5.6 The *conjugate* of a signed formula is defined as follows: The conjugate of a formula TA is FA, and the conjugate of FA is TA.

3.5.5 Hintikka Sets

A Hintikka set is a set of signed formulae satisfying certain downward closure conditions that ensure that such a set is satisfiable.

Definition 3.5.7 A set S of signed formulae is a *Hintikka set* iff the following conditions hold:

(H1) No signed propositional letter and its conjugate are both in S.

(H2) If a signed *a-formula* A is in S then both A_1 and A_2 are in S.

(H3) If a signed *b-formula* B is in S then either B_1 is in S or B_2 is in S.

The following lemma shows that Hintikka sets arise when the *search* procedure does not produce a closed tree.

Lemma 3.5.2 Whenever the tree T constructed by the *search* procedure is not a closed tree, a Hintikka set can be extracted from T.

Proof: If T is not a closed tree, then either it is finite and some leaf is not an axiom, or it is infinite. If T is infinite, by lemma 3.5.1, there is an infinite path in T. In the first case, consider a path to some nonaxiom leaf, and in the second consider an infinite path. Let

$$S = \{TA \mid A \in U\} \cup \{FB \mid B \in V\}$$

be the set of signed formulae such that U is the union of all propositions occurring in the premise of each sequent in the chosen path, and V is is the union of all propositions occurring in the conclusion of each sequent in the chosen path. S is a Hintikka set.

(1) H1 holds. Since every atomic formula occurring in a sequent occurs in every path having this sequent as source, if S contains both TP and FP for some propositional letter P, some sequent in the path is an axiom. This contradicts the fact that either the path is finite and ends in a non-axiom, or is an infinite path.

(2) H2 and H3 hold. This is true because the definition of *a-components* and *b-components* mirrors the inference rules. Since all nonatomic propositions in a sequent $\Gamma \rightarrow \Delta$ on the chosen path are considered during the expansion phase, and since every proposition in the input sequent is eventually considered (as $head(L)$ or $head(R)$):

(i) For every proposition A in U, if A belongs to $\Gamma \rightarrow \Delta$ and TA is of type a, A_1 and A_2 are added to the successor of $\Gamma \rightarrow \Delta$ during the expansion step. More precisely, if A_1 (or A_2) is of the form TC_1 (or TC_2), C_1 (C_2) is added to the premise of the successor of $\Gamma \rightarrow \Delta$; if A_1 (A_2) is of the form FC_1 (FC_2), C_1 (C_2) is added to the conclusion of the successor of $\Gamma \rightarrow \Delta$. In both cases, A_1 and A_2 belong to S.

(ii) If A belongs to $\Gamma \rightarrow \Delta$ and TA is of type b, A_1 is added to the left successor of $\Gamma \rightarrow \Delta$, and A_2 is added to the right successor of $\Gamma \rightarrow \Delta$, during the expansion step. As in (i), more precisely, if A_1 (or A_2) is of the form TC_1 (TC_2), C_1 (C_2) is added to the premise of the left successor (right successor) of $\Gamma \rightarrow \Delta$; if A_1 (A_2) is of the form FC_1 (FC_2), C_1 (C_2) is added to the conclusion of the left successor (right successor) of $\Gamma \rightarrow \Delta$. Hence, either B_1 or B_2 belongs to S.

Properties (i) and (ii) also apply to the set V. This proves that S is a Hintikka set. $\square$

The following lemma establishes the fundamental property of Hintikka sets.

Lemma 3.5.3 Every Hintikka set S is satisfiable.

Proof: We define a valuation v satisfying S as follows: For every signed propositional symbol TP in S let $v(P) = \mathbf{T}$; for every signed propositional symbol FP in S let $v(P) = \mathbf{F}$; for every propositional symbol P such that neither TP nor FP is in S, set arbitrarily $v(P) = \mathbf{T}$. By clause (H1) of a Hintikka set, v is well defined. It remains to show that v makes every signed formula TX or FX true (that is, in the first case X true and in the second case X false). This is shown by induction on the number of logical connectives in X. Since every signed formula is either of type a or of type b, there are two cases.

(1) If A of type a is in S, by (H2) both A_1 and A_2 are in S. But A_1 and A_2 have fewer connectives than A and so, the induction hypothesis applies. Hence, v makes both A_1 and A_2 true. This implies that v makes A true.

(2) If B of type b is in S, by (H3) either B_1 or B_2 is in S. Without loss of generality assume that B_1 is in S. Since B_1 has fewer connectives than B, the induction hypothesis applies. Hence, v makes B_1 true. This implies that v makes B true. $\square$

3.5.6 Extended Completeness of the Gentzen System G'

We are now ready to prove the generalization of theorem 3.4.1.

Theorem 3.5.1 Given a sequent $\Gamma \to \Delta$, either

(1) $\Gamma \to \Delta$ is falsifiable and

    (i) If $\Gamma \to \Delta$ is infinite, then *search* runs forever building an infinite tree T, or

    (ii) If $\Gamma \to \Delta$ is finite, *search* produces a finite tree T with some non-axiom leaf. In both cases, a falsifying valuation can be obtained from some path in the tree produced by procedure *search*; or

(2) $\Gamma \to \Delta$ is valid and *search* terminates with a finite closed tree T. In this case, there exist finite subsequences $C_1, ..., C_m$ and $D_1, ..., D_n$ of Γ and Δ respectively such that the sequent $C_1, ..., C_m \to D_1, ..., D_n$ is provable.

Proof: First, observe that if a subsequent of a sequent is valid, the sequent itself is valid. Also, if $\Gamma \to \Delta$ is infinite, *search* terminates if and only if the tree is closed. This last statement holds because any node that is not an axiom is not finished, since otherwise L and R would be empty, contradicting the fact that $\Gamma \to \Delta$ is infinite. Hence, at every step of the procedure *search*, some node is unfinished, and since the procedure *expand* adds at least one new node to the tree (when $head(L)$ is added to Γ and $head(R)$ is added to Δ), *search* builds an infinite tree. Consequently, if $\Gamma \to \Delta$ is infinite, either *search* halts with a closed tree, or it builds an infinite tree.

If *search* halts with a closed tree, let $C_1, ...C_m$ be the initial subsequence of propositions in Γ that were deleted from Γ to obtain L, and $D_1, ..., D_n$ the initial subsequence of propositions in Δ which were deleted from Δ to obtain R. A proof tree for a the finite sequent $C_1, ..., C_m \to D_1, ..., D_n$ can easily be obtained from T, using the technique illustrated in example 3.5.1.

First, starting from the root and proceeding bottom-up, for each node

$$\Gamma, head(L) \to \Delta, head(R)$$

at depth k created at the end of a call to procedure *expand*, add $head(L)$ after the rightmost proposition in the premise of every sequent at depth less than k, and add $head(R)$ after the rightmost proposition in the conclusion of every sequent at depth less than k, obtaining the tree T'. Then, a proof tree T'' for $C_1, ..., C_m \to D_1, ..., D_n$ is constructed from T' by deleting all duplicate nodes. The tree T'' is a proof tree because the same inference rules that have been used in T are used in T''. A proof similar to that of theorem 3.4.1 shows that $C_1, ..., C_m \to D_1, ..., D_n$ is valid and consequently that $\Gamma \to \Delta$ is valid.

Hence, if the *search* procedure halts with a closed tree, a subsequent of $\Gamma \to \Delta$ is provable, which implies that $\Gamma \to \Delta$ is provable (and consequently valid). Hence, if $\Gamma \to \Delta$ is falsifiable, either the *search* procedure halts with a finite nonclosed tree if $\Gamma \to \Delta$ is finite, or else *search* must go on forever if $\Gamma \to \Delta$ is infinite. If the tree is finite, some leaf is not an axiom, and consider the path to this leaf. Otherwise, let T be the infinite tree obtained in the limit. This tree is well defined since for every integer k, *search* will produce the subtree of depth k of T. Since T is infinite and finite branching, by König's lemma, there is an infinite path $u_0, u_1, ..., u_n,...$ in T. By lemma 3.5.1, the set

$$S = \{TA \mid A \in U\} \cup \{FB \mid B \in V\}$$

of signed formulae such that U is the union of all propositions occurring in the premise of each sequent in the chosen path, and V is is the union of all propositions occurring in the conclusion of each sequent in the chosen path, is a Hintikka set. By lemma 3.5.2, S is satisfiable. But any valuation satisfying S falsifies $\Gamma \to \Delta$, and $\Gamma \to \Delta$ is falsifiable.

To summarize, if the *search* procedure halts with a closed tree, $\Gamma \to \Delta$ is provable, and therefore valid. Otherwise $\Gamma \to \Delta$ is falsifiable.

Conversely, if $\Gamma \to \Delta$ is valid, *search* must halt with a closed tree, since otherwise the above reasoning shows that a falsifying valuation can be found. But then, we have shown that $\Gamma \to \Delta$ is provable. If $\Gamma \to \Delta$ is falsifiable, *search* cannot halt with a closed tree, since otherwise $\Gamma \rightarrow \Delta$ would be provable, and consequently valid. But then, we have shown that a falsifying valuation can be found from the tree T. This concludes the proof of the theorem. $\square$

We now derive some easy consequences of the main theorem. Since a provable sequent is valid, the following is an obvious corollary.

Theorem 3.5.2 (Extended completeness theorem for G') For every (possibly infinite) sequent $\Gamma \rightarrow \Delta$, $\Gamma \rightarrow \Delta$ is valid if and only if $\Gamma \rightarrow \Delta$ is provable.

3.5.7 Compactness, Model Existence, Consistency

Recall that a proposition A is satisfiable if some valuation makes it **true**.

Definition 3.5.8 A set Γ of propositions is *satisfiable* iff some valuation makes all propositions in Γ **true**.

Theorem 3.5.3 (Compactness theorem for G') For any (possibly infinite) set Γ of propositions, if every finite (nonempty) subset of Γ is satisfiable then Γ is satisfiable.

Proof: Assume Γ is not satisfiable. Viewing Γ as a sequence of proposition, it is clear that the sequent

$$\Gamma \rightarrow$$

is valid, and by theorem 3.5.1 there is a finite subsequence $A_1, ..., A_p$ of Γ such that

$$A_1, ..., A_p \rightarrow$$

is provable. But then, by lemma 3.4.3 $A_1, ..., A_p \rightarrow$ is valid, which means that $A_1, ..., A_p$ is not satisfiable contrary to the hypothesis. Hence Γ is satisfiable. $\square$

Definition 3.5.9 A set Γ of propositions is *consistent* if there exists some proposition B such that the sequent $\Gamma \rightarrow B$ is not provable (that is, $A_1, ..., A_m \rightarrow B$ is not provable for any finite subsequence $A_1, ..., A_m$ of Γ). Otherwise, we say that Γ is *inconsistent*.

Theorem 3.5.4 (Model existence theorem for G') If a set of propositions Γ is consistent then it is satisfiable.

Proof: Assume Γ unsatisfiable. Hence, for every proposition B, the sequent

$$\Gamma \rightarrow B$$

is valid. By theorem 3.5.1, for every such B, there is a finite subsequence $A_1, ..., A_p$ of Γ such that the sequent

$$A_1, ..., A_p \rightarrow B$$

is provable. But then, Γ is not consistent, contrary to the hypothesis. $\square$

The converse of theorem 3.5.4 is also true.

Lemma 3.5.4 (Consistency lemma for G') If a set Γ of propositions is satisfiable then it is consistent.

Proof: Let v be a valuation such that

$$v \models A$$

for every proposition in Γ. Assume that Γ is inconsistent. Then,

$$\Gamma \to B$$

is provable for every proposition B, and in particular, there is a finite subsequence $A_1, ..., A_m$ of Γ such that

$$A_1, ..., A_m \to P \wedge \neg P$$

is provable (for some propositional symbol P). By lemma 3.4.3, $A_1, ..., A_m \to P \wedge \neg P$ is valid and, since the valuation v makes all propositions in Γ true, v should make $P \wedge \neg P$ true, which is impossible. Hence, Γ is consistent. $\square$

Note that if a set Γ of propositions is consistent, theorem 3.5.4 shows that the sequent $\Gamma \to$ is falsifiable. Hence, by theorem 3.5.1, a falsifying valuation can be obtained (in fact, all falsifying valuations can be obtained by considering all infinite paths in the counter-example tree).

One may view the goal of procedure *search* as the construction of Hintikka sets. If this goal fails, the original sequent was valid and otherwise, any Hintikka set yields a falsifying valuation. The decomposition of propositions into *a-components* or *b-components* is the basis of a variant of Gentzen systems called the tableaux system. For details, see Smullyan, 1968.

3.5.8 Maximal Consistent Sets

We conclude this section by discussing briefly the concept of maximal consistent sets. This concept is important because it can be used to give another proof of the completeness theorem (theorem 3.5.2).

Definition 3.5.10 A set Γ of propositions is *maximally consistent* (or a *maximal consistent set*) iff, for every consistent set Δ, if Γ is a subset of Δ, then $\Gamma = \Delta$. Equivalently, every proper superset of Γ is inconsistent.

The importance of maximal consistent sets lies in the following lemma.

Lemma 3.5.5 Every consistent set Γ is a subset of some maximal consistent set Δ.

Proof: If Γ is a consistent set, by theorem 3.5.4, it is satisfiable. Let v be a valuation satisfying Γ. Let Δ be the set

$$\{A \mid v \models A\}$$

of all propositions satisfied by v. Clearly, Γ is a subset of Δ. We claim that Δ is a maximal consistent set. First, by lemma 3.5.4, Δ is consistent since it is satisfied by v. It remains to show that it is maximally consistent. We proceed by contradiction. Assume that there is a consistent set Λ such that Δ is a proper subset of Λ. Since Λ is consistent, by theorem 3.5.4, it is satisfied by a valuation v'. Since Δ is a proper subset of Λ, there is a proposition A which is in Λ but not in Δ. Hence,

$$v \not\models A,$$

since otherwise A would be in Δ. But then,

$$v \models \neg A,$$

and $\neg A$ is in Δ. Since Δ is a subset of Λ, v' satisfies every proposition in Δ, and in particular

$$v' \models \neg A.$$

But since v' satisfies Λ, we also have

$$v' \models A,$$

which is impossible. Hence, Δ is indeed maximally consistent. $\square$

The above lemma was shown using theorem 3.5.4, but it can be shown more directly and without theorem 3.5.4. Actually, theorem 3.5.4 can be shown from lemma 3.5.5, and in turn, the completeness theorem can be shown from theorem 3.5.4. Such an approach to the completeness theorem is more traditional, but not as constructive, in the sense that it does not provide a procedure for constructing a deduction tree.

There is also a close relationship between maximally consistent sets and Hintikka sets. Indeed, by reformulating Hintikka sets as unsigned sets of propositions, it can be shown that every maximal consistent set is a Hintikka set. However, the converse is not true. Hintikka sets are more general (and in a sense more economical) than maximal consistent sets. For details, we refer the reader to the problems.

PROBLEMS

3.5.1. (i) Show that the infinite sequent $\Gamma \to \Delta$ where

$$\Gamma = < P_0, (P_0 \supset P_1), (P_1 \supset P_2), ..., (P_i \supset P_{i+1}), ... >$$

and

$$\Delta = < (P_1 \supset Q) >$$

is falsifiable.

(ii) Prove that for every $i > 0$, the sequent $\Gamma \to \Delta'$, where Γ is as above and $\Delta' = < (P_0 \supset P_i) >$ is provable.

3.5.2. The *cut rule* is the following inference rule:

$$\frac{\Gamma \to \Delta, A \qquad A, \Lambda \to \Theta}{\Gamma, \Lambda \to \Delta, \Theta}$$

A is called the *cut formula* of this inference.

Let $G' + \{cut\}$ be the formal system obtained by adding the cut rule to G'. The notion of a deduction tree is extended to allow the cut rule as an inference. A proof in G' is called a *cut-free* proof.

(i) Prove that for every valuation v, if v satisfies the premises of the cut rule, then it satisfies its conclusion.

(ii) Prove that if a sequent is provable in the system $G' + \{cut\}$, then it is valid.

(iii) Prove that if a sequent is provable in $G' + \{cut\}$, then it has a cut-free proof.

3.5.3. (i) Prove solely in terms of proofs in $G' + \{cut\}$ that a set Γ of propositions is inconsistent if and only if there is a proposition A such that both $\Gamma \to A$ and $\Gamma \to \neg A$ are provable in $G' + \{cut\}$. (For inspiration see Section 3.6.)

(ii) Prove solely in terms of proofs in $G' + \{cut\}$ that, $\Gamma \to A$ is not provable in $G' + \{cut\}$ if and only if $\Gamma \cup \{\neg A\}$ is consistent. (For inspiration see Section 3.6.)

Note: Properties (i) and (ii) also hold for the proof system G', but the author does not know of any proof not involving a proof-theoretic version of Gentzen's *cut elimination theorem*. The cut elimination theorem states that any proof in the system $G' + \{cut\}$ can be transformed to a proof in G' (without cut). The completeness theorem for G' provides a semantic proof of the cut elimination theorem. However, in order to show (i) and (ii) without using semantic arguments, it appears that one has to mimic Gentzen's original proof. (See Szabo, 1969.)

* **3.5.4.** A set Γ of propositions is said to be *complete* if, for every proposition A, either $\Gamma \to A$ or $\Gamma \to \neg A$ is provable, but not both. Prove that for any set Γ of propositions, the following are equivalent:

(i) The set

$$\{A \mid \vdash \Gamma \to A \text{ in } G'\}$$

is a maximal consistent set.

(ii) Γ is complete.

(iii) There is a single valuation v satisfying Γ.

(iv) There is a valuation v such that for all A,

$\Gamma \to A$ is provable (in G') if and only if $v \models A$.

3.5.5. Let Γ be a consistent set. Let $A_1, A_2, ..., A_n, ...$ be an enumeration of *all* propositions in $PROP$. Define the sequence Γ_n inductively as follows:

$$\Gamma_0 = \Gamma,$$

$$\Gamma_{n+1} = \begin{cases} \Gamma_n \cup \{A_n\} & \text{if } \Gamma_n \cup \{A_n\} \text{ is consistent;} \\ \Gamma_n & \text{otherwise.} \end{cases}$$

Let

$$\Delta = \bigcup_{n \geq 0} \Gamma_n.$$

Prove the following:

(a) Each Γ_n is consistent.

(b) Δ is consistent.

(c) Δ is maximally consistent.

Note that this exercise provides another proof of lemma 3.5.5, not using the completeness theorem.

3.5.6. Prove that if a proposition A over the language using the logical connectives $\{\vee, \wedge, \supset, \neg\}$ is a tautology, then A contains some occurrence of either $\neg$ or $\supset$.

* **3.5.7.** Given a proposition A, its *immediate descendants* A_1 and A_2 are given by the following table:

Type-a formulae

A	A_1	A_2
$(X \wedge Y)$	X	Y
$\neg(X \vee Y)$	$\neg X$	$\neg Y$
$\neg(X \supset Y)$	X	$\neg Y$
$\neg(\neg X)$	X	X

Type-b formulae

B	B_1	B_2
$\neg(X \wedge Y)$	$\neg X$	$\neg Y$
$(X \vee Y)$	X	Y
$(X \supset Y)$	$\neg X$	Y

Note that neither propositional letters nor negations of propositional letters have immediate descendants.

Given a set S of propositions, let $Des(S)$ be the set of immediate descendants of propositions in S, and define S^n by induction as follows:

$$S^0 = S;$$

$$S^{n+1} = Des(S^n)$$

Let

$$S^* = \bigcup_{n \geq 0} S^n$$

be the union of all the S^n. Hintikka sets can also be defined without using signed formulae, in terms of immediate descendants:

A set S of propositions is a *Hintikka set* if the following conditions hold:

(H1) No propositional letter and its negation are both in S.

(H2) If an *a-formula* A is in S then both A_1 and A_2 are in S.

(H3) If a *b-formula* B is in S then either B_1 is in S or B_2 is in S.

In this problem and some of the following problems, given any set S of propositions and any propositions $A_1,...,A_n$, the set $S \cup \{A_1, ... A_n\}$ will also be denoted by $\{S, A_1, ..., A_n\}$.

Assume that S is consistent.

(a) Using a modification of the construction given in problem 3.5.5, show that S can be extended to a maximal consistent subset U of S^* (that is, to a consistent subset U of S^* containing S, such that U is not a proper subset of any consistent subset of S^*).

(b) Prove that consistent sets satisfy the following properties:

C_0: No set S containing a propositional letter and its negation is consistent.

C_1: If $\{S, A\}$ is consistent, so is $\{S, A_1, A_2\}$, where A is a proposition of type a.

C_2: If $\{S, B\}$ is consistent, then either $\{S, B_1\}$ or $\{S, B_2\}$ is consistent, where B is a proposition of type b.

(c) Prove that U is a Hintikka set.

(d) Show that U is not necessarily a maximal consistent subset of $PROP$, the set of all propositions.

* **3.5.8.**   The purpose of this problem is to prove the compactness theorem for propositional logic without using the completeness theorem. The proof developed in the following questions is in a sense more constructive than the proof given by using the completeness theorem, since if we are given a set Γ such that every finite subset of Γ is satisfiable, we will actually construct a valuation that satisfies Γ. However, the existence of ultrafilters requires Zorn's Lemma, and so this proof is not constructive in the recursion-theoretic sense.

For this problem, you may use the following result stated below and known as Zorn's lemma. For details, the reader should consult a text on set Theory, such as Enderton, 1977; Suppes, 1972; or Kuratowski and Mostowski, 1976.

We recall the following concepts from Subsection 2.1.9. A chain in a poset $(P, \leq)$ is a totally ordered subset of P. A chain C is bounded if there exists an element $b \in P$ such that for all $p \in C$, $p \leq b$. A maximal element of P is some $m \in P$ such that for any $m' \in P$, if $m \leq m'$ then $m = m'$.

Zorn's lemma: Given a partially ordered set S, if every chain in S is bounded, then S has a maximal element.

(1) Let E be a nonempty set, and F a class of subsets of E. We say that F is a *filter* on E iff:

1. E is in F;

2. if u and v are in F, then $u \cap v$ is in F;

3. if u is in F and v is any subset of E, if $u \subseteq v$, then v is also in F.

A filter F is a *proper filter* if $\emptyset$ (the empty set) is not in F. A proper filter F is *maximal* if, for any other proper filter D, if F is a subset of D, then $D = F$.

A (nonempty) class C of subsets of a nonempty set E has the *finite intersection property* (*f.i.p.*) iff the intersection of every finite number of sets in C is nonempty. Let C be any nonempty class of subsets of a nonempty set E. The *filter generated by* C is the intersection D of all filters over E which include C.

Prove the following properties:

(i) The filter D generated by C is indeed a filter over E.

(ii) D is equal to the set of all subsets X of E such that either $X = E$, or for some $Y_1, ..., Y_n \in C$,

$$Y_1 \cap ... \cap Y_n \subseteq X.$$

(iii) D is a proper filter if and only if C has the finite intersection property.

(2) A maximal proper filter is called an *ultrafilter*.

Prove that a nonempty collection U of sets with the finite intersection property is an ultrafilter over E if and only if, for every subset X of E,

$$X \in U \quad \text{if and only if} \quad (E - X) \notin U.$$

Hint: Assume that $(E - X) \notin U$. Let $D = U \cup \{X\}$, and let F be the filter generated by D (as in question 1). Show that F is a proper filter including U. Hence, $U = F$ and D is a subset of U, so that $X \in U$.

(3) Use Zorn's lemma to show that if a class C of subsets of a nonempty set E has the finite intersection property, then it is contained in some ultrafilter.

Hint: Show that the union of a chain of proper filters is a proper filter that bounds the chain.

* **3.5.9.** Let I be a nonempty set and $V = \{v_i \mid i \in I\}$ be a set of valuations. Let U be a proper filter on I. Define the valuation v such that for each propositional symbol $P \in \mathbf{PS}$,

$$v(P) = \mathbf{T} \quad \text{iff} \quad \{i \mid v_i(P) = \mathbf{T}\} \in U.$$

(Such a valuation v is called a *reduced product*).

(a) Show that

$$v(P) = \mathbf{F} \quad \text{iff} \quad \{i \mid v_i(P) = \mathbf{T}\} \notin U$$

or

$$\{i \mid v_i(P) = \mathbf{T}\} = \emptyset.$$

If U is an ultrafilter, show that for all propositions A,

$$v \models A \quad \text{iff} \quad \{i \mid v_i \models A\} \in U.$$

Such a valuation v is called the *ultraproduct* of V with respect to U.

(b) Show that for any Horn formula A (see problem 3.3.18), whenever U is a proper filter, if

$$\{i \mid v_i \models A\} \in U \quad \text{then} \quad v \models A.$$

As a consequence, show that for every Horn formula A,

$$\text{if } v_i \models A \text{ for all } i \in I, \text{ then } v \models A.$$

(c) Let $I = \{1, 2\}$. Give all the filters on I. Give an example showing that there exists a proper filter U on $\{1, 2\}$, a set of valuations $\{v_1, v_2\}$, and a proposition A, such that $v \models A$, but $\{i \mid v_i \models A\} \notin U$.

(d) Consider the proper filter $U = \{\{1, 2\}\}$ on $I = \{1, 2\}$, and let $A = P_1 \vee P_2$. Find two valuations v_1 and v_2 such that $v_1 \models A$ and $v_2 \models A$, but the reduced product v of v_1 and v_2 with respect to U does not satisfy A. Conclude that not every proposition is logically equivalent to a Horn formula.

* **3.5.10.** (a) Let Γ be a set of propositions such that every finite subset of Γ is satisfiable. Let I be the set of all finite subsets of Γ, and for each $i \in I$, let v_i be a valuation satisfying i. For each proposition $A \in \Gamma$, let

$$A^* = \{i \in I \mid A \in i\}.$$

Let

$$\mathcal{C} = \{A^* \mid A \in \Gamma\}.$$

Note that $\mathcal{C}$ has the finite intersection property since

$$\{A_1, ..., A_n\} \in A_1^* \cap ... \cap A_n^*.$$

By problem 3.5.8, let U be an ultrafilter including $\mathcal{C}$, so that every A^* is in U. If $i \in A^*$, then $A \in i$, and so

$$v_i \models A.$$

Thus, for every A in Γ, A^* is a subset of $\{i \in I \mid v_i \models A\}$.

Show that each set $\{i \in I \mid v_i \models A\}$ is in U.

(b) Show that the ultraproduct v (defined in problem 3.5.9) of the set of valuations $\{v_i \mid i \in I\}$ with respect to the ultrafilter U satisfies Γ.

* **3.5.11.** Recall the definition of a Horn formula given in problem 3.3.18. Given a countable set $\{v_i \mid i \geq 0\}$ of truth assignments, the product v of $\{v_i \mid i \geq 0\}$ is the truth assignment such that for every propositional symbol P_j,

$$v(P_j) = \begin{cases} \mathbf{T} & \text{if } v_i(P_j) = \mathbf{T}, \text{ for all } v_i, \\ \mathbf{F} & \text{otherwise.} \end{cases}$$

(a) Show that if X is a set of propositional Horn formulae and every truth assignment in $\{v_i \mid i \geq 0\}$ satisfies X, then the product v satisfies X.

(b) Let
$$X^* = \{\neg P \mid P \text{ is atomic and } X \nvdash P\}.$$

Show that if X is a consistent set of basic Horn formulas, then $X \cup X^*$ is consistent.

Hint: Using question (a), show that there is a truth assignment v satisfying $X \cup X^*$.

∗ **3.5.12.** In this problem, we are using the definitions given in problem 3.5.7. Given a set S, a property P about subsets of S (P is a subset of 2^S) is a *property of finite character* iff the following hold:

Given any subset U of S, P holds for U if and only if P holds for all finite subsets of U.

A property P about sets of propositions is a *consistency property* if P is of finite character and the following hold:

C_0: No set S containing a propositional letter and its negation satisfies P.

C_1: If $\{S, A\}$ satisfies P, so does $\{S, A_1, A_2\}$, where A is a proposition of type a.

C_2: If $\{S, B\}$ satisfies P, then either $\{S, B_1\}$ or $\{S, B_2\}$ satisfies P, where B is a proposition of type b.

(a) Using Zorn's lemma (see problem 3.5.7), show that for any set S, for any property P about subsets of S, if P is of finite character, then any subset U of S for which P holds is a subset of some maximal subset of S for which P holds.

(b) Prove that if P is a consistency property and P satisfies a set U of propositions, then U can be extended to a Hintikka set.

Hint: Use the technique described in problem 3.5.6.

(c) Prove that if P is a consistency property and P satisfies a set U of propositions, then U is satisfiable.

∗ **3.5.13.** Using the definitions given in problem 3.5.6, show that a maximal consistent set S is a Hintikka set satisfying the additional property:

M_0 : For every proposition A, $\quad A \in S \quad$ if and only if $\quad \neg A \notin S$.

∗ **3.5.14.** In this problem, we also use the definitions of problem 3.5.7. A set S of propositions is *downward closed* iff the following conditions hold:

D_1: For every proposition A of type a, if $A \in S$, then both A_1 and A_2 are in S.

D_2: For every proposition B of type b, if $B \in S$, then either B_1 is in S or B_2 is in S.

A set S of propositions is *upward closed* iff:

U_1: For every proposition A of type a, if A_1 and A_2 are both in S, then A is in S.

U_2: For every proposition B of type b, if either B_1 is in S or B_2 is in S, then B is in S.

(a) Prove that any downward closed set satisfying condition M_0 (given in problem 3.5.13) is a maximal consistent set.

(b) Prove that any upward closed set satisfying condition M_0 is a maximal consistent set.

Note: Conditions D_1 and D_2 are conditions $H2$ and $H3$ for Hintikka sets. Furthermore, U_1 and U_2 state the converse of D_1 and D_2. Hence, the above problem shows that a maximal consistent set is a set satisfying condition M_0 and the "if and only if" version of $H2$ and $H3$. Consequently, this reproves that a maximal consistent set is a Hintikka set.

In the next problems, some connections between logic and the theory of boolean algebras are explored.

* **3.5.15.** Recall from Section 3.3 that a *boolean algebra* is a structure $\mathbf{A} =< A, +, *, \neg, 0, 1 >$, where A is a nonempty set, $+$ and $*$ are binary functions on A, $\neg$ is a unary function on A, and 0 and 1 are distinct elements of A, such that the following axioms hold:

Associativity rules:
$$((A + B) + C) = (A + (B + C)) \quad ((A * B) * C) = (A * (B * C))$$
Commutativity rules:
$$(A + B) = (B + A) \quad (A * B) = (B * A)$$
Distributivity rules:
$$(A + (B * C)) = ((A + B) * (A + C))$$
$$(A * (B + C)) = ((A * B) + (A * C))$$
De Morgan's rules:
$$\neg(A + B) = (\neg A * \neg B) \quad \neg(A * B) = (\neg A + \neg B)$$
Idempotency rules:
$$(A + A) = A \quad (A * A) = A$$
Double negation rule:
$$\neg\neg A = A$$
Absorption rules:
$$(A + (A * B)) = A \quad (A * (A + B)) = A$$
Laws of zero and one:

$$(A + 0) = A \quad (A * 0) = 0$$
$$(A + 1) = 1 \quad (A * 1) = A$$
$$(A + \neg A) = 1 \quad (A * \neg A) = 0$$

When dealing with boolean algebras, $\neg A$ is also denoted as $\overline{A}$. Given a boolean algebra $\mathbf{A}$, a partial ordering $\leq$ is defined on A as follows:

$$a \leq b \quad \text{if and only if} \quad a + b = b.$$

A *filter* D is a subset of A such that D is nonempty, for all $x, y \in D$, $x * y \in D$, and for all $z \in A$ and $x \in D$, if $x \leq z$, then $z \in D$. A *proper filter* is a filter such that $0 \notin D$ (equivalently, $D \neq A$). (Note that this is a generalization of the notion defined in problem 3.5.8.)

An *ideal* is a nonempty subset I of A such that, for all $x, y \in I$, $x + y \in I$, and for all $z \in A$ and $x \in I$, $x * z \in I$.

(a) Show that for any (proper) filter D, the set

$$\{\overline{x} \mid x \in D\}$$

is an ideal.

Given a filter D, the relation (D) defined by

$$x(D)y \quad \text{if and only if} \quad x * y + \overline{x} * \overline{y} \in D.$$

is a congruence relation on $\mathbf{A}$. The set of equivalences classes modulo (D) is a boolean algebra denoted by $\mathbf{A}/D$, whose 1 is D, and whose 0 is $\{\overline{x} \mid x \in D\}$.

(b) Prove that $x(D)y$ if and only if there is some $z \in D$ such that

$$x * z = y * z,$$

if and only if

$$x * \overline{y} + \overline{x} * y \in \{\overline{x} \mid x \in D\},$$

if and only if

$$(\overline{x} + y) * (\overline{y} + x) \in D.$$

Note: Intuitively speaking, $(\overline{x} + y)$ corresponds to the proposition $(x \supset y)$ and $(\overline{x} + y) * (\overline{y} + x)$ to $(x \equiv y)$.

A subset E of A has the *finite intersection property* iff for any finite number of elements $x_1, ..., x_n \in E$,

$$x_1 * ... * x_n \neq 0.$$

(c) Prove that every subset E with the finite intersection property is contained in a smallest proper filter. (See problem 3.5.8.)

A filter D is *principal* iff for some $a \neq 0$ in A, $x \in D$ if and only if $a \leq x$. A proper filter is an *ultrafilter* iff it is maximal.

(d) Prove that any set with the finite intersection property can be extended to an ultrafilter.

(e) If D is an ultrafilter, then

$$x + y \in D \quad \text{iff} \quad \text{either } x \in D \text{ or } y \in D,$$

$$x \in D \quad \text{iff} \quad \overline{x} \notin D.$$

Prove that D is an ultrafilter if and only if the quotient boolean algebra $\mathbf{A}/D$ is isomorphic to the two-element boolean algebra $BOOL$.

* **3.5.16.** Let $\simeq$ be the proof-theoretic version of the equivalence relation on $PROP$ defined in problem 3.4.6, so that for any two propositions A and B,

$$A \simeq B \text{ if and only if } \vdash (A \equiv B) \text{ in } G'.$$

(a) Show that the set $\mathbf{B}_0$ of equivalence classes modulo $\simeq$ is a boolean algebra if we define the operations $+$, $*$ and $\overline{}$ on $\mathbf{B}_0$ as follows:

$$[A] + [B] = [A \vee B],$$
$$[A] * [B] = [A \wedge B],$$
$$\overline{[A]} = [\neg A].$$

Also, let $0 = [\bot]$ and $1 = [\top]$. Observe that $0 \neq 1$. The algebra $\mathbf{B}_0$ is called the *Lindenbaum algebra* of $PROP$.

Hint: Use problems 3.4.6 and 3.4.7.

(b) Prove that the following statements are equivalent:

(1) Every consistent set can be extended to a maximal consistent set.

(2) Every filter on $\mathbf{B}_0$ can be extended to an ultrafilter.

* **3.5.17.** Let T be any subset of propositions in $PROP$. We say that T is *finitely axiomatizable* if there is a finite set S of propositions such that for every proposition A in $PROP$,

$$\vdash T \rightarrow A \text{ in } G' + \{cut\} \quad \text{if and only if} \quad \vdash S \rightarrow A \text{ in } G' + \{cut\}.$$

Let

$$D_T = \{[A] \mid \vdash T \rightarrow A \text{ in } G' + \{cut\}\},$$

where $[A]$ denotes the equivalence class of A modulo $\simeq$. Prove the following statements:

(i) T is consistent iff D_T is a proper filter on $\mathbf{B}_0$.

(ii) T is consistent and finitely axiomatizable iff D_T is a principal filter on $\mathbf{B}_0$.

(iii) T is complete iff D_T is an ultrafilter on $\mathbf{B}_0$. (For the definition of a complete set of propositions, See problem 3.5.4).

(iv) T is complete and finitely axiomatizable iff D_T is a principal ultrafilter on $\mathbf{B}_0$.

Given a subset D of $\mathbf{B}_0$, let

$$T_D = \{A \in PROP \mid [A] \in D\}.$$

Show that the converses of (i) to (iv) each hold, with T replaced by T_D and D_T by D.

Say that a set T of propositions is *closed* if, for every $A \in PROP$, $\vdash T \to A$ implies that $A \in T$. Show that there is a one-to-one correspondence between complete closed extensions of T and ultrafilters in $\mathbf{B}_0/D_T$.

Note: In this problem the cut rule seems necessary to prove that D_T is a filter, specifically, that if $\vdash T \to A$ and $\vdash T \to (A \supset B)$ in $G' + \{cut\}$, then $\vdash T \to B$ in $G' + \{cut\}$. To prove this in G', a form of Gentzen's cut elimination theorem seems necessary.

* **3.5.18.** (a) Let $\mathbf{A}_1$ and $\mathbf{A}_2$ be two boolean algebras. A function $h : \mathbf{A}_1 \to \mathbf{A}_2$ is a homomorphism if, for all $x, y \in \mathbf{A}_1$,

$$h(x \vee_1 y) = h(x) \vee_2 h(y),$$
$$h(x \wedge_1 y) = h(x) \wedge_2 h(y) \text{ and}$$
$$h(\overline{x}) = \overline{h(x)}.$$

Show that $h(0) = 0$ and $h(1) = 1$.

(b) Given a boolean algebra $\mathbf{A}$ and a proper filter D, show that the mapping $h_D : \mathbf{A} \to \mathbf{A}/D$ that maps every element a of $\mathbf{A}$ to its equivalence class $[a]$ modulo (D) is a homomorphism.

(c) Let T be a consistent set of propositions. The equivalence relation $\simeq_T$ on $PROP$ is defined as follows:

$$A \simeq_T B \quad \text{if and only if} \quad \vdash T \to (A \equiv B) \text{ in } G' + \{cut\}.$$

Show that the set $\mathbf{B}_T$ of equivalence classes modulo $\simeq_T$ is a boolean algebra if we define the operations $+$, $*$ and $^-$ on $\mathbf{B}_T$ as follows:

$$[A]_T + [B]_T = [A \vee B]_T,$$
$$[A]_T * [B]_T = [A \wedge B]_T,$$
$$\overline{[A]_T} = [\neg A]_T.$$

($[A]_T$ denotes the equivalence class of A modulo $\simeq_T$.) Furthermore, the element 1 is the equivalence class

$$\{A \mid \ \vdash T \to A \text{ in } G' + \{cut\}\},$$

and the element 0 is the class

$$\{A \mid \ \vdash T \to \neg A \text{ in } G' + \{cut\}\}.$$

The boolean algebra $\mathbf{B}_T$ is called the *Lindenbaum algebra of T*. Note that the equivalence class $[A]_T$ of A modulo $\simeq_T$ is the set

$$\{B \mid \ \vdash T \to (A \equiv B) \text{ in } G' + \{cut\}\}.$$

For any homomorphism $h : \mathbf{B}_T \to BOOL$, let $v : \mathbf{PS} \to BOOL$ be defined such that for every propositional letter P,

$$v(P) = h([P]_T).$$

Show that v is a valuation satisfying T such that $\widehat{v}(A) = h([A]_T)$ for all $A \in PROP$.

(d) There is a correspondence between valuations satisfying T and ultrafilters U in $\mathbf{B}_T$ defined as follows: For every ultrafilter U in $\mathbf{B}_T$, the quotient algebra $\mathbf{B}_T/U$ is isomorphic to the boolean algebra $BOOL$ (see problem 3.5.15(e)). By questions 3.5.18(a) and 3.5.18(b), there is a valuation v_U satisfying T induced by the homomorphism from $\mathbf{B}_T$ to $\mathbf{B}_T/U$. Conversely, if v is a valuation satisfying T, show that

$$U_v = \{[A]_T \mid \widehat{v}(A) = \mathbf{T}\}$$

is an ultrafilter in $\mathbf{B}_T$.

(e) Prove the extended completeness theorem for $G' + \{cut\}$.

Hint: Assume that $T \to A$ is valid, but that A is not provable from T. Then, in the Lindenbaum algebra $\mathbf{B}_T$, $[A]_T \neq 1$, and so $[\neg A]_T \neq 0$. Using problem 3.5.15(d), there is an ultrafilter U in $\mathbf{B}_T$ containing $[\neg A]_T$. Since $\mathbf{B}_T/U$ is isomorphic to $BOOL$, by questions 3.5.18(c) and 3.5.18(d), there is a valuation v satisfying T such that

$$\widehat{v}(\neg A) = h([\neg A]_T),$$

where h is the homormophism from $\mathbf{B}_T$ to $\mathbf{B}_T/U$. Since $[\neg A]_T$ is in U,

$$h([\neg A]_T) = \mathbf{T}.$$

Hence, there is a valuation satisfying T such that $\widehat{v}(A) = \mathbf{F}$. This contradicts the validity of $T \to A$.

3.5.19. Write a computer program (preferably in PASCAL or C) implementing the extended *search* procedure of definition 3.5.1.

3.6 More on Gentzen Systems: The Cut Rule

The rules of the Gentzen system G' given in definition 3.4.2 were chosen so as to give ourselves as few choices as possible at each step *upward* in searching systematically for a falsifying valuation. The use of other Gentzen-type systems may afford simpler proofs, especially working *downward*. One such system is the system LK' due to Gentzen. The system LK' contains a rule called the *cut rule*, which is important from a historical point of view, but also from a mathematical point of view. Indeed, even though it is possible to find complete proof systems not using the cut rule, we will discover some unexpected complications when we study first-order logic with equality. Indeed, the system for first-order logic with equality not using the cut rule is not very natural, and the cut rule cannot be dispensed with easily.

3.6.1 Using Auxiliary Lemmas in Proofs

There are also "pragmatic" reasons for considering the cut rule. The cut rule is the following:

$$\frac{\Gamma \to \Delta, A \qquad A, \Lambda \to \Theta}{\Gamma, \Lambda \to \Delta, \Theta}$$

A is called the *cut formula* of this inference.

Notice that this rule formalizes the technique constantly used in practice to use an *auxiliary lemma* in a proof. This is more easily visualized if we assume that Δ is empty. Then, $\Gamma \to A$ is the auxiliary lemma, which can be assumed to belong to a catalogue of already-proven results. Now, using A as an assumption, if we can show that using other assumptions Λ, that Θ is provable, we can conclude that $\Gamma, \Lambda \to \Theta$ is provable. The conclusion does not refer to A.

One might say that a proof using the cut rule is not as "direct" and consequently, not as perspicuous as a proof not using the cut rule. On the other hand, if we already have a vast catalogue of known results, and we can use them to give short and "easy" proofs of other results, why force ourselves not to use the convenience afforded by the cut rule? We shall not try to answer these questions of a philosophical nature. Let us just make a few remarks.

Let us call a proof not using the cut rule a cut-free proof. Cut-free proofs are important in investigations regarding consistency results. The object of such investigations is to establish constructively the consistency of mathematical theories such as arithmetic or set theory, the ultimate goal beeing to show that mathematics formalized as a logical theory is free of contradictions. First-order logic is simple enough that Gentzen's cut elimination theorem holds constructively. This means that for every proof using the cut rule, another proof not using the cut rule can effectively be constructed. We shall give a (nonconstructive) semantic proof of this result in this chapter, and a constructive proof for a simpler system in Chapter 6. From a mathematical point of view, this shows that the cut rule can be dispensed with. For richer logics, such as second-order logic, the cut-elimination theorem also holds, but not constructively, in the sense that there is no effective method for converting a proof with cut to a proof without cut.

Another interesting issue is to examine the relative complexity of proofs with or without cut. Proofs with cuts can be much shorter than cut-free proofs. This will be shown in Chapter 6. However, from the point of view of automatic theorem proving, cut-free proofs are easier to find. For more on cut-free proofs and the cut rule, the reader is referred to Takeuti, 1975, and Pfenning's paper in Shostack, 1984a.

We now present the system LK'.

3.6.2 The Gentzen System LK'

The system LK' consists of *structural rules*, the *cut rule*, and of *logical rules*.

Definition 3.6.1 Gentzen system LK'. The letters $\Gamma, \Delta, \Lambda, \Theta$ stand for arbitrary (possibly empty) sequences of propositions and A, B for arbitrary propositions.

(1) Structural rules:

(i) Weakening:

$$\frac{\Gamma \to \Delta}{A, \Gamma \to \Delta} \ (left) \qquad \frac{\Gamma \to \Delta}{\Gamma \to \Delta, A} \ (right)$$

A is called the *weakening formula*.

(ii) Contraction:

$$\frac{A, A, \Gamma \to \Delta}{A, \Gamma \to \Delta} \ (left) \qquad \frac{\Gamma \to \Delta, A, A}{\Gamma \to \Delta, A} \ (right)$$

(iii) Exchange:

$$\frac{\Gamma, A, B, \Delta \to \Lambda}{\Gamma, B, A, \Delta \to \Lambda} \ (left) \qquad \frac{\Gamma \to \Delta, A, B, \Lambda}{\Gamma \to \Delta, B, A, \Lambda} \ (right)$$

(2) Cut rule:

$$\frac{\Gamma \to \Delta, A \qquad A, \Lambda \to \Theta}{\Gamma, \Lambda \to \Delta, \Theta}$$

A is called the *cut formula* of this inference.

(3) Logical rules:

$$\frac{A, \Gamma \to \Delta}{A \wedge B, \Gamma \to \Delta} \ (\wedge : left) \quad \text{and} \quad \frac{B, \Gamma \to \Delta}{A \wedge B, \Gamma \to \Delta} \ (\wedge : left)$$

$$\frac{\Gamma \to \Delta, A \qquad \Gamma \to \Delta, B}{\Gamma \to \Delta, A \wedge B} \ (\wedge : right)$$

$$\frac{A, \Gamma \to \Delta \qquad B, \Gamma \to \Delta}{A \vee B, \Gamma \to \Delta} \ (\vee : left)$$

$$\frac{\Gamma \to \Delta, A}{\Gamma \to \Delta, A \vee B} \ (\vee : right) \quad \text{and} \quad \frac{\Gamma \to \Delta, B}{\Gamma \to \Delta, A \vee B} \ (\vee : right)$$

$$\frac{\Gamma \to \Delta, A \qquad B, \Lambda \to \Theta}{A \supset B, \Gamma, \Lambda \to \Delta, \Theta} \ (\supset: left) \qquad \frac{A, \Gamma \to \Delta, B}{\Gamma \to \Delta, A \supset B} \ (\supset: right)$$

$$\frac{\Gamma \to \Delta, A}{\neg A, \Gamma \to \Delta} \ (\neg : left) \qquad \frac{A, \Gamma \to \Delta}{\Gamma \to \Delta, \neg A} \ (\neg : right)$$

In the rules above, the propositions $A \vee B$, $A \wedge B$, $A \supset B$ and $\neg A$ are called the *principal formulae* and the propositions A, B the *side formulae*.

The *axioms* of the system LK' are all sequents of the form

$$A \to A.$$

Note that in view of the exchange rule, the order of propositions in a sequent is really irrelevant, and the system LK' could be defined using multisets as defined in problem 2.1.8.

Proof trees are defined inductively as in definition 3.4.5, but with the rules of the system LK' given in definition 3.6.1. If a sequent has a proof in the system G' we say that it is G'-*provable* and similarly, if it is provable in the system LK', we say that it is LK'-*provable*. The system obtained by removing the cut rule from LK' will be denoted as $LK' - \{cut\}$. We also say that a sequent is LK'-provable without a cut if it has a proof tree using the rules of the system $LK' - \{cut\}$. We now show that the systems G' and LK' are logically equivalent. We will in fact prove a stronger result, namely that G', $LK' - \{cut\}$ and LK' are equivalent. First, we show that the system LK' is sound.

Lemma 3.6.1 (Soundness of LK') Every axiom of LK' is valid. For every rule of LK', for every valuation v, if v makes all the premises of a rule true then v makes the conclusion of the rule true. Every LK'-provable sequent is valid.

Proof: The proof uses the induction principle for proofs and is straightforward. □

Note that lemma 3.6.1 differs from lemma 3.4.3 in the following point: It is not true that if v makes the conclusion of a rule true then v makes all premises of that rule true. This reveals a remarkable property of the system G'. The system G' is a "two way" system, in the sense that the rules can be used either from top-down or from bottom-up. However, LK' is a top-down system. In order to ensure that the inferences are sound, the rules must be used from top-down.

3.6.3 Logical Equivalence of G', LK', and $LK' - \{cut\}$

The following theorem yields a semantic version of the cut elimination theorem.

Theorem 3.6.1 Logical equivalence of G', LK', and $LK' - \{cut\}$. There is an algorithm to convert any LK'-proof of a sequent $\Gamma \rightarrow \Delta$ into a G'-proof. There is an algorithm to convert any G'-proof of a sequent $\Gamma \rightarrow \Delta$ into a proof using the rules of $LK' - \{cut\}$.

Proof: If $\Gamma \rightarrow \Delta$ has an LK'-proof, by lemma 3.6.1, $\Gamma \rightarrow \Delta$ is valid. By theorem 3.5.2, $\Gamma \rightarrow \Delta$ has a G'-proof given by the algorithm *search*. Note that if $\Gamma \rightarrow \Delta$ is infinite, then the *search* procedure gives a proof for a finite subsequent of $\Gamma \rightarrow \Delta$, but by definition 3.6.2, it is a proof of $\Gamma \rightarrow \Delta$. Conversely, using the induction principle for G'-proofs we show that every G'-proof can be converted to an $(LK' - \{cut\})$-proof. This argument also applies to infinite sequents, since a proof of an infinite sequent is in fact a proof of some finite subsequent of it.

First, every G'-axiom $\Gamma \rightarrow \Delta$ contains some common proposition A, and by application of the weakening and the exchange rules, an $(LK' - \{cut\})$-proof of $\Gamma \rightarrow \Delta$ can be obtained from the axiom $A \rightarrow A$. Next, we have to show that every application of a G'-rule can be replaced by a sequence of $(LK' - \{cut\})$-rules. There are eight cases to consider. Note that the G'-rules $\wedge : right$, $\vee : left$, $\supset : right$, $\supset : left$, $\neg : right$ and $\neg : left$ can easily be simulated in $LK' - \{cut\}$ using the exchange, contraction, and corresponding $(LK'-\{cut\}) - rules$. We show how the G'-rule $\wedge : left$ can be transformed to a sequence of $(LK' - \{cut\})$-rules, leaving the transformation of the G'-rule $\vee : right$ as an exercise. The following is an $(LK' - \{cut\})$-derivation from $\Gamma, A, B, \Delta \to \Lambda$ to $\Gamma, A \wedge B, \Delta \to \Lambda$.

$$
\begin{array}{c}
\Gamma, A, B, \Delta \to \Lambda \\
\hline
\text{(several exchanges)} \\
\hline
A, B, \Gamma, \Delta \to \Lambda \\
\hline
A \wedge B, B, \Gamma, \Delta \to \Lambda \\
\hline
B, A \wedge B, \Gamma, \Delta \to \Lambda \\
\hline
A \wedge B, A \wedge B, \Gamma, \Delta \to \Lambda \\
\hline
A \wedge B, \Gamma, \Delta \to \Lambda \\
\hline
\text{(several exchanges)} \\
\hline
\Gamma, A \wedge B, \Delta \to \Lambda
\end{array}
\qquad
\begin{array}{l}
(\wedge : left\ (A)) \\[1.2em]
\text{(exchange)} \\[1.2em]
(\wedge : left\ (B)) \\[1.2em]
\text{(contraction)}
\end{array}
$$

□

3.6.4 Gentzen's Hauptsatz for LK' (Cut elimination theorem for LK')

Theorem 3.6.1 has the following important corollary.

Corollary (Gentzen's Hauptsatz for LK') A sequent is LK'-provable if and only if it is LK'-provable without a cut.

Note that the *search* procedure together with the above procedure provides an algorithm to construct a cut-free LK'-proof from an LK'-proof with cut. Gentzen proved the above result by a very different method in which an LK'-proof is (recursively) transformed into an LK'-proof without cut. Gentzen's proof is more structural and syntactical than ours, since we completely forget about the LK'-proof and start from scratch using the procedure *search*. Also, Gentzen's proof generalizes to the first-order predicate calculus LK, providing an algorithm for transforming any LK-proof with cut to an LK-proof without cut. The *search* procedure will also provide a cut-free proof, but the argument used in justifying the correctness of the *search* procedure is not constructive. The nonconstructive step arises when we show that the *search* procedure terminates for a valid sequent. Gentzen's proof is difficult and can be found in either Takeuti, 1975; Kleene, 1952; or in Gentzen's original paper in Szabo, 1969. A constructive proof for a simpler system (sequents of formulae in NNF) will also be given in Chapter 6.

3.6.5 Characterization of Consistency in LK'

The following lemma gives a characterization of consistency in the system LK'.

Lemma 3.6.2 (1) A set Γ of propositions is inconsistent if and only if there is some proposition A such that both $\Gamma \to A$ and $\Gamma \to \neg A$ are LK'-provable.

(2) For any proposition A, the sequent $\Gamma \to A$ is not LK'-provable if and only if $\Gamma \cup \{\neg A\}$ is consistent.

Proof: In this proof, we will abbreviate LK'-provable as provable.

(1) If Γ is inconsistent then $\Gamma \to B$ is provable for any proposition B, showing that (1) is necessary. Conversely, assume that for some A, both $\vdash \Gamma \to A$ and $\vdash \Gamma \to \neg A$ in LK', with proofs T_1 and T_2. The following is a proof of $\Gamma \to B$ for any given B.

$$
\begin{array}{cc}
\dfrac{
\dfrac{
\dfrac{
\begin{array}{c} T_1 \\ \hline \Gamma \to A \end{array}
}{\neg A, \Gamma \to} \ (\neg : left)
}{\Gamma \to \neg\neg A} \ (\neg : right)
\qquad
\dfrac{
\dfrac{
\begin{array}{c} T_2 \\ \hline \Gamma \to \neg A \end{array}
}{\neg\neg A, \Gamma \to} \ (\neg : left)
}{\neg\neg A, \Gamma \to B} \ (weakening)
}{\Gamma, \Gamma \to B} & (cut\ (\neg\neg A))
\end{array}
$$

$$
\dfrac{\Gamma, \Gamma \to B}{\Gamma \to B} \ (contractions\ and\ exchanges)
$$

(2) Assume that $\Gamma \to A$ is not provable. If $\Gamma \cup \{\neg A\}$ was inconsistent, then $\neg A, \Gamma \to A$ would be provable with proof T. The following is a proof of $\Gamma \to A$, contradicting the hypothesis.

$$
\dfrac{
\dfrac{
\dfrac{
\begin{array}{c} T \\ \hline \neg A, \Gamma \to A \end{array}
}{\Gamma \to A, \neg\neg A} \ (\neg : right)
\qquad
\dfrac{
\dfrac{A \to A}{\to A, \neg A} \ (\neg : right)
}{\neg\neg A \to A} \ (\neg : left)
}{\Gamma \to A, A} \ (cut\ (\neg\neg A))
}{\Gamma \to A} \ (contraction)
$$

Conversely, assume that $\Gamma \cup \{\neg A\}$ is consistent. If $\Gamma \to A$ is provable, a fortiori $\Gamma, \neg A \to A$ is provable. But $\neg A \to \neg A$ is also provable since it is an axiom, and so $\Gamma, \neg A \to \neg A$ is provable. By (1), $\Gamma \cup \{\neg A\}$ is inconsistent. $\square$

Remark: Recall that for an infinite set of propositions Γ, $\Gamma \to A$ is provable if $\Delta \to A$ is provable for a *finite* subsequence Δ of Γ. Hence, the above proofs should really be modified to refer to finite subsequences of Γ. Using the exchange, weakening and contraction rules, we can ensure that the antecedent in the conclusion of each proof is a subsequence of Γ. We leave the details as an exercise. Also, note that the above characterizations of consistency (or inconsistency) in LK' are purely syntactic (proof theoretic), and that the cut rule was used in a crucial way.

PROBLEMS

3.6.1. Give LK'-proof trees for the following tautologies:

$$A \supset (B \supset A)$$
$$(A \supset B) \supset ((A \supset (B \supset C)) \supset (A \supset C))$$
$$A \supset (B \supset (A \wedge B))$$
$$A \supset (A \vee B) \quad B \supset (A \vee B)$$
$$(A \supset B) \supset ((A \supset \neg B) \supset \neg A)$$
$$(A \wedge B) \supset A \quad (A \wedge B) \supset B$$
$$(A \supset C) \supset ((B \supset C) \supset ((A \vee B) \supset C))$$
$$\neg\neg A \supset A$$

3.6.2. Show that the cut rule is not a two-way rule, that is, if a valuation v satisfies the conclusion of the cut rule, it does not necessarily satisfy its premises. Find the other rules of LK' that are not two-way rules.

* **3.6.3.** Recall that a set Γ of propositions is *maximally consistent* if Γ is consistent and for any other set Δ, if Γ is a proper subset of Δ then Δ is inconsistent.

    (a) Show that if Γ is maximally consistent, for any proposition A such that $\Gamma \to A$ is provable in LK' (with cut), A is in Γ.

    (b) Show that

$$\Delta = \{A \mid \vdash \Gamma \to A \text{ in } LK' \text{ (with cut)}\}$$

    is maximally consistent if and only if for every proposition A, either $\vdash \Gamma \to A$ or $\vdash \Gamma \to \neg A$ in LK', but not both.

    (c) Show that Γ is maximally consistent iff there is a single valuation v satisfying Γ.

    (d) Show that Γ is maximally consistent iff there is a valuation v such that $v \models A$ if and only if A is in Γ.

3.6.4. Using the technique of problem 3.5.5, prove in LK' $(+\{cut\})$ that every consistent set can be extended to a maximal consistent set.

* **3.6.5.** In this problem, we are adopting the definition of a Hintikka set given in problem 3.5.6. Let Δ be a maximally consistent set. To cut down on the number of cases, in this problem, assume that $(A \supset B)$ is an abbreviation for $(\neg A \vee B)$, so that the set of connectives is $\{\wedge, \vee, \neg\}$.

(a) Show that Δ is a Hintikka set.

(b) Recall that in LK', $\Gamma \to A$ is not provable if and only if $\Gamma \cup \{\neg A\}$ is consistent. Using problem 3.6.4, prove that if $\Gamma \to A$ is valid, then it is provable in LK' (with cut).

Remark: This provides another proof of the completeness of LK' (with cut). Note that a proof tree for $\Gamma \to A$ is not produced (compare with theorem 3.4.1).

3.6.6. Prove that the extended completeness theorem and the model existence theorem are equivalent for LK'. (This is also true for $LK' - \{cut\}$, but apparently requires the cut elimination theorem).

3.6.7. Implement a computer program (preferably in PASCAL or C) converting an LK-proof into a cut-free LK-proof. Compare and investigate the relative lengths of proofs.

Notes and Suggestions for Further Reading

We have chosen Gentzen systems as the main vehicle for presenting propositional logic because of their algorithmic nature and their conceptual simplicity. Our treatment is inspired from Kleene, 1967 and Kleene, 1952. For more on Gentzen systems, the reader should consult Takeuti, 1975; Szabo, 1969; or Smullyan, 1968.

We believe that the use of Hintikka sets improves the clarity of the proof of the completeness theorem. For more details on Hintikka sets and related concepts such as consistency properties, the reader is referred to Smullyan, 1968.

There are other proof systems for propositional logic. The Hilbert system H discussed in problems 3.4.9 to 3.4.12 is from Kleene, 1967, as well as the natural deduction system used in problems 3.4.11 and 3.4.12. For more on natural deduction systems, the reader is referred to Van Dalen, 1982; Prawitz 1965; or Szabo, 1969. A variant of Gentzen systems called tableaux systems is discussed at length in Smullyan, 1968.

The relationship between boolean algebra and logic was investigated by Tarski, Lindenbaum, Rasiowa, and Sikorski. For more details, the reader is referred to Chang and Keisler, 1973, or Bell and Slomson, 1974. Exercise 3.5.18 is adapted from Bell and Slomson, 1974.

The proof of Gentzen's cut elimination theorem can be found in Kleene, 1952; Takeuti, 1975; and Szabo, 1969.

Chapter 4

Resolution in Propositional Logic

4.1 Introduction

In Chapter 3, a procedure for showing whether or not a given proposition is valid was given. This procedure, which uses a Gentzen system, yields a formal proof in the Gentzen system when the input proposition is valid. In this chapter, another method for deciding whether a proposition is valid is presented. This method due to J. Robinson (Robinson, 1965) and called *resolution*, has become quite popular in automatic theorem proving, because it is simple to implement. The essence of the method is to prove the validity of a proposition by establishing that the negation of this proposition is unsatisfiable.

The main attractive feature of the resolution method is that it has a single inference rule (the resolution rule). However, there is a price to pay: The resolution method applies to a proposition in conjunctive normal form.

In this chapter, the resolution method will be presented as a variant of a special kind of Gentzen-like system. A similar approach is followed in Robinson, 1969, but the technical details differ. We shall justify the completeness of the resolution method by showing that Gentzen proofs can be recursively transformed into resolution proofs, and conversely. Such transformations are interesting not only because they show constructively the equivalence of the two proof systems, but because they also show that the complexity of proofs of a proposition is essentially the same in both systems, provided that the systems are defined the right way. Indeed, we shall show that if resolution proofs are represented as directed acyclic graphs (DAGs) and if we allow Gentzen rules to apply to more than two propositions, the number of axioms in a Gentzen proof tree is linearly related to the number of resolution steps in the corresponding resolution DAG.

The essence of the connection between Gentzen proofs and resolution proofs is the following:

Given a proposition A, if B is a conjunctive normal form of $\neg A$, A is valid iff B is unsatisfiable, iff the sequent $B \rightarrow$ is valid.

We will show how a Gentzen proof tree for $B \rightarrow$ can be transformed into a resolution DAG for B (and conversely). The first step is to design an efficient Gentzen system for proving sequents of the form $B \rightarrow$, where B is in conjunctive normal form.

4.2 A Special Gentzen System

The system $GCNF'$ is obtained by restricting the system G' to sequents of the form $A \rightarrow$, where A is a proposition in conjunctive normal form.

4.2.1 Definition of the System $GCNF'$

Recall from definition 3.4.7 that a proposition A is in conjunctive normal form iff it is a conjunction $C_1 \wedge ... \wedge C_m$ of disjunctions $C_i = B_{i,1} \vee ... \vee B_{i,n_i}$, where each $B_{i,j}$ is either a propositional symbol P or the negation $\neg P$ of a propositional symbol.

Definition 4.2.1 A *literal* is either a propositional symbol or its negation. Given a literal L, its *conjugate* $\overline{L}$ is defined as follows: If $L = P$ then $\overline{L} = \neg P$, and if $L = \neg P$, then $\overline{L} = P$. A proposition of the form

$$C_i = B_{i,1} \vee ... \vee B_{i,n_i},$$

where each $B_{i,j}$ is a literal is called a *clause*.

By lemma 3.3.6, since both $\wedge$ and $\vee$ are commutative, associative and idempotent, it can be assumed that the clauses C_i are distinct, and each clause can be viewed as the set of its distinct literals. Also, since the semantics of sequents implies that a sequent

$$C_1 \wedge ... \wedge C_n \rightarrow \quad \text{is valid iff}$$
$$\text{the sequent } C_1, ..., C_n \rightarrow \text{ is valid,}$$

it will be assumed in this section that we are dealing with sequents of the form $C_1, ..., C_n \rightarrow$, where $\{C_1, ..., C_n\}$ is a *set* of clauses, and each clause C_i is a set of literals. We will also view a sequent $C_1, ..., C_n \rightarrow$ as the set of clauses $\{C_1, ..., C_n\}$. For simplicity, the clause $\{L\}$ consisting of a single literal will often be denoted by L. If Γ and Δ denote sets of propositions, then Γ, Δ denotes the set $\Gamma \cup \Delta$. Similarly, if Γ is a set of propositions and A is a proposition, then Γ, A denotes the set $\Gamma \cup \{A\}$. Consequently, $\Gamma, A_1, ..., A_m \rightarrow$ denotes the sequent $\Gamma \cup \{A_1, ..., A_m\} \rightarrow$.

It should be noted that when a sequent $C_1, ..., C_n \rightarrow$ is viewed as the set of clauses $\{C_1, ..., C_n\}$, the comma is interpreted as the connective *and* ($\wedge$), but that when we consider a single clause $C_i = \{L_1, ..., L_m\}$ as a set of literals, the comma is interpreted as the connective *or* ($\vee$).

For sequents of the above form, note that only the $\vee : left$ rule and the $\neg : left$ rule are applicable. Indeed, since each proposition in each C_i is a literal, for every application of the $\neg : left$ rule resulting in a literal of the form $\neg P$ in the antecedent of the conclusion of the rule, the propositional letter P belongs to the right-hand side of the sequent which is the premise of the rule. Since the $\vee : left$ rule does not add propositions to the right-hand side of sequents, only propositional letters can appear on the right-hand side of sequents. Hence, only the $\vee : left$ rule and the $\neg : left$ rule are applicable to sequents of the form defined above. But then, by redefining the axioms to be sequents of the form

$$\Gamma, P, \neg P \rightarrow ,$$

we obtain a proof system in which the only inference rule is $\vee : left$.

A further improvement is achieved by allowing several $\vee : left$ rules to be performed in a single step. If a sequent of the form

$$\Gamma, (A_1 \vee B), ..., (A_m \vee B) \rightarrow$$

is provable, the proof may contain m distinct applications of the $\vee : left$ rule to $(A_1 \vee B),...,(A_m \vee B)$, and the proof may contain $2^m - 1$ copies of the proof tree for the sequent $\Gamma, B \rightarrow$. We can avoid such redundancies if we introduce a rule of the form

$$\frac{\Gamma, A_1, ..., A_m \rightarrow \qquad \Gamma, B \rightarrow}{\Gamma, (A_1 \vee B), ..., (A_m \vee B) \rightarrow}$$

We obtain a proof system in which proofs are even more economical if we introduce a rule of the form

$$\frac{\Gamma_1, C_1, ..., C_k \rightarrow \qquad \Gamma_2, B \rightarrow}{\Gamma, (A_1 \vee B), ..., (A_m \vee B) \rightarrow}$$

where Γ_1 and Γ_2 are arbitrary subsets of $\Gamma \cup \{(A_1 \vee B), ..., (A_m \vee B)\}$ (not necessarily disjoint), and $\{C_1, ..., C_k\}$ is any *nonempty* subset of $\{A_1, ..., A_m\}$. In this fashion, we obtain a system similar to LK' with implicit weakenings, in which every nontrivial sequent (see definition below) has a proof in which the axioms are of the form $P, \neg P \rightarrow$.

Definition 4.2.2 A sequent of the form $C_1, ..., C_n \rightarrow$ is *trivial* if $C_i = P$ and $C_j = \neg P$ for some letter P and some $i, j, 1 \leq i, j \leq n$. Otherwise, we say that the sequent is *nontrivial*.

The system $GCNF'$ is defined as follows.

Definition 4.2.3 (The system $GCNF'$). Let $\Gamma, \Gamma_1, \Gamma_2$ denote sets of propositions, $A_1, ..., A_m$ denote propositions, B denote a literal, and P denote a propositional letter.

Axioms: All trivial sequents, that is, sequents of the form

$$\Gamma, P, \neg P \rightarrow$$

Inference Rule: All instances of the rule

$$\frac{\Gamma_1, C_1, ..., C_k \rightarrow \qquad \Gamma_2, B \rightarrow}{\Gamma, (A_1 \vee B), ..., (A_m \vee B) \rightarrow}$$

where Γ_1 and Γ_2 are arbitrary subsets of $\Gamma \cup \{(A_1 \vee B), ..., (A_m \vee B)\}$ (possibly empty and not necessarily disjoint), $\{C_1, ..., C_k\}$ is any *nonempty* subset of $\{A_1, ..., A_m\}$, and B is a *literal*.

Remark: In the above rule, B is a *literal* and not an arbitrary *proposition*. This is not a restriction because the system $GCNF'$ is complete. We could allow arbitrary propositions, but this would complicate the transformation of a proof tree into a resolution refutation (see the problems). The system obtained by restricting Γ_1 and Γ_2 to be subsets of Γ is also complete. However, if such a rule was used, this would complicate the transformation of a resolution refutation into a proof tree.

Deduction trees and proof trees are defined in the obvious way as in definition 3.4.5.

EXAMPLE 4.2.1

The following is a proof tree in the system $GCNF'$.

$$\frac{\neg S, S \rightarrow \qquad \dfrac{S, \neg S \rightarrow \qquad \neg Q, Q \rightarrow}{\neg Q, S, \{\neg S, Q\} \rightarrow}}{\{P, \neg Q\}, \{\neg S, \neg Q\}, S, \{\neg S, Q\} \rightarrow}$$

In the above proof, the $\vee$: *left* rule is applied to $\{P, \neg Q\}, \{\neg S, \neg Q\}$. The literal $\neg Q$, which occurs in both clauses, goes into the right premise of the lowest inference, and from the set $\{P, \neg S\}$, only $\neg S$ goes into the left premise. The above proof in $GCNF'$ is more concise than the following proof in G'. This is because rules of $GCNF'$ can apply to several propositions in a single step, avoiding the duplication of subtrees.

$$\frac{T_1 \qquad\qquad T_2}{\{P, \neg Q\}, \{\neg S, \neg Q\}, S, \{\neg S, Q\} \rightarrow}$$

where T_1 is the tree

$$\frac{P, \neg S, S, \{\neg S, Q\} \rightarrow \qquad \dfrac{P, \neg Q, S, \neg S \rightarrow \qquad P, \neg Q, S, Q \rightarrow}{P, \neg Q, S, \{\neg S, Q\} \rightarrow}}{P, \{\neg S, \neg Q\}, S, \{\neg S, Q\} \rightarrow}$$

and T_2 is the tree

$$\frac{\neg Q, \neg S, S, \{\neg S, Q\} \rightarrow \qquad \dfrac{\neg Q, S, \neg S \rightarrow \qquad \neg Q, S, Q \rightarrow}{\neg Q, S, \{\neg S, Q\} \rightarrow}}{\neg Q, \{\neg S, \neg Q\}, S, \{\neg S, Q\} \rightarrow}$$

In order to prove the soundness and completeness of the System $GCNF'$, we need the following lemmas.

4.2.2 Soundness of the System $GCNF'$

First, we prove the following lemma.

Lemma 4.2.1 Every axiom is valid. For every valuation v, if v satisfies both premises of a rule of $GCNF'$, then v satisfies the conclusion.

Proof: It is obvious that every axiom is valid. For the second part, it is equivalent to show that if v falsifies $\Gamma, (A_1 \vee B), ..., (A_m \vee B) \rightarrow$, then either v falsifies $\Gamma_1, C_1, ..., C_k \rightarrow$, or v falsifies $\Gamma_2, B \rightarrow$. But v falsifies $\Gamma, (A_1 \vee B), ..., (A_m \vee B) \rightarrow$ if v satisfies all propositions in Γ and v satisfies all of $(A_1 \vee B), ..., (A_m \vee B)$. If v satisfies B, then v satisfies Γ and B, and so v falsifies $\Gamma_2, B \rightarrow$ for every subset Γ_2 of $\Gamma, (A_1 \vee B), ..., (A_m \vee B)$. If v does not satisfy B, then it must satisfy all of $A_1, ..., A_m$. But then, v satisfies Γ_1 and $C_1, ..., C_k$ for any subset Γ_1 of $\Gamma, (A_1 \vee B), ..., (A_m \vee B)$ and nonempty subset $\{C_1, ..., C_k\}$ of $\{A_1, ..., A_m\}$. This concludes the proof. $\square$

Using induction on proof trees as in lemma 3.4.3, we obtain the following corollary:

Corollary (Soundness of $GCNF'$) Every sequent provable in $GCNF'$ is valid.

In order to prove the completeness of $GCNF'$, we will also need the following normal form lemma.

Lemma 4.2.2 Every proof in G' for a sequent of the form $C_1, ..., C_m \rightarrow$, where each C_i is a clause is equivalent to a G'-proof in which all instances of the $\neg : left$ rule precede all instances of the $\vee : left$ rule, in the sense that on every path from a leaf to the root, all $\neg : left$ rules precede all $\vee : left$ rules.

Proof: First, observe that every proof tree in G' of the form

Subtree of type (1)

$$\frac{\dfrac{T_1}{\Gamma, A \rightarrow P} \qquad \dfrac{T_2}{\Gamma, B \rightarrow P}}{\dfrac{\Gamma, (A \vee B) \rightarrow P}{\Gamma, (A \vee B), \neg P \rightarrow}}$$

can be transformed to a proof tree of the same depth of the form

Subtree of type (2)

$$\frac{\dfrac{\dfrac{T_1}{\Gamma, A \rightarrow P}}{\Gamma, A, \neg P \rightarrow} \qquad \dfrac{\dfrac{T_2}{\Gamma, B \rightarrow P}}{\Gamma, B, \neg P \rightarrow}}{\Gamma, (A \vee B), \neg P \rightarrow}$$

Next, we prove the lemma by induction on proof trees. Let T be a proof tree. If T is a one-node tree, it is an axiom and the lemma holds trivially. Otherwise, either the inference applied at the root is the $\vee : left$ rule, or it is the $\neg : left$ rule. If it is the $\vee : left$ rule, T has two subtrees T_1 and T_2 with root S_1 and S_2, and by the induction hypothesis, there are proof trees T_1' and T_2' with root S_1 and S_2 satisfying the conditions of the lemma. The tree obtained from T by replacing T_1 by T_1' and T_2 by T_2' satisfies the conditions of the lemma. Otherwise, the inference applied at the root of T is the $\neg : left$ rule. Let T_1 be the subtree of T having $\Gamma, (A \vee B) \rightarrow P$ as its root. If T_1 is an axiom, the lemma holds trivially. If the rule applied at the root of T_1 is a $\neg : left$ rule, by the induction hypothesis, there is a tree T_1' with the same depth as T_1 satisfying the condition of the lemma. If T_1' contains only $\neg : left$ rules, the lemma holds since for some letter Q, both Q and $\neg Q$ must belong to $\Gamma, (A \vee B), \neg P \rightarrow$. Otherwise, the tree T' obtained by replacing T_1 by T_1' in T is a tree of type (1), and $depth(T') = depth(T)$. By the the remark at the beginning of the proof, this tree can be replaced by a tree of type (2) having the same depth as T' (and T). We conclude by applying the induction hypothesis to the two subtrees of T'. Finally, if the rule applied at the root of the subtree T_1 is a $\vee : left$ rule, T is a tree of type (1). We conclude as in the previous case. $\square$

4.2.3 Completeness of the System $GCNF'$

Having this normal form, we can prove the completeness of $GCNF'$.

Theorem 4.2.1 (Completeness of $GCNF'$). If a sequent

$$C_1, ..., C_m \rightarrow$$

is valid, then it is provable in $GCNF'$. Equivalently, if

$$C_1 \wedge ... \wedge C_m$$

is unsatisfiable, the sequent

$$C_1, ..., C_m \rightarrow$$

is provable in $GCNF'$. Furthermore, every nontrivial valid sequent has a proof in $GCNF'$ in which the axioms are of the form $P, \neg P \rightarrow$.

Proof: Since $\vee$ is associative, commutative and idempotent, a clause $\{L_1, ..., L_k\}$ is equivalent to the formula

$$(((...(L_1 \vee L_2) \vee ...) \vee L_{k-1}) \vee L_k).$$

By the completeness of G' (theorem 3.4.1), the sequent $C_1,...,C_m \rightarrow$ has a proof in which the $\vee : left$ rules are instances of the rule of definition 4.2.3. From lemma 4.2.2, the sequent $C_1, ..., C_m \rightarrow$ has a G'-proof in which all the $\neg : left$ rules precede all the $\vee : left$ rules. We prove that every nontrivial valid sequent $C_1, ..., C_m \rightarrow$ has a proof in $GCNF'$ by induction on proof trees in G'.

Let T be a G'-proof of $C_1, ..., C_m \rightarrow$ in G'. Since it is assumed that $C_1, ..., C_m \rightarrow$ is nontrivial, some C_i, say C_1, is of the form $(A \vee B)$, and the bottom inference must be the $\vee : left$ rule. So, T is of the form

$$
\frac{
\begin{array}{cc}
\dfrac{T_1}{A, C_2, ..., C_m \rightarrow} & \dfrac{T_2}{B, C_2, ..., C_m \rightarrow}
\end{array}
}{(A \vee B), C_2, ..., C_m \rightarrow}
$$

If both T_1 and T_2 only contain applications of the $\neg : left$ rule, A must be a literal and some C_i is its conjugate, since otherwise, $C_2, ..., C_m$ would be trivial. Assume that C_i is the conjugate of A. Similarly, B is a literal, and some C_j is its conjugate.

Then, we have the following proof in $GCNF'$:

$$\frac{A, \overline{A} \to \qquad B, \overline{B} \to}{(A \vee B), C_2, ..., C_m \to}$$

In this proof, $\Gamma_1 = \{\overline{A}\} = C_i$, and $\Gamma_2 = \{\overline{B}\} = C_j$.

Otherwise, either $A, C_2, ..., C_m$ or $B, C_2, ..., C_m$ is nontrivial. If both are nontrivial, by the induction hypothesis, they have proofs S_1 and S_2 in $GCNF'$, and by one more application of the $\vee : left$ rule, we obtain the following proof in $GCNF'$:

$$\frac{\dfrac{S_1}{A, C_2, ..., C_m \to} \qquad\qquad \dfrac{S_2}{B, C_2, ..., C_m \to}}{(A \vee B), C_2, ..., C_m \to}$$

If $A, C_2, ..., C_m$ is trivial, then A must be a literal and some C_i is its conjugate, since otherwise, $C_1, ..., C_m$ would be trivial. Assume that $C_i = \overline{A}$. Since $B, C_2, ..., C_m$ is nontrivial, by the induction hypothesis it has a proof S_2 in $GCNF'$. Then, the following is a proof of $C_1, ..., C_m \to$ in $GCNF'$:

$$\frac{A, \overline{A} \to \qquad\qquad \dfrac{S_2}{B, C_2, ..., C_m \to}}{(A \vee B), C_2, ..., C_m \to}$$

The case in which $A, C_2, ..., C_m$ is nontrivial and $B, C_2, ..., C_m$ is trivial is symmetric. This concludes the proof of the theorem. $\square$

The completeness of the system $GCNF'$ has been established by showing that it can simulate the system G'. However, the system $GCNF'$ is implicitly more nondeterministic than the system G'. This is because for a given sequent of the form

$$\Gamma, (A_1 \vee B), ..., (A_m \vee B) \to ,$$

one has the choice to pick the number of propositions $(A_1 \vee B), ..., (A_m \vee B)$, and to pick the subsets Γ_1, Γ_2 and $C_1, ..., C_k$. A consequence of this nondeterminism is that smaller proofs can be obtained, but that the process of finding proofs is not easier in $GCNF'$ than it is in G'.

PROBLEMS

4.2.1. Show that the set of clauses

$$\{\{A, B, \neg C\}, \{A, B, C\}, \{A, \neg B\}, \{\neg A\}\}$$

is unsatisfiable.

4.2.2. Show that the following sets of clauses are unsatisfiable:

(a) $\{\{A, \neg B, C\}, \{B, C\}, \{\neg A, C\}, \{B, \neg C\}, \{\neg B\}\}$

(b) $\{\{A, \neg B\}, \{A, C\}, \{\neg B, C\}, \{\neg A, B\}, \{B, \neg C\}, \{\neg A, \neg C\}\}$

4.2.3. Which of the following sets of clauses are satisfiable? Give satisfying valuations for those that are satisfiable, and otherwise, give a proof tree in $GCNF'$:

(a) $\{\{A, B\}, \{\neg A, \neg B\}, \{\neg A, B\}\}$

(b) $\{\{\neg A\}, \{A, \neg B\}, \{B\}\}$

(c) $\{\{A, B\}, \bot\}$

4.2.4. Consider the following algorithm for converting a proposition A into a proposition B in conjunctive normal form, such that A is satisfiable if and only if B is satisfiable. First, express A in terms of the connectives $\vee$, $\wedge$ and $\neg$. Then, let $A_1, ..., A_n$ be all the subformulae of A, with $A_n = A$. Let $P_1, ..., P_n$ be distinct propositional letters. The proposition B is the conjunction of P_n with the conjunctive normal forms of the following propositions:

$$((P_i \vee P_j) \equiv P_k) \quad \text{whenever} \quad A_k \text{ is } (A_i \vee A_j)$$
$$((P_i \wedge P_j) \equiv P_k) \quad \text{whenever} \quad A_k \text{ is } (A_i \wedge A_j)$$
$$(\neg P_i \equiv P_k) \quad \text{whenever} \quad A_k \text{ is } \neg A_i.$$

(a) Prove that A is satisfiable if and only if B is satisfiable, by showing how to obtain a valuation satisfying B from a valuation satisfying A and vice versa.

(b) Prove that the above algorithm runs in polynomial time in the length of the proposition A.

Hint: Use problem 3.2.4.

4.2.5. Prove that the system $GCNF'$ is still complete if we require all leaf nodes of a proof tree to contain only literals.

4.2.6. Design a new system $GDNF'$ with a single inference rule $\wedge$: *right*, for sequents of the form

$$\rightarrow D_1, ..., D_n,$$

where $D_1 \vee ... \vee D_n$ is in disjunctive normal form. Prove the completeness of such a system.

4.2.7. Prove that the system $GCNF'$ is complete for infinite sets of clauses.

4.2.8. Write a computer program for testing whether a set of clauses is unsatisfiable, using the system $GCNF'$.

4.3 The Resolution Method for Propositional Logic

We will now present the resolution method, and show how a proof in the system $GCNF'$ can be transformed into a proof by resolution (and vice versa). In the resolution method, it is convenient to introduce a notation for the clause $\{\perp\}$ consisting of the constant *false* ($\perp$). This clause is denoted by $\square$ and is called the *empty clause*. Then, one attempts to establish the unsatisfiability of a set $\{C_1, ..., C_m\}$ of clauses by constructing a kind of tree whose root is the empty clause. Such trees usually contain copies of identical subtrees, and the resolution method represents them as collapsed trees. Technically, they are represented by directed acyclic graphs (DAGs).

Proof trees in $GCNF'$ can also be represented as DAGs and in this way, more concise proofs can be obtained. We could describe the algorithms for converting a proof in $GCNF'$ into a proof by resolution and vice versa in terms of DAGs, but this is not as clear and simple for the system $GCNF'$ using DAGs than it is for the standard system $GCNF'$ using trees. Furthermore, the second algorithm converts a resolution DAG into a proof tree that has at most twice the number of nodes in the original DAG and so, the system $GCNF'$ is essentially as efficient as the resolution method. Hence, for clarity and simplicity, we shall present our results in terms of proof trees and resolution DAGs. Some remarks on proof DAGs will be made at the end of this section.

4.3.1 Resolution DAGs

For our purposes, it is convenient to define DAGs as pairs (t, R) where t is a tree and R is an equivalence relation on the set of tree addresses of t satisfying certain conditions. Roughly speaking, R specifies how common subtrees are shared (collapsed into the same DAG).

EXAMPLE 4.3.1

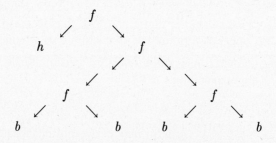

The above tree t contains two copies of the subtree

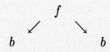

which itself contains two copies of the one-node subtree b. If we define the equivalence relation R on the domain $dom(t) = \{e, 1, 2, 21, 22, 211, 212, 221, 222\}$ as the least equivalence relation containing the set of pairs

$$\{(21, 22), (211, 221), (212, 222), (211, 212)\},$$

the equivalence classes modulo R are the nodes of a graph representing the tree t as a collapsed tree. This graph can be represented as follows:

EXAMPLE 4.3.2

The equivalence relation R is such that two equivalent tree addresses must be the roots of identical subtrees. If two addresses u, v are equivalent, then they must be independent (as defined in Subsection 2.2.2), so that there are no cycles in the graph.

The formal definition of a DAG is as follows.

Definition 4.3.1 A *directed acyclic graph* (for short, a *DAG*) is a pair (t, R), where t is a tree labeled with symbols from some alphabet Δ, and R is an equivalence relation on $dom(t)$ satisfying the following conditions:

For any pair of tree addresses $u, v \in dom(t)$, if $(u, v) \in R$, then either $u = v$ or u and v are independent (as defined in Subsection 2.2.2) and, for all $i > 0$, we have:

(1) $ui \in dom(t)$ iff $vi \in dom(t)$;

(2) If $ui \in dom(t)$, then $(ui, vi) \in R$;

(3) $t(u) = t(v)$.

The tree t is called the *underlying tree* of the DAG. The equivalence classes modulo R are called the *nodes* of the DAG. Given any two equivalence classes S and T, an *edge* from S to T is any pair $([u], [ui])$ of equivalence classes of tree addresses modulo R, such that $u \in S$, and $ui \in T$. By conditions (1)-(3), the definition of an edge makes sense.

The concept of root, descendant, ancestor, and path are also well defined for DAGs: They are defined on the underlying tree.

This definition only allows connected rooted DAGs, but this is enough for our purpose. Indeed, DAGs arising in the resolution method are sets of DAGs as defined above.

We now describe the resolution method.

4.3.2 Definition of the Resolution Method for Propositional Logic

The resolution method rests on the fact that the proposition

$$((A \vee P) \wedge (B \vee \neg P)) \equiv ((A \vee P) \wedge (B \vee \neg P) \wedge (A \vee B)) \qquad (*)$$

is a tautology. Indeed, since the above is a tautology, the set of clauses

$$\{C_1, ..., C_m, \{A, P\}, \{B, \neg P\}\}$$

is logically equivalent to the set

$$\{C_1, ..., C_m, \{A, P\}, \{B, \neg P\}, \{A, B\}\}$$

obtained by adding the clause $\{A, B\}$. Consequently, the set

$$\{C_1, ..., C_m, \{A, P\}, \{B, \neg P\}\}$$

is unsatisfiable if and only if the set

$$\{C_1, ..., C_m, \{A, P\}, \{B, \neg P\}, \{A, B\}\}$$

is unsatisfiable.

The clause $\{A, B\}$ is a *resolvent* of the clauses $\{A, P\}$ and $\{B, \neg P\}$. The resolvent of the clauses $\{P\}$ and $\{\neg P\}$ is the empty clause $\square$. The process of adding a resolvent of two clauses from a set S to S is called a *resolution step*. The resolution method attempts to build a sequence of sets of clauses obtained by successive resolution steps, and ending with a set containing the empty clause. When this happens, we know that the original clause is unsatisfiable, since resolution steps preserve unsatisfiability, and a set of clauses containing the empty clause is obviously unsatisfiable.

There are several ways of recording the resolution steps. A convenient and space-efficient way to do so is to represent a sequence of resolution steps as a DAG. First, we show that the proposition $(*)$ defined above is a tautology.

Lemma 4.3.1 The proposition

$$((A \vee P) \wedge (B \vee \neg P)) \equiv ((A \vee P) \wedge (B \vee \neg P) \wedge (A \vee B))$$

is a tautology, even when either A or B is empty. (When A is empty, $(A \vee P)$ reduces to P and $(A \vee B)$ to B, and similarly when B is empty. When both A and B are empty, we have the tautology $(P \wedge \neg P) \equiv (P \wedge \neg P)$.)

Proof: We prove that

$$((A \vee P) \wedge (B \vee \neg P)) \supset ((A \vee P) \wedge (B \vee \neg P) \wedge (A \vee B))$$

is valid and that

$$((A \vee P) \wedge (B \vee \neg P) \wedge (A \vee B)) \supset ((A \vee P) \wedge (B \vee \neg P))$$

is valid. The validity of the second proposition is immediate. For the first one, it is sufficient to show that

$$((A \vee P) \wedge (B \vee \neg P)) \supset (A \vee B)$$

is valid. We have the following proof tree (in G'):

$$
\cfrac{
A, (B \vee \neg P) \to A, B
\qquad
\cfrac{
P, B \to A, B
\qquad
\cfrac{P \to P, A, B}{P, \neg P \to A, B}
}{
P, (B \vee \neg P) \to A, B
}
}{
\cfrac{
\cfrac{
\cfrac{
(A \vee P), (B \vee \neg P) \to A, B
}{
(A \vee P), (B \vee \neg P) \to (A \vee B)
}
}{
(A \vee P) \wedge (B \vee \neg P) \to (A \vee B)
}
}{
\to ((A \vee P) \wedge (B \vee \neg P)) \supset (A \vee B))
}
}
$$

$\square$

Definition 4.3.2 Given two clauses C_1 and C_2, a clause C is a *resolvent* of C_1 and C_2 iff, for some literal L, $L \in C_1$, $\overline{L} \in C_2$, and

$$C = (C_1 - \{L\}) \cup (C_2 - \{\overline{L}\}).$$

In other words, a resolvent of two clauses is any clause obtained by striking out a literal and its conjugate, one from each, and merging the remaining literals into a single clause.

EXAMPLE 4.3.3

The clauses $\{A, B\}$ and $\{\neg A, \neg B\}$ have the two resolvents $\{A, \neg A\}$ and $\{B, \neg B\}$. The clauses $\{P\}$ and $\{\neg P\}$ have the empty clause as their resolvent.

Observe that by lemma 4.3.1, any set S of clauses is logically equivalent to the set $S \cup \{C\}$ obtained by adding any resolvent of clauses in S. However, it is not true that the set

$$(S \cup \{C\}) - \{C_1, C_2\}$$

obtained by adding a resolvent of two clauses C_1 and C_2 and deleting C_1 and C_2 from S is equivalent to S. If $S = \{\{P, Q\}, \{\neg Q\}\}$, then $\{P\}$ is a resolvent of $\{P, Q\}$ and $\{\neg Q\}$, but S is not equivalent to $\{\{P\}\}$. Indeed, the valuation v which satisfies both P and Q satisfies $\{\{P\}\}$ but does not satisfy S since it does not satisfy $\{\neg Q\}$.

We now define resolution DAGs. We believe that it is more appropriate to call resolution DAGs having the empty clause at their root *resolution refutations* rather than *resolutions proofs*, since they are used to show the unsatisfiability of a set of clauses.

Definition 4.3.3 Given a set $S = \{C_1, ..., C_n\}$ of clauses, a *resolution DAG* for S is any finite set

$$D = \{(t_1, R_1), ..., (t_m, R_m)\}$$

of distinct DAGs labeled with clauses and such that:

(1) The leaf nodes of each underlying tree t_i are labeled with clauses in S.

(2) For every DAG (t_i, R_i), for every non-leaf node u in t_i, u has exactly two successors $u1$ and $u2$, and if $u1$ is labeled with a clause C_1 and $u2$ is labeled with a clause C_2 (not necessarily distinct from C_1), then u is labeled with a resolvent of C_1 and C_2.

A resolution DAG is a *resolution refutation* iff it consists of a single DAG (t, R) whose root is labeled with the empty clause. The nodes of a DAG which are not leaves are also called *resolution steps*.

EXAMPLE 4.3.4

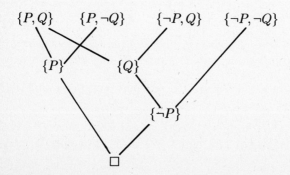

The DAG of example 4.3.4 is a resolution refutation.

4.3.3 Soundness of the Resolution Method

The soundness of the resolution method is given by the following lemma.

Lemma 4.3.2 If a set S of clauses has a resolution refutation DAG, then S is unsatisfiable.

Proof: Let (t_1, R_1) be a resolution refutation for S. The set S' of clauses labeling the leaves of t_1 is a subset of S.

First, we prove by induction on the number of nodes in a DAG (t, R) that the set of clauses S' labeling the leaves of t is equivalent to the set of clauses labeling all nodes in t.

If t has one node, the property holds trivially. If t has more than one node, let l and r be the left subtree and right subtree of t respectively. By the induction hypothesis, the set S_1 of clauses labeling the nodes of l is equivalent to the set L_1 of clauses labeling the leaves of l, and the set S_2 of clauses labeling the nodes of r is equivalent to the set L_2 of clauses labeling the leaves of r. Let C_1 and C_2 be the clauses labeling the root of l and r respectively. By definition 4.3.3, the root of t is a resolvent R of C_1 and C_2. By lemma 4.3.1, the set $S_1 \cup S_2$ is equivalent to the set $S_1 \cup S_2 \cup \{R\}$. But then, $S_1 \cup S_2 \cup \{R\}$ is equivalent to $S' = L_1 \cup L_2$, since $S_1 \cup S_2$ is equivalent to $L_1 \cup L_2$. This concludes the induction proof.

Applying the above property to t_1, S' is equivalent to the set of clauses labeling all nodes in t_1. Since the resolvent D labeling the root of t_1 is the empty clause, the set S' of clauses labeling the leaves of t_1 is unsatisfiable, which implies that S is unsatisfiable since S' is a subset of S. $\square$

4.3.4 Converting $GCNF'$-proofs into Resolution Refutations and Completeness

The completeness of the resolution method is proved by showing how to transform a proof in the system $GCNF'$ into a resolution refutation.

Theorem 4.3.1 There is an algorithm for constructing a resolution refutation D from a proof tree T in the system $GCNF'$. Furthermore, the number of resolution steps in D (nonleaf nodes) is less than or equal to the number of axioms in T.

Proof: We give a construction and prove its correctness by induction on proof trees in $GCNF'$. Let $S = \{C_1, ..., C_m\}$. If the proof tree of the sequent $C_1, ..., C_m \rightarrow$ is a one-node tree labeled with an axiom, there must a literal L such that some $C_i = L$ and some $C_j = \overline{L}$. Hence, we have the resolution refutation consisting of the DAG

This resolution DAG has one resolution step, so the base step of the induction holds.

Otherwise, the proof tree is of the form

$$\frac{\dfrac{T_1}{\Gamma_1, C_1, ..., C_k \rightarrow} \qquad \dfrac{T_2}{\Gamma_2, B \rightarrow}}{\Gamma, \{A_1, B\}, ..., \{A_n, B\} \rightarrow}$$

where $\Gamma_1, \Gamma_2 \subseteq \Gamma \cup \{\{A_1, B\}, ..., \{A_n, B\}\}$, $\{C_1, ..., C_k\} \subseteq \{A_1, ... A_n\}$, and $\Gamma \cup \{\{A_1, B\}, ..., \{A_n, B\}\} = S$. By the induction hypothesis, there is a resolution refutation D_1 obtained from the tree T_1 with root $\Gamma_1, C_1, ..., C_k \rightarrow$, and a resolution refutation D_2 obtained from the tree T_2 with root $\Gamma_2, B \rightarrow$. The leaves of D_1 are labeled with clauses in $\Gamma_1 \cup \{C_1, ..., C_k\}$, and the leaves of D_2 are labeled with clauses in $\Gamma_2 \cup \{B\}$. Let $\{C'_1, ..., C'_p\}$ be the subset of clauses in $\{C_1, ..., C_k\}$ that label leaves of D_1.

If $\{C'_1, ..., C'_p\} \subseteq S$, then D_1 is also resolution refutation for S. Similarly, if the set of clauses labeling the leaves of D_2 is a subset of S, D_2 is a resolution refutation for S. In both cases, by the induction hypothesis, the number of resolution steps in D_1 (resp. D_2) is less than or equal to the number of leaves of T_1 (resp. T_2), and so, it is less than the number of leaves of T.

Otherwise, let $\{C''_1, ..., C''_q\} \subseteq \{C_1, ..., C_k\}$ be the set of clauses *not* in S labeling the leaves of D_1. Also, $B \notin S$ and B labels some leaf of D_2, since otherwise the clauses labeling the leaves of D_2 would be in S. Let D'_1 be the resolution DAG obtained from D_1 by replacing $C''_1, ..., C''_q$ by $\{C''_1, B\}, ..., \{C''_q, B\}$, and applying the same resolution steps in D'_1 as the resolution steps applied in D_1. We obtain a resolution DAG that may be a resolution refutation, or in which the clause B is the label of the root node.

If D'_1 is a resolution refutation, we are done.

Otherwise, since $B \notin S$ and D'_1 is not a resolution refutation, the root of D'_1 is labeled with B and no leaf of D'_1 is labeled with B. By the induction hypothesis, the number of resolution steps n_1 in D_1 (and D'_1) is less than or equal to the number of axioms in T_1, and the number n_2 of resolution steps in D_2 is less than or equal to the number of axioms in T_2. We construct a resolution refutation for a subset of $\Gamma, \{A_1, B\}, ..., \{A_n, B\}$ by combining D'_1 and D_2 as follows: Identify the root labeled with B in D'_1 with the leaf labeled with B in D_2, and identify all leaf nodes u, v with $u \in D'_1$ and $v \in D_2$, iff u and v are labeled with the same clause. (Since $B \notin S$, B cannot be such a clause.) The number of resolution steps in D is $n_1 + n_2$, which is less than or equal to the number of axioms in T. This concludes the construction and the proof. □

Theorem 4.3.2 (Completeness of the resolution method) Given a finite set S of clauses, S is unsatisfiable if and only if there is a resolution refutation for S.

   Proof: By lemma 4.3.2, if S has a resolution refutation, S is unsatisfiable. If S is unsatifiable, then the sequent $S \to$ is valid. Since the system $GCNF'$ is complete (theorem 4.2.1), there is a proof tree T for $S \to$. Finally, by theorem 4.3.1, a resolution refutation D is obtained from T. □

EXAMPLE 4.3.5

   Consider the following proof tree in $GCNF'$:

$$\frac{\dfrac{P, \neg P \to \qquad\qquad S, \neg S \to}{P, S, \{\neg P, \neg S\} \to} \qquad\qquad \neg Q, Q \to}{\{P, \neg Q\}, \{S, \neg Q\}, Q, \{\neg P, \neg S\} \to}$$

The leaves $P, \neg P \to$ and $S, \neg S \to$ are mapped into the following DAGs:

$$D_1 \qquad\qquad\qquad D_2$$

$$\{P\} \quad \{\neg P\} \qquad\qquad \{S\} \quad \{\neg S\}$$

Let D_3 be the DAG obtained from D_1 by adding $\neg S$ to $\{\neg P\}$:

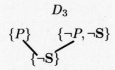

$$D_3$$

$$\{P\} \quad \{\neg P, \neg S\}$$
$$\{\neg S\}$$

We now combine D_3 and D_2 to obtain D_4:

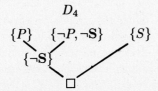

$$D_4$$

$$\{P\} \quad \{\neg P, \neg S\} \qquad \{S\}$$
$$\{\neg S\}$$

The leaf $Q, \neg Q \rightarrow$ is mapped into the DAG D_5:

D_5

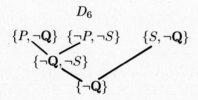

Let D_6 be the DAG obtained from D_4 by adding $\neg Q$ to $\{P\}$ and $\{S\}$:

D_6

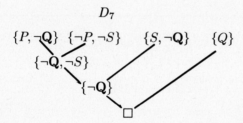

Finally, let D_7 be obtained by combining D_6 and D_5:

D_7

The theorem actually provides an algorithm for constructing a resolution refutation. Since the number of resolution steps in D is less than or equal to the number of axioms in T, the number of nodes in the DAG D cannot be substantially larger than the number of nodes in T. In fact, the DAG D has at most two more nodes than the tree, as shown by the following lemma.

Lemma 4.3.3 If T is a proof tree in $GCNF'$ and T has m nodes, the number n of nodes in the DAG D given by theorem 4.3.1 is such that, $r+1 \leq n \leq m+2$, where r is the number of resolution steps in D.

Proof: Assume that T has k leaves. Since this tree is binary branching, it has $m = 2k - 1$ nodes. The number r of resolution steps in D is less than or equal to the number k of leaves in T. But the number n of nodes in the DAG is at least $r + 1$ and at most $2r + 1$, since every node has at most two successors. Hence, $n \leq 2r + 1 \leq 2k + 1 = m + 2$. $\square$

In example 4.3.5, the number of resolution steps in the DAG is equal to the number of leaves in the tree. The following example shows that the number of resolution steps can be strictly smaller.

EXAMPLE 4.3.6

Consider the following proof tree T:

$$
\cfrac{\neg R, R \to \quad \cfrac{\cfrac{\neg P, P \to \quad \neg Q, Q \to}{\neg Q, \neg P, \{P, Q\} \to}}{\neg R, \neg Q, \{R, \neg P\}, \{P, Q\} \to} \quad \neg R, R \to}{\neg R, \neg Q, \{R, \neg P\}, \{P, Q, R\} \to}
$$

The result of recursively applying the algorithm to T yields the two DAGs D_1 and D_2:

corresponding to the axioms $\neg P, P \to$ and $\neg Q, Q \to$. Adding **Q** to P in the DAG D_1, we obtain the DAG D_3:

D_3

$$\neg P \quad \{P, \mathbf{Q}\}$$
$$\diagdown \!\! \diagup$$
$$\mathbf{Q}$$

Identifying the root of DAG D_3 with the leaf labeled Q of DAG D_2, we obtain the DAG D_4 corresponding to the subtree

$$
\cfrac{\neg P, P \to \quad \neg Q, Q \to}{\neg Q, \neg P, \{P, Q\} \to}
$$

of T:

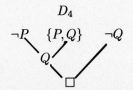

Adding **R** to $\neg P$ in D_4, we obtain the DAG D_5:

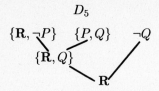

The axiom $\neg R, R \rightarrow$ yields the DAG D_6:

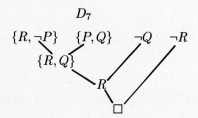

and merging the root of D_5 with the leaf labeled R in D_6, we obtain the DAG D_7 corresponding to the left subtree

$$
\frac{
\begin{array}{cc}
\dfrac{\neg P, P \rightarrow \qquad \neg Q, Q \rightarrow}{}
\end{array}
}{}
$$

$$
\frac{\neg R, R \rightarrow \qquad \dfrac{\neg P, P \rightarrow \qquad \neg Q, Q \rightarrow}{\neg Q, \neg P, \{P, Q\} \rightarrow}}{\neg R, \neg Q, \{R, \neg P\}, \{P, Q\} \rightarrow}
$$

of the original proof tree:

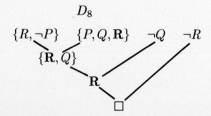

At this stage, since the right subtree of the proof tree T is the axiom $\neg R, R \rightarrow$, we add **R** to $\{P, Q\}$ in D_7. However, since **R** is also in $\{R, \neg P\}$, the resulting DAG D_8 is a resolution refutation:

The DAG D_8 has three resolution steps, and the original proof tree T has four axioms. This is because the original proof tree is not optimal. If we had applied the $\vee : rule$ to $\{\neg P, R\}$ and $\{P, Q, R\}$, we would obtain a proof tree with three axioms, which yields a DAG with three resolution steps.

4.3.5 From Resolution Refutations to $GCNF'$-proofs

We will now prove that every resolution refutation can be transformed into a proof tree in $GCNF'$, and that the size of the resulting tree is at most twice the size of the DAG. This will show that the resolution method and the Gentzen system $GCNF'$ are equivalent in terms of the complexity of proofs (measured by their size).

Theorem 4.3.3 There is an algorithm which transforms any resolution refutation D into a proof tree T in the system $GCNF'$. Furthermore, the number of axioms (leaves) of T is less than or equal to the number of resolution steps in D.

Proof: We construct the tree and prove the theorem by induction on the number of resolution steps in D. If D contains a single resolution step, D is a DAG of the form

$$\{P\} \qquad \{\neg P\}$$
$$\searrow \quad \swarrow$$
$$\square$$

Hence, for some clauses C_i and C_j in the set of clauses $\{C_1, ..., C_m\}$, $C_i = \{P\}$ and $C_j = \{\neg P\}$. Hence, the one-node tree labeled with the axiom $C_1, ..., C_m \to$ is a proof tree.

If D has more than one resolution step, it is a DAG (t, R) whose root is labeled with the empty clause. The descendants n_1 and n_2 of the root r of (t, R) are labeled with clauses $\{P\}$ and $\{\neg P\}$ for some propositional letter P. There are two cases: Either one of n_1, n_2 is a leaf, or neither is a leaf.

Case 1: Assume that n_1 is a leaf, so that $C_1 = \{P\}$, the other case being similar. Let
$$\{A_1, \neg P\}, ..., \{A_n, \neg P\}$$
be the terminal nodes of all paths from the node n_2, such that every node in each of these paths contains $\neg P$. Let D_2 be the resolution DAG obtained by deleting r and n_1 (as well as the edges having these nodes as source) and by deleting $\neg P$ from every node in every path from n_2 such that every node in such a path contains $\neg P$. The root of D_2 is labeled with the empty clause. By the induction hypothesis, there is a proof tree T_2 in $GCNF'$ for

$$\Gamma, A_1, ..., A_n \to, \quad \text{where}$$
$$\Gamma = \{C_2, ..., C_m\} - \{\{A_1, \neg P\}, ..., \{A_n, \neg P\}\}.$$

If $\{A_1, ..., A_n\}$ is a subset of Γ, then the tree obtained from T_2 by replacing Γ by $\{C_1, ..., C_m\}$ is also a proof tree. By the induction hypothesis the number of leaves k in T_2 is less than or equal to the number of resolution steps in D_2, which is less than the number of resolution steps in D.

If $\{A_1, ..., A_n\}$ is not a subset of Γ, then, the following is a proof tree T in $GCNF'$ for $C_1, ..., C_m \to$.

$$\frac{\dfrac{T_2}{\Gamma, A_1, ..., A_n \to} \qquad P, \neg P \to}{P, \Gamma, \{A_1, \neg P\}, ..., \{A_n, \neg P\} \to}$$

The number of axioms in T is $k - 1 + 1 = k$, which is less than or equal to the number of resolution steps in D. In the special case where $\Gamma, A_1, ..., A_n \to$ is an axiom, since $C_1, ..., C_m \to$ is not a trivial sequent (the case of a trivial sequent being covered by the base case of a DAG with a single resolution step), there is some proposition in $\Gamma, A_1, ..., A_n$ which is the conjugate of one of the A_i. Then, T_2 is simply the axiom $A_i, \overline{A_i} \to$.

Case 2: Neither n_1 nor n_2 is a leaf. Let

$$\{A_1, P\}, ..., \{A_n, P\}$$

be the set of terminal nodes of all paths from the node n_1, such that every node in each of these paths contains P. Let D_1 be the resolution DAG consisting of the descendant nodes of n_1 (and n_1 itself), and obtained by deleting P from every node in every path from n_1 such that every node in such a path contains P. The root of D_1 is labeled with the empty clause. By the induction hypothesis, there is a proof tree T_1 in $GCNF'$ for $\Gamma_1, A_1, ..., A_n \to$, where Γ_1 is some subset of $\{C_1, ...C_m\}$. By the induction hypothesis the number of axioms in T_1 is less than or equal to the number m_1 of resolution steps in D_1. Similarly, let D_2 be the DAG obtained by deleting all descendants of n_1 that are not descendants of n_2, and deleting all edges having these nodes as source, including the edges from n_1, so that n_1 is now a leaf. By the induction hypothesis, there is a proof tree T_2 in $GCNF'$ with root labeled with $\Gamma_2, P \to$, where Γ_2 is some subset of $\{C_1, ..., C_m\}$. Also, the number of leaves in T_2 is less than or equal to the number m_2 of resolution steps in D_2. Note than the number of resolution steps in D is $m_1 + m_2$.

If $\{A_1, ..., A_n\}$ is a subset of Γ_1, then the tree T_1 is also a proof tree for $\{C_1, ..., C_m\}$. Similarly, if P belongs to Γ_2, the tree T_2 is a proof tree for $\{C_1, ..., C_m\}$. Otherwise, we have the following proof tree T in $GCNF'$:

$$\frac{\dfrac{T_1}{\Gamma_1, A_1, ..., A_n \rightarrow} \qquad\qquad \dfrac{T_2}{\Gamma_2, P \rightarrow}}{\Gamma, \{A_1, P\}, ..., \{A_n, P\} \rightarrow}$$

The number of axioms in T is the sum of the number of axioms in T_1 and T_2, which is less than or equal to $m_1 + m_2$, the number of resolution steps in D. The special cases in which either $\Gamma_1, A_1, ..., A_n \rightarrow$ or $\Gamma_2, P \rightarrow$ is an axiom is handled as in case 1. The details are left as an exercise. This completes the proof. $\square$

EXAMPLE 4.3.7

Let D be the following resolution refutation:

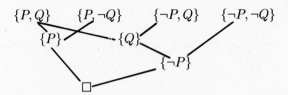

Let D_1 be the set of descendants of $\{P\}$:

$$
\begin{array}{ccc}
\{\mathbf{P}, Q\} & & \{\mathbf{P}, \neg Q\} \\
 & \{\mathbf{P}\} &
\end{array}
$$

Let D_2 be the resolution refutation obtained by deleting P:

$$
\begin{array}{ccc}
\{Q\} & & \{\neg Q\} \\
 & \square &
\end{array}
$$

Let D_3 be the DAG obtained from D by deleting the descendants of $\{P\}$ that are not descendants of $\{\neg P\}$. Note that since $\{P, Q\}$ is a descendant of $\{\neg P\}$, it is retained. However, $\{P, \neg Q\}$ is deleted.

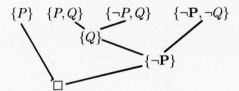

Let D_4 be the resolution refutation obtained by deleting $\neg P$ from all paths from $\{\neg P\}$ containing it:

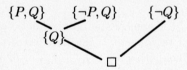

Let D_5 be the set of descendants of $\{Q\}$:

$$\{P, \mathbf{Q}\} \quad \{\neg P, \mathbf{Q}\}$$
$$\{\mathbf{Q}\}$$

Let D_6 be the resolution refutation obtained from D_5 by deleting Q:

$$\{P\} \quad \{\neg P\}$$
$$\square$$

The method of theorem 4.3.3 yields the following proof trees: The tree T_1:

$$Q, \neg Q \rightarrow$$

corresponds to D_2;

The tree T_2:

$$\frac{Q, \neg Q \rightarrow \qquad P, \neg P \rightarrow}{\{P, Q\}, \{\neg P, Q\}, \neg Q \rightarrow}$$

corresponds to D_4;

The tree T_3:

$$\frac{\dfrac{Q, \neg Q \rightarrow \qquad P, \neg P \rightarrow}{\{P, Q\}, \{\neg P, Q\}, \neg \mathbf{Q} \rightarrow} \qquad P, \neg P \rightarrow}{P, \{P, Q\}, \{\neg P, Q\}, \{\neg \mathbf{P}, \neg \mathbf{Q}\} \rightarrow}$$

corresponds to D_3;

The tree T_4:

$$\cfrac{Q,\neg Q\to \qquad \cfrac{P,\neg P\to}{\cfrac{\{P,Q\},\{\neg P,Q\},\neg Q\to \qquad \cfrac{P,\neg P\to}{P,\{P,Q\},\{\neg P,Q\},\{\neg P,\neg Q\}\to}}{\ }}}{\{\mathbf{P},\mathbf{Q}\},\{\mathbf{P},\neg\mathbf{Q}\},\{\neg P,Q\},\{\neg P,\neg Q\}\to}$$

corresponds to D. The number of axioms of T_4 is four, which is the number of resolution steps in D. The next example shows that it is possible that the proof tree constructed from a resolution DAG has fewer leaves than the number of resolution steps.

EXAMPLE 4.3.8

Consider the following resolution refutation D:

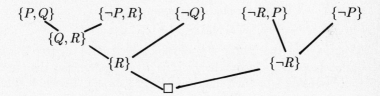

Let D_1 be the DAG consisting of the descendants of $\{R\}$:

Let D_2 be the DAG obtained from D_1 by deleting $\mathbf{R}$:

Note that D_2 is a resolution for a subset of the set of clauses

$$\{\{P,Q\},\{\neg P,R\},\{\neg Q\},\{\neg R,P\},\{\neg P\}\}.$$

The construction recursively applied to D_2 yields the following proof tree:

$$\frac{P, \neg P \rightarrow \qquad Q, \neg Q \rightarrow}{\{P, Q\}, \neg P, \neg Q \rightarrow}$$

From the above tree, we obtain the following proof tree for the original set of clauses:

$$\frac{P, \neg P \rightarrow \qquad Q, \neg Q \rightarrow}{\{P, Q\}, \{\neg P, R\}, \neg Q, \{\neg R, P\}, \neg P \rightarrow}$$

This last tree has two leaves, whereas the DAG D has four resolution steps. The following lemma shows that the size of the tree T given by the method of theorem 4.3.3 is linearly related to the size of D.

Lemma 4.3.4 Given a resolution refutation D having m nodes, the proof tree T constructed by the method of theorem 4.3.3 has a number of nodes n such that, $n \leq 2m - 3$.

Proof: Assume that D has r resolution steps. Since the tree T is binary branching and has $k \leq r$ leaves, it has $n = 2k - 1$ nodes. But the number m of nodes in the DAG D is such that, $r + 1 \leq m \leq 2r + 1$. Hence, $n \leq 2m - 3$. □

This last lemma shows that for every resolution refutation D, there is a proof tree whose size is at most twice the size of D. Hence, the Gentzen system $GCNF'$ is basically as efficient as the resolution method.

In the proof tree of example 4.3.7, there are two leaves labeled with $P, \neg P \rightarrow$, and two leaves labeled with $Q, \neg Q \rightarrow$. Hence, if this tree is represented as a DAG, it will have five nodes instead of seven. The original DAG has eight nodes. This suggests that the system $GCNF'$ using DAGs instead of trees is just as efficient as the resolution method. However, we do not have a proof of this fact at this time.

PROBLEMS

4.3.1. Show that the set of clauses

$$\{\{A, B, \neg C\}, \{A, B, C\}, \{A, \neg B\}, \{\neg A\}\}$$

is unsatisfiable using the resolution method.

4.3.2. Show that the following sets of clauses are unsatisfiable using the resolution method:

(a) $\{\{A, \neg B, C\}, \{B, C\}, \{\neg A, C\}, \{B, \neg C\}, \{\neg B\}\}$

(b) $\{\{A, \neg B\}, \{A, C\}, \{\neg B, C\}, \{\neg A, B\}, \{B, \neg C\}, \{\neg A, \neg C\}\}$

4.3.3. Construct a proof tree in $GCNF'$ for the set of clauses of problem 4.3.1, and convert it into a resolution DAG using the algorithm of theorem 4.3.1.

4.3.4. Construct proof trees in $GCNF'$ for the sets of clauses of problem 4.3.2, and convert them into resolution DAGs using the algorithm of theorem 4.3.1.

4.3.5. Convert the resolution DAG for the clause of problem 4.3.1 into a proof tree using the algorithm of theorem 4.3.3.

4.3.6. Convert the resolution DAGs for the clause of problem 4.3.2 into proof trees using the algorithm of theorem 4.3.3.

4.3.7. Find all resolvents of the following pairs of clauses:

(a) $\{A, B\}, \{\neg A, \neg B\}$

(b) $\{A, \neg B\}, \{B, C, D\}$

(c) $\{\neg A, B, \neg C\}, \{B, C\}$

(d) $\{A, \neg A\}, \{A, \neg A\}$

4.3.8. Construct a resolution refutation for the following set of clauses:

$$\{\{P, Q\}, \{\neg P, Q\}, \{P, \neg Q\}, \{\neg P, \neg Q\}\}.$$

Convert this resolution DAG into a proof tree using the algorithm of theorem 4.3.3

4.3.9. Another way of presenting the resolution method is as follows. Given a (finite) set S of clauses, let

$$R(S) = S \cup \{C \mid C \text{ is a resolvent of two clauses in } S\}.$$

Also, let

$$R^0(S) = S,$$
$$R^{n+1}(S) = R(R^n(S)), n \geq 0, \text{ and let}$$
$$R^*(S) = \bigcup_{n \geq 0} R^n(S).$$

(a) Prove that S is unsatisfiable if and only if $R^*(S)$ is unsatisfiable.

(b) Prove that if S is finite, there is some $n \geq 0$ such that

$$R^*(S) = R^n(S).$$

(c) Prove that there is a resolution refutation for S if and only if the empty clause $\square$ is in $R^*(S)$.

(d) Prove that S is unsatisfiable if and only if $\square$ belongs to $R^*(S)$.

4.3.10. Find $R(S)$ for the following sets of clauses:

(a) $\{\{A, \neg B\}, \{A, B\}, \{\neg A\}\}$

(b) $\{\{A, B, C\}, \{\neg B, \neg C\}, \{\neg A, \neg C\}\}$

(c) $\{\{\neg A, \neg B\}, \{B, C\}, \{\neg C, A\}\}$

(d) $\{\{A, B, C\}, \{A\}, \{B\}\}$

4.3.11. Prove that the resolution method is still complete if the resolution rule is restricted to clauses that are not tautologies (that is, clauses not containing both P and $\neg P$ for some propositional letter P.)

4.3.12. We say that a clause C_1 *subsumes* a clause C_2 if C_1 is a proper subset of C_2. In the version of the resolution method described in problem 4.3.8, let

$$R_1(S) = R(S) - \{C \mid C \text{ is subsumed by some clause in } R(S)\}.$$

Let

$$R_1^0 = S,$$
$$R_1^{n+1}(S) = R_1(R_1^n(S)) \text{ and}$$
$$R_1^*(S) = \bigcup_{n \geq 0} \{R_1^n(S)\}.$$

Prove that S is unsatisfiable if and only if $\square$ belongs to $R_1^*(S)$.

4.3.13. Prove that the resolution method is also complete for infinite sets of clauses.

4.3.14. Write a computer program implementing the resolution method.

Notes and Suggestions for Further Reading

The resolution method was discovered by J.A. Robinson (Robinson, 1965). An earlier method for testing the unsatisfiability of a set of clauses is the Davis-Putnam procedure (Davis and Putnam, 1960). Improvements due to Prawitz (Prawitz, 1960) and Davis (Davis, 1963) led to the resolution method. An exposition of resolution based on Gentzen sequents has also been given by J. Robinson (Robinson, 1969), but the technical details are different. To the best of our knowledge, the constructive method used in this chapter for proving the completeness of the resolution method by transforming a Gentzen-like proof is original.

With J.A. Robinson, this author believes that a Gentzen-sequent based exposition of resolution is the best from a pedagogical point of view and for theoretical understanding.

The resolution method and many of its refinements are discussed in Robinson, 1969; Loveland, 1978; and in Chang and Lee, 1973.

Chapter 5

First-Order Logic

5.1 INTRODUCTION

In propositional logic, it is not possible to express assertions about elements
of a structure. The weak expressive power of propositional logic accounts for
its relative mathematical simplicity, but it is a very severe limitation, and it
is desirable to have more expressive logics. First-order logic is a considerably
richer logic than propositional logic, but yet enjoys many nice mathematical properties. In particular, there are finitary proof systems complete with
respect to the semantics.

In first-order logic, assertions about elements of structures can be expressed. Technically, this is achieved by allowing the propositional symbols
to have arguments ranging over elements of structures. For convenience, we
also allow symbols denoting functions and constants.

Our study of first-order logic will parallel the study of propositional logic
conducted in Chapter 3. First, the *syntax* of first-order logic will be defined.
The syntax is given by an inductive definition. Next, the *semantics* of first-
order logic will be given. For this, it will be necessary to define the notion of
a *structure*, which is essentially the concept of an algebra defined in Section
2.4, and the notion of *satisfaction*. Given a structure $\mathbf{M}$ and a formula A, for
any assignment s of values in $\mathbf{M}$ to the variables (in A), we shall define the
satisfaction relation $\models$, so that

$$\mathbf{M} \models A[s]$$

expresses the fact that the assignment s satisfies the formula A in **M**.

The satisfaction relation $\models$ is defined recursively on the set of formulae. Hence, it will be necessary to prove that the set of formulae is freely generated by the atomic formulae.

A formula A is said to be *valid in a structure* **M** if

$$\mathbf{M} \models A[s]$$

for every assignment s. A formula A is *valid* (or universally valid) if A is valid in every structure **M**.

Next, we shall attempt to find an algorithm for deciding whether a formula is valid. Unfortunately, there is no such algorithm. However, it is possible to find a procedure that will construct a proof for a valid formula. Contrary to the propositional case, this procedure may run forever if the input formula is not valid. We will set up a *proof system* that is a generalization of the Gentzen system of Section 3.4 and extend the *search* procedure to first-order formulae. Then, some fundamental theorems will be proved, including the completeness theorem, compactness theorem, and model existence theorem.

The two main concepts in this chapter are the *search* procedure to build proof trees and Hintikka sets. The main theme is that the *search* procedure is a *Hintikka set constructor*. If the *search* procedure fails to find a counter example, a proof is produced. Otherwise, the Hintikka set yields a counter example. This approach yields the completeness theorem in a very direct fashion. Another theme that will emerge in this chapter is that the *search* procedure is a very inefficient proof procedure. The purpose of the next chapters is to try to find more efficient proof procedures.

5.2 FIRST-ORDER LANGUAGES

First, we define alphabets for first-order languages.

5.2.1 Syntax

In contrast with propositional logic, first-order languages have a fixed part consisting of logical connectives, variables, and auxiliary symbols, and a part that depends on the intended application of the language, consisting of predicate, function, and constant symbols, called the *non logical part*.

Definition 5.2.1 The *alphabet* of a first-order language consists of the following sets of symbols:

Logical connectives: $\wedge$ (and), $\vee$ (or), $\neg$ (not), $\supset$ (implication), $\equiv$ (equivalence), $\perp$ (falsehood); quantifiers: $\forall$ (for all), $\exists$ (there exists); the equality symbol $\doteq$.

Variables: A countably infinite set $\mathbf{V} = \{x_0, x_1, x_2, ...\}$.

Auxiliary symbols: "(" and ")".

A set $\mathbf{L}$ of *nonlogical symbols* consisting of:

(i) *Function symbols*: A (countable, possibly empty) set $\mathbf{FS}$ of symbols $f_0, f_1, ...$, and a *rank function* r assigning a positive integer $r(f)$ (called *rank* or *arity*) to every function symbol f.

(ii) *Constants*: A (countable, possibly empty) set $\mathbf{CS}$ of symbols $c_0, c_1, ...$, each of rank zero.

(iii) *Predicate symbols*: A (countable, possibly empty) set $\mathbf{PS}$ of symbols $P_0, P_1, ...$, and a *rank function* r assigning a nonnegative integer $r(P)$ (called rank or arity) to each predicate symbol P.

It is assumed that the sets $\mathbf{V}$, $\mathbf{FS}$, $\mathbf{CS}$ and $\mathbf{PS}$ are disjoint. We will refer to a first-order language with set of nonlogical symbols $\mathbf{L}$ as the language $\mathbf{L}$. First-order languages obtained by omitting the equality symbol are referred to as *first-order languages without equality*. Note that predicate symbols of rank zero are in fact propositional symbols. Note also that we are using the symbol $\doteq$ for equality in the object language in order to avoid confusion with the symbol $=$ used for equality in the meta language.

We now give inductive definitions for the sets of terms and formulae. Note that a first-order language is in fact a two-sorted ranked alphabet in the sense of Subsection 2.5.1. The sorts are *term* and *formula*. The symbols in $\mathbf{V}$, $\mathbf{CS}$ and $\mathbf{FS}$ are of sort *term*, and the symbols in $\mathbf{PS}$ and $\perp$ are of sort *formula*.

Definition 5.2.2   Given a first-order language $\mathbf{L}$, let Γ be the union of the sets $\mathbf{V}$, $\mathbf{CS}$ and $\mathbf{FS}$. For every function symbol f of rank $n > 0$, let C_f be the function $C_f : (\Gamma^*)^n \to \Gamma^*$ such that, for all strings $t_1, ..., t_n \in \Gamma^*$,

$$C_f(t_1, ..., t_n) = f t_1 ... t_n.$$

The set $TERM_\mathbf{L}$ of $\mathbf{L}$-*terms* (for short, terms) is the inductive closure of the union of the set $\mathbf{V}$ of variables and the set $\mathbf{CS}$ of constants under the constructors C_f.

A more informal way of stating definition 5.2.2 is the following:

(i) Every constant and every variable is a term.

(ii) If $t_1, ..., t_n$ are terms and f is a function symbol of rank $n > 0$, then $f t_1 ... t_n$ is a term.

EXAMPLE 5.2.1

Let **L** be the following first-order language for arithmetic where, **CS** = $\{0\}$, **FS** = $\{S, +, *\}$, and **PS** = $\{<\}$. The symbol S has rank 1, and the symbols $+$, $*$ and $<$ have rank 2. Then, the following are terms:

$$S0$$

$$+S0SS0$$

$$*x_1 S + Sx_1 x_2.$$

Definition 5.2.3 Given a first-order language **L**, let Δ be the union of the sets Γ and **PS** (where Γ is the union of the sets **V**, **CS**, **FS**). For every predicate symbol P of rank $n > 0$, let C_P be the function

$$C_P : (\Gamma^*)^n \to \Delta^*$$

such that, for all strings $t_1, ..., t_n \in \Gamma^*$,

$$C_P(t_1, ..., t_n) = Pt_1...t_n.$$

Also, let $C_{\doteq}$ be the function $C_{\doteq} : (\Gamma^*)^2 \to \Delta^*$ such that, for all strings $t_1, t_2 \in \Gamma^*$,

$$C_{\doteq}(t_1, t_2) = \doteq t_1 t_2.$$

The set of **L**-*atomic formulae* (for short, atomic formulae) is the inductive closure of the pair of sets $TERM_{\mathbf{L}}$ (of sort *term*) and $\{P \mid P \in \mathbf{PS}, r(P) = 0\} \cup \{\bot\}$ (of sort *formula*), under the functions C_P and $C_{\doteq}$.

A less formal definition is the following:

(i) Every predicate symbol of rank 0 is an atomic formula, and so is $\bot$.

(ii) If $t_1, ..., t_n$ are terms and P is a predicate symbol of rank $n > 0$, then $Pt_1...t_n$ is an atomic formula, and so is $\doteq t_1 t_2$.

Let Σ be the union of the sets **V**, **CS**, **FS**, **PS**, and $\{\wedge, \vee, \neg, \supset, \equiv, \forall, \exists, \doteq, (,), \bot\}$. The functions $C_\wedge$, $C_\vee$, $C_\supset$, $C_\equiv$, $C_\neg$ are defined (on Σ^*) as in definition 3.2.2, and the functions A_i and E_i are defined such that, for any string $A \in \Sigma^*$,

$$A_i(A) = \forall x_i A, \text{ and } E_i(A) = \exists x_i A.$$

The set $FORM_{\mathbf{L}}$ of **L**-*formulae* (for short, formulae) is the inductive closure of the set of atomic formulae under the functions $C_\wedge$, $C_\vee$, $C_\supset$, $C_\equiv$, $C_\neg$ and the functions A_i and E_i.

A more informal way to state definition 5.2.3 is to define a formula as follows:

(i) Every atomic formula is a formula.

(ii) For any two formulae A and B, $(A \wedge B)$, $(A \vee B)$, $(A \supset B)$, $(A \equiv B)$ and $\neg A$ are also formulae.

(iii) For any variable x_i and any formula A, $\forall x_i A$ and $\exists x_i A$ are also formulae.

We let the letters x, y, z subscripted or not range over variables. We also omit parentheses whenever possible, as in the propositional calculus.

EXAMPLE 5.2.2

Using the first-order language of example 5.2.1, the following are atomic formulae:

$$< 0S0$$

$$\doteq ySx$$

The following are formulae:

$$\forall x \forall y (< xy \supset \exists z \doteq y + xz)$$

$$\forall x \forall y ((< xy \vee < yx) \vee \doteq xy)$$

Next, we will show that terms and formulae are freely generated.

5.2.2 Free Generation of the Set of Terms

We define a function K such that for a symbol s (variable, constant or function symbol), $K(s) = 1 - n$, where n is the number of terms that must follow s to obtain a term (n is the "tail deficiency").

$$K(x) = 1 - 0 = 1, \text{ for a variable } x;$$
$$K(c) = 1 - 0 = 1, \text{ for a constant } c;$$
$$K(f) = 1 - n, \text{ for a } n\text{-ary function symbol } f.$$

We extend K to strings composed of variables, constants and function symbols as follows: if $w = w_1 ... w_m$ then

$$K(w) = K(w_1) + ... + K(w_m),$$

and the following property holds.

Lemma 5.2.1 For any term t, $K(t) = 1$.

Proof: We use the induction principle for the set of terms. The basis of the induction holds trivially for the atoms. For a term $ft_1...t_n$ where f is of arity $n > 0$ and $t_1, ..., t_n$ are terms, since by definition $K(ft_1...t_n) = K(f) +$

$K(t_1) + ... + K(t_n)$ and by the induction hypothesis $K(t_1) = ... = K(t_n) = 1$, we have $K(ft_1...t_n) = 1 - n + (1 + ... + 1) = 1$. $\square$

Lemma 5.2.2 Every nonempty suffix of a term is a concatenation of one or more terms.

Proof: We use the induction principle for terms. The basis is obvious for the atoms. For a term $ft_1...t_n$, any proper suffix w must be of the form $st_{k+1}...t_n$, where $k \leq n$, and s is a suffix of t_k. By the induction hypothesis, s is a concatenation of terms $s_1, ..., s_m$, and $w = s_1...s_m t_{k+1}...t_n$, which is a concatenation of terms. $\square$

Lemma 5.2.3 No proper prefix of a term is a term.

Proof: Assume a term t is divided into a proper prefix t_1 and a (proper) suffix t_2. Then, $1 = K(t) = K(t_1) + K(t_2)$. By lemma 5.2.2, $K(t_2) \geq 1$. Hence $K(t_1) \leq 0$ and t_1 cannot be a term by lemma 5.2.1. $\square$

Theorem 5.2.1 The set of **L**-terms is freely generated from the variables and constants as atoms and the functions C_f as operations.

Proof: First, $f \neq g$ clearly implies that C_f and C_g have disjoint ranges and these ranges are disjoint from the set of variables and the set of constants. Since the constructors increase the length of strings, condition (3) for free generation holds. It remains to show that the restrictions of the functions C_f to $TERM_{\mathbf{L}}$ are injective. If $ft_1...t_m = fs_1...s_n$, we must have $t_1...t_m = s_1...s_n$. Then either $t_1 = s_1$, or t_1 is a proper prefix of s_1, or s_1 is a proper prefix of t_1. In the first case, $s_2...s_m = t_2...t_n$. The other two cases are ruled out by lemma 5.2.3 since both t_1 and s_1 are terms. By repeating the above reasoning, we find that $m = n$ and $t_i = s_i$, $1 \leq i \leq n$. Hence the set of terms is freely generated. $\square$

5.2.3 Free Generation of the Set of Formulae

To extend this argument to formulae, we define K on the other symbols with the following idea in mind: $K(s)$ should be $1 - n$, where n is the number of things (right parentheses, terms or formulae) required to go along with s in order to form a formula (n is the "tail deficiency").

$$K(``(") = -1;$$
$$K(``)") = 1;$$
$$K(\forall) = -1;$$
$$K(\exists) = -1;$$
$$K(\wedge) = -1;$$
$$K(\vee) = -1;$$
$$K(\supset) = -1;$$

$$K(\equiv) = -1;$$
$$K(\neg) = 0;$$
$$K(\dot{=}) = -1;$$
$$K(P) = 1 - n \quad \text{for any } n\text{-ary predicate symbol } P;$$
$$K(\bot) = 1.$$

We also extend K to strings as usual:

$$K(w_1...w_m) = K(w_1) + ... + K(w_m).$$

A lemma analogous to lemma 5.2.1 holds for formulae.

Lemma 5.2.4 For any formula A, $K(A) = 1$.

Proof: The proof uses the induction principle for formulae. Let Y be the subset of the set of formulae A such that $K(A) = 1$. First, we show that Y contains the atomic formulae. If A is of the form $Pt_1...t_m$ where P has rank m, since by lemma 5.2.1, $K(t_i) = 1$, and by definition $K(P) = 1 - m$, $K(A) = 1 - m + m = 1$, as desired. If A is of the form $\dot{=} t_1 t_2$, since $K(t_1) = K(t_2) = 1$ and $K(\dot{=}) = -1$, $K(A) = -1 + 1 + 1 = 1$. By definition $K(\bot) = 1$. Next, if A is of the form $\neg B$, by the induction hypothesis, $K(B) = 1$. Since $K(\neg) = 0$, $K(A) = 0 + 1 = 1$. If A is of the form $(B * C)$ where $* \in \{\wedge, \vee, \supset, \equiv\}$, by the induction hypothesis, $K(B) = 1$ and $K(C) = 1$, and

$$K(A) = K(``(") + K(B) + K(*) + K(C) + K(``)") = -1 + 1 + -1 + 1 + 1 = 1.$$

Finally if A is of the form $\forall x_i B$ (or $\exists x_i B$), by the induction hypothesis $K(B) = 1$, and $K(A) = K(\forall) + K(x_i) + K(B) = -1 + 1 + 1 = 1$. Hence Y is closed under the connectives, which concludes the proof by induction. $\square$

Lemma 5.2.5 For any proper prefix w of a formula A, $K(w) \leq 0$.

Proof: The proof uses the induction principle for formulae and is similar to that of lemma 5.2.4. It is left as an exercise. $\square$

Lemma 5.2.6 No proper prefix of a formula is a formula.

Proof: Immediate from lemma 5.2.4 and lemma 5.2.5. $\square$

Theorem 5.2.2 The set of **L**-formulae is freely generated by the atomic formulae as atoms and the functions C_X (X a logical connective), A_i and E_i.

Proof: The proof is similar to that of theorem 3.2.1 and rests on the fact that no proper prefix of a formula is a formula. The argument for atomic formulae is similar to the arguments for terms. The propositional connectives are handled as in the proof of theorem 3.2.1. For the quantifiers, observe that given two formulae, one in the range of a function A_i or E_i, the other in the range of another function, either the leftmost characters differ (for $C_\wedge$, $C_\vee$, $C_\supset$, $C_\equiv$ and $C_\neg$, "(" and "¬" are different from "∀" and "∃"), or the substrings consisting of the two leftmost symbols differ ($\forall x_i$ is different from $\forall x_j$ for $j \neq i$ and different from $\exists x_j$ for any x_j, and similarly $\exists x_i$ is different from $\exists x_j$ for $x_j \neq x_i$ and different from $\forall x_j$ for any x_j).

The fact that the restrictions of the functions $C_\wedge$, $C_\vee$, $C_\supset$, $C_\equiv$ and $C_\neg$ to $FORM_\mathbf{L}$ are injective is shown as in lemma 3.2.1, using the property that a proper prefix of a formula is not a formula (and a proper prefix of a term is not a term). For the functions A_i, $A_i(A) = A_i(B)$ iff $\forall x_i A = \forall x_i B$ iff $A = B$, and similarly for E_i. Finally, the constructors increase the number of connectives (or quantifiers). $\square$

Remarks:

(1) Instead of defining terms and atomic formulae in prefix notation, one can define them as follows (using parentheses):

The second clause of definition 5.2.2 is changed to: For every function symbol f of arity n and any terms $t_1,...,t_n$, $f(t_1,...,t_n)$ is a term. Also, atomic formulae are defined as follows: For every predicate symbol P of arity n and any terms $t_1, ..., t_n$, $P(t_1, ..., t_n)$ is an atomic formula; $t_1 \doteq t_2$ is an atomic formula.

One can still show that the terms and formulae are freely generated. In the sequel, we shall use the second notation when it is convenient. For simplicity, we shall also frequently use = instead of $\doteq$ and omit parentheses whenever possible, using the conventions adopted for the propositional calculus.

(2) The sets $TERM_\mathbf{L}$ and $FORM_\mathbf{L}$ are the carriers of a two-sorted algebra $T(\mathbf{L}, \mathbf{V})$ with sorts *term* and *formula*, in the sense of Subsection 2.5.2. The operations are the functions C_f for all function symbols, $C_{\doteq}$, C_P for all predicate symbols, $C_\wedge$, $C_\vee$, $C_\supset$, $C_\equiv$, $C_\neg$, A_i, E_i ($i \geq 0$) for formulae, and the symbols of arity 0 are the propositional symbols in $\mathbf{PS}$, the constants in $\mathbf{CS}$, and $\bot$. This two-sorted algebra $T(\mathbf{L}, \mathbf{V})$ is free on the set of variables $\mathbf{V}$ (as defined in Section 2.5), and is isomorphic to the tree algebra $T_\mathbf{L}(\mathbf{V})$ (in $T_\mathbf{L}(\mathbf{V})$, the term $A_i(A)$ is used instead of $\forall x_i A$, and $E_i(A)$ instead of $\exists x_i A$).

5.2.4 Free and Bound Variables

In first-order logic, variables may occur *bound* by quantifiers. Free and bound occurrences of variables are defined by recursion as follows.

Definition 5.2.4 Given a term t, the set $FV(t)$ of *free variables* of t is defined by recursion as follows:

$$FV(x_i) = \{x_i\}, \text{ for a variable } x_i;$$
$$FV(c) = \emptyset, \text{ for a constant } c;$$

$$FV(ft_1...t_n) = FV(t_1) \cup ... \cup FV(t_n),$$
for a function symbol f of rank n.

For a formula A, the set $FV(A)$ of *free variables* of A is defined by:

$$FV(Pt_1...t_n) = FV(t_1) \cup ... \cup FV(t_n),$$
for a predicate symbol P of rank n;
$$FV(\doteq t_1t_2) = FV(t_1) \cup FV(t_2);$$
$$FV(\neg A) = FV(A);$$
$$FV((A * B)) = FV(A) \cup FV(B), \text{ where } * \in \{\wedge, \vee, \supset, \equiv\};$$
$$FV(\bot) = \emptyset;$$
$$FV(\forall x_i A) = FV(A) - \{x_i\};$$
$$FV(\exists x_i A) = FV(A) - \{x_i\}.$$

A term t or a formula A is *closed* if, respectively $FV(t) = \emptyset$, or $FV(A) = \emptyset$. A closed formula is also called a *sentence*. A formula without quantifiers is called *open*.

Definition 5.2.5 Given a formula A, the set $BV(A)$ of *bound variables* in A is given by:

$$BV(Pt_1...t_n) = \emptyset;$$
$$BV(\doteq t_1t_2) = \emptyset;$$
$$BV(\neg A) = BV(A);$$
$$BV((A * B)) = BV(A) \cup BV(B), \text{ where } * \in \{\wedge, \vee, \supset, \equiv\};$$
$$BV(\bot) = \emptyset;$$
$$BV(\forall x_i A) = BV(A) \cup \{x_i\};$$
$$BV(\exists x_i A) = BV(A) \cup \{x_i\}.$$

In a formula $\forall x_i A$ (or $\exists x_i A$), we say that the variable x_i *is bound by the quantifier* $\forall$ (or $\exists$).

For a formula A, the intersection of $FV(A)$ and $BV(A)$ need not be empty, that is, the same variable may have both a free and a bound occurrence in A. As we shall see later, every formula is equivalent to a formula where the free and bound variables are disjoint.

EXAMPLE 5.2.3

Let

$$A = (\forall x(Rxy \supset Px) \wedge \forall y(\neg Rxy \wedge \forall x Px)).$$

Then
$$FV(A) = \{x, y\}, \ BV(A) = \{x, y\}.$$

For
$$B = \forall x (Rxy \supset Px),$$

we have
$$FV(B) = \{y\}, \ BV(B) = \{x\}.$$

For
$$C = \forall y (\neg Rxy \land \forall x Px),$$

we have
$$FV(C) = \{x\} \text{ and } BV(C) = \{x, y\}.$$

5.2.5 Substitutions

We will also need to define the substitution of a term for a free variable in a term or a formula.

Definition 5.2.6 Let s and t be terms. The result of substituting t in s for a variable x, denoted by $s[t/x]$ is defined recursively as follows:

$$y[t/x] = \ if \ y \neq x \ then \ y \ else \ t, \text{ when } s \text{ is a variable } y;$$
$$c[t/x] = c, \text{ when } s \text{ is a constant } c;$$
$$ft_1...t_n[t/x] = ft_1[t/x]...t_n[t/x], \text{ when } s \text{ is a term } ft_1...t_n.$$

For a formula A, $A[t/x]$ is defined recursively as follows:

$$\bot \ [t/x] = \bot, \text{ when } A \text{ is } \bot;$$
$$Pt_1...t_n[t/x] = Pt_1[t/x]...t_n[t/x], \text{ when } A = Pt_1...t_n;$$
$$\doteq t_1t_2[t/x] = \ \doteq t_1[t/x]t_2[t/x], \text{ when } A = \ \doteq t_1t_2;$$
$$(B * C)[t/x] = (B[t/x] * C[t/x]), \text{ when } A = (B * C), \ * \in \{\land, \lor, \supset, \equiv\}.$$
$$(\neg B)[t/x] = \neg B[t/x], \text{ when } A = \neg B;$$
$$(\forall y B)[t/x] = \ if \ x \neq y \ then \ \forall y B[t/x] \ else \ \forall y B, \text{ when } A = \forall y B;$$
$$(\exists y B)[t/x] = \ if \ x \neq y \ then \ \exists y B[t/x] \ else \ \exists y B, \text{ when } A = \exists y B.$$

EXAMPLE 5.2.4

Let
$$A = \forall x (P(x) \supset Q(x, f(y))), \text{ and let } t = g(y).$$

We have
$$f(x)[t/x] = f(g(y)), \quad A[t/y] = \forall x (P(x) \supset Q(x, f(g(y)))),$$

$$A[t/x] = \forall x(P(x) \supset Q(x, f(y))) = A,$$

since x is a bound variable.

The above definition prevents substitution of a term for a bound variable. When we give the semantics of the language, certain substitutions will not behave properly in the sense that they can change the truth value in a wrong way. These are substitutions in which some variable in the term t becomes bound in $A[t/x]$.

EXAMPLE 5.2.5

Let

$$A = \exists x(x < y)[x/y] = \exists x(x < x).$$

The sentence

$$\exists x(x < x)$$

is false in an ordered structure, but

$$\exists x(x < y)$$

may well be satisfied.

Definition 5.2.7 A term t is *free for x in A* if either:

(i) A is atomic or

(ii) $A = (B * C)$ and t is free for x in B and C, $* \in \{\land, \lor, \supset, \equiv\}$ or

(iii) $A = \neg B$ and t is free for x in B or

(iv) $A = \forall y B$ or $A = \exists y B$ and either

$x = y$ or

$x \neq y$, $y \notin FV(t)$ and t is free for x in B.

EXAMPLE 5.2.6

Let

$$A = \forall x(P(x) \supset Q(x, f(y))).$$

Then $g(y)$ is free for y in A, but $g(x)$ is not free for y in A.
If

$$B = \forall x(P(x) \supset \forall z Q(z, f(y))),$$

then $g(z)$ is not free for y in B, but $g(z)$ is free for z in B (because the substitution of $g(z)$ for z will not take place).

From now on we will assume that all substitutions satisfy the conditions of definition 5.2.7. Sometimes, if a formula A contains x as a free variable we write A as $A(x)$ and we abbreviate $A[t/x]$ as $A(t)$. In the next section, we will turn to the semantics of first-order logic.

PROBLEMS

5.2.1. Prove lemma 5.2.6.

5.2.2. Let w be a string consisting of variables, constants and function symbols.

(a) Prove that if $K(v) > 0$ for every proper suffix v of w, then w is a concatenation of $K(w)$ terms.

(b) Prove that w is a term iff $K(w) = 1$ and $K(v) > 0$ for every proper suffix v of w.

5.2.3. Given a formula A and constants a and b, show that for any two distinct variables x_i and x_j,

$$A[a/x_i][b/x_j] = A[b/x_j][a/x_i].$$

5.2.4. Prove that for every formula A and for every constant c,

$$FV(A[c/x]) = FV(A) - \{x\}.$$

5.2.5. Prove that for any formula A and term t, if y is not in $FV(A)$, then $A[t/y] = A$.

5.2.6. Prove that for every formula A and every term t, if t is free for x in A, $x \neq z$ and z is not in $FV(t)$, for every constant d,

$$A[t/x][d/z] = A[d/z][t/x].$$

5.2.7. Prove that for every formula A, if x and y are distinct variables, y is free in A, and z is a variable not occurring in A, for every term t free for x in A,

$$A[t/x][z/y] = A[z/y][t[z/y]/x],$$

and $t[z/y]$ is free for x in $A[z/y]$.

* **5.2.8.** The purpose of this problem is to generalize problem 3.2.6, that is, to give a context-free grammar defining terms and formulae. For this, it is necessary to encode the variables, constants, and the function and predicate symbols, as strings over a finite alphabet. Following Lewis and Papadimitriou, 1981, each variable x_n is encoded as $xI^n\$$, each m-ary predicate symbol P_n is encoded as $PI^m\$I^n\$$ $(m, n \geq 0)$, and each m-ary function symbol f_n is encoded as $fI^m\$I^n\$$.

Then, the set of terms and formulae is the language $L(G)$ defined by the following context-free grammar $G = (V, \Sigma, R, S)$:

$$\Sigma = \{P, I, f, x, \$, \wedge, \vee, \supset, \equiv, \neg, \forall, \exists, \doteq, \bot\},$$

$$V = \Sigma \cup \{S, N, T, U, W\},$$

$$
\begin{aligned}
R = \{ & N \to e, \\
& N \to NI, \\
& W \to xN\$, \\
& T \to W, \\
& T \to fU, \\
& U \to IUT, \\
& U \to \$N\$, \\
& S \to PU, \\
& S \to \perp, \\
& S \to \doteq TT, \\
& S \to (S \vee S), \\
& S \to (S \wedge S), \\
& S \to (S \supset S), \\
& S \to (S \equiv S), \\
& S \to \neg S, \\
& S \to \forall W S, \\
& S \to \exists W S \}
\end{aligned}
$$

Prove that the grammar G is unambiguous.

Note: The above language is actually $SLR(1)$. For details on parsing techniques, consult Aho and Ullman, 1977.

5.3 SEMANTICS OF FIRST-ORDER LANGUAGES

Given a first-order language **L**, the semantics of formulae is obtained by interpreting the function, constant and predicate symbols in **L** and assigning values to the free variables. For this, we need to define the concept of a *structure*.

5.3.1 First-Order Structures

A first-order structure assigns a meaning to the symbols in **L** as explained below.

Definition 5.3.1 Given a first-order language **L**, an **L**-*structure* **M** (for short, a structure) is a pair $\mathbf{M} = (M, I)$ where M is a nonempty set called the *domain* (or *carrier*) of the structure and I is a function called the *interpretation function* and which assigns functions and predicates over M to the symbols in **L** as follows:

(i) For every function symbol f of rank $n > 0$, $I(f) : M^n \to M$ is an n-ary function.

(ii) For every constant c, $I(c)$ is an element of M.

(iii) For every predicate symbol P of rank $n \geq 0$, $I(P) : M^n \to BOOL$ is an n-ary predicate. In particular, predicate symbols of rank 0 are interpreted as truth values (Hence, the interpretation function I restricted to predicate symbols of rank zero is a valuation as in propositional logic.)

We will also use the following notation in which I is omitted: $I(f)$ is denoted as $f_{\mathbf{M}}$, $I(c)$ as $c_{\mathbf{M}}$ and $I(P)$ as $P_{\mathbf{M}}$.

EXAMPLE 5.3.1

Let **L** be the language of arithmetic where, $\mathbf{CS} = \{0\}$ and $\mathbf{FS} = \{S, +, *\}$. The symbol S has rank 1, and the symbols $+, *$ have rank 2. The **L**-structure $\mathcal{N}$ is defined such that its domain is the set **N** of natural numbers, and the constant and function symbols are interpreted as follows: 0 is interpreted as zero, S as the function such that $S(x) = x+1$ for all $x \in \mathbf{N}$ (the successor function), $+$ is interpreted as addition and $*$ as multiplication. The structure $\mathcal{Q}$ obtained by replacing **N** by the set **Q** of rational numbers and the structure $\mathcal{R}$ obtained by replacing **N** by the set **R** of real numbers and interpreting 0, S, $+$ and $*$ in the natural way are also **L**-structures.

Remark: Note that a first-order **L**-structure is in fact a two-sorted algebra as defined in Subsection 2.5.2 (with carrier $BOOL$ of sort *formula* and carrier M of sort *term*).

5.3.2 Semantics of Formulae

We now wish to define the semantics of formulae. Since a formula A may contain free variables, its truth value will generally depend on the specific assignment s of values from the domain M to the variables. Hence, we shall define the semantics of a formula A as a *function* $A_{\mathbf{M}}$ from the set of assignments of values in M to the variables, to the set $BOOL$ of truth values. First, we need some definitions.

Definition 5.3.2 Given a first-order language **L** and an **L**-structure **M**, an *assignment* is any function $s : \mathbf{V} \to M$ from the set of variables **V** to the domain M. The set of all such functions is denoted by $[\mathbf{V} \to M]$.

To define the meaning of a formula, we shall construct the function $A_{\mathbf{M}}$ recursively, using the fact that formulae are freely generated from the atomic formulae, and using theorem 2.4.1. Since the meaning of a formula is a function from $[\mathbf{V} \to M]$ to $BOOL$, let us denote the set of all such functions as $[[\mathbf{V} \to M] \to BOOL]$. In order to apply theorem 2.4.1, it is necessary to extend the connectives to the set of functions $[[\mathbf{V} \to M] \to BOOL]$ to make this set into an algebra. For this, the next two definitions are needed.

Definition 5.3.3 Given any nonempty domain M, given any element $a \in M$ and any assignment $s : \mathbf{V} \to M$, $s[x_i := a]$ denotes the new assignment $s' : \mathbf{V} \to M$ such that

$$s'(y) = s(y) \text{ for } y \neq x_i \text{ and } s'(x_i) = a.$$

For all $i \geq 0$, define the function $(A_i)_{\mathbf{M}}$ and $(E_i)_{\mathbf{M}}$ from $[[\mathbf{V} \to M] \to BOOL]$ to $[[\mathbf{V} \to M] \to BOOL]$ as follows: For every function $f \in [[\mathbf{V} \to M] \to BOOL]$, $(A_i)_{\mathbf{M}}(f)$ is the function such that: For every assignment $s \in [\mathbf{V} \to M]$,

$$(A_i)_{\mathbf{M}}(f)(s) = \mathbf{F} \quad \text{iff} \quad f(s[x_i := a]) = \mathbf{F} \quad \text{for some } a \in M;$$

The function $(E_i)_{\mathbf{M}}(f)$ is the function such that: For every assignment $s \in [\mathbf{V} \to M]$,

$$(E_i)_{\mathbf{M}}(f)(s) = \mathbf{T} \quad \text{iff} \quad f(s[x_i := a]) = \mathbf{T} \quad \text{for some } a \in M.$$

Note that $(A_i)_{\mathbf{M}}(f)(s) = \mathbf{T}$ iff the function $g_s : M \to BOOL$ such that $g_s(a) = f(s[x_i := a])$ for all $a \in M$ is the constant function whose value is $\mathbf{T}$, and that $(E_i)_{\mathbf{M}}(f)(s) = \mathbf{F}$ iff the function $g_s : M \to BOOL$ defined above is the constant function whose value is $\mathbf{F}$.

Definition 5.3.4 Given a nonempty domain M, the functions $\wedge_{\mathbf{M}}$, $\vee_{\mathbf{M}}$, $\supset_{\mathbf{M}}$ and $\equiv_{\mathbf{M}}$ from $[[\mathbf{V} \to M] \to BOOL] \times [[\mathbf{V} \to M] \to BOOL]$ to $[[\mathbf{V} \to M] \to BOOL]$, and the function $\neg_{\mathbf{M}}$ from $[[\mathbf{V} \to M] \to BOOL]$ to $[[\mathbf{V} \to M] \to BOOL]$ are defined as follows: For every two functions f and g in $[[\mathbf{V} \to M] \to BOOL]$, for all s in $[\mathbf{V} \to M]$,

$$\wedge_{\mathbf{M}}(f, g)(s) = H_\wedge(f(s), g(s));$$
$$\vee_{\mathbf{M}}(f, g)(s) = H_\vee(f(s), g(s));$$
$$\supset_{\mathbf{M}}(f, g)(s) = H_\supset(f(s), g(s));$$
$$\equiv_{\mathbf{M}}(f, g)(s) = H_\equiv(f(s), g(s));$$
$$\neg_{\mathbf{M}}(f)(s) = H_\neg(f(s)).$$

Using theorem 2.4.1, we can now define the meaning $t_{\mathbf{M}}$ of a term t and the meaning $A_{\mathbf{M}}$ of a formula A. We begin with terms.

Definition 5.3.5 Given an **L**-structure **M**, the function

$$t_{\mathbf{M}} : [\mathbf{V} \to M] \to M$$

defined by a term t is the function such that for every assignment $s \in [\mathbf{V} \to M]$, the value $t_{\mathbf{M}}[s]$ is defined recursively as follows:

(i) $\qquad\qquad\qquad\qquad x_{\mathbf{M}}[s] = s(x)$, for a variable x;

(ii) $\qquad\qquad\qquad\qquad c_{\mathbf{M}}[s] = c_{\mathbf{M}}$, for a constant c;

(iii) $\qquad\qquad\qquad (ft_1...t_n)_{\mathbf{M}}[s] = f_{\mathbf{M}}((t_1)_{\mathbf{M}}[s], ..., (t_n)_{\mathbf{M}}[s])$.

The recursive definition of the function $A_{\mathbf{M}} : [\mathbf{V} \to M] \to BOOL$ is now given.

Definition 5.3.6 The function

$$A_{\mathbf{M}} : [\mathbf{V} \to M] \to BOOL$$

is defined recursively by the following clauses:

(1) For atomic formulae: $A_{\mathbf{M}}$ is the function such that, for every assignment $s \in [\mathbf{V} \to M]$,

(i) $\qquad (Pt_1...t_n)_{\mathbf{M}}[s] = P_{\mathbf{M}}((t_1)_{\mathbf{M}}[s], ..., (t_n)_{\mathbf{M}}[s])$;

(ii) $\qquad (\doteq t_1 t_2)_{\mathbf{M}}[s] = if \ (t_1)_{\mathbf{M}}[s] = (t_2)_{\mathbf{M}}[s] \ then \ \mathbf{T} \ else \ \mathbf{F}$;

(iii) $\qquad (\bot)_{\mathbf{M}}[s] = \mathbf{F}$.

(2) For nonatomic formulae:

(i) $\qquad\qquad\qquad\qquad (A * B)_{\mathbf{M}} = *_{\mathbf{M}}(A_{\mathbf{M}}, B_{\mathbf{M}})$,

where $* \in \{\wedge, \vee, \supset, \equiv\}$ and $*_{\mathbf{M}}$ is the corresponding function defined in definition 5.3.4;

(ii) $\qquad\qquad\qquad\qquad (\neg A)_{\mathbf{M}} = \neg_{\mathbf{M}}(A_{\mathbf{M}})$;

(iii) $\qquad\qquad\qquad\qquad (\forall x_i A)_{\mathbf{M}} = (A_i)_{\mathbf{M}}(A_{\mathbf{M}})$;

(iv) $\qquad\qquad\qquad\qquad (\exists x_i A)_{\mathbf{M}} = (E_i)_{\mathbf{M}}(A_{\mathbf{M}})$.

Note that by definitions 5.3.3, 5.3.4, 5.3.5, and 5.3.6, for every assignment $s \in [\mathbf{V} \to M]$,

$$(\forall x_i A)_{\mathbf{M}}[s] = \mathbf{T} \ \text{iff} \ A_{\mathbf{M}}[s[x_i := m]] = \mathbf{T} \ \text{for all} \ m \in M,$$

and

$$(\exists x_i A)_{\mathbf{M}}[s] = \mathbf{T} \ \text{iff} \ A_{\mathbf{M}}[s[x_i := m]] = \mathbf{T} \ \text{for some} \ m \in M.$$

Hence, if M is infinite, evaluating $(\forall x_i A)_{\mathbf{M}}[s]$ (or $(\exists x_i A)_{\mathbf{M}}[s]$) requires testing infinitely many values (all the truth values $A_{\mathbf{M}}[s[x_i := m]]$, for $m \in$

M). Hence, contrary to the propositional calculus, there does not appear to be an algorithm for computing the truth value of $(\forall x_i A)_\mathbf{M}[s]$ (or $(\exists x_i A)_\mathbf{M}[s]$) and in fact, no such algorithm exists in general.

5.3.3 Satisfaction, Validity, and Model

We can now define the notions of satisfaction, validity and model.

Definition 5.3.7 Let **L** be a first-order language and **M** be an L-structure.

(i) Given a formula A and an assignment s, we say that **M** *satisfies A with s* iff

$$A_\mathbf{M}[s] = \mathbf{T}.$$

This is also denoted by

$$\mathbf{M} \models A[s].$$

(ii) A formula A is *satisfiable in* **M** iff there is some assignment s such that

$$A_\mathbf{M}[s] = \mathbf{T};$$

A is *satisfiable* iff there is some **M** in which A is satisfiable.

(iii) A formula A is *valid in* **M** (or *true* in **M**) iff

$$A_\mathbf{M}[s] = \mathbf{T} \quad \text{for every assignment } s.$$

This is denoted by

$$\mathbf{M} \models A.$$

In this case, **M** is called a *model* of A. A formula A is *valid* (or *universally valid*) iff it is valid in every structure **M**. This is denoted by

$$\models A.$$

(iv) Given a set Γ of formulae, Γ is *satisfiable* iff there exists a structure **M** and an assignment s such that

$$\mathbf{M} \models A[s] \quad \text{for every formula} \quad A \in \Gamma;$$

A structure **M** is a *model* of Γ iff **M** is a model of every formula in Γ. This is denoted by

$$\mathbf{M} \models \Gamma.$$

The set Γ is *valid* iff $\mathbf{M} \models \Gamma$ for every structure **M**. This is denoted by

$$\models \Gamma.$$

(v) Given a set Γ of formulae and a formula B, B is a *semantic consequence* of Γ, denoted by

$$\Gamma \models B$$

iff, for every **L**-structure **M**, for every assignment s,

> if $\mathbf{M} \models A[s]$ for every formula $A \in \Gamma$ then $\mathbf{M} \models B[s]$.

EXAMPLE 5.3.2

Consider the language of arithmetic defined in example 5.3.1. The following formulae are valid in the model $\mathcal{N}$ with domain the set of natural numbers defined in example 5.3.1. This set A_P is known as the axioms of *Peano's arithmetic*:

$$\forall x \neg (S(x) \doteq 0)$$
$$\forall x \forall y (S(x) \doteq S(y) \supset x \doteq y)$$
$$\forall x (x + 0 \doteq x)$$
$$\forall x \forall y (x + S(y) \doteq S(x + y))$$
$$\forall x (x * 0 \doteq 0)$$
$$\forall x \forall y (x * S(y) \doteq x * y + x)$$

For every formula A with one free variable x,

$$(A(0) \wedge \forall x (A(x) \supset A(S(x)))) \supset \forall y A(y)$$

This last axiom scheme is known as an *induction axiom*. The structure **N** is a model of these formulae. These formulae are not valid in all structures. For example, the first formula is not valid in the structure whose domain has a single element.

As in propositional logic, formulae that are universally valid are particularly interesting. These formulae will be characterized in terms of a proof system in the next section.

5.3.4 A More Convenient Semantics

It will be convenient in the sequel, especially in proofs involving Hintikka sets, to have a slightly different definition of the truth of a quantified formula. The idea behind this definition is to augment the language **L** with a set of constants naming the elements in the structure **M**, so that elements in M can be treated as constants in formulae. Instead of defining the truth value of $(\forall x_i A)_{\mathbf{M}}[s]$ using the modified assignments $s[x_i := m]$ so that

$$(\forall x_i A)_{\mathbf{M}}[s] = \mathbf{T} \text{ iff } A_{\mathbf{M}}[s[x_i := m]] = \mathbf{T} \text{ for all } m \in M,$$

we will define it using the modified formula $A[\mathbf{m}/x_i]$, so that

$$(\forall x_i A)_{\mathbf{M}}[s] = \mathbf{T} \text{ iff } A[\mathbf{m}/x_i]_{\mathbf{M}}[s] = \mathbf{T} \text{ for all } m \in M,$$

where **m** is a constant naming the element $m \in M$. Instead of computing the truth value of A in the modified assignment $s[x_i := m]$, we compute the truth value of the modified formula $A[\mathbf{m}/x_i]$ obtained by substituting the constant $\mathbf{m}$ for the variable x_i, in the assignment s itself.

Definition 5.3.8 Given a first-order language **L** and an **L**-structure **M**, the *extended language* **L(M)** is obtained by adjoining to the set **CS** of constants in **L** a set

$$\{\mathbf{m} \mid m \in M\}$$

of new constants, one for each element of M. The interpretation function I of the first-order structure **M** is extended to the constants in the set $\{\mathbf{m} \mid m \in M\}$ by defining $I(\mathbf{m}) = m$. Hence, for any assignment $s : \mathbf{V} \to M$,

$$(\mathbf{m})_{\mathbf{M}}[s] = m.$$

The resulting alphabet is not necessarily countable and the set of formulae may also not be countable. However, since a formula contains only finitely many symbols, there is no problem with inductive definitions or proofs by induction.

Next, we will prove a lemma that shows the equivalence of definition 5.3.6 and the more convenient definition of the truth of a quantified formula sketched before definition 5.3.8. First, we need the following technical lemma.

Lemma 5.3.1 Given a first-order language **L** and an **L**-structure **M**, the following hold:

(1) For any term t, for any assignment $s \in [\mathbf{V} \to M]$, any element $m \in M$ and any variable x_i,

$$t_{\mathbf{M}}[s[x_i := m]] = (t[\mathbf{m}/x_i])_{\mathbf{M}}[s].$$

(2) For any formula A, for any assignment $s \in [\mathbf{V} \to M]$, any element $m \in M$ and any variable x_i,

$$A_{\mathbf{M}}[s[x_i := m]] = (A[\mathbf{m}/x_i])_{\mathbf{M}}[s].$$

Proof: This proof is a typical induction on the structure of terms and formulae. Since we haven't given a proof of this kind for first-order formulae, we will give it in full. This will allow us to omit similar proofs later. First, we prove (1) by induction on terms.

If t is a variable $x_j \neq x_i$, then

$$(x_j)_{\mathbf{M}}[s[x_i := m]] = s(x_j)$$

and

$$(x_j[\mathbf{m}/x_i])_{\mathbf{M}}[s] = (x_j)_{\mathbf{M}}[s] = s(x_j),$$

establishing (1).

If t is the variable x_i, then

$$(x_i)_{\mathbf{M}}[s[x_i := m]] = m$$

and

$$(x_i[\mathbf{m}/x_i])_{\mathbf{M}}[s] = (\mathbf{m})_{\mathbf{M}}[s] = m,$$

establishing (1).

If t is a constant c, then

$$(c)_{\mathbf{M}}[s[x_i := m]] = c_{\mathbf{M}} = c_{\mathbf{M}}[s] = (c[\mathbf{m}/x_i])_{\mathbf{M}}[s],$$

establishing (1).

If t is a term $ft_1...t_k$, then

$$(ft_1...t_k)_{\mathbf{M}}[s[x_i := m]] = f_{\mathbf{M}}((t_1)_{\mathbf{M}}[s[x_i := m]], ..., (t_k)_{\mathbf{M}}[s[x_i := m]]).$$

By the induction hypothesis, for every j, $1 \leq j \leq k$,

$$(t_j)_{\mathbf{M}}[s[x_i := m]] = (t_j[\mathbf{m}/x_i])_{\mathbf{M}}[s].$$

Hence,

$$
\begin{aligned}
(ft_1...t_k)_{\mathbf{M}}[s[x_i := m]] &= f_{\mathbf{M}}((t_1)_{\mathbf{M}}[s[x_i := m]], ..., (t_k)_{\mathbf{M}}[s[x_i := m]]) \\
&= f_{\mathbf{M}}((t_1[\mathbf{m}/x_i])_{\mathbf{M}}[s], ..., (t_k[\mathbf{m}/x_i])_{\mathbf{M}}[s]) \\
&= (ft_1[\mathbf{m}/x_i]...t_k[\mathbf{m}/x_i])_{\mathbf{M}}[s] \\
&= (t[\mathbf{m}/x_i])_{\mathbf{M}}[s],
\end{aligned}
$$

establishing (1).

This concludes the proof of (1). Next, we prove (2) by induction on formulae.

If A is an atomic formula of the form $Pt_1...t_k$, then

$$(Pt_1...t_k)_{\mathbf{M}}[s[x_i := m]] = P_{\mathbf{M}}((t_1)_{\mathbf{M}}[s[x_i := m]], ..., (t_k)_{\mathbf{M}}[s[x_i := m]]).$$

By part (1) of the lemma, for every j, $1 \leq j \leq k$,

$$(t_j)_{\mathbf{M}}[s[x_i := m]] = (t_j[\mathbf{m}/x_i])_{\mathbf{M}}[s].$$

Hence,

$$
\begin{aligned}
(Pt_1...t_k)_{\mathbf{M}}[s[x_i := m]] &= P_{\mathbf{M}}((t_1)_{\mathbf{M}}[s[x_i := m]], ..., (t_k)_{\mathbf{M}}[s[x_i := m]]) \\
&= P_{\mathbf{M}}((t_1[\mathbf{m}/x_i])_{\mathbf{M}}[s], ..., (t_k[\mathbf{m}/x_i])_{\mathbf{M}}[s]) \\
&= (Pt_1[\mathbf{m}/x_i]...t_k[\mathbf{m}/x_i])_{\mathbf{M}}[s] \\
&= (A[\mathbf{m}/x_i])_{\mathbf{M}}[s],
\end{aligned}
$$

establishing (2).

It is obvious that (2) holds for the constant $\perp$.

If A is an atomic formula of the form $\doteq t_1 t_2$, then

$$(\doteq t_1 t_2)_{\mathbf{M}}[s[x_i := m]] = \mathbf{T} \text{ iff } (t_1)_{\mathbf{M}}[s[x_i := m]] = (t_2)_{\mathbf{M}}[s[x_i := m]].$$

By part (1) of the lemma, for $j = 1, 2$, we have

$$(t_j)_{\mathbf{M}}[s[x_i := m]] = (t_j[\mathbf{m}/x_i])_{\mathbf{M}}[s].$$

Hence,

$$(\doteq t_1 t_2)_{\mathbf{M}}[s[x_i := m]] = \mathbf{T} \text{ iff}$$
$$(t_1)_{\mathbf{M}}[s[x_i := m]] = (t_2)_{\mathbf{M}}[s[x_i := m]] \text{ iff}$$
$$(t_1[\mathbf{m}/x_i])_{\mathbf{M}}[s] = (t_2[\mathbf{m}/x_i])_{\mathbf{M}}[s] \text{ iff}$$
$$((\doteq t_1 t_2)[\mathbf{m}/x_i])_{\mathbf{M}}[s] = \mathbf{T},$$

establishing (2).

If A is a formula of the form $(B * C)$ where $* \in \{\wedge, \vee, \supset, \equiv\}$, we have

$$(B * C)_{\mathbf{M}}[s[x_i := m]] = *_{\mathbf{M}}(B_{\mathbf{M}}, C_{\mathbf{M}})[s[x_i := m]]$$
$$= H_*(B_{\mathbf{M}}[s[x_i := m]], C_{\mathbf{M}}[s[x_i := m]]).$$

By the induction hypothesis,

$$B_{\mathbf{M}}[s[x_i := m]] = (B[\mathbf{m}/x_i])_{\mathbf{M}}[s] \text{ and}$$
$$C_{\mathbf{M}}[s[x_i := m]] = (C[\mathbf{m}/x_i])_{\mathbf{M}}[s].$$

Hence,

$$(B * C)_{\mathbf{M}}[s[x_i := m]] = H_*(B_{\mathbf{M}}[s[x_i := m]], C_{\mathbf{M}}[s[x_i := m]])$$
$$= H_*(B[\mathbf{m}/x_i])_{\mathbf{M}}[s], C[\mathbf{m}/x_i])_{\mathbf{M}}[s]) = ((B * C)[\mathbf{m}/x_i])_{\mathbf{M}}[s],$$

establishing (2).

If A is of the form $\neg B$, then

$$(\neg B)_{\mathbf{M}}[s[x_i := m]] = \neg_{\mathbf{M}}(B_{\mathbf{M}})[s[x_i := m]] = H_\neg(B_{\mathbf{M}}[s[x_i := m]]).$$

By the induction hypothesis,

$$B_{\mathbf{M}}[s[x_i := m]] = (B[\mathbf{m}/x_i])_{\mathbf{M}}[s].$$

Hence,

$$(\neg B)_{\mathbf{M}}[s[x_i := m]] = H_{\neg}(B_{\mathbf{M}}[s[x_i := m]])$$
$$= H_{\neg}(B[\mathbf{m}/x_i])_{\mathbf{M}}[s] = ((\neg B)[\mathbf{m}/x_i])_{\mathbf{M}}[s],$$

establishing (2).

If A is of the form $\forall x_j B$, there are two cases. If $x_i \neq x_j$, then

$$(\forall x_j B)_{\mathbf{M}}[s[x_i := m]] = \mathbf{T} \text{ iff}$$
$$B_{\mathbf{M}}[s[x_i := m][x_j := a]] = \mathbf{T} \text{ for all } a \in M.$$

By the induction hypothesis,

$$B_{\mathbf{M}}[s[x_i := m][x_j := a]] = (B[\mathbf{a}/x_j])_{\mathbf{M}}[s[x_i := m]],$$

and by one more application of the induction hypothesis,

$$(B[\mathbf{a}/x_j])_{\mathbf{M}}[s[x_i := m]] = (B[\mathbf{a}/x_j][\mathbf{m}/x_i])_{\mathbf{M}}[s].$$

By problem 5.2.3, since $x_i \neq x_j$ and $\mathbf{m}$ and $\mathbf{a}$ are constants,

$$B[\mathbf{a}/x_j][\mathbf{m}/x_i] = B[\mathbf{m}/x_i][\mathbf{a}/x_j].$$

Hence,

$$B_{\mathbf{M}}[s[x_i := m][x_j := a]] = \mathbf{T} \text{ for all } a \in M \text{ iff}$$
$$(B[\mathbf{a}/x_j][\mathbf{m}/x_i])_{\mathbf{M}}[s] = \mathbf{T} \text{ for all } a \in M \text{ iff}$$
$$((B[\mathbf{m}/x_i])[\mathbf{a}/x_j])_{\mathbf{M}}[s] = \mathbf{T} \text{ for all } a \in M.$$

By the induction hypothesis,

$$((B[\mathbf{m}/x_i])[\mathbf{a}/x_j])_{\mathbf{M}}[s] = (B[\mathbf{m}/x_i])_{\mathbf{M}}[s[x_j := a]].$$

Hence,

$$((B[\mathbf{m}/x_i])[\mathbf{a}/x_j])_{\mathbf{M}}[s] = \mathbf{T} \text{ for all } a \in M \text{ iff}$$
$$(B[\mathbf{m}/x_i])_{\mathbf{M}}[s[x_j := a]] = \mathbf{T} \text{ for all } a \in M \text{ iff}$$
$$((\forall x_j B)[\mathbf{m}/x_i])_{\mathbf{M}}[s] = \mathbf{T},$$

establishing (2).

If $x_i = x_j$, then

$$s[x_i := m][x_j := a] = s[x_i := a] \text{ and } (\forall x_j B)[\mathbf{m}/x_i] = \forall x_i B,$$

and so

$$(\forall x_j B)_{\mathbf{M}}[s[x_i := m]] = \mathbf{T} \text{ iff}$$
$$B_{\mathbf{M}}[s[x_i := m][x_j := a]] = \mathbf{T} \text{ for all } a \in M \text{ iff}$$
$$B_{\mathbf{M}}[s[x_i := a]] = \mathbf{T} \text{ for all } a \in M \text{ iff}$$
$$(\forall x_i B)_{\mathbf{M}}[s] = \mathbf{T} \text{ iff}$$
$$((\forall x_j B)[\mathbf{m}/x_i])_{\mathbf{M}}[s] = \mathbf{T},$$

establishing (2).

The case in which A is of the form $\exists x_j B$ is similar to the previous case and is left as an exercise. This concludes the induction proof for (2). $\square$

We can now prove that the new definition of the truth of a quantified formula is equivalent to the old one.

Lemma 5.3.2 For any formula B, for any assignment $s \in [\mathbf{V} \to M]$ and any variable x_i, the following hold:

(1) $(\forall x_i B)_{\mathbf{M}}[s] = \mathbf{T} \text{ iff } (B[\mathbf{m}/x_i])_{\mathbf{M}}[s] = \mathbf{T} \text{ for all } m \in M;$

(2) $(\exists x_i B)_{\mathbf{M}}[s] = \mathbf{T} \text{ iff } (B[\mathbf{m}/x_i])_{\mathbf{M}}[s] = \mathbf{T} \text{ for some } m \in M;$

Proof: Recall that

$$(\forall x_i B)_{\mathbf{M}}[s] = \mathbf{T} \text{ iff } B_{\mathbf{M}}[s[x_i := m]] = \mathbf{T} \text{ for all } m \in M.$$

By lemma 5.3.1,
$$B_{\mathbf{M}}[s[x_i := m]] = (B[\mathbf{m}/x_i])_{\mathbf{M}}[s].$$

Hence,

$$(\forall x_i B)_{\mathbf{M}}[s] = \mathbf{T} \text{ iff}$$
$$(B[\mathbf{m}/x_i])_{\mathbf{M}}[s] = \mathbf{T} \text{ for all } m \in M,$$

proving (1).

Also recall that

$$(\exists x_i B)_{\mathbf{M}}[s] = \mathbf{T} \text{ iff } B_{\mathbf{M}}[s[x_i := m]] = \mathbf{T} \text{ for some } m \in M.$$

By lemma 5.3.1,
$$B_{\mathbf{M}}[s[x_i := m]] = (B[\mathbf{m}/x_i])_{\mathbf{M}}[s].$$

Hence,

$$(\exists x_i B)_{\mathbf{M}}[s] = \mathbf{T} \text{ iff}$$
$$(B[\mathbf{m}/x_i])_{\mathbf{M}}[s] = \mathbf{T} \text{ for some } m \in M,$$

proving (2). $\square$

In view of lemma 5.3.2, the recursive clauses of the definition of satisfaction can also be stated more informally as follows:

$$\mathbf{M} \models (\neg A)[s] \text{ iff } \mathbf{M} \not\models A[s],$$
$$\mathbf{M} \models (A \wedge B)[s] \text{ iff } \mathbf{M} \models A[s] \text{ and } \mathbf{M} \models B[s],$$
$$\mathbf{M} \models (A \vee B)[s] \text{ iff } \mathbf{M} \models A[s] \text{ or } \mathbf{M} \models B[s],$$
$$\mathbf{M} \models (A \supset B)[s] \text{ iff } \mathbf{M} \not\models A[s] \text{ or } \mathbf{M} \models B[s],$$
$$\mathbf{M} \models (A \equiv B)[s] \text{ iff } (\mathbf{M} \models A[s] \text{ } iff \text{ } \mathbf{M} \models B[s]),$$
$$\mathbf{M} \models (\forall x_i)A[s] \text{ iff } \mathbf{M} \models (A[\mathbf{a}/x_i])[s] \text{ for every } a \in M,$$
$$\mathbf{M} \models (\exists x_i)A[s] \text{ iff } \mathbf{M} \models (A[\mathbf{a}/x_i])[s] \text{ for some } a \in M.$$

5.3.5 Free Variables and Semantics of Formulae

If A is a formula and the set $FV(A)$ of variables free in A is $\{y_1, ..., y_n\}$, for every assignment s, the truth value $A_{\mathbf{M}}[s]$ only depends on the restriction of s to $\{y_1, ..., y_n\}$. The following lemma makes the above statement precise.

Lemma 5.3.3 Given a formula A with set of free variables $\{y_1, ..., y_n\}$, for any two assignments s_1, s_2 such that $s_1(y_i) = s_2(y_i)$ for $1 \leq i \leq n$,

$$A_{\mathbf{M}}[s_1] = A_{\mathbf{M}}[s_2].$$

Proof: The lemma is proved using the induction principle for terms and formulae. First, let t be a term, and assume that s_1 and s_2 agree on $FV(t)$.

If $t = c$ (a constant), then

$$t_{\mathbf{M}}[s_1] = c_{\mathbf{M}} = t_{\mathbf{M}}[s_2].$$

If $t = y$ (a variable), then

$$t_{\mathbf{M}}[s_1] = s_1(y) = s_2(y) = t_{\mathbf{M}}[s_2],$$

since $FV(t) = \{y\}$ and s_1 and s_2 agree on $FV(t)$.

If $t = ft_1...t_n$, then

$$t_{\mathbf{M}}[s_1] = f_{\mathbf{M}}((t_1)_{\mathbf{M}}[s_1], ..., (t_n)_{\mathbf{M}}[s_1]).$$

Since every $FV(t_i)$ is a subset of $FV(t)$, $1 \leq i \leq n$, and s_1 and s_2 agree on $FV(t)$, by the induction hypothesis,

$$(t_i)_{\mathbf{M}}[s_1] = (t_i)_{\mathbf{M}}[s_2],$$

for $i = 1, ..., n$. Hence,

$$t_{\mathbf{M}}[s_1] = f_{\mathbf{M}}((t_1)_{\mathbf{M}}[s_1], ..., (t_n)_{\mathbf{M}}[s_1])$$
$$= f_{\mathbf{M}}((t_1)_{\mathbf{M}}[s_2], ..., (t_n)_{\mathbf{M}}[s_2]) = t_{\mathbf{M}}[s_2].$$

Now, let A be a formula, and assume that s_1 and s_2 agree on $FV(A)$.

If $A = Pt_1...t_n$, since $FV(t_i)$ is a subset of $FV(A)$ and s_1 and s_2 agree on $FV(A)$, we have

$$(t_i)_{\mathbf{M}}[s_1] = (t_i)_{\mathbf{M}}[s_2],$$

$1 \leq i \leq n$. But then, we have

$$A_{\mathbf{M}}[s_1] = P_{\mathbf{M}}((t_1)_{\mathbf{M}}[s_1], ..., (t_n)_{\mathbf{M}}[s_1])$$
$$= P_{\mathbf{M}}((t_1)_{\mathbf{M}}[s_2], ..., (t_n)_{\mathbf{M}}[s_2]) = A_{\mathbf{M}}[s_2].$$

If $A = \doteq t_1 t_2$, since $FV(t_1)$ and $FV(t_2)$ are subsets of $FV(t)$ and s_1 and s_2 agree on $FV(A)$,

$$(t_1)_{\mathbf{M}}[s_1] = (t_1)_{\mathbf{M}}[s_2] \text{ and } (t_2)_{\mathbf{M}}[s_1] = (t_2)_{\mathbf{M}}[s_2],$$

and so

$$A_{\mathbf{M}}[s_1] = A_{\mathbf{M}}[s_2].$$

The cases in which A is of the form $(B \vee C)$, or $(B \wedge C)$, or $(B \supset C)$, or $(A \equiv B)$, or $\neg B$ are easily handled by induction as in lemma 3.3.1 (or 5.3.1), and the details are left as an exercise. Finally, we treat the case in which A is of the form $\forall x B$ (the case $\exists x B$ is similar).

If $A = \forall x B$, then $FV(A) = FV(B) - \{x\}$. First, recall the following property shown in problem 5.2.4: For every constant c,

$$FV(B[c/x]) = FV(B) - \{x\}.$$

Recall that from lemma 5.3.2,

$$A_{\mathbf{M}}[s_1] = \mathbf{T} \text{ iff } (B[a/x])_{\mathbf{M}}[s_1] = \mathbf{T} \text{ for every } a \in M.$$

Since s_1 and s_2 agree on $FV(A) = FV(B) - \{x\}$ and $FV(B[a/x]) = FV(B) - \{x\}$, by the induction hypothesis,

$$(B[a/x])_{\mathbf{M}}[s_1] = (B[a/x])_{\mathbf{M}}[s_2] \text{ for every } a \in M.$$

This shows that

$$(B[a/x])_{\mathbf{M}}[s_1] = \mathbf{T} \text{ iff } (B[a/x])_{\mathbf{M}}[s_2] = \mathbf{T},$$

that is,
$$A_\mathbf{M}[s_1] = A_\mathbf{M}[s_2].$$

□

As a consequence, if A is a sentence (that is $FV(A) = \emptyset$) the truth value of A in $\mathbf{M}$ is the same for all assignments. Hence for a *sentence* A, for every structure $\mathbf{M}$,
$$\text{either } \mathbf{M} \models A \text{ or } \mathbf{M} \models \neg A.$$

5.3.6 Subformulae and Rectified Formulae

In preparation for the next section in which we present a Gentzen system that is sound and complete with respect to the semantics, we need the following definitions and lemmas.

Definition 5.3.9 Given a formula A, a formula X is a *subformula* of A if either:

(i) $A = Pt_1...t_n$ and $X = A$, or

(ii) $A = \doteq t_1 t_2$ and $X = A$, or

(iii) $A = \perp$ and $X = A$, or

(iv) $A = (B * C)$ and $X = A$, or X is a subformula of B, or X is a subformula of C, where $* \in \{\wedge, \vee, \supset, \equiv\}$, or

(v) $A = \neg B$ and $X = A$, or X is a subformula of B, or

(vi) $A = \forall x B$ and $X = A$, or X is a subformula of $B[t/x]$ for any term t free for x in B, or

(vii) $A = \exists x B$ and $X = A$, or X is a subformula of $B[t/x]$ for any term t free for x in B.

Definition 5.3.10 A quantifier $\forall$ (or $\exists$) *binds* an occurrence of a variable x in A iff A contains a subformula of the form $\forall x B$ (or $\exists x B$).

A formula A is *rectified* if

(i) $FV(A)$ and $BV(A)$ are disjoint and

(ii) Distinct quantifiers in A bind occurrences of distinct variables.

The following lemma will be needed later.

Lemma 5.3.4 (i) For every formula A,
$$\models \forall x A \equiv \forall y A[y/x]$$

and
$$\models \exists x A \equiv \exists y A[y/x],$$

for every variable y free for x in A and not in $FV(A) - \{x\}$.

(ii) There is an algorithm such that, for any input formula A, the algorithm produces a rectified formula A' such that

$$\models A \equiv A'$$

(that is, A' is semantically equivalent to A).

Proof: We prove (i), leaving (ii) as an exercise. First, we show that for every term t and every term r,

$$t[r/x] = t[y/x][r/y], \text{ if } y \text{ is not in } FV(t) - \{x\}.$$

The proof is by induction on the structure of terms and it is left as an exercise.

Now, let A be a formula, $\mathbf{M}$ be a structure, s an assignment, and a any element in M. We show that

$$(A[\mathbf{a}/x])_{\mathbf{M}}[s] = (A[y/x][\mathbf{a}/y])_{\mathbf{M}}[s],$$

provided that y is free for x in A, and that $y \notin FV(A) - \{x\}$.

This proof is by induction on formulae. We only treat the case of quantifiers, leaving the other cases as an exercise. We consider the case $A = \forall z B$, the case $\exists z B$ being similar. Recall that the following property was shown in problem 5.2.5:

If y is not in $FV(A)$, then $A[t/y] = A$ for every term t. Also recall that

$$FV(A) = FV(B) - \{z\}.$$

If $z = x$, since $y \notin FV(A) - \{x\} = FV(A) - \{z\}$ and z is not free in A, we know that x and y are not free in A, and

$$A[\mathbf{a}/x] = A, \ A[y/x] = A, \text{ and } A[\mathbf{a}/y] = A.$$

If $z \neq x$, then
$$A[\mathbf{a}/x] = \forall z B[\mathbf{a}/x].$$

Since y is free for x in A, we know that $y \neq z$, y is free for x in B, and

$$A[y/x][\mathbf{a}/y] = \forall z B[y/x][\mathbf{a}/y]$$

(a constant is always free for a substitution). Furthermore, since $y \notin FV(A) - \{x\}$, $FV(A) = FV(B) - \{z\}$ and $z \neq y$, we have $y \notin FV(B) - \{x\}$. By the induction hypothesis,

$$(B[\mathbf{a}/x])_{\mathbf{M}}[s] = (B[y/x][\mathbf{a}/y])_{\mathbf{M}}[s] \text{ for every } a \in M,$$

which proves that

$$(A[\mathbf{a}/x])_\mathbf{M}[s] = (A[y/x][\mathbf{a}/y])_\mathbf{M}[s] \text{ for every } a \in M.$$

(ii) This is proved using the induction principle for formulae and repeated applications of (i). $\square$

EXAMPLE 5.3.3

Let
$$A = (\forall x(Rxy \supset Px) \land \forall y(\neg Rxy \land \forall x Px)).$$

A rectified formula equivalent to A is

$$B = (\forall u(Ruy \supset Pu) \land \forall v(\neg Rxv \land \forall z Pz)).$$

From now on, we will assume that we are dealing with rectified formulae.

5.3.7 Valid Formulae Obtained by Substitution in Tautologies

As in propositional logic we will be particularly interested in those formulae that are valid. An easy way to generate valid formulae is to substitute formulae for the propositional symbols in a propositional formula.

Definition 5.3.11 Let $\mathbf{L}$ be a first-order language. Let $\mathbf{PS_0}$ be the subset of $\mathbf{PS}$ consisting of all predicate symbols of rank 0 (the propositional letters). Consider the set $PROP_\mathbf{L}$ of all $\mathbf{L}$-formulae obtained as follows: $PROP_\mathbf{L}$ consists of all $\mathbf{L}$-formulae A such that for some proposition B and some substitution

$$\sigma : \mathbf{PS_0} \to FORM_\mathbf{L}$$

assigning an arbitrary formula to each propositional letter in $\mathbf{PS_0}$,

$$A = \widehat{\sigma}(B),$$

the result of performing the substitution σ on B.

EXAMPLE 5.3.4

Let
$$B = (P \equiv Q) \equiv ((P \supset Q) \land (Q \supset P)).$$

If $\sigma(P) = \forall x C$ and $\sigma(Q) = \exists y(C \land D)$, then

$$A = (\forall x C \equiv \exists y(C \land D)) \equiv ((\forall x C \supset \exists y(C \land D)) \land (\exists y(C \land D) \supset \forall x C))$$

is of the above form.

However, note that not all **L**-formulae can be obtained by substituting formulae for propositional letters in tautologies. For example, the formulae $\forall x C$ and $\exists y(C \wedge D)$ cannot be obtained in this fashion.

The main property of the formulae in $PROP_{\mathbf{L}}$ is that those obtained by substitution into tautologies are valid. Note that not all valid formulae can be obtained in this fashion. For example, we will prove shortly that the formula $\neg \forall x A \equiv \exists x \neg A$ is valid.

Lemma 5.3.5 Let A be a formula obtained by substitution into a proposition B as explained in definition 5.3.11. If B is a tautology then A is valid.

Proof: Let $\sigma : \mathbf{PS}_0 \to FORM_{\mathbf{L}}$ be the substitution and B the proposition such that $\hat{\sigma}(B) = A$. First, we prove by induction on propositions that for every **L**-structure **M** and every assignment s, for every formula A in $PROP_{\mathbf{L}}$, if v is the valuation defined such that for every propositional letter P,

$$v(P) = (\sigma(P))_{\mathbf{M}}[s],$$

then

$$\hat{v}(B) = A_{\mathbf{M}}[s].$$

The base case $B = P$ is obvious by definition of v. If B is of the form $(C * D)$ with $* \in \{\vee, \wedge, \supset, \equiv\}$, we have

$$\hat{v}((C * D)) = H_*(\hat{v}(C), \hat{v}(D)),$$
$$\hat{\sigma}((C * D)) = (\hat{\sigma}(C) * \hat{\sigma}(D)),$$
$$(\hat{\sigma}((C * D)))_{\mathbf{M}}[s] = (\hat{\sigma}(C) * \hat{\sigma}(D))_{\mathbf{M}}[s]$$
$$= *_{\mathbf{M}}(\hat{\sigma}(C)_{\mathbf{M}}, \hat{\sigma}(D)_{\mathbf{M}})[s]$$
$$= H_*(\hat{\sigma}(C)_{\mathbf{M}}[s], \hat{\sigma}(D)_{\mathbf{M}}[s]).$$

By the induction hypothesis,

$$\hat{v}(C) = (\hat{\sigma}(C))_{\mathbf{M}}[s] \text{ and } \hat{v}(D) = (\hat{\sigma}(D))_{\mathbf{M}}[s].$$

Hence,

$$\hat{v}((C * D)) = H_*(\hat{v}(C), \hat{v}(D))$$
$$= H_*((\hat{\sigma}(C))_{\mathbf{M}}[s], (\hat{\sigma}(D))_{\mathbf{M}}[s]) = (\hat{\sigma}((C * D)))_{\mathbf{M}}[s],$$

as desired.

If B is of the form $\neg C$, then we have

$$\hat{v}(\neg C) = H_\neg(\hat{v}(C)), \ \hat{\sigma}(\neg C) = \neg \hat{\sigma}(C),$$

and

$$(\hat{\sigma}(\neg C))_{\mathbf{M}}[s] = (\neg \hat{\sigma}(C))_{\mathbf{M}}[s] = \neg_{\mathbf{M}}((\hat{\sigma}(C))_{\mathbf{M}})[s] = H_\neg((\hat{\sigma}(C))_{\mathbf{M}}[s]).$$

By the induction hypothesis,

$$\hat{v}(C) = (\hat{\sigma}(C))_\mathbf{M}[s].$$

Hence,

$$\hat{v}(\neg C) = H_\neg(\hat{v}(C)) = H_\neg((\hat{\sigma}(C))_\mathbf{M}[s]) = (\hat{\sigma}(\neg C))_\mathbf{M}[s],$$

as desired. This completes the induction proof. To conclude the lemma, observe that if B is a tautology,

$$\hat{v}(B) = \mathbf{T}.$$

Hence, by the above property, for every structure $\mathbf{M}$ and every assignment s, there is a valuation v such that $\hat{v}(B) = A_\mathbf{M}[s]$, and so $A_\mathbf{M}[s] = \mathbf{T}$. $\square$

5.3.8 Complete Sets of Connectives

As in the propositional case, the logical connectives are not independent. The following lemma shows how they are related.

Lemma 5.3.6 The following formulae are valid for all formulae A,B.

$$(A \equiv B) \equiv ((A \supset B) \wedge (B \supset A)) \tag{1}$$
$$(A \supset B) \equiv (\neg A \vee B) \tag{2}$$
$$(A \vee B) \equiv (\neg A \supset B) \tag{3}$$
$$(A \vee B) \equiv \neg(\neg A \wedge \neg B) \tag{4}$$
$$(A \wedge B) \equiv \neg(\neg A \vee \neg B) \tag{5}$$
$$\neg A \equiv (A \supset \bot) \tag{6}$$
$$\bot \equiv (A \wedge \neg A) \tag{7}$$
$$\neg \forall x A \equiv \exists x \neg A \tag{8}$$
$$\neg \exists x A \equiv \forall x \neg A \tag{9}$$
$$\forall x A \equiv \neg \exists x \neg A \tag{10}$$
$$\exists x A \equiv \neg \forall x \neg A \tag{11}$$

Proof: The validity of (1) to (7) follows from lemma 5.3.5, since these formulae are obtained by substitution in tautologies. To prove (8), recall that for any formula B,

$$(\forall x B)_\mathbf{M}[s] = \mathbf{T} \text{ iff } B_\mathbf{M}[s[x := m]] = \mathbf{T} \text{ for all } m \in M,$$

and

$$(\exists x B)_\mathbf{M}[s] = \mathbf{T} \text{ iff } B_\mathbf{M}[s[x := m]] = \mathbf{T} \text{ for some } m \in M.$$

Furthermore,

$$(\neg B)_{\mathbf{M}}[s] = \mathbf{T} \text{ iff } B_{\mathbf{M}}[s] = \mathbf{F}.$$

Hence,

$$(\neg \forall x A)_{\mathbf{M}}[s] = \mathbf{T} \text{ iff}$$
$$(\forall x A)_{\mathbf{M}}[s] = \mathbf{F} \text{ iff}$$
$$A_{\mathbf{M}}[s[x := m]] = \mathbf{F} \text{ for some } m \in M \text{ iff}$$
$$(\neg A)_{\mathbf{M}}[s[x := m]] = \mathbf{T} \text{ for some } m \in M \text{ iff}$$
$$(\exists x \neg A)_{\mathbf{M}}[s] = \mathbf{T}.$$

The proof of (9) is similar.

To prove (10), observe that

$$(\neg \forall x \neg A)_{\mathbf{M}}[s] = \mathbf{T} \text{ iff}$$
$$(\forall x \neg A)_{\mathbf{M}}[s] = \mathbf{F} \text{ iff}$$
$$(\neg A)_{\mathbf{M}}[s[x := m]] = \mathbf{F} \text{ for some } m \in M, \text{ iff}$$
$$A_{\mathbf{M}}[s[x := m]] = \mathbf{T} \text{ for some } m \in M, \text{ iff}$$
$$(\exists x A)_{\mathbf{M}}[s] = \mathbf{T}.$$

The proof of (11) is similar. $\square$

The above lemma shows that, as in the propositional case, we can restrict our attention to any functionally complete set of connectives. But it also shows that we can either dispense with $\exists$ or with $\forall$, taking $\exists x A$ as an abbreviation for $\neg \forall x \neg A$, or $\forall x A$ as an abbreviation for $\neg \exists x \neg A$.

In the rest of this text, we shall use mostly the set $\{\wedge, \vee, \neg, \supset, \forall, \exists\}$ even though it is not minimal, because it is particularly convenient and natural. Hence, $(A \equiv B)$ will be viewed as an abbreviation for $((A \supset B) \wedge (B \supset A))$ and $\bot$ as an abbreviation for $(P \wedge \neg P)$.

5.3.9 Logical Equivalence and Boolean Algebras

As in the propositional case, we can also define the notion of logical equivalence for formulae.

Definition 5.3.12 The relation $\simeq$ on $FORM_{\mathbf{L}}$ is defined so that for any two formulae A and B,

$$A \simeq B \quad \text{if and only if} \quad \models (A \equiv B).$$

We say that A and B are *logically equivalent*, or for short, *equivalent*.

It is immediate to show that $\simeq$ is an equivalence relation. The following additional properties show that it is a congruence in the sense of Subsection 2.4.6.

Lemma 5.3.7 For all formulae A, A', B, B', the following properties hold: If $A \simeq A'$ and $B \simeq B'$ then, for $* \in \{\wedge, \vee, \supset, \equiv\}$,

$$(A * B) \simeq (A' * B'),$$
$$\neg A \simeq \neg A',$$
$$\forall x A \simeq \forall x A' \text{ and}$$
$$\exists x A \simeq \exists x A'.$$

Proof: First, we note that a formula $(A \equiv B)$ is valid iff for every structure $\mathbf{M}$ and every assignment s,

$$A_{\mathbf{M}}[s] = B_{\mathbf{M}}[s].$$

Then, the proof is similar to the proof of the propositional case (lemma 3.3.5) for formulae of the form $(A * B)$ or $\neg A$.

Since

$$(\forall x A)_{\mathbf{M}}[s] = \mathbf{T} \text{ iff } A_{\mathbf{M}}[s[x := m]] = \mathbf{T} \text{ for all } m \in M,$$

and $A \simeq A'$ implies that

$$A_{\mathbf{M}}[s] = A'_{\mathbf{M}}[s] \text{ for all } \mathbf{M} \text{ and } s,$$

then

$$A_{\mathbf{M}}[s[x := m]] = \mathbf{T} \text{ for all } m \in M \text{ iff}$$
$$A'_{\mathbf{M}}[s[x := m]] = \mathbf{T} \text{ for all } m \in M \text{ iff}$$
$$(\forall x A')_{\mathbf{M}}[s] = \mathbf{T}.$$

Hence $\forall x A$ and $\forall x A'$ are equivalent. The proof that if $A \simeq A'$ then $\exists x A \simeq \exists x A'$ is similar. $\square$

In the rest of this section, it is assumed that the constant symbol $\top$ is added to the alphabet of definition 5.2.1, and that $\top$ is interpreted as $\mathbf{T}$. By lemma 5.3.5 and lemma 3.3.6, the following identities hold.

Lemma 5.3.8 The following identities hold.

Associativity rules:
$$((A \vee B) \vee C) \simeq (A \vee (B \vee C)) \quad ((A \wedge B) \wedge C) \simeq (A \wedge (B \wedge C))$$
Commutativity rules:
$$(A \vee B) \simeq (B \vee A) \quad (A \wedge B) \simeq (B \wedge A)$$
Distributivity rules:
$$(A \vee (B \wedge C)) \simeq ((A \vee B) \wedge (A \vee C))$$

$$(A \wedge (B \vee C)) \simeq ((A \wedge B) \vee (A \wedge C))$$

De Morgan's rules:

$$\neg(A \vee B) \simeq (\neg A \wedge \neg B) \quad \neg(A \wedge B) \simeq (\neg A \vee \neg B)$$

Idempotency rules:

$$(A \vee A) \simeq A \quad (A \wedge A) \simeq A$$

Double negation rule:

$$\neg\neg A \simeq A$$

Absorption rules:

$$(A \vee (A \wedge B)) \simeq A \quad (A \wedge (A \vee B)) \simeq A$$

Laws of zero and one:

$$(A \vee \perp) \simeq A \quad (A \wedge \perp) \simeq \perp$$
$$(A \vee \top) \simeq \top \quad (A \wedge \top) \simeq A$$
$$(A \vee \neg A) \simeq \top \quad (A \wedge \neg A) \simeq \perp$$

$\square$

Let us denote the equivalence class of a proposition A modulo $\simeq$ as $[A]$, and the set of all such equivalence classes as $\mathbf{B_L}$. As in Chapter 3, we define the operations $+$, $*$ and $\overline{}$ on $\mathbf{B_L}$ as follows:

$$[A] + [B] = [A \vee B],$$
$$[A] * [B] = [A \wedge B],$$
$$\overline{[A]} = [\neg A].$$

Also, let $0 = [\perp]$ and $1 = [\top]$.

By lemma 5.3.7, the above functions (and constants) are independent of the choice of representatives in the equivalence classes, and the properties of lemma 5.3.8 are identities valid on the set $\mathbf{B_L}$ of equivalence classes modulo $\simeq$. The structure $\mathbf{B_L}$ is a *boolean algebra* called the *Lindenbaum algebra* of $\mathbf{L}$. This algebra is important for studying algebraic properties of formulae. Some of these properties will be investigated in the problems of the next section.

Remark: Note that properties of the quantifiers and of equality are not captured in the axioms of a boolean algebra. There is a generalization of the notion of a boolean algebra, the *cylindric algebra*, due to Tarski. Cylindric algebras have axioms for the existential quantifier and for equality. However, this topic is beyond the scope of this text. The interested reader is referred to Henkin, Monk, and Tarski, 1971.

PROBLEMS

5.3.1.   Prove that the following formulae are valid:

$$\forall x A \supset A[t/x], \quad A[t/x] \supset \exists x A,$$
$$\text{where } t \text{ is free for } x \text{ in } A.$$

5.3.2.   Let x, y be any distinct variables. Let A, B be any formulae, C, D any formulae not containing the variable x free, and let E be any formula such that x is free for y in E. Prove that the following formulae are valid:

$$\forall x C \equiv C \qquad\qquad \exists x C \equiv C$$
$$\forall x \forall y A \equiv \forall y \forall x A \qquad\qquad \exists x \exists y A \equiv \exists y \exists x A$$
$$\forall x \forall y E \supset \forall x E[x/y] \qquad\qquad \exists x E[x/y] \supset \exists x \exists y E$$
$$\forall x A \supset \exists x A$$
$$\exists x \forall y A \supset \forall y \exists x A$$

5.3.3.   Let A, B be any formulae, and C any formula not containing the variable x free. Prove that the following formulae are valid:

$$\neg \exists x A \equiv \forall x \neg A \qquad\qquad \neg \forall x A \equiv \exists x \neg A$$
$$\exists x A \equiv \neg \forall x \neg A \qquad\qquad \forall x A \equiv \neg \exists x \neg A$$
$$\forall x A \wedge \forall x B \equiv \forall x (A \wedge B) \qquad \exists x A \vee \exists x B \equiv \exists x (A \vee B)$$
$$C \wedge \forall x A \equiv \forall x (C \wedge D) \qquad C \vee \exists x A \equiv \exists x (C \vee A)$$
$$C \wedge \exists x A \equiv \exists x (C \wedge D) \qquad C \vee \forall x A \equiv \forall x (C \vee A)$$
$$\exists x (A \wedge B) \supset \exists x A \wedge \exists x B \qquad \forall x A \vee \forall x B \supset \forall x (A \vee B)$$

5.3.4.   Let A, B be any formulae, and C any formula not containing the variable x free. Prove that the following formulae are valid:

$$(C \supset \forall x A) \equiv \forall x (C \supset A) \qquad (C \supset \exists x A) \equiv \exists x (C \supset A)$$
$$(\forall x A \supset C) \equiv \exists x (A \supset C) \qquad (\exists x A \supset C) \equiv \forall x (A \supset C)$$
$$(\forall x A \supset \exists x B) \equiv \exists x (A \supset B)$$
$$(\exists x A \supset \forall x B) \supset \forall x (A \supset B)$$

5.3.5.   Prove that the following formulae are not valid:

$$A[t/x] \supset \forall x A, \quad \exists x A \supset A[t/x],$$
$$\text{where } t \text{ is free for } x \text{ in } A.$$

5.3.6.   Show that the following formulae are not valid:

$$\exists x A \supset \forall x A$$
$$\forall y \exists x A \supset \exists x \forall y A$$
$$\exists x A \wedge \exists x B \supset \exists x (A \wedge B)$$
$$\forall x (A \vee B) \supset \forall x A \vee \forall x B$$

5.3.7. Prove that the formulae of problem 5.3.4 are not necessarily valid if x is free in C.

5.3.8. Let A be any formula and B any formula in which the variable x does not occur free.

(a) Prove that

$$\text{if } \models (B \supset A), \text{ then } \models (B \supset \forall x A),$$

and

$$\text{if } \models (A \supset B), \text{ then } \models (\exists x A \supset B).$$

(b) Prove that the above may be false if x is free in B.

5.3.9. Let $\mathbf{L}$ be a first-order language, and $\mathbf{M}$ be an $\mathbf{L}$-structure. For any formulae A, B, prove that:

(a) $\mathbf{M} \models A \supset B$ implies that $(\mathbf{M} \models A$ implies $\mathbf{M} \models B)$,

but not vice versa.

(b) $A \models B$ implies that $(\models A$ implies $\models B)$,

but not vice versa.

5.3.10. Given a formula A with set of free variables $FV(A) = \{x_1, ..., x_n\}$, the *universal closure* A' of A is the sentence $\forall x_1 ... \forall x_n A$. Prove that A is valid iff $\forall x_1 ... \forall x_n A$ is valid.

5.3.11. Let $\mathbf{L}$ be a first-order language, Γ a set of formulae, and B some formula. Let Γ' be the set of universal closures of formulae in Γ.

Prove that
$$\Gamma \models B \text{ implies that } \Gamma' \models B,$$

but not vice versa.

5.3.12. Let $\mathbf{L}$ be the first-order language with equality consisting of one unary function symbol f. Write formulae asserting that for every $\mathbf{L}$-structure $\mathbf{M}$:

(a) f is injective

(b) f is surjective

(c) f is bijective

5.3.13. Let $\mathbf{L}$ be a first-order language with equality.

Find a formula asserting that any $\mathbf{L}$-structure $\mathbf{M}$ has at least n elements.

5.3.14. Prove that the following formulae are valid:

$$\forall x \exists y (x \doteq y)$$
$$A[t/x] \equiv \forall x((x \doteq t) \supset A), \text{ if } x \text{ is not free in } t.$$
$$A[t/x] \equiv \exists x((x \doteq t) \wedge A), \text{ if } x \text{ is not free in } t.$$

$*$ **5.3.15.** Let $\mathbf{A}$ and $\mathbf{B}$ be two **L**-structures. A surjective function $h : A \to B$ is a *homomorphism* of $\mathbf{A}$ onto $\mathbf{B}$ if:

(i) For every n-ary function symbol f, for every $(a_1, ..., a_n) \in A^n$,

$$h(f_{\mathbf{A}}(a_1, ..., a_n)) = f_{\mathbf{B}}(h(a_1), ..., h(a_n));$$

(ii) For every constant symbol c,

$$h(c_{\mathbf{A}}) = c_{\mathbf{B}};$$

(iii) For every n-ary predicate symbol P, for every $(a_1, ..., a_n) \in A^n$,

$$\text{if } \quad P_{\mathbf{A}}(a_1, ..., a_n) = \mathbf{T} \quad \text{then} \quad P_{\mathbf{B}}(h(a_1), ..., h(a_n)) = \mathbf{T}.$$

Let t be a term containing the free variables $\{y_1, ..., y_n\}$, and A be a formula with free variables $\{y_1, ..., y_n\}$. For any assignment s whose restriction to $\{y_1, ..., y_n\}$ is given by $s(y_i) = a_i$, the notation $t[a_1, ..., a_n]$ is equivalent to $t[s]$, and $A[a_1, ..., a_n]$ is equivalent to $A[s]$.

(a) Prove that for any term t, if t contains the variables $\{y_1, ..., y_n\}$, for every $(a_1, ..., a_n) \in A^n$, if h is a homomorphism from $\mathbf{A}$ onto $\mathbf{B}$, then

$$h(t_{\mathbf{A}}[a_1, ..., a_n]) = t_{\mathbf{B}}[h(a_1), ..., h(a_n)],$$

even if h is not surjective.

A *positive formula* is a formula built up from atomic formulae (excluding $\perp$), using only the connectives $\wedge$, $\vee$ and the quantifiers $\forall$, $\exists$.

(b) Prove that for any positive formula X, for every $(a_1, ..., a_n) \in A^n$,

$$\text{if } \quad \mathbf{A} \models X[a_1, ..., a_n] \quad \text{then} \quad \mathbf{B} \models X[h(a_1), ..., h(a_n)],$$

where h is a homomorphism from $\mathbf{A}$ onto $\mathbf{B}$ (we say that positive formulae are preserved under homomorphisms).

An *isomorphism* between $\mathbf{A}$ and $\mathbf{B}$ is a bijective homomorphism $h : \mathbf{A} \to \mathbf{B}$ such that for any n-ary predicate P and any $(a_1, ..., a_n) \in A^n$,

$$P_{\mathbf{A}}(a_1, ..., a_n) = \mathbf{T} \quad \text{iff} \quad P_{\mathbf{B}}(h(a_1), ..., h(a_n)) = \mathbf{T}.$$

(c) If **L** is a first-order language without equality, prove that for any formula X with free variables $\{y_1, ..., y_n\}$, for any $(a_1, ..., a_n) \in A^n$, if h is a homomorphism of **A** onto **B** then

$$\mathbf{A} \models X[a_1, ..., a_n] \quad \text{iff} \quad \mathbf{B} \models X[h(a_1), ..., h(a_n)].$$

(d) If **L** is a first-order language with or without equality, prove that for any formula X with free variables $\{y_1, ..., y_n\}$, for any $(a_1, ..., a_n) \in A^n$, if h is an isomorphism between **A** and **B** then

$$\mathbf{A} \models X[a_1, ..., a_n] \quad \text{iff} \quad \mathbf{B} \models X[h(a_1), ..., h(a_n)].$$

(e) Find two structures, a homomorphism h between them, and a nonpositive formula X that is not preserved under h.

5.3.16. Let A be the sentence:

$$\forall x \neg R(x, x) \wedge \forall x \forall y \forall z (R(x, y) \wedge R(y, z) \supset R(x, z)) \wedge$$
$$\forall x \exists y R(x, y).$$

Give an infinite model for A and prove that A has *no* finite model.

* **5.3.17.** The *monadic predicate calculus* is the language **L** with no equality having only unary predicate symbols, and no function or constant symbols. Let **M** be an **L**-structure having n distinct unary predicates $Q_1,...,Q_n$ on M. Define the relation $\cong$ on M as follows:

$$a \cong b \quad \text{iff}$$
$$Q_i(a) = \mathbf{T} \quad \text{iff} \quad Q_i(b) = \mathbf{T}, \text{ for all } i, 1 \leq i \leq n.$$

(a) Prove that $\cong$ is an equivalence relation having at most 2^n equivalence classes.

Let $\mathbf{M}/\cong$ be the structure having the set $M/\cong$ of equivalence classes modulo $\cong$ as its domain, and the unary predicates $R_1,...,R_n$ such that, for every equivalence class $\bar{x}$,

$$R_i(\bar{x}) = \mathbf{T} \quad \text{iff} \quad Q_i(x) = \mathbf{T}.$$

(b) Prove that for every **L**-formula A containing only predicate symbols in the set $\{P_1, ..., P_n\}$,

$$\mathbf{M} \models A \quad \text{iff} \quad \mathbf{M}/\cong \models A,$$

and that

$$\models A \quad \text{iff}$$
$$\mathbf{M} \models A \quad \text{for all } \mathbf{L}\text{-structures } \mathbf{M} \text{ with at most } 2^n \text{ elements.}$$

(c) Using part (b), outline an algorithm for deciding validity in the monadic predicate calculus.

5.3.18. Give a detailed rectification algorithm (see lemma 5.3.4(ii)).

5.3.19. Let **A** and **B** be two **L**-structures. The structure **B** is a *substructure* of the structure **A** iff the following conditions hold:

(i) The domain B is a subset of the domain A;

(ii) For every constant symbol c,

$$c_\mathbf{B} = c_\mathbf{A};$$

(iii) For every function symbol f of rank $n > 0$, $f_\mathbf{B}$ is the restriction of $f_\mathbf{A}$ to B^n, and for every predicate symbol P of rank $n \geq 0$, $P_\mathbf{B}$ is the restriction of $P_\mathbf{A}$ to B^n.

A universal sentence is a sentence of the form

$$\forall x_1 ... \forall x_m B,$$

where B is quantifier free. Prove that if **A** is a model of a set Γ of universal sentences, then **B** is also a model of Γ.

* **5.3.20.** In this problem, some properties of reduced products are investigated. We are considering first-order languages with or without equality.

The notion of a reduced product depends on that of a filter, and the reader is advised to consult problem 3.5.8 for the definition of filters and ultrafilters.

Let I be a nonempty set which will be used as an index set, and let D be a proper filter over I. Let $(A_i)_{i \in I}$ be an I-indexed family of nonempty sets. The Cartesian product C of these sets, denoted by

$$\prod_{i \in I} A_i$$

is the set of all I-indexed sequences

$$f : I \to \bigcup_{i \in I} A_i$$

such that, for each $i \in I$, $f(i) \in A_i$. Such I-sequences will also be denoted as $< f(i) \mid i \in I >$.

The relation $=_D$ on C is defined as follows:

For any two functions $f, g \in C$,

$$f =_D g \quad \text{iff} \quad \{i \in I \mid f(i) = g(i)\} \in D.$$

In words, f and g are related iff the set of indices on which they "agree" belongs to the proper filter D.

(a) Prove that $=_D$ is an equivalence relation.

(b) Let $\mathbf{L}$ be a first-order language (with or without equality). For each $i \in I$, let $\mathbf{A}_i$ be an $\mathbf{L}$-structure. We define the *reduced product* $\mathbf{B}$ of the $(A_i)_{i \in I}$ modulo D, as the $\mathbf{L}$-structure defined as follows:

(i) The domain of $\mathbf{B}$ the set of equivalence classes of

$$\prod_{i \in I} A_i$$

modulo $=_D$.

(ii) For every constant symbol c, c is interpreted in $\mathbf{B}$ as the equivalence class

$$[< c_{\mathbf{A}_i} \mid i \in I >]_D.$$

(iii) For every function symbol f of rank $n > 0$, f is interpreted as the function such that, for any n equivalence classes $G^1 = [< g^1(i) \mid i \in I >]_D,...,G^n = [< g^n(i) \mid i \in I >]_D$,

$$f_{\mathbf{B}}(G^1,...,G^n) = [< f_{\mathbf{A}_i}(g^1(i),...,g^n(i)) \mid i \in I >]_D.$$

(iv) For every predicate symbol P of rank $n \geq 0$, P is interpreted as the predicate such that, for any n equivalence classes $G^1 = [< g^1(i) \mid i \in I >]_D,...,G^n = [< g^n(i) \mid i \in I >]_D$,

$$P_{\mathbf{B}}(G^1,...,G^n) = \mathbf{T} \quad \text{iff} \quad \{i \in I \mid P_{\mathbf{A}_i}(g^1(i),...,g^n(i)) = \mathbf{T}\} \in D.$$

The reduced product $\mathbf{B}$ is also denoted by

$$\prod_D (A_i)_{i \in I}.$$

(c) Prove that $=_D$ is a congruence; that is, that definitions (ii) to (iv) are independent of the representatives chosen in the equivalence classes $G^1,...,G^n$.

** **5.3.21.** Let $\mathbf{B}$ be the reduced product

$$\prod_D (A_i)_{i \in I}$$

as defined in problem 5.3.20. When D is an ultrafilter, $\mathbf{B}$ is called an *ultraproduct*.

(a) Prove that for every term t with free variables $\{y_1, ..., y_n\}$, for any n equivalence classes $G^1 = [< g^1(i) \mid i \in I >]_D, ..., G^n = [< g^n(i) \mid i \in I >]_D$ in $\mathbf{B}$,

$$t_{\mathbf{B}}[G^1, ..., G^n] = [< t_{\mathbf{A}_i}[g^1(i), ..., g^n(i)] \mid \ \in I >]_D.$$

(b) Let D be an ultrafilter. Prove that for any formula A with free variables $FV(A) = \{y_1, ..., y_n\}$, for any n equivalence classes $G^1 = [< g^1(i) \mid i \in I >]_D, ..., G^n = [< g^n(i) \mid i \in I >]_D$ in $\mathbf{B}$,

$$\mathbf{B} \models A[G^1, ..., G^n] \quad \text{iff} \quad \{i \in I \mid \mathbf{A}_i \models A[g^1(i), ..., g^n(i)]\} \in D.$$

(c) Prove that if D is an ultrafilter, for every sentence A,

$$\mathbf{B} \models A \quad \text{iff} \quad \{i \in I \mid \mathbf{A}_i \models A\} \in D.$$

Hint: Proceed by induction on formulae. The fact that D is an ultrafilter is needed in the case where A is of the form $\neg B$.

* **5.3.22.** This problem generalizes problem 3.5.10 to first-order logic. It provides a proof of the compactness theorem for first-order languages of any cardinality.

(a) Let Γ be a set of sentences such that every finite subset of Γ is satisfiable. Let I be the set of all finite subsets of Γ, and for each $i \in I$, let $\mathbf{A}_i$ be a structure satisfying i. For each sentence $A \in \Gamma$, let

$$A^* = \{i \in I \mid A \in i\}.$$

Let

$$\mathcal{C} = \{A^* \mid A \in \Gamma\}.$$

Note that $\mathcal{C}$ has the finite intersection property since

$$\{A_1, ..., A_n\} \in A_1^* \cap ... \cap A_n^*.$$

By problem 3.5.8, there is an ultrafilter U including $\mathcal{C}$, so that every A^* is in U. If $i \in A^*$, then $A \in i$, and so

$$\mathbf{A}_i \models A.$$

Thus, for every A in Γ, A^* is a subset of the set $\{i \in I \mid \mathbf{A}_i \models A\}$.

Show that

$$\{i \in I \mid \mathbf{A}_i \models A\} \in U.$$

(b) Show that the ultraproduct $\mathbf{B}$ (defined in problem 5.3.20) satisfies Γ.

** **5.3.23.** Given a first-order language **L**, a *basic Horn formula* is a disjunction of atomic formulae, in which at most one atomic formula is negated. The class of *Horn formulae* is the least class of formulae containing the basic Horn formulae, and such that:

(i) If A and B are Horn formulae, so is $A \wedge B$;

(ii) If A is a Horn formula, so is $\forall x A$;

(iii) If A is a Horn formula, so is $\exists x A$.

A Horn sentence is a closed Horn formula.

Let A be a Horn formula, and let $(\mathbf{A}_i)_{i \in I}$ be an I-indexed family of structures satisfying A. Let D be a proper filter over D, and let $\mathbf{B}$ be the reduced product of $(\mathbf{A}_i)_{i \in I}$ modulo $=_D$.

(a) Prove that for any n equivalence classes $G^1 = [< g^1(i) \mid i \in I >]_D,...,G^n = [< g^n(i) \mid i \in I >]_D$ in $\mathbf{B}$,

$$\text{if} \quad \{i \in I \mid \mathbf{A}_i \models A[g^1(i), ..., g^n(i)]\} \in D \quad \text{then}$$
$$\mathbf{B} \models A[G^1, ..., G^n].$$

(b) Given a Horn sentence A,

$$\text{if} \quad \mathbf{A}_i \models A \quad \text{for all} \in I, \text{ then} \quad \mathbf{B} \models A.$$

(We say that Horn sentences are preserved under reduced products.)

(c) If we let the filter D be the family of sets $\{I\}$, the reduced product $\mathbf{B}$ is called the *direct product* of $(A_i)_{i \in I}$.

Show that the formula A below (which is not a Horn formula) is preserved under direct products, but is not preserved under reduced products:

A is the conjunction of the boolean algebra axioms plus the sentence

$$\exists x \forall y ((x \neq 0) \wedge (x * y = y \supset (y = x \vee y = 0))).$$

* **5.3.24.** Show that in the proof of the compactness theorem given in problem 5.3.22, if Γ is a set of Horn formulae, it is not necessary to extend C to an ultrafilter, but simply to take the filter generated by C.

* **5.3.25.** Let **L** be a first-order language. Two **L**-structures $\mathbf{M}_1$ and $\mathbf{M}_2$ are *elementary equivalent* iff for every sentence A,

$$\mathbf{M}_1 \models A \quad \text{iff} \quad \mathbf{M}_2 \models A.$$

Prove that if there is an isomorphism between $\mathbf{M}_1$ and $\mathbf{M}_2$, then $\mathbf{M}_1$ and $\mathbf{M}_2$ are elementary equivalent.

Note: The converse of the above statement holds if one of the structures has a finite domain. However, there are infinite structures that are elementary equivalent but are not isomorphic. For example, if the language **L** has a unique binary predicate $<$, the structure $\mathcal{Q}$ of the rational numbers and the structure $\mathcal{R}$ of the real numbers (with $<$ interpreted as the strict order of each structure) are elementary equivalent, but nonisomorphic since $\mathcal{Q}$ is countable but $\mathcal{R}$ is not.

5.4 Proof Theory of First-Order Languages

In this section, the Gentzen system G' is extended to first-order logic.

5.4.1 The Gentzen System G for Languages Without Equality

We first consider the case of a first-order language **L** *without equality*. In addition to the rules for the logical connectives, we need four new rules for the quantifiers. The rules presented below are designed to make the search for a counter-example as simple as possible. In the following sections, sequents are defined as before but are composed of formulae instead of propositions. Furthermore, it is assumed that the set of connectives is $\{\wedge, \vee, \supset, \neg, \forall, \exists\}$. By lemma 5.3.7, there is no loss of generality. Hence, when $\equiv$ is used, $(A \equiv B)$ is considered as an abbreviation for $((A \supset B) \wedge (B \supset A))$.

Definition 5.4.1 (Gentzen system G for languages without equality) The symbols Γ, Δ, Λ will be used to denote arbitrary sequences of formulae and A, B to denote formulae. The rules of the sequent calculus G are the following:

$$\frac{\Gamma, A, B, \Delta \to \Lambda}{\Gamma, A \wedge B, \Delta \to \Lambda} \; (\wedge : left) \qquad \frac{\Gamma \to \Delta, A, \Lambda \quad \Gamma \to \Delta, B, \Lambda}{\Gamma \to \Delta, A \wedge B, \Lambda} \; (\wedge : right)$$

$$\frac{\Gamma, A, \Delta \to \Lambda \quad \Gamma, B, \Delta \to \Lambda}{\Gamma, A \vee B, \Delta \to \Lambda} \; (\vee : left) \qquad \frac{\Gamma \to \Delta, A, B, \Lambda}{\Gamma \to \Delta, A \vee B, \Lambda} \; (\vee : right)$$

$$\frac{\Gamma, \Delta \to A, \Lambda \quad B, \Gamma, \Delta \to \Lambda}{\Gamma, A \supset B, \Delta \to \Lambda} \; (\supset : left) \qquad \frac{A, \Gamma \to B, \Delta, \Lambda}{\Gamma \to \Delta, A \supset B, \Lambda} \; (\supset : right)$$

$$\frac{\Gamma, \Delta \to A, \Lambda}{\Gamma, \neg A, \Delta \to \Lambda} \; (\neg : left) \qquad \frac{A, \Gamma \to \Delta, \Lambda}{\Gamma \to \Delta, \neg A, \Lambda} \; (\neg : right)$$

In the quantifier rules below, x is any variable and y is any variable free for x in A and not free in A, unless $y = x$ ($y \notin FV(A) - \{x\}$). The term t is any term free for x in A.

$$\frac{\Gamma, A[t/x], \forall x A, \Delta \to \Lambda}{\Gamma, \forall x A, \Delta \to \Lambda} \ (\forall : left) \qquad \frac{\Gamma \to \Delta, A[y/x], \Lambda}{\Gamma \to \Delta, \forall x A, \Lambda} \ (\forall : right)$$

$$\frac{\Gamma, A[y/x], \Delta \to \Lambda}{\Gamma, \exists x A, \Delta \to \Lambda} \ (\exists : left) \qquad \frac{\Gamma \to \Delta, A[t/x], \exists x A, \Lambda}{\Gamma \to \Delta, \exists x A, \Lambda} \ (\exists : right)$$

Note that in both the $(\forall : right)$-rule and the $(\exists : left)$-rule, the variable y does *not* occur free in the lower sequent. In these rules, the variable y is called the *eigenvariable* of the inference. The condition that the eigenvariable does not occur free in the conclusion of the rule is called the *eigenvariable condition*. The formula $\forall x A$ (or $\exists x A$) is called the *principal formula* of the inference, and the formula $A[t/x]$ (or $A[y/x]$) the *side formula* of the inference.

The *axioms* of G are all sequents $\Gamma \to \Delta$ such that Γ and Δ contain a common formula.

5.4.2 Deduction Trees for the System G

First, we define when a sequent is falsifiable or valid.

Definition 5.4.2 (i) A sequent $A_1, ..., A_m \to B_1, ..., B_n$ is *falsifiable* iff for some structure **M** and some assignment s:

$$\mathbf{M} \models (A_1 \wedge ... \wedge A_m \wedge \neg B_1 \wedge ... \wedge \neg B_n)[s].$$

Note that when $m = 0$ the condition reduces to

$$\mathbf{M} \models (\neg B_1 \wedge ... \wedge \neg B_n)[s]$$

and when $n = 0$ the condition reduces to

$$\mathbf{M} \models (A_1 \wedge ... \wedge A_m)[s].$$

(ii) A sequent $A_1, ..., A_m \to B_1, ..., B_n$ is *valid* iff for every structure **M** and every assignment s:

$$\mathbf{M} \models (\neg A_1 \vee ... \vee \neg A_m \vee B_1 \vee ... \vee B_n)[s].$$

This is denoted by

$$\models A_1, ..., A_m \to B_1, ..., B_n.$$

Note that a sequent is valid if and only if it is not falsifiable.

Deduction trees and *proof trees* for the system G are defined as in definition 3.4.5, but for formulae and with the rules given in definition 5.4.1.

EXAMPLE 5.4.1

Let $A = \exists x(P \supset Q(x)) \supset (P \supset \exists z Q(z))$, where P is a propositional symbol and Q a unary predicate symbol.

$$
\cfrac{
    P \to P, \exists z Q(z) \qquad
    \cfrac{
        \cfrac{Q(y_1), P \to Q(y_1), \exists z Q(z)}{Q(y_1), P \to \exists z Q(z)}
    }{
    }
}{
\cfrac{
\cfrac{
\cfrac{
\cfrac{P, P \supset Q(y_1) \to \exists z Q(z)}{P \supset Q(y_1) \to P \supset \exists z Q(z)}
}{\exists x(P \supset Q(x)) \to P \supset \exists z Q(z)}
}{\to \exists x(P \supset Q(x)) \supset (P \supset \exists z Q(z))}
}{}
}
$$

The above is a proof tree for A. Note that $P[y_1/x] = P$.

EXAMPLE 5.4.2

Let $A = (\forall x P(x) \wedge \exists y Q(y)) \supset (P(f(v)) \wedge \exists z Q(z))$, where P and Q are unary predicate symbols, and f is a unary function symbol. The tree below is a proof tree for A.

$$
\Pi \qquad
\cfrac{
\cfrac{
\cfrac{
\cfrac{P(f(v)), \forall x P(x), Q(u) \to Q(u), \exists z Q(z)}{P(f(v)), \forall x P(x), Q(u) \to \exists z Q(z)}
}{P(f(v)), \forall x P(x), \exists y Q(y) \to \exists z Q(z)}
}{P(f(v)), \forall x P(x), \exists y Q(y) \to P(f(v)) \wedge \exists z Q(z)}
}{
\cfrac{
\cfrac{
\cfrac{\forall x P(x), \exists y Q(y) \to P(f(v)) \wedge \exists z Q(z)}{\forall x P(x) \wedge \exists y Q(y) \to P(f(v)) \wedge \exists z Q(z)}
}{\to (\forall x P(x) \wedge \exists y Q(y)) \supset (P(f(v)) \wedge \exists z Q(z))}
}{}
}
$$

where $\Pi = P(f(v)), \forall x P(x), \exists y Q(y) \to P(f(v))$.

5.4.3 Soundness of the System G

In order to establish the soundness of the system G, the following lemmas will be needed.

Lemma 5.4.1 (i) For any two terms t and r,

$$
(r[t/x])_{\mathbf{M}}[s] = (r[\mathbf{a}/x])_{\mathbf{M}}[s], \text{ where } a = t_{\mathbf{M}}[s].
$$

(ii) If t is free for x in A, then

$$(A[t/x])_{\mathbf{M}}[s] = (A[\mathbf{a}/x])_{\mathbf{M}}[s], \text{ where } a = t_{\mathbf{M}}[s].$$

Proof: The lemma is shown using the induction principle for terms and formulae as in lemma 5.3.1. The proof of (i) is left as an exercise. Since the proof of (ii) is quite similar to the proof of lemma 5.3.1, we only treat one case, leaving the others as an exercise.

Assume $A = \forall z B$. If $z = x$, then $A[t/x] = A$, and $A[\mathbf{a}/x] = A$. If $x \neq z$, since t is free for x in A, $z \notin FV(t)$, t is free for x in B,

$$A[t/x] = \forall z B[t/x] \text{ and } A[\mathbf{a}/x] = \forall z B[\mathbf{a}/x].$$

By problem 5.2.6, the following property holds: If t is free for x in B, $x \neq z$ and z is not in $FV(t)$, for every constant d,

$$B[t/x][d/z] = B[d/z][t/x].$$

Then, for every $c \in M$, by the induction hypothesis,

$$(B[t/x][\mathbf{c}/z])_{\mathbf{M}}[s] = (B[\mathbf{c}/z][t/x])_{\mathbf{M}}[s]$$
$$= (B[\mathbf{c}/z][\mathbf{a}/x])_{\mathbf{M}}[s] = (B[\mathbf{a}/x][\mathbf{c}/z])_{\mathbf{M}}[s],$$

with $\mathbf{a} = t_{\mathbf{M}}[s]$. But this proves that

$$(\forall z B[t/x])_{\mathbf{M}}[s] = (\forall z B[\mathbf{a}/x])_{\mathbf{M}}[s],$$

as desired. $\square$

Lemma 5.4.2 For any rule stated in definition 5.4.1, the conclusion is falsifiable in some structure $\mathbf{M}$ if and only if one of the premises is falsifiable in the structure $\mathbf{M}$. Equivalently, the conclusion is valid if and only if all premises are valid.

Proof: We prove the lemma for the rules $\forall : right$ and $\forall : left$, leaving the others as an exercise. In this proof, the following abbreviating conventions will be used: If Γ (or Δ) is the antecedent of a sequent,

$$\mathbf{M} \models \Gamma[s]$$

means that

$$\mathbf{M} \models A[s]$$

for every formula $A \in \Gamma$, and if Δ (or Λ) is the succedent of a sequent,

$$\mathbf{M} \not\models \Delta[s]$$

means that

$$\mathbf{M} \not\models B[s]$$

for every formula $B \in \Delta$.

(i) Assume that $\Gamma \rightarrow \Delta, A[y/x], \Lambda$ is falsifiable. This means that there is a structure $\mathbf{M}$ and an assignment s such that

$$\mathbf{M} \models \Gamma[s], \ \mathbf{M} \not\models \Delta[s],$$
$$\mathbf{M} \not\models (A[y/x])[s] \text{ and } \mathbf{M} \not\models \Lambda[s].$$

Recall that

$$\mathbf{M} \models (\forall x A)[s]$$

iff

$$\mathbf{M} \models (A[\mathbf{a}/x])[s] \text{ for every } a \in M.$$

Since y is free for x in A, by lemma 5.4.1, for $a = s(y)$, we have

$$(A[y/x])_{\mathbf{M}}[s] = (A[\mathbf{a}/x])_{\mathbf{M}}[s].$$

Hence,

$$\mathbf{M} \not\models (\forall x A)[s],$$

which implies that $\Gamma \rightarrow \Delta, \forall x A, \Lambda$ is falsified by $\mathbf{M}$ and s.

Conversely, assume that $\mathbf{M}$ and s falsify $\Gamma \rightarrow \Delta, \forall x A, \Lambda$. In particular, there is some $a \in M$ such that

$$\mathbf{M} \not\models (A[\mathbf{a}/x])[s].$$

Let s' be the assignment $s(y := a)$. Since $y \notin FV(A) - \{x\}$ and y is free for x in A, by lemma 5.3.3 and lemma 5.4.1, then

$$\mathbf{M} \not\models (A[y/x])[s'].$$

Since y does not appear free in Γ, Δ or Λ, by lemma 5.3.3, we also have

$$\mathbf{M} \models \Gamma[s'], \ \mathbf{M} \not\models \Delta[s'] \text{ and } \mathbf{M} \not\models \Lambda[s'].$$

Hence, $\mathbf{M}$ and s' falsify $\Gamma \rightarrow \Delta, A[y/x], \Lambda$.

(ii) Assume that $\mathbf{M}$ and s falsify $\Gamma, A[t/x], \forall x A, \Delta \rightarrow \Lambda$. Then, it is clear that $\mathbf{M}$ and s also falsify $\Gamma, \forall x A, \Delta \rightarrow \Lambda$.

Conversely, assume that $\mathbf{M}$ and s falsify $\Gamma, \forall x A, \Delta \rightarrow \Lambda$. Then,

$$\mathbf{M} \models \Gamma[s], \ \mathbf{M} \models (\forall x A)[s],$$
$$\mathbf{M} \models \Delta[s] \text{ and } \mathbf{M} \not\models \Lambda[s].$$

In particular,

$$\mathbf{M} \models (A[\mathbf{a}/x])[s] \text{ for every } a \in M.$$

By lemma 5.4.1, for $a = t_\mathbf{M}[s]$, we have

$$(A[t/x])_\mathbf{M}[s] = (A[\mathbf{a}/x])_\mathbf{M}[s],$$

and so

$$\mathbf{M} \models (A[t/x])[s].$$

Hence, $\mathbf{M}$ and s also falsify $\Gamma, A[t/x], \forall x A, \Delta \rightarrow \Lambda$. $\square$

As a consequence, we obtain the soundness of the system G.

Lemma 5.4.3 (Soundness of G) Every sequent provable in G is valid.

Proof: Use the induction principle for G-proofs and the fact that the axioms are obviously valid. $\square$

We shall now turn to the completeness of the system G. For that, we shall modify the procedure *expand*. The additional complexity comes from the quantifier rules and the handling of terms, and we need to revise the definition of a Hintikka set. We shall define the concept of a Hintikka set with respect to a term algebra H defined in the next section. First, we extend the notation for signed formulae given in definition 3.5.3 to quantified formulae.

5.4.4 Signed Formulae and Term Algebras (no Equality Symbol)

We begin with signed formulae.

Definition 5.4.3 *Signed formulae of type a, type b, type c, and type d* and their *components* are defined in the tables below.

Type-a formulae

A	A_1	A_2
$T(X \wedge Y)$	TX	TY
$F(X \vee Y)$	FX	FY
$F(X \supset Y)$	TX	FY
$T(\neg X)$	FX	FX
$F(\neg X)$	TX	TX

Type-b formulae

B	B_1	B_2
$F(X \wedge Y)$	FX	FY
$T(X \vee Y)$	TX	TY
$T(X \supset Y)$	FX	TY

Type-c formulae

C	C_1	C_2
$T\forall xY$	$TY[t/x]$	$TY[t/x]$
$F\exists xY$	$FY[t/x]$	$FY[t/x]$

where t is free for x in Y

Type-d formulae

D	D_1	D_2
$T\exists xY$	$TY[t/x]$	$TY[t/x]$
$F\forall xY$	$FY[t/x]$	$FY[t/x]$

where t is free for x in Y

For a formula C of type c, we let $C(t)$ denote $TY[t/x]$ if $C = T\forall xY$, $FY[t/x]$ if $C = F\exists xY$, and for a formula D of type d, we let $D(t)$ denote $TY[t/x]$ if $D = T\exists xY$, $FY[t/x]$ if $D = F\forall xY$.

We define satisfaction for signed formulae as follows.

Definition 5.4.4 Given a structure $\mathbf{M}$, an assignment s, and a formula A,

$$\mathbf{M} \models (TA)[s] \quad \text{iff} \quad \mathbf{M} \models A[s],$$

and

$$\mathbf{M} \models (FA)[s] \quad \text{iff} \quad \mathbf{M} \models (\neg A)[s]$$

(or equivalently, $\mathbf{M} \not\models A[s]$).

Lemma 5.4.4 For any structure $\mathbf{M}$ and any assignment s:

(i) For any formula C of type c,

$$\mathbf{M} \models C[s] \quad \text{iff} \quad \mathbf{M} \models C(\mathbf{a})[s] \text{ for every } a \in M;$$

(ii) For any formula D of type d,

$$\mathbf{M} \models D[s] \quad \text{iff} \quad \mathbf{M} \models D(\mathbf{a})[s] \text{ for at least one } a \in M.$$

Proof: The lemma follows immediately from lemma 5.3.1. $\square$

In view of lemma 5.4.4, formulae of type c are also called formulae of *universal type*, and formulae of type d are called formulae of *existential type*. In order to define Hintikka sets with respect to a term algebra H, some definitions are needed.

Definition 5.4.5 Given a first-order language **L**, a nonempty set H of terms in $TERM_\mathbf{L}$ (with variables from **V**) is a *term algebra* iff:

(i) For every n-ary function symbol f in **L** $(n > 0)$, for any terms $t_1, ..., t_n \in H$, $ft_1...t_n$ is also in H.

(ii) Every constant symbol c in **L** is also in H.

Note that this definition is consistent with the definition of an algebra given in Section 2.4. Indeed, every function symbol and every constant in **L** receives an interpretation. A term algebra is simply an algebra whose carrier H is a nonempty set of terms, and whose operations are the *term constructors*. Note also that the terms in the carrier H may contain variables from the countable set **V**. However, variables are not in **L** and are not treated as constants. Also, observe that if **L** has at least one function symbol, by condition (i) any term algebra on **L** is infinite. Hence, a term algebra is finite only if **L** does not contain any function symbols.

5.4.5 Reducts, Expansions

Given a set S of signed **L**-formulae, not all symbols in **L** need occur in formulae in S. Hence, we shall define the *reduct* of **L** to S.

Definition 5.4.6 (i) Given two first-order languages (with or without equality) **L** and **L'**, if **L** is a subset of **L'** (that is, the sets of constant symbols, function symbols, and predicate symbols of **L** are subsets of the corresponding sets of **L'**), we say that **L** is a *reduct* of **L'** and that **L'** is an *expansion* of **L**. (The set of variables is neither in **L** nor in **L'** and is the given countable set **V**.)

(ii) If **L** is a reduct of **L'**, an **L**-structure $\mathbf{M} = (M, I)$ is a *reduct* of an **L'**-structure $\mathbf{M'} = (M', I')$ if $M' = M$ and I is the restriction of I' to **L**. $\mathbf{M'}$ is called an *expansion* of **M**.

(iii) Given a set S of signed **L**-formulae, the *reduct* of **L** with respect to S, denoted by $\mathbf{L}_S$, is the subset of **L** consisting of all constant, function, and predicate symbols occurring in formulae in S.

Note: If **L** is a reduct of **L'**, any **L**-structure **M** can be expanded to an **L'**-structure $\mathbf{M'}$ (in many ways). Furthermore, for any **L**-formula A and any assigment s,

$$\mathbf{M} \models A[s] \quad \text{if and only if} \quad \mathbf{M'} \models A[s].$$

5.4.6 Hintikka Sets (Languages Without Equality)

The definition of a Hintikka set is generalized to first-order languages without equality as follows.

Definition 5.4.7 A *Hintikka set* S (over a language **L** without equality) *with respect to a term algebra* H (over the reduct $\mathbf{L}_S$) is a set of signed **L**-formulae such that the following conditions hold for all signed formulae A, B, C, D of type a, b, c, d:

H0: No atomic formula and its conjugate are both in S (TA or FA is atomic iff A is).

H1: If a type-a formula A is in S, then both A_1 and A_2 are in S.

H2: If a type-b formula B is in S, then either B_1 is in S or B_2 is in S.

H3: If a type-c formula C is in S, then for every term $t \in H$, $C(t)$ is in S (where t is free for x in C).

H4: If a type-d formula D is in S, then for at least one term $t \in H$, $D(t)$ is in S (where t is free for x in D).

H5: Every variable x occurring free in some formula of S is in H.

Observe that condition H5 and the fact that H is a term algebra imply that, for every term occurring in some formula in S, if that term is closed or contains only variables free in S, then it is in H.

We can now prove the generalization of lemma 3.5.3 for first-order logic (without equality). From lemma 5.3.4, we can assume without loss of generality that the set of variables occurring free in formulae in S is disjoint from the set of variables occurring bound in formulae in S.

Lemma 5.4.5 Every Hintikka set S (with respect to a term algebra H) is satisfiable in a structure $\mathbf{H_S}$ with domain H.

Proof: The $\mathbf{L}_S$-structure $\mathbf{H_S}$ is defined as follows. The domain of $\mathbf{H_S}$ is H.

Every constant c in $\mathbf{L}_S$ is interpreted as the term c;

Every function symbol f of rank n in $\mathbf{L}_S$ is interpreted as the function such that, for any terms $t_1, ..., t_n \in H$,

$$f_{\mathbf{H_S}}(t_1, ..., t_n) = ft_1...t_n.$$

For every predicate symbol P of rank n in $\mathbf{L}_S$, for any terms $t_1, ..., t_n \in H$,

$$P_{\mathbf{H_S}}(t_1, ..., t_n) = \begin{cases} \mathbf{F} & \text{if } FPt_1...t_n \in S, \\ \mathbf{T} & \text{if } TPt_1...t_n \in S, \\ & \text{or neither } TPt_1...t_n \text{ nor } FPt_1...t_n \text{ is in } S. \end{cases}$$

By conditions H0 and the fact that H is an algebra, this definition is proper. Let s be any assignment that is the identity on the variables belonging to H.

We now prove using the induction principle for formulae that

$$\mathbf{H_S} \models X[s]$$

for every signed formula $X \in S$. We will need the following claim that is proved using the induction principle for terms. (For a proof of a more general version of this claim, see claim 2 in lemma 5.6.1.)

Claim: For every term t (in H),

$$t_{\mathbf{H_S}}[s] = t.$$

Assume that $TPt_1...t_n$ is in S. By H5, the variables in $t_1,...,t_n$ are in H, and since H is a term algebra, $t_1,...,t_n$ are in H. By definition of $P_{\mathbf{H_S}}$ and the above claim,

$$\begin{aligned}(Pt_1...t_n)_{\mathbf{H_S}}[s] &= P_{\mathbf{H_S}}((t_1)_{\mathbf{H_S}}[s], ..., (t_n)_{\mathbf{H_S}}[s]) \\ &= P_{\mathbf{H_S}}(t_1,...,t_n) = \mathbf{T}.\end{aligned}$$

Hence,

$$\mathbf{H_S} \models Pt_1...t_n[s].$$

Similarly, it is shown that if $FPt_1...t_n$ is in S then

$$(Pt_1...t_n)_{\mathbf{H_S}}[s] = \mathbf{F}.$$

The propositional connectives are handled as in lemma 3.5.3.

If a signed formula C of type c is in S, $C(t)$ is in S for every term t in H by H3. Since $C(t)$ contains one less quantifier than C, by the induction hypothesis,

$$\mathbf{H_S} \models C(t)[s] \text{ for every } t \in H.$$

By lemma 5.4.1, for any formula A, any term t free for x in A, any structure $\mathbf{M}$ and any assignment v, we have

$$(A[t/x])_{\mathbf{M}}[v] = (A[\mathbf{a}/x])_{\mathbf{M}}[v],$$

where $a = t_{\mathbf{M}}[v]$. Since

$$t_{\mathbf{H_S}}[s] = t,$$

we have

$$(C[t/x])_{\mathbf{H_S}}[s] = (C[\mathbf{t}/x])_{\mathbf{H_S}}[s].$$

Hence,

$$\mathbf{H_S} \models C(\mathbf{t})[s] \text{ for every } t \in H,$$

which implies that

$$\mathbf{H_S} \models C[s]$$

by lemma 5.4.4.

If a signed formula D of type d is in S, $D(t)$ is in S for some t in H by H4. Since $D(t)$ contains one less quantifier than D, by the induction hypothesis,

$$\mathbf{H_S} \models D(t)[s].$$

As above, it can be shown that

$$\mathbf{H_S} \models D(\mathbf{t})[s] \text{ for some } t \in H,$$

which implies that

$$\mathbf{H_S} \models D[s]$$

(by lemma 5.4.4). Finally, using the note before definition 5.4.7, $\mathbf{H_S}$ can be expanded to an **L**-structure satisfying S. $\square$

The domain H of the structure $\mathbf{H_S}$ is also called a *Herbrand universe*.

In order to prove the completeness of the system G (in case of a first-order language **L** without equality) we shall extend the methods used in sections 3.4 and 3.5 for the propositional calculus to first-order logic. Given a (possibly infinite) sequent $\Gamma \to \Delta$, our goal is to attempt to falsify it. For this, we design a *search* procedure with the following properties:

(1) If the original sequent is valid, the *search* procedure stops after a finite number of steps, yielding a proof tree.

(2) If the original sequent is falsifiable, the *search* procedure constructs a possibly infinite tree, and along some (possibly infinite) path in the tree it can be shown that a Hintikka set exists, which yields a counter example for the sequent.

The problem is to modify the *expand* procedure to deal with quantifier rules and terms. Clauses H3 and H4 in the definition of a Hintikka set suggest that a careful procedure must be designed in dealing with quantified formulae.

We first treat the *special case* in which we are dealing with a first-order language without equality, without function symbols and with a finite sequent $\Gamma \to \Delta$.

5.4.7 Completeness: Special Case of Languages Without Function Symbols and Without Equality

Given a sequent $\Gamma \to \Delta$, using lemma 5.3.4, we can assume that the set of all variables occurring free in some formula in the sequent is disjoint from the set of all variables occurring bound in some formula in the sequent. This condition ensures that terms occurring in formulae in the sequent are free for susbtitutions. Even though it is not strictly necessary to assume that all the formulae in the sequent are rectified, it is convenient to assume that they are.

First, it is convenient for proving the correctness of the *search* procedure to give a slightly more general version of the quantifier rules $\forall : left$ and $\exists : right$.

Definition 5.4.8 The *extended rules* $\forall : left$ and $\exists : right$ are the following:

$$\frac{\Gamma, A[t_1/x], ..., A[t_k/x], \forall x A, \Delta \to \Lambda}{\Gamma, \forall x A, \Delta \to \Lambda} \ (\forall : left)$$

$$\frac{\Gamma \to \Delta, A[t_1/x], ..., A[t_k/x], \exists x A, \Lambda}{\Gamma \to \Delta, \exists x A, \Lambda} \ (\exists : right)$$

where $t_1, ..., t_k$ are any k terms $(k \geq 1)$ free for x in A.

It is clear that an inference using this new version of the $\forall : left$ rule (resp. $\exists : right$ rule) can be simulated by k applications of the old $\forall : left$ rule (resp. $\exists : right$ rule). Hence, there is no gain of generality. However, these new rules may reduce the size of proof trees. Consequently, we will assume from now on that the rules of definition 5.4.8 are used as the $\forall : left$ and $\exists : right$ rules of the Gentzen system G.

In order to fulfill conditions H3 and H4 we build lists of variables and constants as follows.

Let $TERM_0$ be a nonempty list of terms and variables defined as follows. If no free variables and no constants occur in any of the formulae in $\Gamma \to \Delta$,

$$TERM_0 = < y_0 >,$$

where y_0 is the first variable in **V** not occurring in any formula in $\Gamma \to \Delta$. Otherwise,

$$TERM_0 = < u_0, ..., u_p >,$$

a list of all free variables and constants occurring in formulae in $\Gamma \to \Delta$. Let

$$AVAIL_0 = < y_1, ..., y_n, ... >$$

be a countably infinite list disjoint from $TERM_0$ and consisting of variables not occurring in any formula in $\Gamma \to \Delta$.

The terms in $TERM_0$ and the variables in $AVAIL_0$ will be used as the t's and y's for our applications of quantifier rules $\forall : right$, $\forall : left$, $\exists : right$ and $\exists : left$. Because the variables in $TERM_0$ do not occur bound in $\Gamma \to \Delta$ and the variables in $AVAIL_0$ are new, the substitutions with results $A[t/x]$ and $A[y/x]$ performed using the quantifier rules will be free.

As the search for a counter example for $\Gamma \to \Delta$ progresses, we keep track step by step of which of $u_0, ..., u_p, y_1, y_2, y_3, ...$ have been thus far activated. The list of activated terms is kept in $TERM_0$ and the list of available variables in $AVAIL_0$.

Every time a rule $\forall : right$ or $\exists : left$ is applied, as the variable y we use the head of the list $AVAIL_0$, we append y to the end of $TERM_0$ and delete y from the head of $AVAIL_0$.

When a rule $\forall : left$ or $\exists : right$ is applied, we use as the terms t each of the terms $u_0,...,u_q$ in $TERM_0$ that have not previously served as a term t for that rule with the same principal formula.

To handle the $\forall : left$ rule and the $\exists : right$ rule correctly, it is necessary to keep track of the formulae $\forall x A$ (or $\exists x A$) for which the term u_i was used as a term for the rule $\forall : left$ (or $\exists : right$) with principal formula $\forall x A$ (or $\exists x A$). The first reason is economy, but the second is more crucial:

If a sequent has the property that all formulae in it are either atomic or of the form $\forall x A$ (or $\exists x A$) such that all the terms $u_0,...,u_q$ in $TERM_0$ have already been used as terms for the rule $\forall : left$ (or $\exists : right$), and if this sequent is not an axiom, then it will never become an axiom and we can stop expanding it.

Hence, we structure $TERM_0$ as a list of records where every record $< u_i, FORM_0(i) >$ contains two fields: u_i is a term and $FORM_0(i)$ a list of the formulae $\forall x A$ (or $\exists x A$) for which u_i was used as a term t for the rule $\forall : left$ (or $\exists : right$) with principal formula $\forall x A$ (or $\exists x A$). Initially, each list $FORM_0(i)$ is the null list. The lists $FORM_0(i)$ are updated each time a term t is used in a rule $\forall : left$ or $\exists : right$. We also let $t(TERM_0)$ denote the set of terms $\{u_i \mid < u_i, FORM_0(i) > \in TERM_0\}$.

Finally, we need to take care of another technical detail: These lists must be updated only at the end of a round, so that the same substitutions are performed for all occurrences of a formula. Hence, we create another variable $TERM_1$ local to the procedure $search$. During a round, $TERM_1$ is updated but $TERM_0$ is not, and at the end of the round, $TERM_0$ is set to its updated version $TERM_1$.

A leaf of the tree constructed by procedure $search$ is *finished* iff either:

(1) The sequent labeling it is an axiom, or

(2) The sequent contains only atoms or formulae $\forall x A$ (or $\exists x A$) belonging to all of the lists $FORM_0(i)$ for all $< u_i, FORM_0(i) >$ in $TERM_0$.

The $search$ procedure is obtained by modifying the procedure given in definition 3.4.6 by adding the initialization of $TERM_0$ and $AVAIL_0$.

Definition 5.4.9 The $search$ procedure. The input to $search$ is a one-node tree labeled with a sequent $\Gamma \rightarrow \Delta$. The output is a possibly infinite tree T called a *systematic deduction tree*.

Procedure Search

procedure $search(\Gamma \to \Delta : sequent;$ **var** $T : tree);$
  begin
    let T be the one-node tree labeled with $\Gamma \to \Delta$;
    Let $TERM_0 := << u_0, nil >, ..., < u_p, nil >>$
    and let $AVAIL_0 := < y_1, y_2, y_3... >$
    with $u_0, ..., u_p$ as explained above.
    while not *all leaves of T are finished* **do**
      $TERM_1 := TERM_0; T_0 := T;$
      for each *leaf node of T_0*
      (in lexicographic order) **do**
        if not $finished(node)$ **then**
          $expand(node, T)$
        endif
      endfor;
      $TERM_0 := TERM_1$
    endwhile;
    if *all leaves are closed*
    then
      write (*'T is a proof of $\Gamma \to \Delta$'*)
    else
      write (*'$\Gamma \to \Delta$ is falsifiable'*)
    endif
  end

Procedure Expand

procedure $expand(node : tree\text{-}address;$ **var** $T : tree);$
  begin
    let $A_1, ..., A_m \to B_1, ..., B_n$ be the label of node;
    let S be the one-node tree labeled with
      $A_1, ..., A_m \to B_1, ..., B_n;$
    for $i := 1$ **to** m **do**
      if $nonatomic(A_i)$ **then**
        $grow\text{-}left(A_i, S)$
      endif
    endfor;
    for $i := 1$ **to** n **do**
      if $nonatomic(B_i)$ **then**
        $grow\text{-}right(B_i, S)$
      endif
    endfor;
    $T := dosubstitution(T, node, S)$
  end

Procedure Grow-Left

procedure *grow-left*(A : *formula*; **var** S : *tree*);
  begin
    case A **of**
    $B \wedge C$, $B \vee C$,
    $B \supset C$, $\neg B$: *extend every nonaxiom leaf of S using the*
                          left rule corresponding to the main
                          propositional connective;
    $\forall x B$: **for** *every term* $u_k \in t(TERM_0)$
                          such that A is not in $FORM_0(k)$ **do**
                              extend every nonaxiom leaf of S by applying
                              the $\forall : left$ rule using the term u_k
                              as one of the terms of the rule;
                              $FORM_1(k) := Append(FORM_1(k), A)$
                          endfor;
    $\exists x B$: *extend every nonaxiom leaf of S by applying*
                          the $\exists : left$ rule using $y = head(AVAIL_0)$
                          as the new variable;
                          $TERM_1 := append(TERM_1, < y, nil >)$;
                          $AVAIL_0 := tail(AVAIL_0)$
    endcase
  end

Procedure Grow-Right

procedure *grow-right*(A : *formula*; **var** S : *tree*);
  begin
    case A **of**
    $B \wedge C$, $B \vee C$,
    $B \supset C$, $\neg B$: *extend every nonaxiom leaf of S using the*
                          right rule corresponding to the main
                          propositional connective;
    $\exists x B$: **for** *every term* $u_k \in t(TERM_0)$
                          such that A is not in $FORM_0(k)$ **do**
                              extend every nonaxiom leaf of S by applying
                              the $\exists : right$ rule using the term u_k
                              as one of the terms of the rule;
                              $FORM_1(k) := Append(FORM_1(k), A)$
                          endfor;
    $\forall x B$: *extend every nonaxiom leaf of S by applying*
                          the $\forall : right$ rule using $y = head(AVAIL_0)$
                          as the new variable;
                          $TERM_1 := append(TERM_1, < y, nil >)$;
                          $AVAIL_0 := tail(AVAIL_0)$
    endcase
  end

EXAMPLE 5.4.3

Let
$$A = \exists x(P \supset Q(x)) \supset (P \supset \exists z Q(z)),$$

where P is a propositional symbol and Q a unary predicate symbol. Let us trace the construction of the proof tree constructed by the *search* procedure. Since A does not have any free variables, $TERM_0 = <<$ $y_0, nil >>$. After the first round, we have the following tree, and $TERM_0$ and $AVAIL_0$ have not changed.

$$\frac{\exists x(P \supset Q(x)) \rightarrow P \supset \exists z Q(z)}{\rightarrow \exists x(P \supset Q(x)) \supset (P \supset \exists z Q(z))}$$

The $\exists : left$ rule is applied using the head y_1 of $AVAIL_0$, and the following tree is obtained:

$$\frac{\dfrac{P, P \supset Q(y_1) \rightarrow \exists z Q(z)}{P \supset Q(y_1) \rightarrow P \supset \exists z Q(z)}}{\dfrac{\exists x(P \supset Q(x)) \rightarrow P \supset \exists z Q(z)}{\rightarrow \exists x(P \supset Q(x)) \supset (P \supset \exists z Q(z))}}$$

At the end of this round,

$$TERM_0 = << y_0, nil >, < y_1, nil >>,$$

and
$$AVAIL_0 = < y_2, y_3, ... > .$$

During the next round, the $\exists : right$ rule is applied to the formula $\exists z Q(z)$ with the terms y_0 and y_1. The following tree is obtained:

$$\frac{P \rightarrow P, \exists z Q(z) \qquad \dfrac{Q(y_1), P \rightarrow Q(y_0), Q(y_1), \exists z Q(z)}{Q(y_1), P \rightarrow \exists z Q(z)}}{\dfrac{P, P \supset Q(y_1) \rightarrow \exists z Q(z)}{\dfrac{P \supset Q(y_1) \rightarrow P \supset \exists z Q(z)}{\dfrac{\exists x(P \supset Q(x)) \rightarrow P \supset \exists z Q(z)}{\rightarrow \exists x(P \supset Q(x)) \supset (P \supset \exists z Q(z))}}}}$$

At the end of the round,

$$TERM_0 = << y_0, < \exists z Q(z) >>, < y_1, < \exists z Q(z) >>> .$$

This last tree is a proof tree for A.

EXAMPLE 5.4.4

Let

$$A = \exists x(P \supset Q(x)) \supset (P \supset \forall z Q(z)),$$

where P is a propositional letter and Q is a unary predicate symbol. Initially, $TERM_0 = <<y_0, nil>>$. After the first round, the following tree is obtained and $TERM_0$ and $AVAIL_0$ are unchanged:

$$\frac{\exists x(P \supset Q(x)) \rightarrow P \supset \forall z Q(z)}{\rightarrow \exists x(P \supset Q(x)) \supset (P \supset \forall z Q(z))}$$

During the second round, the $\exists : left$ rule is applied and the following tree is obtained:

$$\frac{\dfrac{\dfrac{P, P \supset Q(y_1) \rightarrow \forall z Q(z)}{P \supset Q(y_1) \rightarrow P \supset \forall z Q(z)}}{\exists x(P \supset Q(x)) \rightarrow P \supset \forall z Q(z)}}{\rightarrow \exists x(P \supset Q(x)) \supset (P \supset \forall z Q(z))}$$

At the end of this round,

$$TERM_0 = <<y_0, nil>, <y_1, nil>>,$$

and

$$AVAIL_0 = <y_2, y_3, \dots>.$$

During the next round, the $\forall : right$ rule is applied to the formula $\forall z Q(z)$ with the new variable y_2. The following tree is obtained:

$$\frac{\dfrac{P \rightarrow P, \forall z Q(z) \qquad \dfrac{\dfrac{Q(y_1), P \rightarrow Q(y_2)}{Q(y_1), P \rightarrow \forall z Q(z)}}{\;}}{\dfrac{P, P \supset Q(y_1) \rightarrow \forall z Q(z)}{\dfrac{P \supset Q(y_1) \rightarrow P \supset \forall z Q(z)}{\dfrac{\exists x(P \supset Q(x)) \rightarrow P \supset \forall z Q(z)}{\rightarrow \exists x(P \supset Q(x)) \supset (P \supset \forall z Q(z))}}}}}{}$$

At the end of this round,

$$TERM_0 = <<y_0, nil>, <y_1, nil>, <y_2, nil>>,$$

and
$$AVAIL_0 = <y_3, y_4, \ldots > .$$

Since all formulae in the sequent $Q(y_1), P \rightarrow Q(y_2)$ are atomic and it is not an axiom, this last tree yields a counter example with domain $\{y_0, y_1, y_2\}$, by making Q arbitrarily **true** for y_0 and y_1, **false** for y_2, and P **true**. Note that $\{y_1, y_2\}$ is also the domain of a counter example.

The following theorem is a generalization of theorem 3.4.1 to finite sequents in first-order languages without function symbols (and without equality).

Theorem 5.4.1 (i) If the input sequent $\Gamma \rightarrow \Delta$ is valid, the procedure *search* halts with a finite closed tree T which is a proof tree for $\Gamma \rightarrow \Delta$.

(ii) If the input sequent $\Gamma \rightarrow \Delta$ is falsifiable, either *search* halts with a finite counter-example tree T and $\Gamma \rightarrow \Delta$ can be falsified in a finite structure, or *search* generates an infinite tree T and $\Gamma \rightarrow \Delta$ can be falsified in a countably infinite structure.

Proof: First, assume that the sequent $\Gamma \rightarrow \Delta$ is falsifiable. If the tree T was finite and closed, by lemma 5.4.3, $\Gamma \rightarrow \Delta$ would be valid, a contradiction. Hence, either T is finite and contains some path to a nonaxiom leaf, or T is infinite and by König's lemma contains an infinite path. In either case, we show as in theorem 3.4.1 that a Hintikka set can be found along that path. Let U be the union of all formulae occurring in the left-hand side of each sequent along that path, and V the union of all formulae occurring in the right-hand side of any such sequent. Let

$$S = \{TA \mid A \in U\} \cup \{FB \mid B \in V\}.$$

We prove the following claim:

Claim: S is a Hintikka set with respect to the term algebra consisting of the set H of terms in $t(TERM_0)$.

Conditions H0, H1, and H2 are proved as in the propositional case (proof of lemma 3.5.2), and we only have to check that $t(TERM_0)$ is a term algebra, and conditions H3, H4, and H5. Since **L** does not contain function symbols, $t(TERM_0)$ is trivially closed under the operations (there are none). Since the constants occurring in the input sequent are put in $t(TERM_0)$, $t(TERM_0)$ is a term algebra. Since $t(TERM_0)$ is initialized with the list of free variables and constants occurring in the input sequent (or y_0 if this list is empty), and since every time a variable y is removed from the head of $AVAIL_0$, $<y, <nil>>$ is added to $TERM_0$, H5 holds. Every time a formula C of type c is expanded, all substitution instances $C(t)$ for all t in $t(TERM_0)$ that have not already been used with C are added to the upper sequent. Hence, H3 is satisfied. Every time a formula D of type d is expanded, y is added to $t(TERM_0)$ and the substitution instance $D(y)$ is added to the upper sequent. Hence, H4 is satisfied, and the claim holds. $\square$

By lemma 5.4.5, some assignment s satisfies S in the structure $\mathbf{H_S}$. This implies that $\Gamma \rightarrow \Delta$ is falsified by $\mathbf{H_S}$ and s.

Note that H must be infinite if the tree T is infinite. Otherwise, since we start with a finite sequent, every path would be finite and would end either with an axiom or a finished sequent.

If the sequent $\Gamma \rightarrow \Delta$ is valid the tree T must be finite and closed since otherwise, the above argument shows that $\Gamma \rightarrow \Delta$ is falsifiable. $\square$

As a corollary, we obtain a version of Gödel's completeness theorem for first-order languages without function symbols or equality.

Corollary   If a sequent (over a first-order language without function symbols or equality) is valid then it is G-provable. $\square$

In Section 5.5, we shall modify the *search* procedure to handle function symbols and possibly infinite sequents. Finally in Section 5.6, we will adapt the procedure to deal with languages with equality.

PROBLEMS

5.4.1. Give proof trees for the following formulae:

$$\forall x A \supset A[t/x], \quad A[t/x] \supset \exists x A,$$
$$\text{where } t \text{ is free for } x \text{ in } A.$$

5.4.2. Let x, y be any distinct variables. Let A, B be any formulae, C, D any formulae not containing the variable x free, and let E be any formula such that x is free for y in E. Give proof trees for the following formulae:

$$\forall x C \equiv C \qquad\qquad \exists x C \equiv C$$
$$\forall x \forall y A \equiv \forall y \forall x A \qquad\qquad \exists x \exists y A \equiv \exists y \exists x A$$
$$\forall x \forall y E \supset \forall x E[x/y] \qquad\qquad \exists x E[x/y] \supset \exists x \exists y E$$
$$\forall x A \supset \exists x A$$
$$\exists x \forall y A \supset \forall y \exists x A$$

5.4.3. Let A, B be any formulae, and C any formula not containing the variable x free. Give proof trees for the following formulae:

$$\neg \exists x A \equiv \forall x \neg A \qquad\qquad \neg \forall x A \equiv \exists x \neg A$$
$$\exists x A \equiv \neg \forall x \neg A \qquad\qquad \forall x A \equiv \neg \exists x \neg A$$
$$\forall x A \wedge \forall x B \equiv \forall x (A \wedge B) \qquad\qquad \exists x A \vee \exists x B \equiv \exists x (A \vee B)$$
$$C \wedge \forall x A \equiv \forall x (C \wedge D) \qquad\qquad C \vee \exists x A \equiv \exists x (C \vee A)$$
$$C \wedge \exists x A \equiv \exists x (C \wedge D) \qquad\qquad C \vee \forall x A \equiv \forall x (C \vee A)$$
$$\exists x (A \wedge B) \supset \exists x A \wedge \exists x B \qquad\qquad \forall x A \vee \forall x B \supset \forall x (A \vee B)$$

5.4.4. Let A, B be any formulae, and C any formula not containing the variable x free. Give proof trees for the following formulae:

$$(C \supset \forall x A) \equiv \forall x (C \supset A) \qquad (C \supset \exists x A) \equiv \exists x (C \supset A)$$
$$(\forall x A \supset C) \equiv \exists x (A \supset C) \qquad (\exists x A \supset C) \equiv \forall x (A \supset C)$$
$$(\forall x A \supset \exists x B) \equiv \exists x (A \supset B)$$
$$(\exists x A \supset \forall x B) \supset \forall x (A \supset B)$$

5.4.5. Give a proof tree for the following formula:

$$\neg \forall x \exists y (\neg P(x) \wedge P(y)).$$

5.4.6. Show that the rules $\forall : right$ and $\exists : left$ are not necessarily sound if y occurs free in the lower sequent.

* **5.4.7.** Let **L** be a first-order language without function symbols and without equality. A first-order formula A is called *simple* if, for every subformula of the form $\forall x B$ or $\exists x B$, B is quantifier free.

  Prove that the *search* procedure always terminates for simple formulae. Conclude that there is an algorithm for deciding validity of simple formulae.

5.4.8. Let $\Gamma \to A$ be a finite sequent, let c be a constant not occurring in $\Gamma \to A$, and let $\Gamma[c/x] \to A[c/x]$ be the result of substituting the constant c for x in all formulae in Γ and in A.

  Prove that if $\Gamma[c/x] \to A[c/x]$ is provable, then $\Gamma[y/x] \to A[y/x]$ is provable for every variable y not occurring in Γ or in A. Conclude that if x does not occur in Γ, if $\Gamma \to A[c/x]$ is provable, then $\Gamma \to \forall x A$ is provable.

* **5.4.9.** The language of a binary relation (simple directed graphs) consists of one binary predicate symbol R. The axiom for simple directed graphs is:

$$\forall x \forall y (R(x,y) \supset \neg R(y,x)).$$

  This expresses that a simple directed graph has no cycles of length ≤ 2. Let T_1 contain the above axiom together with the axiom

$$\forall x \exists y R(x,y)$$

asserting that every node has an outgoing edge.

(a) Find a model for T_1 having three elements.

Let T_2 be T_1 plus the density axiom:

$$\forall x \forall y (R(x,y) \supset \exists z (R(x,z) \wedge R(z,y))).$$

(b) Find a model for T_2. Find a seven-node graph model of T_2.

5.5 Completeness for Languages with Function Symbols and no Equality

In order to deal with (countably) infinite sequents, we shall use the technique for building a tree used in Section 3.5. First, we need to deal with function symbols.

5.5.1 Organizing the Terms for Languages with Function Symbols and no Equality

Function symbols are handled in the following way. First, we can assume without loss of generality that the set of variables $\mathbf{V}$ is the disjoint union of two countably infinite sets $\{x_0, x_1, x_2, ...\}$ and $\{y_0, y_1, y_2, ...\}$ and that only variables in the first set (the $x's$) are used to build formulae. In this way, $\{y_0, y_1, y_2, ...\}$ is an infinite supply of new variables. Let $\Gamma \to \Delta$ be a possibly infinite sequent. As before, using lemma 5.3.4, we can assume that the set of variables occurring free in $\Gamma \to \Delta$ is disjoint from the set of variables occurring bound in $\Gamma \to \Delta$.

Let $\mathbf{L'}$ be the reduct of $\mathbf{L}$ consisting of the constant, function and predicate symbols occurring in formulae in $\Gamma \to \Delta$. If the set of constants and free variables occurring in $\Gamma \to \Delta$ is nonempty, let

$$TERMS = < u_0, u_1, ..., u_k, ... >$$

be an enumeration of all $\mathbf{L'}$-terms constructed from the variables occurring free in formulae in $\Gamma \to \Delta$, and the constant and function symbols in $\mathbf{L'}$. Otherwise, let

$$TERMS = < y_0, u_1, ..., u_k, ... >$$

be an enumeration of all the terms constructed from the variable y_0 and the function symbols in $\mathbf{L'}$.

Let

$$TERM_0 = << head(TERMS), nil >>$$

and let

$$AVAIL_0 = tail(TERMS).$$

For any $i \geq 1$, let $AVAIL_i$ be an enumeration of the set of all $\mathbf{L'}$-terms *actually containing* some occurrence of y_i and constructed from the variables occurring free in formulae in $\Gamma \to \Delta$, the constant and function symbols occurring in $\Gamma \to \Delta$, and the variables $y_1, ..., y_i$. We assume that such an enumeration begins with y_i and is of the form

$$AVAIL_i = < y_i, u_{i,1}, ..., u_{i,j}, .. > .$$

EXAMPLE 5.5.1

Assume that $\mathbf{L'}$ contains a constant symbol a, and the function symbols f of rank 2, and g of rank 1. Then, a possible enumeration of $TERMS$ is the following:

$$TERMS = < a, g(a), f(a,a), g(g(a)), g(f(a,a)), f(g(a),a),$$
$$f(a,g(a)), f(g(a),g(a)), \ldots > .$$

Hence, $TERM_0 = << a, nil >>$, and

$$AVAIL_0 = < g(a), f(a,a), g(g(a)), g(f(a,a)),$$
$$f(g(a),a), f(a,g(a)), f(g(a),g(a)), \ldots > .$$

For $i = 1$, a possible enumeration of $AVAIL_1$ is:

$$AVAIL_1 = < y_1, g(y_1), f(a,y_1), f(y_1,a), f(y_1,y_1),$$
$$g(g(y_1)), g(f(a,y_1)), g(f(y_1,a)), g(f(y_1,y_1)), \ldots > .$$

Each time a rule $\forall : right$ or $\exists : left$ is applied, we use as the variable y the first y_i of $y_1, y_2, \ldots$ not yet activated, append y_i to $t(TERM_0)$, and delete y_i from the list $AVAIL_i$. We use a counter $NUMACT$ to record the number of variables $y_1, y_2, \ldots$ thus far activated. Initially $NUMACT = 0$, and every time a new y_i is activated $NUMACT$ is incremented by 1.

In a $\forall : left$ or a $\exists : right$ step, all the terms in $t(TERM_0)$ thus far activated are available as terms t. Furthermore, at the end of every round, the head of every available list $AVAIL_i$ thus far activated (with $0 \le i \le NUMACT$) is appended to $t(TERM_0)$ and deleted from $AVAIL_i$. Thus, along any path in the tree that does not close, once any term in an $AVAIL_i$ list is activated, every term in the list is eventually activated.

Note that if the sets $\mathbf{FS}$ and $\mathbf{CS}$ are effective (recursive), recursive functions enumerating the lists $t(TERM_0)$ and $AVAIL_i$ $(i \ge 0)$ can be written (these lists are in fact recursive).

The definition of a *finished leaf* is the following: A leaf of the tree is finished if either

(1) It is an axiom, or

(2) L and R are empty, and the sequent only contains atoms or formulae $\forall x A$ (or $\exists x A$) belonging to all lists $FORM_0(i)$, for all $< u_i, FORM_0(i) >$ in $TERM_0$.

5.5.2 The Search Procedure for Languages with Function Symbols and no Equality

The *search* procedure (and its subprograms) are revised as follows.

Definition 5.5.1 The *search* procedure.

Procedure Search

```
procedure search(Γ₀ → Δ₀ : sequent; var T : tree);
  begin
    L := tail(Γ₀); Γ := head(Γ₀);
    R := tail(Δ₀); Δ := head(Δ₀);
    let T be the one-node tree labeled with Γ → Δ;
    Let TERM₀ and AVAILᵢ, 0 ≤ i,
    be initialized as explained above.
    NUMACT := 0;
    while not all leaves of T are finished do
      TERM₁ := TERM₀; T₀ := T;
      for each leaf node of T₀
      (in lexicographic order) do
        if not finished(node) then
          expand(node, T)
        endif
      endfor;
      TERM₀ := TERM₁; L := tail(L); R := tail(R);
      for i := 0 to NUMACT do
        TERM₀ := append(TERM₀, < head(AVAILᵢ), nil >);
        AVAILᵢ := tail(AVAILᵢ)
      endfor
    endwhile;
    if all leaves are closed
    then
      write ('T is a proof of  Γ₀ → Δ₀')
    else
      write ('Γ₀ → Δ₀ is falsifiable')
    endif
  end
```

The Procedures *expand, grow-left,* and *grow-right* appear on the next two pages.

procedure *expand*(*node* : *tree-address*; **var** *T* : *tree*);
  begin
    let $A_1, ..., A_m \rightarrow B_1, ..., B_n$ *be the label of node*;
    let S be the one-node tree labeled with
      $A_1, ..., A_m \rightarrow B_1, ..., B_n$;
      for $i := 1$ **to** m **do**
        if *nonatomic*(A_i) **then**
          grow-left(A_i, S)
        endif
      endfor;
      for $i := 1$ **to** n **do**
        if *nonatomic*(B_i) **then**
          grow-right(B_i, S)
        endif
      endfor;
      for *each leaf u of S* **do**
        let $\Gamma \rightarrow \Delta$ *be the label of u*;
        $\Gamma' := \Gamma, head(L)$;
        $\Delta' := \Delta, head(R)$;
        create a new node u1 labeled with $\Gamma' \rightarrow \Delta'$
      endfor; $T := dosubstitution(T, node, S)$
  end

procedure *grow-left*(A : *formula*; **var** S : *tree*);
  begin
    case A **of**
    $B \wedge C, B \vee C,$
    $B \supset C, \neg B$: *extend every nonaxiom leaf of S using the*
                        left rule corresponding to the main
                        propositional connective;
    $\forall x B$: **for** *every term* $u_k \in t(TERM_0)$
                        such that A is not in $FORM_0(k)$ **do**
                          extend every nonaxiom leaf of S by applying
                          the $\forall$: *left rule using the term* u_k
                          as one of the terms of the rule;
                          $FORM_1(k) := Append(FORM_1(k), A)$
                        endfor;
    $\exists x B$: $NUMACT := NUMACT + 1$;
                        extend every nonaxiom leaf of S by applying
                        the $\exists$: *left rule using* $y = head(AVAIL_{NUMACT})$
                        as the new variable;
                        $TERM_1 := append(TERM_1, < y, nil >)$;
                        $AVAIL_{NUMACT} := tail(AVAIL_{NUMACT})$
    endcase
  end

```
procedure grow-right(A : formula; var S : tree);
  begin
    case A of
    B ∧ C, B ∨ C,
    B ⊃ C, ¬B    : extend every nonaxiom leaf of S using the
                   right rule corresponding to the main
                   propositional connective;
    ∃xB          : for every term u_k ∈ t(TERM_0)
                   such that A is not in FORM_0(k) do
                       extend every nonaxiom leaf of S by applying
                       the ∃ : right rule using the term u_k
                       as one of the terms of the rule;
                       FORM_1(k) := Append(FORM_1(k), A)
                   endfor;
    ∀xB          : NUMACT := NUMACT + 1;
                   extend every nonaxiom leaf of S by applying
                   the ∀ : right rule using y = head(AVAIL_NUMACT)
                   as the new variable;
                   TERM_1 := append(TERM_1, < y, nil >);
                   AVAIL_NUMACT := tail(AVAIL_NUMACT)
    endcase
  end
```

Let us give an example illustrating this new version of the *search* procedure.

EXAMPLE 5.5.2

Let

$$A = (\forall x P(x) \wedge \exists y Q(y)) \supset (P(f(v)) \wedge \exists z Q(z)),$$

where P and Q are unary predicate symbols, and f is a unary function symbol. The variable v is free in A. Initially,

$$TERM_0 = << v, nil >>,$$
$$AVAIL_0 = < f(v), f(f(v)), ..., f^n(v), ... >,$$

and for $i \geq 1$,

$$AVAIL_i = < y_i, f(y_i), ..., f^n(y_i), ... > .$$

After the first round, we have the following tree:

$$\frac{\forall x P(x) \wedge \exists y Q(y) \rightarrow P(f(v)) \wedge \exists z Q(z)}{\rightarrow \forall x P(x) \wedge \exists y Q(y) \supset P(f(v)) \wedge \exists z Q(z)}$$

At the end of this round,

$$TERM_0 = << v, nil >, < f(v), nil >>,$$

$$AVAIL_0 = < f^2(v), ..., f^n(v), ... >,$$

and for $i \geq 1$, $AVAIL_i$ is unchanged. After the second round, we have the following tree:

$$
\cfrac{
\cfrac{
\cfrac{
\cfrac{\forall x P(x), \exists y Q(y) \to P(f(v)) \qquad \forall x P(x), \exists y Q(y) \to \exists z Q(z)}
{\forall x P(x), \exists y Q(y) \to P(f(v)) \land \exists z Q(z)}
}{\forall x P(x) \land \exists y Q(y) \to P(f(v)) \land \exists z Q(z)}
}{\to \forall x P(x) \land \exists y Q(y) \supset P(f(v)) \land \exists z Q(z)}
}{}
$$

At the end of this round,

$$TERM_0 = << v, nil >, < f(v), nil >, < f(f(v)), nil >>,$$

$$AVAIL_0 = < f^3(v), ..., f^n(v), ... >,$$

and for $i \geq 1$, $AVAIL_i$ is unchanged. After the third round, we have the following tree:

$$
\cfrac{
\cfrac{
\cfrac{
\cfrac{
\cfrac{T_1}{\forall x P(x), \exists y Q(y) \to P(f(v))} \qquad
\cfrac{T_2}{\forall x P(x), \exists y Q(y) \to \exists z Q(z)}
}{\forall x P(x), \exists y Q(y) \to P(f(v)) \land \exists z Q(z)}
}{\forall x P(x) \land \exists y Q(y) \to P(f(v)) \land \exists z Q(z)}
}{\to \forall x P(x) \land \exists y Q(y) \supset P(f(v)) \land \exists z Q(z)}
}{}
$$

where the tree T_1 is the proof tree

$$\cfrac{P(v), P(f(v)), P(f(f(v))), \forall x P(x), \exists y Q(y) \to P(f(v))}{\forall x P(x), \exists y Q(y) \to P(f(v))}$$

and the tree T_2 is the tree

$$
\cfrac{
\cfrac{
\cfrac{\Gamma, \forall x P(x), Q(y_1) \to Q(v), Q(f(v)), Q(f(f(v))), \exists z Q(z)}
{P(v), P(f(v)), P(f(f(v))), \forall x P(x), Q(y_1) \to \exists z Q(z)}
}{P(v), P(f(v)), P(f(f(v))), \forall x P(x), \exists y Q(y) \to \exists z Q(z)}
}{\forall x P(x), \exists y Q(y) \to \exists z Q(z)}
$$

where
$$\Gamma = P(v), P(f(v)), P(f(f(v))).$$

At the end of this round,

$$TERM_0 =<< v, < \forall x P(x), \exists z Q(z) >>, < f(v), < \forall x P(x), \exists z Q(z) >>,$$
$$< f(f(v)), < \forall x P(x), \exists z Q(z) >>, < y_1, nil >, < f^3(y), nil >,$$
$$< f(y_1), nil >>,$$

$$AVAIL_0 =< f^4(v), ..., f^n(v), ... >,$$
$$AVAIL_1 =< f^2(y_1), ..., f^n(y_1), ... >,$$

and for $i > 1$, $AVAIL_i$ is unchanged. At the end of the fourth round, T_2 expands into the following proof tree:

$$\frac{\Gamma, \forall x P(x), Q(y_1) \to Q(v), Q(f(v)), Q(f(f(v))), Q(y_1), \exists z Q(z)}{\dfrac{\Gamma, \forall x P(x), Q(y_1) \to Q(v), Q(f(v)), Q(f(f(v))), \exists z Q(z)}{\dfrac{P(v), P(f(v)), P(f(f(v))), \forall x P(x), Q(y_1) \to \exists z Q(z)}{\dfrac{P(v), P(f(v)), P(f(f(v))), \forall x P(x), \exists y Q(y) \to \exists z Q(z)}{\forall x P(x), \exists y Q(y) \to \exists z Q(z)}}}}$$

where
$$\Gamma = P(v), P(f(v)), P(f(f(v))).$$

EXAMPLE 5.5.3

Let
$$A = (\exists x P(x) \wedge Q(a)) \supset \forall y P(f(y)).$$

Initially, $TERM_0 =<< a, nil >>$, and

$$AVAIL_0 =< f(a), f^2(a), ..., f^n(a), ... > .$$

At the end of the first round, the following tree is obtained:

$$\frac{\exists x P(x) \wedge Q(a) \to \forall y P(f(y))}{\to \exists x P(x) \wedge Q(a) \supset \forall y P(f(y))}$$

At the end of the round,

$$TERM_0 =<< a, nil >, < f(a), nil >>,$$

and
$$AVAIL_0 =< f^2(a), ..., f^n(a), ... > .$$

At the end of the second round, the following tree is obtained:

$$\frac{\exists x P(x), Q(a) \to P(f(y_1))}{\frac{\exists x P(x) \wedge Q(a) \to \forall y P(f(y))}{\to \exists x P(x) \wedge Q(a) \supset \forall y P(f(y))}}$$

At the end of this round,

$$TERM_0 = <<a, nil>, <f(a), nil>, <y_1, nil>, <f(y_1), nil>,$$
$$<f^2(a), nil>>,$$
$$AVAIL_0 = <f^3(a), ..., f^n(a), ...>, \text{ and}$$
$$AVAIL_1 = <f^2(y_1), ..., f^n(y_1), ...>.$$

At the end of the third round, the following tree is obtained:

$$\frac{P(y_2), Q(a) \to P(f(y_1))}{\frac{\exists x P(x), Q(a) \to P(f(y_1))}{\frac{\exists x P(x) \wedge Q(a) \to \forall y P(f(y))}{\to \exists x P(x) \wedge Q(a) \supset \forall y P(f(y))}}}$$

This tree is a finished nonclosed tree, since all formulae in the top sequent are atomic. A counter example is given by the structure having

$$H = <a, f(a), ..., f^n(a), ..., y_1, f(y_1), ..., f^n(y_1), ...,$$
$$y_2, f(y_2), ..., f^n(y_2), ...>$$

as its domain, and by interpreting P as taking the value **T** for y_2, **F** for $f(y_1)$, and Q taking the value **T** for a.

5.5.3 Completeness of the System G (Languages Without Equality)

We can now prove the following fundamental theorem.

Theorem 5.5.1 (i) If the input sequent $\Gamma_0 \to \Delta_0$ is valid, then the procedure *search* halts with a finite closed tree T, from which a proof tree for a finite subsequent $C_1, ..., C_m \to D_1, ..., D_n$ of $\Gamma_0 \to \Delta_0$ can be constructed.

(ii) If the input sequent $\Gamma_0 \to \Delta_0$ is falsifiable, then there is a Hintikka set S such that either

(a) Procedure *search* halts with a finite counter-example tree T and, if we let

$$H_S = t(TERM_0) \cup \{AVAIL_i \mid 0 \le i \le NUMACT\},$$

$\Gamma_0 \to \Delta_0$ is falsifiable in a structure with countable domain H_S; or

(b) Procedure *search* generates an infinite tree T and, if we let $H_S = t(TERM_0)$, then $\Gamma_0 \to \Delta_0$ is falsifiable in a structure with countable domain H_S.

Proof: The proof combines techniques from the proof of theorem 5.4.1 and the proof of theorem 3.5.1. The difference with the proof of theorem 5.4.1 is that a closed tree T is not exactly a proof tree, and that it is necessary to modify the tree T in order to obtain a proof tree.

Assume that *search* produces a closed tree, and let $C_1, ... C_m$ be the initial subsequence of formulae in Γ which were deleted from Γ to obtain L, and $D_1, ..., D_n$ the initial subsequence of formulae in Δ which were deleted from Δ to obtain R. A proof tree for a the finite sequent $C_1, ..., C_m \to D_1, ..., D_n$ can easily be obtained from T, using the following technique:

First, starting from the root and proceeding bottom-up, for each node $\Gamma, head(L) \to \Delta, head(R)$ at depth k created at the end of a call to procedure *expand*, add $head(L)$ after the rightmost formula in the premise of every sequent at depth less than k, and add $head(R)$ after the rightmost formula in the conclusion of every sequent at depth less than k, obtaining the tree T'. Then, a proof tree T'' for $C_1, ..., C_m \to D_1, ..., D_n$ is constructed from T' by deleting all duplicate nodes. The tree T'' is a proof tree because the same inference rules that have been used in T are used in T''.

If T is not a closed tree, along a finite or infinite path we obtain a set S of signed formulae as in the proof of theorem 5.4.1. We show the following claim:

Claim: The set H_S of terms defined in clause (ii) of theorem 5.5.1 is a term algebra (over the reduct $\mathbf{L}'$), and S is a Hintikka set with respect to H_S.

To prove the claim, as in theorem 5.4.1, we only need to prove that H_S is a term algebra, and that H3, H4, and H5 hold. We can assume that $\mathbf{L}'$ contains function symbols since the other case has been covered by theorem 5.4.1. Since $\mathbf{L}'$ contains function symbols, all the sets $AVAIL_i$ ($i \geq 0$) are countably infinite.

First, observe that every variable free in S is either a variable free in the input sequent or one of the activated variables. Since

(i) $t(TERM_0)$ and $AVAIL_0$ are initialized in such a way that the variables free in the input sequent belong to the union of $t(TERM_0)$ and $AVAIL_0$,

(ii) Whenever a new variable y_i is removed from the head of the list $AVAIL_i$ it is added to $t(TERM_0)$ and,

(iii) At the end of every round the head of every activated list (list $AVAIL_i$, for $0 \leq i \leq NUMACT$) is removed from that list and added to $t(TERM_0)$, it follows from (i)-(iii) that all variables free in S are in $t(TERM_0) \cup \{AVAIL_i \mid 0 \le i \le NUMACT\}$ if T is finite or in $t(TERM_0)$ if T is infinite, and condition H5 holds.

Observe that if T is finite, from (i), (ii), and (iii) it also follows that $t(TERM_0) \cup \{AVAIL_i \mid 0 \le i \le NUMACT\}$ contains the set of all terms built up from the variables free in the input sequent, the variables activated during applications of $\exists : left$ or $\forall : right$ rules, and the constant and function symbols occurring in the input sequent. Hence, H_S is closed under the function symbols in $\mathbf{L}'$, and it is a term algebra.

If T is infinite, all the terms in all the activated lists $AVAIL_i$ are eventually transferred to $t(TERM_0)$, and conditions (i) to (iii) also imply that $t(TERM_0)$ contains the set of all terms built up from the variables free in the input sequent, the variables activated during applications of $\exists : left$ or $\forall : right$ rules, and the constant and function symbols occurring in the input sequent. Hence, $H_S = t(TERM_0)$ is a term algebra.

If T is infinite, condition H5 also follows from (i) to (iii).

Every time a formula C of type c is expanded, all substitution instances $C(t)$ for all t in $t(TERM_0)$ which have not already been used with C are added to the upper sequent. Hence, H3 is satisfied. Every time a formula D of type d is expanded, the new variable y_i removed from $AVAIL_i$ is added to $t(TERM_0)$ and the substitution instance $D(y_i)$ is added to the upper sequent. Hence, H4 is satisfied, and the claim holds.

Since S is a Hintikka set, by lemma 5.4.5, some assignment s satisfies S in a structure $\mathbf{H_S}$ with domain H_S, and so, $\Gamma_0 \to \Delta_0$ is falsified by $\mathbf{H_S}$ and s. $\square$

Corollary   (A version of Gödel's extended completeness theorem for G) A sequent (even infinite) is valid iff it is G-provable. $\square$

As a second corollary, we obtain the following useful result.

Corollary   There is an algorithm for deciding whether a finite sequent consisting of quantifier-free formulae is valid.

Proof: Observe that for quantifier-free formulae, the *search* procedure never uses the quantifier rules. Hence, it behaves exactly as in the propositional case, and the result follows from theorem 3.4.1. $\square$

Unfortunately, this second corollary does not hold for all formulae. Indeed, it can be shown that there is no algorithm for deciding whether any given first-order formula is valid. This is known as *Church's theorem*. A proof of Church's theorem can be found in Enderton, 1972, or Kleene, 1952. A particularly concise and elegant proof due to Floyd is also given in Manna, 1974.

The completeness theorem only provides a semidecision procedure, in the sense that if a formula is valid, this can be demonstrated in a finite number of steps, but if it is falsifiable, the procedure may run forever.

Even though the *search* procedure provides a rather natural proof procedure which is theoretically complete, in practice it is extremely inefficient in terms of the number of steps and the amount of memory needed.

This is illustated by the very simple formulae of example 5.5.2 and example 5.5.3 for which it is already laborious to apply the *search* procedure. The main difficulty is the proper choice of terms in applications of $\forall : left$ and $\exists : right$ rules. In particular, note that the ordering of the terms in the lists $t(TERM_0)$ and $AVAIL_i$ can have a drastic influence on the length of proofs.

After having tried the *search* procedure on several examples, the following fact emerges: It is highly desirable to perform all the quantifier rules first, in order to work as soon as possible on quantifier-free formulae. Indeed, by the second corollary to the completeness theorem, proving quantifier-free formulae is a purely mechanical process.

It will be shown in the next chapter that provided that the formulae in the input sequent have a certain form, if such a sequent is provable, then it has a proof in which all the quantifier inferences are performed below all the propositional inferences. This fact will be formally established by Gentzen's sharpened Hauptsatz, proved in Chapter 7. Hence, the process of finding a proof can be viewed as a two-step procedure:

(1) In the first step, one attempts to "guess" the right terms used in $\forall : left$ and $\exists : right$ inferences;

(2) In the second step, one checks that the quantifier-free formula obtained in step 1 is a tautology.

A rigorous justification of this method will be given by Herbrand's theorem, proved in Chapter 7. The resolution method for first-order logic presented in Chapter 8 can be viewed as an improvement of the above method.

For the time being, we consider some applications of theorem 5.5.1. The following theorems are easily obtained as consequences of the main theorem.

5.5.4 Löwenheim-Skolem, Compactness, and Model Existence Theorems for Languages Without Equality

The following result known as Löwenheim-Skolem's theorem is often used in model theory.

Theorem 5.5.2 (Löwenheim-Skolem) If a set of formulae Γ is satisfiable in some structure **M**, then it is satisfiable in a structure whose domain is at most countably infinite.

Proof: It is clear that $\Gamma \rightarrow$ is falsifiable in **M**. By theorem 5.5.1, the *search* procedure yields a tree from which a Hintikka set S can be obtained. But the domain H of $\mathbf{H_S}$ is at most countably infinite since it consists of terms built from countable sets. Hence, $\mathbf{H_S}$ is a countable structure in which Γ is satisfiable. $\square$

The other results obtained as consequences of theorem 5.5.1 are counterparts of theorem 3.5.3, theorem 3.5.4, and lemma 3.5.4. The proofs are similar and use structures instead of valuations.

Theorem 5.5.3 (Compactness theorem) For any (possibly countably infinite) set Γ of formulae, if every nonempty finite subset of Γ is satisfiable then Γ is satisfiable.

Proof: Similar to that of theorem 3.5.3. $\square$

Recall that a set Γ of formulae is *consistent* if there exists some formula A such that $C_1, ..., C_m \rightarrow A$ is *not* G-provable for any $C_1, ..., C_m$ in Γ.

Theorem 5.5.4 (Model existence theorem) If a set Γ of formulae is consistent then it is satisfiable.

Proof: Similar to that of theorem 3.5.4. $\square$

Note that the *search* procedure will actually yield a structure $\mathbf{H_S}$ for each Hintikka set S arising along each nonclosed path in the tree T, in which S and Γ are satisfiable.

Lemma 5.5.1 (Consistency lemma) If a set Γ of formulae is satisfiable then it is consistent.

Proof: Similar to that of lemma 3.5.4. $\square$

5.5.5 Maximal Consistent Sets

The concept of a maximal consistent set introduced in definition 3.5.9 is generalized directly to sets of first-order formulae:

A set Γ of first-order formulae (without equality) is *maximally consistent* (or a *maximal consistent set*) iff, for every consistent set Δ, if Γ is a subset of Δ, then $\Gamma = \Delta$. Equivalently, every proper superset of Γ is inconsistent. lemma 3.5.5 can be easily generalized to the first-order case.

Lemma 5.5.2 Given a first-order language (without equality), every consistent set Γ is a subset of some maximal consistent set Δ.

Proof: Almost identical to that of lemma 3.5.5, but using structures instead of valuations. $\square$

It is possible as indicated in Section 3.5 for propositional logic to prove lemma 5.5.2 directly, and use lemma 5.5.2 to prove the model existence theorem (theorem 5.5.4). From the model existence theorem, the extended Gödel completeness theorem can be shown (corollary to theorem 5.5.1). Using this approach, it is necessary to use a device due to Henkin known as *adding witnesses*. Roughly speaking, this is necessary to show that a Henkin maximal consistent set is a Hintikka set with respect to a certain term algebra. The addition of witnesses is necessary to ensure condition H4 of Hintikka sets. Such an approach is explored in the problems and for details, we refer the reader to Enderton, 1972; Chang and Keisler, 1973; or VanDalen, 1982.

PROBLEMS

5.5.1. Using the *search* procedure, find counter examples for the formulae:

$$\exists x A \wedge \exists x B \supset \exists x (A \wedge B)$$
$$\forall y \exists x A \supset \exists x \forall y A$$
$$\exists x A \supset \forall x A$$

5.5.2. Using the *search* procedure, prove that the following formula is valid:

$$\neg \exists y \forall x (S(y, x) \equiv \neg S(x, x))$$

5.5.3. Using the *search* procedure, prove that the following formulae are valid:

$$\forall x A \supset A[t/x],$$
$$A[t/x] \supset \exists x A,$$
where t is free for x in A.

5.5.4. This problem is a generalization of the Hilbert system H of problem 3.4.9 to first-order logic. For simplicity, we first treat the case of first-order languages without equality. The *Hilbert system* H for first-order logic (without equality) over the language using the connectives, $\wedge, \vee, \supset, \neg, \forall$ and $\exists$ is defined as follows:

The *axioms* are all formulae given below, where A, B, C denote arbitrary formulae.

$$A \supset (B \supset A)$$
$$(A \supset B) \supset ((A \supset (B \supset C)) \supset (A \supset C))$$

$$A \supset (B \supset (A \wedge B))$$
$$A \supset (A \vee B), \qquad B \supset (A \vee B)$$
$$(A \supset B) \supset ((A \supset \neg B) \supset \neg A)$$
$$(A \wedge B) \supset A, \qquad (A \wedge B) \supset B$$
$$(A \supset C) \supset ((B \supset C) \supset ((A \vee B) \supset C))$$
$$\neg\neg A \supset A$$
$$\forall x A \supset A[t/x]$$
$$A[t/x] \supset \exists x A$$

where in the last two axioms, t is free for x in A.

There are three inference rules:

(i) The rule *modus ponens* given by:

$$\frac{A \qquad (A \supset B)}{B}$$

(ii) The *generalization rules* given by:

$$\frac{(B \supset A)}{(B \supset \forall x A)} \; \forall : rule \qquad \frac{(A \supset B)}{(\exists x A \supset B)} \; \exists : rule$$

where x does not occur free in B.

Let $\{A_1, ..., A_m\}$ be any set of formulae. The concept of a *deduction tree* of a formula B from the set $\{A_1, ..., A_m\}$ in the system H is defined inductively as follows:

(i) Every one-node tree labeled with an axiom B or a formula B in $\{A_1, ..., A_m\}$ is a deduction tree of B from $\{A_1, ..., A_m\}$.

(ii) If T_1 is a deduction tree of A from $\{A_1, ..., A_m\}$ and T_2 is a deduction tree of $(A \supset B)$ from $\{A_1, ..., A_m\}$, then the following tree is a deduction tree of B from $\{A_1, ..., A_m\}$:

$$\frac{\dfrac{T_1}{A} \qquad \dfrac{T_2}{(A \supset B)}}{B}$$

(iii) If T_1 is a deduction tree of $(B \supset A)$ from $\{A_1, ..., A_m\}$ (or a deduction tree of $(A \supset B)$ from $\{A_1, ..., A_m\}$), and x does not occur free in any of the formulae in $\{A_1, ..., A_m\}$, then the following trees are deduction trees of $(B \supset \forall x A)$ from $\{A_1, ..., A_m\}$ (or $(\exists x A \supset B)$ from $\{A_1, ..., A_m\}$).

$$\frac{\begin{array}{c} T_1 \\ \hline (B \supset A) \end{array}}{(B \supset \forall x A)} \qquad \frac{\begin{array}{c} T_1 \\ \hline (A \supset B) \end{array}}{(\exists x A \supset B)}$$

A *proof tree* is a deduction tree whose leaves are labeled with axioms. Given a set $\{A_1, ..., A_m\}$ of formulae and a formula B, we use the notation $A_1, ..., A_m \vdash B$ to denote that there is a deduction tree of B from $\{A_1, ..., A_m\}$. In particular, if the set $\{A_1, ..., A_m\}$ is empty, the tree is a proof tree and we write $\vdash B$.

Prove that the generalization rules are sound rules, in the sense that if the premise is valid, then the conclusion is valid. Prove that the system H is sound; that is, every provable formula is valid.

5.5.5. (i) Show that if $A_1, ..., A_m, A \vdash B$ is a deduction in H not using the generalization rules, then $A_1, ..., A_m \vdash (A \supset B)$.

(ii) Check that the following is a deduction of C from $A \supset (B \supset C)$ and $A \wedge B$:

$$\frac{\begin{array}{ccc} & & \dfrac{A \wedge B \quad A \wedge B \supset A}{A} \end{array}}{\dfrac{\dfrac{A \wedge B \quad A \wedge B \supset B}{B}}{C}} \qquad A \supset (B \supset C)$$

Conclude that $A \supset (B \supset C) \vdash (A \wedge B) \supset C$.

(iii) Check that the following is a deduction of C from $(A \wedge B) \supset C$, A, and B:

$$\frac{\begin{array}{cc} B \quad \dfrac{A \quad A \supset (B \supset (A \wedge B))}{\dfrac{B \supset (A \wedge B)}{A \wedge B}} & (A \wedge B) \supset C \end{array}}{C}$$

Conclude that $(A \wedge B) \supset C \vdash A \supset (B \supset C)$.

* 5.5.6. In this problem, we are also considering the proof system H of problem 5.5.4. The *deduction theorem* states that, for arbitrary formulae $A_1, ..., A_m, A, B$,

$$\text{if } A_1, ..., A_m, A \vdash B, \text{ then } A_1, ..., A_m \vdash (A \supset B).$$

Prove the deduction theorem.

Hint: Use induction on deduction trees. In the case of the generalization rules, use problem 5.5.5.

5.5.7. Prove the following:

$$\text{If } A_1, ..., A_m \vdash B_i \text{ for every } i, \ 1 \leq i \leq m \text{ and}$$
$$B_1, ..., B_m \vdash C, \text{ then } A_1, ..., A_m \vdash C.$$

Hint: Use the deduction theorem.

5.5.8. Prove that for any set Γ of formulae, and any two formulae B and C,

$$\text{if } \vdash C \quad \text{then}$$
$$\Gamma, C \vdash B \quad \text{iff} \quad \Gamma \vdash B.$$

* **5.5.9.** In this problem, we are still considering the proof system H of problem 5.5.4. Prove that the following meta rules hold about deductions in the system H:

For all formulae A, B, C and finite sequence Γ of formulae (possibly empty), we have:

Introduction	*Elimination*
$\supset$ If $\Gamma, A \vdash B$, then $\Gamma \vdash (A \supset B)$	$A, (A \supset B) \vdash B$
$\wedge$ $A, B \vdash (A \wedge B)$	$(A \wedge B) \vdash A$ $(A \wedge B) \vdash B$
$\vee$ $A \vdash (A \vee B)$	If $\Gamma, A \vdash C$ and $\Gamma, B \vdash C$ then $\Gamma, (A \vee B) \vdash C$
$\neg$ If $\Gamma, A \vdash B$ and $\Gamma, A \vdash \neg B$ then $\Gamma \vdash \neg A$ (reductio ad absurdum)	$\neg\neg A \vdash A$ (double negation elimination) $A, \neg A \vdash B$ (weak negation elimination)
$\forall$ If $\Gamma \vdash A$ then $\Gamma \vdash \forall x A$	$\forall x A \vdash A[t/x]$
$\exists$ $A[t/x] \vdash \exists x A$	If $\Gamma, A \vdash C$ then $\Gamma, \exists x A \vdash C$

where t is free for x in A, x does not occur free in Γ, and x does not occur free in C.

Hint: Use problem 5.5.7, problem 5.5.8, and the deduction theorem.

* **5.5.10.** In this problem it is shown that the Hilbert system H is complete, by proving that for every Gentzen proof T of a sequent $\rightarrow A$, where A is any formula, there is a proof in the system H.

(i) Prove that for arbitrary formulae $A_1, ..., A_m, B_1, ..., B_n$,

(a) in H, for $n > 0$,
$$A_1, ..., A_m, \neg B_1, ..., \neg B_n \vdash P \wedge \neg P \quad \text{if and only if}$$
$$A_1, ..., A_m, \neg B_1, ..., \neg B_{n-1} \vdash B_n, \quad \text{and}$$

(b) in H, for $m > 0$,
$$A_1, ..., A_m, \neg B_1, ..., \neg B_n \vdash P \wedge \neg P \quad \text{if and only if}$$
$$A_2, ..., A_m, \neg B_1, ..., \neg B_n \vdash \neg A_1.$$

(ii) Prove that for any sequent $A_1, ..., A_m \to B_1, ..., B_n$,

if $\vdash A_1, ..., A_m \to B_1, ..., B_n$ in the Gentzen system G then
$$A_1, ..., A_m, \neg B_1, ..., \neg B_n \vdash (P \wedge \neg P)$$
is a deduction in the Hilbert system H.

Conclude that H is complete.

Hint: Use problem 5.5.9.

5.5.11. Recall that the cut rule is the rule
$$\frac{\Gamma \to \Delta, A \qquad A, \Lambda \to \Theta}{\Gamma, \Lambda \to \Delta, \Theta}$$

A is called the *cut formula* of this inference.

Let $G + \{cut\}$ be the formal system obtained by adding the cut rule to G. The notion of a deduction tree is extended to allow the cut rule as an inference. A proof in G is called a *cut-free* proof.

(i) Prove that for every structure **A**, if **A** satisfies the premises of the cut rule, then it satisfies its conclusion.

(ii) Prove that if a sequent is provable in the system $G + \{cut\}$, then it is valid.

(iii) Prove that if a sequent is provable in $G + \{cut\}$, then it has a cut-free proof.

5.5.12. (i) Prove solely in terms of proofs in $G + \{cut\}$ that a set Γ of formulae is inconsistent if and only if there is a formula A such that both $\Gamma \to A$ and $\Gamma \to \neg A$ are provable in $G + \{cut\}$.

(ii) Prove solely in terms of proofs in $G + \{cut\}$ that, $\Gamma \to A$ is not provable in $G + \{cut\}$ if and only if $\Gamma \cup \{\neg A\}$ is consistent.

Note: Properties (i) and (ii) also hold for the proof system G, but the author does not know of any proof not involving a proof-theoretic version of Gentzen's *cut elimination theorem*. The completeness theorem for G provides a semantic proof of the cut elimination theorem. However, in order to show (i) and (ii) without using semantic arguments, it appears that one has to mimic Gentzen's original proof. (See Szabo, 1969.)

5.5.13. A set Γ of sentences is said to be *complete* if, for every sentence A, either

$$\vdash \Gamma \rightarrow A \quad \text{or}$$
$$\vdash \Gamma \rightarrow \neg A,$$

but not both. Prove that for any set Γ of *sentences*, the following are equivalent:

(i) The set $\{A \mid \vdash \Gamma \rightarrow A \text{ in } G\}$ is a maximal consistent set.

(ii) Γ is complete.

(iii) Any two models of Γ are elementary equivalent.

(See problem 5.3.25 for the definition of *elementary equivalence*.)

5.5.14. Let Γ be a consistent set (of **L**-formulae). Let $A_1, A_2, ..., A_n, ...$ be an enumeration of *all* **L**-formulae. Define the sequence Γ_n inductively as follows:

$$\Gamma_0 = \Gamma,$$

$$\Gamma_{n+1} = \begin{cases} \Gamma_n \cup \{A_n\} & \text{if } \Gamma_n \cup \{A_n\} \text{ is consistent;} \\ \Gamma_n & \text{otherwise.} \end{cases}$$

Let

$$\Delta = \bigcup_{n \geq 0} \Gamma_n.$$

Prove the following:

(a) Each Γ_n is consistent.

(b) Δ is consistent.

(c) Δ is maximally consistent.

* **5.5.15.** Prove that the results of problem 3.5.16 hold for first-order logic if we change the word *proposition* to *sentence*.

* **5.5.16.** Prove that the results of problem 3.5.17 hold for first-order logic if we change the word *proposition* to *sentence* and work in $G + \{cut\}$.

* **5.5.17.** In this problem and the next four, Henkin's version of the completeness theorem is worked out. The approach is to prove the model existence theorem, and derive the completeness theorem as a consequence. To prove that every consistent set S is satisfiable, first problem 5.5.14 is used to extend S to a maximal consistent set. However, such a maximal consistent set is not necessarily a Hintikka set, because condition H4 may not be satisfied. To overcome this problem, we use Henkin's method, which consists of adding to S formulae of the form $\exists x B \supset B[c/x]$, where c is a new constant called a *witness* of $\exists x B$. However, we have to iterate this process to obtain the desired property.

Technically, we proceed as follows. Let **L** be a first-order language (without equality). Let **L*** be the extension of **L** obtained by adding the set of new constants

$$\{c_D \mid D \text{ is any sentence of the form } \exists x B\}.$$

The constant c_D is called a *witness* of D. Let S be any set of **L**-sentences, and let

$$S^* = S \cup \{\exists x B \supset B[c_D/x] \mid \exists x B \in FORM_{\mathbf{L}},$$
$$c_D \text{ is a witness of the sentence } D = \exists x B\}.$$

(Note that $\exists x B$ is any *arbitrary* existential sentence which need not belong to S).

(a) Prove that for every **L**-sentence A, if $S^* \to A$ is provable in $G + \{cut\}$, then $S \to A$ is provable in $G + \{cut\}$.

Hint: Let Γ be a finite subset of S^* such that $\Gamma \to A$ is provable (in $G + \{cut\}$). Assume that Γ contains some sentence of the form $\exists x B \supset B[c_D/x]$, where $D = \exists x B$. Let $\Delta = \Gamma - \{\exists x B \supset B[c_D/x]\}$. Note that c_D does not occur in Δ, $\exists x B$, or A. Using the result of problem 5.4.8, show that $\Delta, \exists x B \supset B[y/x] \to A$ is provable (in $G + \{cut\}$), where y is a new variable not occurring in $\Gamma \to A$. Next, show that $\to \exists x(\exists x B \supset B)$ is provable (in G). Then, show (using a cut) that $\Delta \to A$ is provable (in $G + \{cut\}$). Conclude by induction on the number of sentences of the form $\exists x B \supset B[c_D/x]$ in Γ.

(b) A set S of **L**-sentences is a *theory* iff it is closed under provability, that is, iff

$$S = \{A \mid \vdash S \to A \text{ in } G + \{cut\}, FV(A) = \emptyset\}.$$

A theory S is a *Henkin theory* iff for any sentence D of the form $\exists x B$ in $FORM_{\mathbf{L}}$ (*not necessarily* in S), there is a constant c in **L** such that $\exists x B \supset B[c/x]$ is also in S.

Let S be any set of **L**-sentences and let T be the theory

$$T = \{A \mid \vdash S \to A \text{ in } G + \{cut\}, FV(A) = \emptyset\}.$$

Define the sequence of languages $\mathbf{L}_n$ and the sequence of theories T_n as follows:

$$\mathbf{L}_0 = \mathbf{L}; \ \mathbf{L}_{n+1} = (\mathbf{L}_n)^*; \text{ and}$$
$$T_0 = T; \ T_{n+1} = \{A \mid \vdash (T_n)^* \to A \text{ in } G + \{cut\}, FV(A) = \emptyset\}.$$

Let

$$\mathbf{L}^H = \bigcup_{n \geq 0} \mathbf{L}_n,$$

and

$$T^H = \bigcup_{n \geq 0} T_n.$$

Prove that T^H is a Henkin theory over the language $\mathbf{L}^H$.

(c) Prove that for every $\mathbf{L}$-sentence C,

$$\text{if } \vdash T^H \to C \text{ in } G + \{cut\},$$
$$\text{then } \vdash T \to C \text{ in } G + \{cut\}.$$

We say that T^H is *conservative* over T.

(d) Prove that T^H is consistent iff T is.

∗ **5.5.18.** Let T be a consistent set of $\mathbf{L}$-sentences.

(a) Show that there exists a maximal consistent extension T' over $\mathbf{L}^H$ of T which is also a Henkin theory.

Hint: Let $S = \{A \mid \vdash T \to A, FV(A) = \emptyset\}$. Show that any maximally consistent extension of S^H is also a Henkin theory which is maximally consistent.

(b) The definition of formulae of type a, b, c, d for unsigned formulae and of their *immediate descendants* given in problem 3.5.7 is extended to the first-order case.

Formulae of types a, b, c, d and their *immediate descendants* are defined by the following table:

Type-a formulae

A	A_1	A_2
$(X \wedge Y)$	X	Y
$\neg(X \vee Y)$	$\neg X$	$\neg Y$
$\neg(X \supset Y)$	X	$\neg Y$
$\neg\neg X$	X	X

Type-b formulae

B	B_1	B_2
$\neg(X \wedge Y)$	$\neg X$	$\neg Y$
$(X \vee Y)$	X	Y
$(X \supset Y)$	$\neg X$	Y

Type-c formulae

C	C_1	C_2
$\forall x Y$	$Y[t/x]$	$Y[t/x]$
$\neg \exists x Y$	$\neg Y[t/x]$	$\neg Y[t/x]$

where t is free for x in Y

Type-d formulae

D	D_1	D_2
$\exists x Y$	$Y[t/x]$	$Y[t/x]$
$\neg \forall x Y$	$\neg Y[t/x]$	$\neg Y[t/x]$

where t is free for x in Y

The definition of a Hintikka set is also adapted in the obvious way to unsigned formulae. Also, in this problem and some of the following problems, given any set S of formulae and any formulae $A_1,...,A_n$, the set $S \cup \{A_1, ...A_n\}$ will also be denoted by $\{S, A_1, ..., A_n\}$.

Prove that a maximal consistent Henkin set T' of sentences is a Hintikka set with respect to the term algebra H consisting of all closed terms built up from the function and constant symbols in $\mathbf{L}^H$.

(c) Prove that every consistent set T of **L**-sentences is satisfiable.

(d) Prove that a formula A with free variables $\{x_1, ..., x_n\}$ is satisfiable iff $A[c_1/x_1, ..., c_n/x_n]$ is satisfiable, where $c_1,...,c_n$ are new constants. Using this fact, prove that every consistent set T of **L**-formulae is satisfiable.

* **5.5.19.** Recall that from problem 5.5.12, in $G + \{cut\}$, $\Gamma \to A$ is not provable if and only if $\Gamma \cup \{\neg A\}$ is consistent. Use the above fact and problem 5.5.18 to give an alternate proof of the extended completeness theorem for $G + \{cut\}$ ($\models \Gamma \to A$ iff $\vdash \Gamma \to A$ in $G + \{cut\}$).

** **5.5.20.** Prove that the results of problem 5.5.17 hold for languages of any cardinality. Using Zorn's lemma, prove that the results of problem 5.5.18 hold for languages of any cardinality. Thus, prove that the extended completeness theorem holds for languages of any cardinality.

* **5.5.21.** For a language **L** of cardinality α, show that the cardinality of the term algebra arising in problem 5.5.18 is at most α. Conclude that any consistent set of **L**-sentences has a model of cardinality at most α.

* **5.5.22.** Let **L** be a countable first-order language without equality. A property P of sets of formulae is an *analytic consistency property* iff the following hold:

(i) P is of finite character (see problem 3.5.12).

(ii) For every set S of formulae for which P is true, the following conditions hold:

A_0: S contains no atomic formula and its negation.

A_1: For every formula A of type a in S, P holds for $\{S, A_1\}$ and $\{S, A_2\}$.

A_2: For every formula B of type b in S, either P holds for $\{S, B_1\}$ or P holds for $\{S, B_2\}$.

A_3: For every formula C of type c in S, P holds for $\{S, C(t)\}$ for every term t.

A_4: For every formula D of type d in S, for some constant c not occurring in S, P holds for $\{S, D(c)\}$.

Prove that if P is an analytic consistency property and P holds for S, then S is satisfiable.

Hint: The *search* procedure can be adapted to work with sets of formulae rather than sequents, by identifying each sequent $\Gamma \to \Delta$ with the set of formulae $\Gamma \cup \{\neg B \mid B \in \Delta\}$, and using the rules corresponding to the definition of descendants given in problem 5.5.18. Using the *search* procedure applied to the set S (that is, to the sequent $S \to$), show that at any stage of the construction of a deduction tree, there is a path such that P holds for the union of S and the union of the formulae along that path. Also, the constant c plays the same role as a new variable.

∗ 5.5.23. Let **L** be a countable first-order language without equality. A set S of formulae is *truth-functionally inconsistent* iff S is the result of substituting first-order formulae for the propositional letters in an unsatisfiable set of propositions (see definition 5.3.11). We say that a formula A is *truth-functionally valid* iff it is obtained by substitution of first-order formulae into a tautology (see definition 5.3.11). We say that a finite set $S = \{A_1, ..., A_m\}$ *truth-functionally implies* B iff the formula $(A_1 \wedge ... \wedge A_m) \supset B$ is truth-functionally valid. A property P of sets of formulae is a *synthetic consistency property* iff the following conditions hold:

(i) P is of finite character (see problem 3.5.12).

(ii) For every set S of formulae, the following conditions hold:

B_0: If S is truth-functionally inconsistent, then P does not hold for S;

B_3: If P holds for S then for every formula C of type c in S, P holds for $\{S, C(t)\}$ for every term t.

B_4: For every formula D of type d in S, for some constant c not occurring in S, P holds for $\{S, D(c)\}$.

B_5: For every formula X, if P does not hold for $\{S, X\}$ or $\{S, \neg X\}$, then P does not hold for S. Equivalently, if P holds for S, then for every formula X, either P holds for $\{S, X\}$ or P holds for $\{S, \neg X\}$

(a) Prove that if P is a synthetic consistency property then the following condition holds:

B_6: If P holds for S and a finite subset of S truth-functionally implies X, then P holds for $\{S, X\}$.

(b) Prove that every synthetic consistency property is an analytic consistency property.

(c) Prove that consistency within the Hilbert system H of problem 5.5.4 is a synthetic consistency property.

∗ **5.5.24.** A set S of formulae is *Henkin-complete* if for every formula $D = \exists x B$ of type d, there is some constant c such that $B(c)$ is also in S.

Prove that if P is a synthetic consistency property and S is a set of formulae that is both Henkin-complete and a maximal set for which P holds, then S is satisfiable.

Hint: Show that S is a Hintikka set for the term algebra consisting of all terms built up from function, constant symbols, and variables occuring free in S.

∗ **5.5.25.** Prove that if P is a synthetic consistency property, then every set S of formulae for which P holds can be extended to a Henkin-complete set which is a maximal set for which P holds.

Hint: Use the idea of problem 5.5.17.

Use the above property to prove the completeness of the Hilbert system H.

∗ **5.5.26.** Let **L** be a countable first-order language without equality. The method of this problem that is due to Henkin (reported in Smullyan, 1968, Chapter 10, page 96) yields one of the most elegant proofs of the model existence theorem.

Using this method, a set that is both Henkin-complete and maximal consistent is obtained from a *single* construction.

Let $A_1, ..., A_n, ...$ be an enumeration of all **L**-formulae. Let P be a synthetic consistency property, and let S be a set of formulae for which P holds. Construct the sequence S_n as follows:

$$S_0 = S;$$

$$S_{n+1} = \begin{cases} S_n \cup \{A_{n+1}\} & \text{If } P \text{ satisfies } S_n \cup \{A_{n+1}\}; \\ S_n \cup \{\neg A_{n+1}\} & \text{otherwise}; \end{cases}$$

In addition, if A_{n+1} is a formula $\exists x D$ of type d, then add $D(c)$ to S_{n+1} for some new constant c not in S_n. (It may be necessary to add countably many new constants to **L**.)

(a) Prove that $S' = \bigcup_{n \geq 0} S_n$ is Henkin-complete, and a maximal set such that P holds for S'.

(b) Let P be an analytic consistency property. Given a set S of formulae, let $Des(S)$ be the set of *immediate descendants* of formulae in S as defined in problem 5.5.18, and define S^n by induction as follows:

$$S^0 = S;$$
$$S^{n+1} = Des(S^n).$$
$$\text{Let } S^* = \bigcup_{n \geq 0} S^n.$$

S^* is called the *set of descendants* of S. Use the construction of (a) to prove that every set S of formulae for which P holds can be extended to a Henkin-complete subset S' of S^* which is a maximal subset of S^* for which P holds.

Hint: Consider an enumeration of S^* and extend S to a subset of S^*, which is Henkin-complete and maximal with respect to P.

Why does such a set generally fail to be the set of formulae satisfied by some structure?

(c) Prove that if M is a subset of S^* which is Henkin-complete and is a maximal subset of S^* for which P holds, then M is a Hintikka set with respect to the term algebra H consisting of all terms built up from function symbols, constants, and variables occurring free in M.

(d) Use (c) to prove that every set S of formulae for which P holds is satisfiable.

5.6 A Gentzen System for First-Order Languages With Equality

Let **L** be a first-order language with equality. The purpose of this section is to generalize the results of Section 5.5 to first-order languages with equality. First, we need to generalize the concept of a Hintikka set and lemma 5.4.5 to deal with languages with equality.

5.6.1 Hintikka Sets for Languages With Equality

The only modification to the definition of a signed formula (definition 5.4.3) is that we allow atomic formulae of the form $(s \doteq t)$ as the formula A in TA or FA.

Definition 5.6.1 A *Hintikka S set with respect to a term algebra H* (over the reduct $\mathbf{L}_S$ *with equality*) is a set of signed L-formulae such that the following conditions hold for all signed formulae A,B,C,D of type a,b,c,d:

H0: No atomic formula and its conjugate are both in S (TA or FA is atomic iff A is).

H1: If a type-a formula A is in S, then both A_1 and A_2 are in S.

H2: If a type-b formula B is in S, then either B_1 is in S or B_2 is in S.

H3: If a type-c formula C is in S, then for every $t \in H$, $C(t)$ is in S (where t is free for x in C).

H4: If a type-d formula D is in S, then for at least one term $t \in H$, $D(t)$ is in S (where t is free for x in D).

H5: Every variable x occurring free in some formula of S is in H.

H6(i): For every term $t \in H$,

$$T(t \doteq t) \in S.$$

(ii): For every n-ary function symbol f in $\mathbf{L}_S$ and any terms $s_1, ..., s_n$, $t_1, ..., t_n \in H$,

$$T((s_1 \doteq t_1) \wedge ... \wedge (s_n \doteq t_n) \supset (fs_1...s_n \doteq ft_1...t_n)) \in S.$$

(iii): For every n-ary predicate symbol P in $\mathbf{L}_S$ (including $\doteq$) and any terms $s_1, ..., s_n, t_1, ..., t_n \in H$

$$T(((s_1 \doteq t_1) \wedge ... \wedge (s_n \doteq t_n) \wedge Ps_1...s_n) \supset Pt_1....t_n) \in S.$$

As in lemma 5.4.5, we can assume without loss of generality that the set of variables occurring free in formulae in S is disjoint from the set of variables occurring bound in formulae in S.

The alert reader will have noticed that conditions H6(i) to H6(iii) do not state explicitly that $\doteq$ is an equivalence relation. However, these conditions will be used in the proof of lemma 5.6.1 to show that a certain relation on terms is a congruence as defined in Subsection 2.4.6. The generalization of lemma 5.4.5 is as follows.

Lemma 5.6.1 Every Hintikka set S (over a language with equality) with respect to a term algebra H is satisfiable in a structure $\mathbf{H_S}$ whose domain H_S is a quotient of H.

Proof: In order to define the $\mathbf{L_S}$-structure $\mathbf{H_S}$, we define the relation $\cong$ on the set of all terms in H as follows:

$$s \cong t \quad \text{if and only if} \quad T(s \doteq t) \in S.$$

First, we prove that $\cong$ is an equivalence relation.

(1) By condition H6(i), $\cong$ is reflexive.

(2) Assume that $T(s \doteq t) \in S$. By H6(iii),

$$T((s \doteq t) \wedge (s \doteq s) \wedge Pss \supset Pts) \in S$$

for any binary predicate symbol P, and in particular when P is $\doteq$. By H2, either $T(t \doteq s) \in S$ or $F((s \doteq t) \wedge (s \doteq s) \wedge (s \doteq s)) \in S$. In the second case, using H2 again either $F(s \doteq t) \in S$, or $F(s \doteq s) \in S$. In each case we reach a contradiction by H0 and H6(i). Hence, we must have $T(t \doteq s) \in S$, establishing that $\cong$ is symmetric.

(3) To prove transitivity we proceed as follows. Assume that $T(r \doteq s)$ and $T(s \doteq t)$ are in S. By (2) above, $T(s \doteq r) \in S$. By H6(iii), we have

$$T((s \doteq r) \wedge (s \doteq t) \wedge (s \doteq s) \supset (r \doteq t)) \in S.$$

(With P being identical to $\doteq$.) A reasoning similar to that used in (2) shows that $T(r \doteq t) \in S$, establishing transitivity.

Hence, $\cong$ is an equivalence relation on H.

We now define the structure $\mathbf{H_S}$ as follows. The domain H_S of this structure is the quotient of the set H modulo the equivalence relation $\cong$, that is, the set of all equivalence classes of terms in H modulo $\cong$ (see Subsection 2.4.7). Given a term $t \in H$, we let $\bar{t}$ denote its equivalence class. The interpretation function is defined as follows:

Every constant c in $\mathbf{L_S}$ is interpreted as the equivalence class $\bar{c}$;

Every function symbol f of rank n in $\mathbf{L_S}$ is interpreted as the function such that, for any equivalence classes $\bar{t_1}, ..., \bar{t_n}$,

$$f_{\mathbf{H_S}}(\bar{t_1}, ..., \bar{t_n}) = \overline{ft_1...t_n};$$

For every predicate symbol P of rank n in $\mathbf{L_S}$, for any equivalence classes $\bar{t_1}, ..., \bar{t_n}$,

$$P_{\mathbf{H_S}}(\bar{t_1}, ..., \bar{t_n}) = \begin{cases} \mathbf{F} & \text{if } FPt_1...t_n \in S, \\ \mathbf{T} & \text{if } TPt_1...t_n \in S, \\ & \text{or neither } TPt_1...t_n \text{ nor } FPt_1...t_n \text{ is in } S. \end{cases}$$

To define a quotient structure as in Subsection 2.4.7, we need to show that the equivalence relation $\cong$ is a congruence (as defined in Subsection 2.4.6) so that the functions $f_{\mathbf{Hs}}$ and the predicates $P_{\mathbf{Hs}}$ are well defined; that is, that their definition is independent of the particular choice of representatives $t_1, ..., t_n$ in the equivalence classes of terms (in H). This is shown using condition H6(ii) for functions and condition H6(iii) for predicates. More specifically, the following claim is proved:

Claim 1: For any terms $s_1, ..., s_n, t_1, ..., t_n \in H$, if $\overline{s_i} = \overline{t_i}$ for $i = 1, .., n$, then:

(i) For each n-ary function symbol f,

$$\overline{fs_1...s_n} = \overline{ft_1...t_n}.$$

(ii) For each n-ary predicate symbol P,

$$\text{if} \quad TPs_1...s_n \in S \quad \text{then} \quad TPt_1...t_n \in S, \quad \text{and}$$
$$\text{if} \quad FPs_1...s_n \in S \quad \text{then} \quad FPt_1...t_n \in S.$$

Proof of claim 1:

To prove (i), assume that $T(s_i \doteq t_i) \in S$, for all i, $1 \leq i \leq n$. Since by H6(ii),

$$T((s_1 \doteq t_1) \wedge ... \wedge (s_n \doteq t_n) \supset (fs_1...s_n \doteq ft_1...t_n)) \in S,$$

by condition H2 of a Hintikka set, either

$$T(fs_1...s_n \doteq ft_1...t_n) \in S,$$

or

$$F((s_1 \doteq t_1) \wedge ... \wedge (s_n \doteq t_n)) \in S.$$

But if $F((s_1 \doteq t_1) \wedge ... \wedge (s_n \doteq t_n)) \in S$, by H2, some $F(s_i \doteq t_i) \in S$, contradicting H0. Hence, $T(fs_1...s_n \doteq ft_1...t_n)$ must be in S, which is just the conclusion of (i) by the definition of the relation $\cong$.

Next, we prove (ii). If $TPs_1...s_n \in S$, since by H6(iii),

$$T((s_1 \doteq t_1) \wedge ... \wedge (s_n \doteq t_n) \wedge Ps_1...s_n \supset Pt_1....t_n) \in S,$$

by condition H2, either

$$TPt_1...t_n \in S,$$

or

$$F((s_1 \doteq t_1) \wedge ... \wedge (s_n \doteq t_n) \wedge Ps_1...s_n) \in S.$$

By condition H2, either

$$FPs_1...s_n \in S,$$

or
$$F((s_1 \doteq t_1) \wedge ... \wedge (s_n \doteq t_n)) \in S.$$

In the first case, H0 is violated, and in the second case, since we have assumed that $T(s_i \doteq t_i) \in S$ for all i, $1 \leq i \leq n$, H0 is also violated as in the proof of (i). Hence, $TPt_1...t_n$ must be in S.

If $FPs_1...s_n \in S$, since by H6(iii),

$$T((t_1 \doteq s_1) \wedge ... \wedge (t_n \doteq s_n) \wedge Pt_1...t_n \supset Ps_1....s_n) \in S,$$

either
$$TPs_1...s_n \in S,$$

or
$$F((t_1 \doteq s_1) \wedge ... \wedge (t_n \doteq s_n) \wedge Pt_1...t_n) \in S.$$

In the first case, H0 is violated. In the second case, either

$$FPt_1...t_n \in S,$$

or
$$F((t_1 \doteq s_1) \wedge ... \wedge (t_n \doteq s_n)) \in S.$$

However, if
$$F((t_1 \doteq s_1) \wedge ... \wedge (t_n \doteq s_n)) \in S,$$

since we have assumed that $Ts_i \doteq t_i \in S$ for all i, $1 \leq i \leq n$, and we have shown in (2) that whenever $Ts_i \doteq t_i \in S$, then $Tt_i \doteq s_i$ is also in S, H0 is also contradicted. Hence, $FPt_1...t_n$ must be in S. This concludes the proof of claim 1. $\square$

Claim (1)(i) shows that $\cong$ is a congruence with respect to function symbols. To show that $\cong$ is a congruence with respect to predicate symbols, we show that:

If $s_i \cong t_i$ for all i, $1 \leq i \leq n$, then

$$\begin{aligned} TPs_1...s_n \in S \quad &\text{iff} \quad TPt_1...t_n \in S \quad \text{and} \\ FPs_1...s_n \in S \quad &\text{iff} \quad FPt_1...t_n \in S. \end{aligned}$$

Proof: If $TPs_1...s_n \in S$, by claim 1(ii), $TPt_1...t_n \in S$. If $TPs_1...s_n \notin S$, then either

$$FPs_1...s_n \in S,$$

or
$$FPs_1...s_n \notin S.$$

If $FPs_1...s_n \in S$, by claim 1(ii), $FPt_1...t_n \in S$, and by H0, $TPt_1...t_n \notin S$. If $FPs_1...s_n \notin S$, then if $TPt_1...t_n$ was in S, then by claim 1(ii), $TPs_1...s_n$

would be in S, contrary to the assumption. The proof that $FPs_1...s_n \in S$ iff $FPt_1...t_n \in S$ is similar. This concludes the proof. $\square$

Having shown that the definition of the structure $\mathbf{H_S}$ is proper, let s be any assignment such that $s(x) = \bar{x}$, for each variable $x \in H$. It remains to show that $\mathbf{H_S}$ and s satisfy S. For this, we need the following claim which is proved using the induction principle for terms:

Claim 2: For every term t (in H),

$$t_{\mathbf{H_S}}[s] = \bar{t}.$$

If t is a variable x, then $x_{\mathbf{H_S}}[s] = s(x) = \bar{x}$, and the claim holds.

If t is a constant c, then

$$c_{\mathbf{H_S}}[s] = c_{\mathbf{H_S}} = \bar{c}$$

by the definition of the interpretation function.

If t is a term $ft_1...t_k$, then

$$(ft_1...t_k)_{\mathbf{H_S}}[s] = f_{\mathbf{H_S}}((t_1)_{\mathbf{H_S}}[s], ..., (t_k)_{\mathbf{H_S}}[s]).$$

By the induction hypothesis, for all i, $1 \le i \le k$,

$$(t_i)_{\mathbf{H_S}}[s] = \bar{t}_i.$$

Hence,

$$(ft_1...t_k)_{\mathbf{H_S}}[s] = f_{\mathbf{H_S}}((t_1)_{\mathbf{H_S}}[s], ..., (t_k)_{\mathbf{H_S}}[s])$$
$$= f_{\mathbf{H_S}}(\bar{t_1}, ..., \bar{t_k}).$$

But by the definition of $f_{\mathbf{H_S}}$,

$$f_{\mathbf{H_S}}(\bar{t_1}, ..., \bar{t_k}) = \overline{ft_1...t_k} = \bar{t}.$$

This concludes the induction and the proof of claim 2. $\square$

Claim 2 is now used to show that for any atomic formula $TPt_1...t_n \in S$, $Pt_1...t_n[s]$ is satisfied in $\mathbf{H_S}$ (and that for any $FPt_1...t_n \in S$, $Pt_1...t_n[s]$ is not satisfied in $\mathbf{H_S}$). Indeed, if $TPt_1...t_n \in S$, by H5 and since H is a term algebra, $t_1, ..., t_n \in H$. Then,

$$(Pt_1...t_n)_{\mathbf{H_S}}[s] = P_{\mathbf{H_S}}((t_1)_{\mathbf{H_S}}[s], ..., (t_n)_{\mathbf{H_S}}[s])$$
$$= \text{(by Claim 2) } P_{\mathbf{H_S}}(\bar{t_1}, ..., \bar{t_n}) = \mathbf{T}.$$

Hence,

$$\mathbf{H_S} \models (Pt_1...t_n)[s].$$

A similar proof applies when $FPt_1....t_n \in S$. Also, if $T(t_1 \doteq t_2) \in S$, by definition of $\cong$ we have $t_1 \cong t_2$, which is equivalent to $\overline{t_1} = \overline{t_2}$, that is, $(t_1)_{\mathbf{Hs}}[s] = (t_2)_{\mathbf{Hs}}[s]$. The rest of the argument proceeds using the induction principle for formulae. The propositional connectives are handled as before.

Let us give the proof for a signed formula of type c. If a formula C of type c is in S, then by H3, $C(t)$ is in S for every term $t \in H$. By the induction hypothesis, we have

$$\mathbf{Hs} \models C(t)[s].$$

By lemma 5.4.1, for any formula A, any term t free for x in A, any structure $\mathbf{M}$ and any assignment v, we have

$$(A[t/x])_{\mathbf{M}}[v] = (A[\mathbf{a}/x])_{\mathbf{M}}[v],$$

where $a = t_{\mathbf{M}}[v]$.

Since $t_{\mathbf{Hs}}[s] = \overline{t}$, we have

$$(C[t/x])_{\mathbf{Hs}}[s] = (C[\overline{t}/x])_{\mathbf{Hs}}[s].$$

Hence,

$$\mathbf{Hs} \models C(t)[s] \quad \text{iff} \quad \mathbf{Hs} \models C(\overline{\mathbf{t}})[s],$$

and since $\overline{t} \in \mathbf{Hs}$, by lemma 5.4.4, this shows that

$$\mathbf{Hs} \models C[s].$$

As in lemma 5.4.5, $\mathbf{Hs}$ is expanded to an $\mathbf{L}$-structure in which S is satisfiable. This concludes the proof that every signed formula in S is satisfied in $\mathbf{Hs}$ by the assignment s. $\square$

5.6.2 The Gentzen System $G_=$ (Languages With Equality)

Let $G_=$ be the Gentzen system obtained from the Gentzen system G (defined in definition 5.4.1) by adding the following rules.

Definition 5.6.2 (Equality rules for $G_=$) Let Γ, Δ, Λ denote arbitrary sequences of formulae (possibly empty) and let $t,s_1,...,s_n$, $t_1,...,t_n$ denote arbitrary $\mathbf{L}$-terms.

For each term t, we have the following rule:

$$\frac{\Gamma, t \doteq t \to \Delta}{\Gamma \to \Delta} \quad reflexivity$$

For each n-ary function symbol f and any terms $s_1, ..., s_n, t_1, ..., t_n$, we have the following rule:

$$\frac{\Gamma, (s_1 \doteq t_1) \wedge ... \wedge (s_n \doteq t_n) \supset (fs_1...s_n \doteq ft_1...t_n) \to \Delta}{\Gamma \to \Delta}$$

For each n-ary predicate symbol P (including $\doteq$) and any terms $s_1, ..., s_n, t_1, ..., t_n$, we have the following rule:

$$\frac{\Gamma, ((s_1 \doteq t_1) \wedge ... \wedge (s_n \doteq t_n) \wedge Ps_1...s_n) \supset Pt_1...t_n \to \Delta}{\Gamma \to \Delta}$$

Note that the reason these formulae are added after the rightmost formula of the premise of a sequent is that in the proof of theorem 5.6.1, this simplifies the reconstruction of a legal proof tree from a closed tree. In fact, as in definition 5.4.8, we generalize the above rules to inferences of the form

$$\frac{\Gamma, A_1, ..., A_k \to \Delta}{\Gamma \to \Delta}$$

where $A_1, ..., A_k$ are any formulae of the form either

$$t \doteq t,$$
$$(s_1 \doteq t_1) \wedge ... \wedge (s_n \doteq t_n) \supset (fs_1...s_n \doteq ft_1...t_n), \quad \text{or}$$
$$((s_1 \doteq t_1) \wedge ... \wedge (s_n \doteq t_n) \wedge Ps_1...s_n) \supset Pt_1...t_n.$$

EXAMPLE 5.6.1

Let

$$A = \forall x (x \doteq f(y) \supset P(x)) \supset P(f(y)),$$

where f is a unary function symbol, and P is a unary predicate symbol.

The following tree is a proof tree for A.

$$\frac{\dfrac{T_1 \quad P(f(y)), \forall x(x \doteq f(y) \supset P(x)) \to P(f(y))}{f(y) \doteq f(y) \supset P(f(y)), \forall x(x \doteq f(y) \supset P(x)) \to P(f(y))}}{\dfrac{\forall x(x \doteq f(y) \supset P(x)) \to P(f(y))}{\to \forall x(x \doteq f(y) \supset P(x)) \supset P(f(y))}}$$

where T_1 is the tree

$$\frac{\forall x(x \doteq f(y) \supset P(x)), f(y) \doteq f(y) \to f(y) \doteq f(y), P(f(y))}{\forall x(x \doteq f(y) \supset P(x)) \to f(y) \doteq f(y), P(f(y))}$$

EXAMPLE 5.6.2

Let

$$A = P(f(y)) \supset \forall x(x \doteq f(y) \supset P(x)).$$

The following tree is a proof tree for A.

$$\cfrac{\cfrac{\cfrac{\cfrac{\cfrac{T_1 \qquad P(z), P(f(y)), z \doteq f(y) \to P(z)}{P(f(y)), z \doteq f(y), (f(y) \doteq z \wedge P(f(y))) \supset P(z) \to P(z)}}{P(f(y)), z \doteq f(y) \to P(z)}}{P(f(y)) \to z \doteq f(y) \supset P(z)}}{P(f(y)) \to \forall x(x \doteq f(y) \supset P(x))}}{\to P(f(y)) \supset \forall x(x \doteq f(y) \supset P(x))}$$

where T_1 is the tree

$$\cfrac{T_2 \qquad P(f(y)), z \doteq f(y) \to P(f(y)), P(z)}{P(f(y)), z \doteq f(y) \to f(y) \doteq z \wedge P(f(y)), P(z)}$$

T_2 is the tree

$$\cfrac{\cfrac{T_3 \qquad f(y) \doteq z, P(f(y)), z \doteq f(y) \to f(y) \doteq z, P(z)}{P(f(y)), z \doteq f(y), (z \doteq f(y) \wedge z \doteq z \wedge z \doteq z) \supset f(y) \doteq z \to f(y) \doteq z, P(z)}}{P(f(y)), z \doteq f(y) \to f(y) \doteq z, P(z)}$$

and T_3 is the tree

$$\cfrac{\cfrac{\cfrac{S_2 \qquad\qquad S_2}{P(f(y)), z \doteq f(y), z \doteq z \to z \doteq z \wedge z \doteq z, f(y) \doteq z, P(z)}}{S_1 \quad P(f(y)), z \doteq f(y) \to z \doteq z \wedge z \doteq z, f(y) \doteq z, P(z)}}{P(f(y)), z \doteq f(y) \to z \doteq f(y) \wedge z \doteq z \wedge z \doteq z, f(y) \doteq z, P(z)}$$

The sequent S_1 is

$$P(f(y)), z \doteq f(y) \to z \doteq f(y), f(y) \doteq z, P(z)$$

and the sequent S_2 is

$$P(f(y)), z \doteq f(y), z \doteq z \to z \doteq z, f(y) \doteq z, P(z).$$

5.6.3 Soundness of the System $G_=$

First, the following lemma is easily shown.

Lemma 5.6.2 For any of the equality rules given in definition 5.6.2 (including their generalization), the premise is falsifiable if and only if the conclusion is falsifiable.

Proof: Straightforward, and left as an exercise. □

Lemma 5.6.3 (Soundness of the System $G_=$) If a sequent is $G_=$-provable then it is valid.

Proof: Use the induction principle for $G_=$-proofs and lemma 5.6.2. □

5.6.4 Search Procedure for Languages With Equality

To prove the completeness of the system $G_=$ we modify the procedure *expand* in the following way. Given a sequent $\Gamma_0 \rightarrow \Delta_0$, let $\mathbf{L}'$ be the reduct of $\mathbf{L}$ consisting of all constant, function, and predicate symbols occurring in formulae in $\Gamma_0 \rightarrow \Delta_0$ (It is also assumed that the set of variables occurring free in $\Gamma_0 \rightarrow \Delta_0$ is disjoint from the set of variables occurring bound in $\Gamma_0 \rightarrow \Delta_0$, which is possible by lemma 5.3.4.)

Let $EQ_{1,0}$ be an enumeration of all $\mathbf{L}'$-formulae of the form

$$(t \doteq t),$$

$EQ_{2,0}$ an enumeration of all $\mathbf{L}'$-formulae of the form

$$(s_1 \doteq t_1) \wedge ... \wedge (s_n \doteq t_n) \supset (f s_1...s_n \doteq f t_1...t_n) \text{ and}$$

$EQ_{3,0}$ an enumeration of all $\mathbf{L}'$-formulae of the form

$$((s_1 \doteq t_1) \wedge ... \wedge (s_n \doteq t_n) \wedge P s_1...s_n) \supset P t_1...t_n,$$

where the terms $t, s_1, ..., s_n, t_1, ...t_n$ are built from the constant and function symbols in $\mathbf{L}'$, and the set of variables free in formulae in $\Gamma_0 \rightarrow \Delta_0$. Note that these terms belong to the set $TERMS$, defined in Subsection 5.5.1.

For $i \geq 1$, we define the sets $EQ_{1,i}$, $EQ_{2,i}$ and $EQ_{3,i}$ as above, except that the terms $t, s_1, ..., s_n, t_1, ...t_n$ belong to the lists $AVAIL_i$ (at the start). During a round, at the end of every expansion step, we shall add each formula that is the head of the list $EQ_{j,i}$, where $j = (RC \bmod 3) + 1$ (RC is a *round counter* initialized to 0 and incremented at the end of every round) for all $i = 0, ..., NUMACT$, to the antecedent of every leaf sequent. At the end of the round, each such formula is deleted from the head of the list $EQ_{j,i}$. In this way, the set S of signed formulae along any (finite or infinite) *nonclosed* path in the tree will be a Hintikka set.

We also add to the procedure *search*, statements to initialize the lists $EQ_{1,i}$, $EQ_{2,i}$ and $EQ_{3,i}$ as explained above.

Note that if the sets **FS**, **PS** and **CS** are effective (recursive), recursive functions enumerating the lists $EQ_{1,i}$, $EQ_{2,i}$ and $EQ_{3,i}$ can be written (these lists are in fact recursive).

In the new version of the *search* procedure for sequents containing the equality symbol, a leaf of the tree is *finished* iff it is an axiom. Hence, the *search* procedure always builds an infinite tree if the input sequent is not valid. This is necessary to guarantee that conditions H6(i) to (iii) are satisfied.

Definition 5.6.3 Procedure *search* for a language with equality.

Procedure Expand for a Language with Equality

```
procedure expand(node : tree-address; var T : tree);
  begin
    let A₁, ..., Aₘ → B₁, ..., Bₙ  be the label of node;
    let S be the one-node tree labeled with
      A₁, ..., Aₘ → B₁, ..., Bₙ;
    for i := 1 to m do
      if nonatomic(Aᵢ) then
        grow-left(Aᵢ, S)
      endif
    endfor;
    for i := 1 to n do
      if nonatomic(Bᵢ) then
        grow-right(Bᵢ, S)
      endif
    endfor;
    for each leaf u of S do
      let Γ → Δ be the label of u;
      Γ' := Γ, head(L);
      Δ' := Δ, head(R);
      for i := 0 to NUMACT₀ do
        Γ' := Γ', head(EQⱼ,ᵢ)
      endfor;
      create a new node u1 labeled with Γ' → Δ'
    endfor;
    T := dosubstitution(T, node, S)
  end
```

Procedure Search for a Language with Equality

procedure $search(\Gamma_0 \to \Delta_0 : sequent;$ **var** $T : tree)$;
  begin
    $L := tail(\Gamma_0); \Gamma := head(\Gamma_0)$;
    $R := tail(\Delta_0); \Delta := head(\Delta_0)$;
    let T be the one-node tree labeled with $\Gamma \to \Delta$;
    Let $TERM_0$, $EQ_{1,i}$, $EQ_{2,i}$, $EQ_{3,i}$ and $AVAIL_i$, $0 \le i$,
    be initialized as explained above.
    $NUMACT := 0; RC := 0$;
    while not *all leaves of T are finished* **do**
      $TERM_1 := TERM_0; T_0 := T; NUMACT_0 := NUMACT$;
      $j := (RC \bmod 3) + 1$;
      for each *leaf node of T_0*
      (in lexicographic order) **do**
        if not $finished(node)$ **then**
          $expand(node, T)$
        endif
      endfor;
      $TERM_0 := TERM_1; L := tail(L); R := tail(R)$;
      for $i := 0$ **to** $NUMACT_0$ **do**
        $EQ_{j,i} := tail(EQ_{j,i})$
      endfor;
      $RC := RC + 1$;
      for $i := 0$ **to** $NUMACT$ **do**
        $TERM_0 := append(TERM_0, < head(AVAIL_i), nil >)$;
        $AVAIL_i := tail(AVAIL_i)$
      endfor
    endwhile;
    if *all leaves are closed*
    then
      write (*'T is a proof of $\Gamma_0 \to \Delta_0$'*)
    else
      write (*'$\Gamma_0 \to \Delta_0$ is falsifiable'*)
    endif
  end

5.6.5 Completeness of the System $G_=$

Using lemma 5.6.1, it is easy to generalize theorem 5.5.1 as follows.

Theorem 5.6.1 Let **L** be a first-order language with equality. For every
sequent $\Gamma_0 \to \Delta_0$ containing the equality symbol the following holds:

   (i) If the input sequent $\Gamma_0 \to \Delta_0$ is valid, the procedure *search* halts with a finite closed tree T, from which a proof tree for a finite subsequent $C_1, ..., C_m \rightarrow D_1, ..., D_n$ of $\Gamma_0 \rightarrow \Delta_0$ can be constructed.

(ii) If the input sequent $\Gamma_0 \rightarrow \Delta_0$ is falsifiable, *search* builds an infinite counter-example tree T and $\Gamma_0 \rightarrow \Delta_0$ can be falsified in a finite or a countably infinite structure which is a quotient of $t(TERM_0)$.

Proof: It is similar to that of theorem 5.5.1 but uses lemma 5.6.1 instead of lemma 5.4.5 in order to deal with equality. What is needed is the following claim:

Claim: If S is the set of signed formulae along any nonclosed path in the tree T generated by *search*, the set $H_S = t(TERM_0)$ is a term algebra (over the reduct $\mathbf{L}'$) and S is a Hintikka set for H_S.

The proof of the claim is essentially the same as the proof given for theorem 5.5.1. The only difference is that it is necessary to check that conditions H6(i) to (iii) hold, but this follows immediately since all formulae in the lists $EQ_{j,0}$ and $EQ_{j,i}$ for all activated variables y_i ($i > 0$) are eventually entered in each infinite path of the tree. $\square$

Corollary (A version of Gödel's extended completeness theorem for $G_=$) Let $\mathbf{L}$ be a first-order language with equality. A sequent (even infinite) is valid iff it is $G_=$-provable. $\square$

Again, as observed in Section 5.5 in the paragraph immediately after the proof of theorem 5.5.1, although constructive and theoretically complete, the *search* procedure is horribly inefficient. But in the case of a language with equality, things are even worse. Indeed, the management of the equality axioms (the lists $EQ_{j,i}$) adds significantly to the complexity of the bookkeeping. In addition, it is no longer true that the *search* procedure always halts for quantifier-free sequents, as it does for languages without equality. This is because the equality rules allow the introduction of new formulae.

It can be shown that there is an algorithm for deciding the validity of quantifier-free formulae with equality by adapting the congruence closure method of Nelson and Oppen, 1980. For a proof, see Chapter 10.

The difficulty is that the *search* procedure no longer preserves the property known as the *subformula property*. For languages without equality, the Gentzen rules are such that given any input sequent $\Gamma \rightarrow \Delta$, the formulae occurring in any deduction tree with root $\Gamma \rightarrow \Delta$ contain only subformulae of the formulae in $\Gamma \rightarrow \Delta$. These formulae are in fact "descendants" of the formulae in $\Gamma \rightarrow \Delta$, which are essentially "signed subformulae" of the formulae in $\Gamma \rightarrow \Delta$ (for a precise definition, see problem 5.5.18). We also say that such proofs are *analytic*. But the equality rules can introduce "brand new" formulae that may be totally unrelated to the formulae in $\Gamma \rightarrow \Delta$. Of course, one could object that this problem arises because the rules of $G_=$ were badly designed, and that this problem may not arise for a better system. However, it is not easy to find a complete proof system for first-order logic with equality, that incorporates a weaker version of the subformula property. It will be shown in the next chapter that there is a Gentzen system LK_e, such that for every sequent $\Gamma \to \Delta$ provable in $G_=$, there is a proof in LK_e in which the formulae occurring in the proof are not too unrelated to the formulae in $\Gamma \to \Delta$. This is a consequence of the cut elimination theorem without essential cuts for LK_e. Nevertherless, theorem proving for languages with equality tends to be harder than theorem proving for languages without equality. We shall come back to this point later, in particular when we discuss the resolution method.

Returning to the completeness theorem, the following classical results apply to first-order languages with equality and are immediate consequences of theorem 5.6.1.

5.6.6 Löwenheim-Skolem, Compactness, and Model Existence Theorems for Languages With Equality

For a language with equality, the structure given by Löwenheim-Skolem theorem may be finite.

Theorem 5.6.2 (Löwenheim-Skolem) Let **L** be a first-order language with equality. If a set of formulae Γ over **L** is satisfiable in some structure **M**, then it is satisfiable in a structure whose domain is at most countably infinite.

Proof: From theorem 5.6.1, Γ is satisfiable in a quotient structure of $\mathbf{H_S}$. Since, $\mathbf{H_S}$ is countable, a quotient of $\mathbf{H_S}$ is countable, but possibly finite. $\square$

Theorem 5.6.3 (Compactness theorem) Let **L** be a first-order language with equality. For any (possibly countably infinite) set Γ of formulae over **L**, if every nonempty finite subset of Γ is satisfiable then Γ is satisfiable. $\square$

Theorem 5.6.4 (Model existence theorem) Let **L** be a first-order language with equality. If a set Γ of formulae over **L** is consistent, then it is satisfiable. $\square$

Note that if Γ is consistent, then by theorem 5.6.4 the sequent $\Gamma \to$ is falsifiable. Hence the *search* procedure run on input $\Gamma \to$ will actually yield a structure $\mathbf{H_S}$ (with domain a quotient of $t(TERM_0)$) for each Hintikka set S arising along each nonclosed path in the tree T, in which S and Γ are satisfiable.

We also have the following lemma.

Lemma 5.6.4 (Consistency lemma) Let **L** be a first-order language with equality. If a set Γ of formulae is satisfiable then it is consistent. $\square$

5.6.7 Maximal Consistent Sets

As in Subsection 5.5.5, a set Γ of first-order formulae (possibly involving equality) is *maximally consistent* (or a *maximal consistent set*) iff, for every consistent set Δ, if Γ is a subset of Δ, then $\Gamma = \Delta$. lemma 5.5.2 can be easily generalized to the first-order languages with equality.

Lemma 5.6.5 Given a first-order language (with or without equality), every consistent set Γ is a subset of some maximal consistent set Δ.

Proof: Almost identical to that of lemma 3.5.5, but using structures instead of valuations. $\square$

5.6.8 Application of the Compactness and Löwenheim-Skolem Theorems: Nonstandard Models of Arithmetic

The compactness and the Löwenhein-Skolem theorems are important tools in model theory. Indeed, they can be used for showing the existence of countable models for certain interesting theories. As an illustration, we show that the theory of example 5.3.2 known as Peano's arithmetic has a countable model nonisomorphic to **N**, the set of natural numbers. Such a model is called a *nonstandard model*.

EXAMPLE 5.6.3

Let **L** be the language of arithmetic defined in example 5.3.1. The following set A_P of formulae is known as the axioms of *Peano's arithmetic*:

$$\forall x \neg (S(x) \doteq 0)$$
$$\forall x \forall y (S(x) \doteq S(y) \supset x \doteq y)$$
$$\forall x (x + 0 \doteq x)$$
$$\forall x \forall y (x + S(y) \doteq S(x + y))$$
$$\forall x (x * 0 \doteq 0)$$
$$\forall x \forall y (x * S(y) \doteq x * y + x)$$

For every formula A with one free variable x,

$$(A(0) \wedge \forall x (A(x) \supset A(S(x)))) \supset \forall y A(y)$$

Let **L**$'$ be the expansion of **L** obtained by adding the new constant c. Let Γ be the union of A_P and the following set I of formulae:

$$I = \{\neg (0 \doteq c), \neg (1 \doteq c), ..., \neg (n \doteq c), ...\},$$

where n is an abbreviation for the term $S(S(...(0)))$, with n occurrences of S.

Observe that **N** can be made into a model of any finite subset of Γ. Indeed, **N** is a model of A_P, and to satisfy any finite subset X of I, it is sufficient to interpret c as any integer strictly larger than the maximum of the finite set $\{n \mid \neg(n \doteq c) \in X\}$. Since every finite subset of Γ is satisfiable, by the compactness theorem, Γ is satisfiable in some model. By the Löwenheim-Skolem theorem, Γ has a countable model, say $\mathbf{M}_0$. Let $\mathbf{M}$ be the reduct of $\mathbf{M}_0$ to the language $\mathbf{L}$. The model $\mathbf{M}$ still contains the element a, which is the interpretation of $c \in \mathbf{L}'$, but since c does not belong to $\mathbf{L}$, the interpretation function of the structure $\mathbf{M}$ does not have c in its domain.

It remains to show that $\mathbf{M}$ and $\mathbf{N}$ are not isomorphic. An *isomorphism* h from $\mathbf{M}$ to $\mathbf{N}$ is a bijection $h : \mathbf{M} \to \mathbf{N}$ such that:

$$h(0_{\mathbf{M}}) = 0 \text{ and, for all } x, y \in \mathbf{M},$$
$$h(S_{\mathbf{M}}(x)) = S(h(x)),$$
$$h(x +_{\mathbf{M}} y) = x + y \text{ and}$$
$$h(x *_{\mathbf{M}} y) = x * y.$$

Assume that there is an isomorphism h between $\mathbf{M}$ and $\mathbf{N}$. Let $k \in \mathbf{N}$ be the natural number $h(a)$ (with $a = c_{\mathbf{M}_0}$ in $\mathbf{M}_0$). Then, for all $n \in \mathbf{N}$,

$$n \neq k.$$

Indeed, since the formula $\neg(n \doteq c)$ is valid in $\mathbf{M}_0$,

$$n_{\mathbf{M}} \neq a \text{ in } \mathbf{M},$$

and since h is injective, $n_{\mathbf{M}} \neq a$ implies $h(n_{\mathbf{M}}) \neq h(a)$, that is, $n \neq k$. But then, we would have $k \neq k$, a contradiction. $\square$

It is not difficult to see that in the model $\mathbf{M}$, $m_{\mathbf{M}} \neq n_{\mathbf{M}}$, whenever $m \neq n$. Hence, the natural numbers are represented in $\mathbf{M}$, and satisfy Peano's axioms. However, there may be other elements in $\mathbf{M}$. In particular, the interpretation a of $c \in \mathbf{L}'$ belongs to $\mathbf{M}$. Intuitively speaking, a is an infinite element of $\mathbf{M}$.

A model of A_P nonisomorphic to $\mathbf{N}$ such as $\mathbf{M}$ is called a *nonstandard model* of Peano's arithmetic.

For a detailed exposition of model theory including applications of the compactness theorem, Löwenheim-Skolem theorem, and other results, the reader is referred to Chang and Keisler, 1973; Monk, 1976; or Bell and Slomson, 1969. The introductory chapter on model theory by J. Keisler in Barwise, 1977, is also highly recommended.

PROBLEMS

5.6.1. Give proof tree for the following formulae:

$$\forall x(x \doteq x)$$

$$\forall x_1...\forall x_n\forall y_1...\forall y_n((x_1 \doteq y_1 \wedge ... \wedge x_n \doteq y_n) \supset$$

$$(f(x_1, ..., x_n) \doteq f(y_1, ..., y_n)))$$

$$\forall x_1...\forall x_n\forall y_1...\forall y_n(((x_1 \doteq y_1 \wedge ... \wedge x_n \doteq y_n) \wedge P(x_1, ..., x_n)) \supset$$

$$P(y_1, ..., y_n))$$

The above formulae are called the *closed equality axioms*.

5.6.2. Let $S = \Gamma \to \Delta$ be a sequent, and let S_e be the set of closed equality axioms for all predicate and function symbols occurring in S. Prove that the sequent $\Gamma \to \Delta$ is provable in $G_=$ iff the sequent $S_e, \Gamma \to \Delta$ is provable in $G_= + \{cut\}$.

5.6.3. Prove that the following formula is provable:

$$f^3(a) \doteq a \wedge f^5(a) \doteq a \supset f(a) \doteq a$$

5.6.4. Let $*$ be a binary function symbol and 1 be a constant. Prove that the following sequent is valid:

$$\forall x(*(x, 1) \doteq x), \forall x(*(1, x) \doteq x),$$

$$\forall x \forall y \forall z(*(x, *(y, z)) \doteq *(*(x, y), z)),$$

$$\forall x(*(x, x) \doteq 1) \to \forall x \forall y(*(x, y) \doteq *(y, x))$$

5.6.5. Give proof trees for the following formulae:

$$\forall x \exists y(x \doteq y)$$

$$A[t/x] \equiv \forall x((x \doteq t) \supset A), \text{ if } x \text{ is not free in } t.$$

$$A[t/x] \equiv \exists x((x \doteq t) \wedge A), \text{ if } x \text{ is not free in } t.$$

5.6.6. (a) Prove that the following formulae are valid:

$$x \doteq x$$

$$x \doteq y \supset (x \doteq z \supset y \doteq z)$$

$$x \doteq y \supset (P(x_1, ..., x_{i-1}, x, x_{i+1}, ..., x_n) \supset$$

$$P(x_1, ..., x_{i-1}, y, x_{i+1}, ..., x_n))$$

$$x \doteq y \supset$$

$$(f(x_1, ..., x_{i-1}, x, x_{i+1}, ..., x_n) \doteq f(x_1, ..., x_{i-1}, y, x_{i+1}, ..., x_n))$$

Consider the extension $H_=$ of the Hilbert system H defined in problem 5.5.4 obtained by adding the above formulae known as the *open equality axioms*.

(b) Prove that

$$\forall x \forall y (x \doteq y \supset y \doteq x)$$

and

$$\forall x \forall y \forall z (x \doteq y \wedge y \doteq z \supset x \doteq z)$$

are provable in $H_=$.

(c) Given a set S of formulae (over a language with equality), let S_e be the set of universal closures (defined in problem 5.3.10) of equality axioms for all function and predicate symbols occuring in S. Prove that

$$S \vdash A \text{ in } H_= \quad \text{iff} \quad S, S_e \vdash A \text{ in } H.$$

(d) Prove the completeness of the Hilbert system $H_=$.

Hint: Use the result of problem 5.5.10 and, given a model for $S \cup S_e$, construct a model of S in which the predicate symbol $\doteq$ is interpreted as equality, using the quotient construction.

* **5.6.7.** In languages with equality, it is possible to eliminate function and constant symbols as shown in this problem.

Let **L** be a language (with or without equality). A *relational version* of **L** is a language **L**$'$ with equality satisfying the following properties:

(1) **L**$'$ has no function or constant symbols;

(2) There is an injection $T : \mathbf{FS} \cup \mathbf{CS} \to \mathbf{PS}'$ called a *translation* such that $r(T(f)) = r(f) + 1$;

(3) $\mathbf{PS}' = \mathbf{PS} \cup T(\mathbf{FS} \cup \mathbf{CS})$, and $T(\mathbf{FS} \cup \mathbf{CS})$ and $\mathbf{PS}$ are disjoint.

We define the *T-translate* A^T of an **L**-formula A as follows:

(1) If A is of the form $(s \doteq x)$, where s is a term and x is a variable, then

   (i) If s is a variable y, then $(y \doteq x)^T = (y \doteq x)$;

   (ii) If s is a constant c, then $(c \doteq x)^T = T(c)(x)$, where $T(c)$ is the unary predicate symbol associated with c;

   (iii) If s is a term of the form $f s_1 ... s_n$, then

$$(s \doteq x)^T = \exists y_1 ... \exists y_n [(s_1 \doteq y_1)^T \wedge ... \wedge (s_n \doteq y_n)^T \wedge T(f)(y_1, ..., y_n, x)],$$

where $y_1, ..., y_n$ are new variables, and $T(f)$ is the $(n+1)$-ary predicate symbol associated with f;

(2) If A is of the form $(s \doteq t)$, where t is not a variable, then

$$(s \doteq t)^T = \exists y((s \doteq y)^T \wedge (t \doteq y)^T),$$

where y is a new variable.

(3) If A is of the form $Ps_1...s_n$, then

(i) If every $s_1, ..., s_n$ is a variable, then $(Ps_1...s_n)^T = Ps_1...s_n$;

(ii) If some s_i is not a variable, then

$$(Ps_1...s_n)^T = \exists y_1...\exists y_n[Py_1...y_n \wedge (s_1 \doteq y_1)^T \wedge ... \wedge (s_n \doteq y_n)^T],$$

where $y_1, ..., y_n$ are new variables.

(4) If A is not atomic, then

$$(\neg B)^T = \neg B^T,$$
$$(B \vee C)^T = B^T \vee C^T,$$
$$(B \wedge C)^T = B^T \wedge C^T,$$
$$(B \supset C)^T = (B^T \supset C^T),$$
$$(\forall x B)^T = \forall x B^T,$$
$$(\exists x B)^T = \exists x B^T.$$

For every function or constant symbol f in **L**, the *existence condition for f* is the following sentence in **L'**:

$$\forall x_1...\forall x_n \exists y T(f)(x_1, ..., x_n, y);$$

The *uniqueness condition for f* is the following sentence in **L'**:

$$\forall x_1...\forall x_n \forall y \forall z (T(f)(x_1, ..., x_n, y) \wedge T(f)(x_1, ..., x_n, z) \supset y \doteq z).$$

The set of *translation conditions* is the set of all existence and uniqueness conditions for all function and constant symbols in **L**.

Given an **L**-structure **A**, the *relational version* **A'** of **A** has the same domain as **A**, the same interpretation for the symbols in **PS**, and interprets each symbol $T(f)$ of rank $n + 1$ as the relation

$$\{(x_1, ..., x_n, y) \in A^{n+1} \mid f_{\mathbf{A}}(x_1, ..., x_n) = y\}.$$

Prove the following properties:

(a) For every formula A of **L**, if A does not contain function or constant symbols, then $A^T = A$.

(b) Let **A** be an **L**-structure, and **A**$'$ its relational version. Prove that **A**$'$ is a model of the translation conditions. Prove that for every **L**-formula A, for every assignment s,

$$\mathbf{A} \models A[s] \quad \text{iff} \quad \mathbf{A}' \models A^T[s].$$

(c) For every **L**$'$-formula A, let A^* be the formula obtained by replacing every atomic formula $T(f)(x_1, ..., x_n, y)$ by $(f(x_1, ..., x_n) \doteq y)$. Prove that

$$\mathbf{A} \models A^*[s] \quad \text{iff} \quad \mathbf{A}' \models A[s].$$

(d) Prove that if **B** is an **L**$'$-structure which is a model of the translation conditions, then $\mathbf{B} = \mathbf{A}'$ for some **L**-structure **A**.

(e) Let Γ be a set of **L**-formulae, and A any **L**-formula. Prove that

$$\Gamma \models A \quad \text{iff}$$
$$\{B^T \mid B \in \Gamma\} \cup \{B \mid B \text{ is a translation condition}\} \models A^T.$$

$*$ **5.6.8.** Let **L** be a first-order language with equality. A *theory* is a set Γ of **L**-sentences such that for every **L**-sentence A, if $\Gamma \models A$, then $A \in \Gamma$. If **L**$'$ is an expansion of **L** and Γ' is a theory over **L**$'$, we say that Γ' is *conservative over* Γ, iff

$$\{A \mid A \text{ is an } \mathbf{L}\text{-sentence in } \Gamma'\} = \Gamma.$$

Let Γ be a theory over **L**, **L**$'$ an expansion of **L**, and Γ' a theory over **L**$'$.

(i) If P is an n-ary predicate symbol in **L**$'$ but not in **L**, a *possible definition of P* over Γ is an **L**-formula A whose set of free variables is a subset of $\{x_1, ..., x_n\}$.

(ii) If f is an n-ary function symbol or a constant in **L**$'$ but not in **L**, a *possible definition of f* over Γ is an **L**-formula A whose set of free variables is a subset of $\{x_1, ..., x_n, y\}$, such that the following *existence* and *uniqueness* conditions are in Γ:

$$\forall x_1...\forall x_n \exists y A,$$
$$\forall x_1...\forall x_n \forall y \forall z (A(x_1, ..., x_n, y) \wedge A(x_1, ..., x_n, z) \supset y \doteq z).$$

(iii) We say that Γ' over **L**$'$ is a *definitional extension* of Γ over **L** iff for every constant, function, or predicate symbol X in **L**$'$ but not in **L**, there is a possible definition A_X over Γ such that the condition stated below holds.

For every possible definition A_X, the sentence A'_X is defined as follows:

For an n-ary predicate symbol P, let A'_P be the sentence

$$\forall x_1...\forall x_n(Px_1...x_n \equiv A_P);$$

For an n-ary function symbol or a constant f, let A'_f be the sentence

$$\forall x_1...\forall x_n \forall y((fx_1...x_n \doteq y) \equiv A_f).$$

Then we require that

$$\Gamma' = \{B \mid B \text{ is an } \mathbf{L}'\text{-formula such that,}$$
$$\Gamma \cup \{A'_X \mid X \text{ is a symbol in } \mathbf{L}' \text{ not in } \mathbf{L}\} \models B\}.$$

(a) Prove that if Γ' is a definitional extension of Γ, then for every $\mathbf{L}'$-formula A, there is an $\mathbf{L}$-formula A^T with the same free variables as A, such that
$$\Gamma' \models (A \equiv A^T).$$

Hint: Use the technique of problem 5.6.7.

(b) Prove that Γ' is conservative over Γ.

5.6.9. Let Γ be a theory over a language $\mathbf{L}$. A *set of axioms* Δ for Γ is any set of $\mathbf{L}$-sentences such that

$$\Gamma = \{B \mid \Delta \models B\}.$$

Given an $\mathbf{L}$-formula A with set of free variables $\{x_1, ..., x_n, y\}$, assume that
$$\Gamma \models \forall x_1...\forall x_n \exists y A(x_1, ..., x_n, y).$$

Let $\mathbf{L}'$ be the expansion of $\mathbf{L}$ obtained by adding the new n-ary function symbol f, and let Γ' be the theory with set of axioms

$$\Gamma \cup \{\forall x_1...\forall x_n A(x_1, ..., x_n, f(x_1, ..., x_n))\}.$$

Prove that Γ' is conservative over Γ.

* **5.6.10.** Let $\mathbf{L}$ be a first-order language with equality. From problem 5.6.1, the closed equality axioms are provable in $G_=$. Recall the concept of a *Henkin theory* from problem 5.5.17.

(a) Prove that a maximally consistent (with respect to $G_=$) Henkin theory T' of sentences is a Hintikka set with respect to the term algebra H consisting of all closed terms built up from the function and constant symbols in $\mathbf{L}^H$.

(b) Prove that every consistent set T of $\mathbf{L}$-sentences is satisfiable.

(c) Prove that a formula A with free variables $\{x_1, ..., x_n\}$ is satisfiable iff $A[c_1/x_1, ..., c_n/x_n]$ is satisfiable, where $c_1, ..., c_n$ are new constants. Using this fact, prove that every consistent set T of $\mathbf{L}$-formulae is satisfiable.

5.6.11. Prove that in $G_= + \{cut\}$,

$$\Gamma \to A \text{ is not provable in } G_= + \{cut\} \text{ iff}$$
$$\Gamma \cup \{\neg A\} \text{ is consistent.}$$

Use the above fact and problem 5.6.10 to give an alternate proof of the extended completeness theorem for $G_= + \{cut\}$.

$*$ **5.6.12.** Prove that the results of problem 5.6.10 hold for languages with equality of any cardinality. Using Zorn's lemma, prove that the results of problem 5.6.10 hold for languages of any cardinality. Thus, prove that the extended completeness theorem holds for languages with equality of any cardinality.

$*$ **5.6.13.** For a language $\mathbf{L}$ with equality of cardinality α, show that the cardinality of the term algebra arising in problem 5.6.10 is at most α. Conclude that any consistent set of $\mathbf{L}$-sentences has a model of cardinality at most α.

$*$ **5.6.14.** Let $\mathbf{L}$ be a countable first-order language with equality. A property P of sets of formulae is an *analytic consistency property* if the following hold:

(i) P is of finite character (see problem 3.5.12).

(ii) For every set S of formulae for which P is true, the following conditions hold:

A_0: S contains no atomic formula and its negation.

A_1: For every formula A of type a in S, P holds for $\{S, A_1\}$ and $\{S, A_2\}$.

A_2: For every formula B of type b in S, P holds either for $\{S, B_1\}$ or for $\{S, B_2\}$.

A_3: For every formula C of type c in S, P holds for $\{S, C(t)\}$ for every term t.

A_4: For every formula D of type d in S, for some constant c not in S, P holds for $\{S, D(c)\}$.

A_5 (i): For every term t, if P holds for S then P holds for $\{S, (t \doteq t)\}$.

(ii): For each n-ary function symbol f, for any terms $s_1, ..., s_n, t_1,$ ..., t_n, if P holds for S then P holds for

$$\{S, (s_1 \doteq t_1) \wedge ... \wedge (s_n \doteq t_n) \supset (f s_1 ... s_n \doteq f t_1 ... t_n)\}.$$

(iii): For each n-ary predicate symbol Q (including $\doteq$), for any terms $s_1, ..., s_n, t_1, ..., t_n$, if P holds for S then P holds for

$$\{S, (s_1 \doteq t_1) \wedge ... \wedge (s_n \doteq t_n) \wedge Q s_1 ... s_n \supset Q t_1 .... t_n\}.$$

Prove that if P is an analytic consistency property and P holds for S, then S is satisfiable.

Hint: See problem 5.5.22.

∗ **5.6.15.** Let **L** be a countable first-order language with equality. A set S of formulae is *truth-functionally inconsistent* iff S is the result of substituting first-order formulae for the propositional letters in an unsatisfiable set of propositions (see definition 5.3.11). We say that a formula A is *truth-functionally valid* iff it is obtained by substitution of first-order formulae into a tautology (see definition 5.3.11). We say that a finite set $S = \{A_1, ..., A_m\}$ *truth-functionally implies* B iff the formula

$$(A_1 \wedge ... \wedge A_m) \supset B$$

is truth-functionally valid. A property P of sets of formulae is a *synthetic consistency property* iff the following conditions hold:

(i) P is of finite character (see problem 3.5.12).

(ii) For every set S of formulae, the following conditions hold:

B_0: If S is truth-functionally inconsistent, then P does not hold for S;

B_3: If P holds for S then for every formula C of type c in S, P holds for $\{S, C(t)\}$ for every term t.

B_4: For every formula D of type d in S, for some constant c not in S, P holds for $\{S, D(c)\}$.

B_5: For every formula X, if P does not hold for $\{S, X\}$ or $\{S, \neg X\}$, then P does not hold for S. Equivalently, if P holds for S, then for every formula X, either P holds for $\{S, X\}$ or P holds for $\{S, \neg X\}$.

B_6 (i): For every term t, if P holds for S then P holds for $\{S, (t \doteq t)\}$.

(ii): For each n-ary function symbol f, for any terms $s_1, ..., s_n, t_1,$ ..., t_n, if P holds for S then P holds for

$$\{S, (s_1 \doteq t_1) \wedge ... \wedge (s_n \doteq t_n) \supset (f s_1 ... s_n \doteq f t_1 ... t_n)\}.$$

(iii): For each n-ary predicate symbol Q (including $\doteq$), for any terms $s_1, ..., s_n, t_1, ..., t_n$, if P holds for S then P holds for

$$\{S, (s_1 \doteq t_1) \land ... \land (s_n \doteq t_n) \land Qs_1...s_n \supset Qt_1....t_n\}.$$

(a) Prove that if P is a synthetic consistency property then the following condition holds:

B_7: If P holds for S and a finite subset of S truth-functionally implies X, then P holds for $\{S, X\}$.

(b) Prove that every synthetic consistency property is an analytic consistency property.

(c) Prove that consistency within the Hilbert system $H_=$ of problem 5.6.6 is a synthetic consistency property.

* **5.6.16.** A set S of formulae is *Henkin-complete* iff for every formula $D = \exists x B$ of type d, there is some constant c such that $B(c)$ is also in S.

Prove that if P is a synthetic consistency property and S is a set of formulae that is both Henkin-complete and a maximally set for which P holds, then S is satisfiable.

Hint: Show that S is a Hintikka set for the term algebra consisting of all terms built up from function, constant symbols, and variables occurring free in S.

* **5.6.17.** Prove that if P is a synthetic consistency property, then every set S of formulae for which P holds can be extended to a Henkin-complete set which is a maximal set for which P holds.

Hint: Use the idea of problem 5.5.17.

Use the above property to prove the completeness of the Hilbert system $H_=$.

* **5.6.18.** Let **L** be a countable first-order language with equality. Prove that the results of problem 5.5.26 are still valid.

* **5.6.19.** Given a first-order language **L** with equality, for any **L**-structure **M**, for any finite or countably infinite sequence X of elements in M (the domain of the structure **M**), the language $\mathbf{L}_X$ is the expansion of **L** obtained by adding new distinct constants ($\{c_1, ..., c_n\}$ if $X =< a_1, ..., a_n >$, $\{c_1, ..., c_n, c_{n+1}, ...\}$ if X is countably infinite) to the set of constants in **L**. The structure $(\mathbf{M}, X)$ is the expansion of **M** obtained by interpreting each c_i as a_i.

An **L**-structure **M** is *countably saturated* if, for every finite sequence $X =< a_1, ..., a_n >$ of elements in M, for every set $\Gamma(x)$ of formulae with at most one free variable x over the expanded language $\mathbf{L}_X$, if every finite subset of $\Gamma(x)$ is satisfiable in $(\mathbf{M}, X)$ then $\Gamma(x)$ is satisfiable in $(\mathbf{M}, X)$.

(1) Prove that every finite structure $\mathbf{M}$ is countably saturated.

Hint: Show that if the conclusion does not hold, a finite unsatisfiable subset of $\Gamma(x)$ can be found.

(2) Two **L**-structures $\mathbf{A}$ and $\mathbf{B}$ are *elementary equivalent* if, for any **L**-sentence D,

$$\mathbf{A} \models D \quad \text{if and only if} \quad \mathbf{B} \models D.$$

(a) Assume that $\mathbf{A}$ and $\mathbf{B}$ are elementary equivalent. Show that for every formula $E(x)$ with at most one free variable x, $E(x)$ is satisfiable in $\mathbf{A}$ if and only if $E(x)$ is satisfiable in $\mathbf{B}$.

(b) Let $X = <a_1, ..., a_n>$ and $Y = <b_1, ..., b_n>$ be two finite sequences of elements in A and B respectively. Assume that $(\mathbf{A}, X)$ and $(\mathbf{B}, Y)$ are elementary equivalent and that $\mathbf{A}$ is countably saturated.

Show that for any set $\Gamma(x)$ of formulae with at most one free variable x over the expansion $\mathbf{L}_Y$ such that every finite subset of $\Gamma(x)$ is satisfiable in $(\mathbf{B}, Y)$, $\Gamma(x)$ is satisfiable in $(\mathbf{A}, X)$. (Note that the languages $\mathbf{L}_X$ and $\mathbf{L}_Y$ are identical. Hence, we will refer to this language as $\mathbf{L}_X$.)

(c) Assume that $\mathbf{A}$ and $\mathbf{B}$ are elementary equivalent, with $\mathbf{A}$ countably saturated. Let $Y = <b_1, ..., b_n, ...>$ be a countable sequence of elements in B.

Prove that there exists a countable sequence $X = <a_1, ..., a_n, ...>$ of elements from A, such that $(\mathbf{A}, X)$ and $(\mathbf{B}, Y)$ are elementary equivalent.

Hint: Proceed in the following way: Define the sequence $X_n = <a_1, ..., a_n>$ by induction so that $(\mathbf{A}, X_n)$ and $(\mathbf{B}, Y_n)$ are elementary equivalent (with $Y_n = <b_1, ..., b_n>$) as follows: Let $\Gamma(x)$ be the set of formulae over $\mathbf{L}_{Y_n}$ satisfied by b_{n+1} in $(\mathbf{B}, Y_n)$.

Show that $\Gamma(x)$ is maximally consistent. Using 2(b), show that $\Gamma(x)$ is satisfied by some a_{n+1} in $(\mathbf{A}, X_n)$ and that it is the set of formulae satisfied by a_{n+1} in $(\mathbf{A}, X_n)$ (recall that $\Gamma(x)$ is maximally consistent). Use these properties to show that $(\mathbf{A}, X_{n+1})$ and $(\mathbf{B}, Y_{n+1})$ are elementary equivalent.

(d) If $(\mathbf{A}, X)$ and $(\mathbf{B}, Y)$ are elementary equivalent as above, show that

$$a_i = a_j \quad \text{if and only if} \quad b_i = b_j \text{ for all } i, j \geq 1.$$

(e) Use (d) to prove that if $\mathbf{A}$ and $\mathbf{B}$ are elementary equivalent and A is finite, then $\mathbf{B}$ is isomorphic to $\mathbf{A}$.

* **5.6.20.** Write a computer program implementing the *search* procedure in the case of a first-order language with equality.

Notes and Suggestions for Further Reading

First-order logic is a rich subject that has been studied extensively. We have presented the basic model-theoretic and proof-theoretic concepts, using Gentzen systems because of their algorithmic nature and their conceptual simplicity. This treatment is inspired from Kleene, 1967. For more on Gentzen systems, the reader should consult Robinson, 1979; Takeuti, 1975; Szabo, 1969; or Smullyan, 1968.

We have focused our attention on a constructive proof of the completeness theorem, using Gentzen systems and Hintikka sets. For more details on Hintikka sets and related concepts such as consistency properties, the reader is referred to Smullyan, 1968.

There are other proof systems for first-order logic. Most texts adopt Hilbert systems. Just to mention a few texts of varying degree of difficulty, Hilbert systems are discussed in Kleene, 1952; Kleene, 1967; Enderton, 1972; Shoenfield, 1967; and Monk, 1976. A more elementary presentation of first-order logic tailored for computer scientists is found in Manna and Waldinger, 1985. Natural deduction systems are discussed in Kleene, 1967; Van Dalen, 1982; Prawitz, 1965; and Szabo, 1969. A variant of Gentzen systems called the tableaux system is discussed at length in Smullyan, 1968.

An illuminating comparison of various approaches to the completeness theorem, including Henkin's method, can be found in Smullyan, 1968.

Since we have chosen to emphasize results and techniques dealing with the foundations of (automatic) theorem proving, model theory is only touched upon lightly. However, this is a very important branch of logic. The reader is referred to Chang and Keisler, 1973; Bell and Slomson, 1974; and Monk, 1976, for thorough and advanced expositions of model theory. The article on fundamentals of model theory by Keisler, and the article by Eklof about ultraproducts, both in Barwise, 1977, are also recommended. Barwise 1977 also contains many other interesting articles on other branches of logic.

We have not discussed the important incompleteness theorems of Gödel. The reader is referred to Kleene,1952; Kleene, 1967; Enderton, 1972; Shoenfield, 1967; or Monk, 1976. One should also consult the survey article by Smorynski in Barwise, 1977.

Chapter 6

Gentzen's Cut Elimination Theorem and Applications

6.1 Introduction

The rules of the Gentzen system G (given in definition 5.4.1) were chosen mainly to facilitate the design of the *search* procedure. The Gentzen system LK' given in Section 3.6 can be extended to a system for first-order logic called LK, which is more convenient for constructing proofs in working downward, and is also useful for proof-theoretical investigations. In particular, the system LK will be used to prove three classical results, Craig's theorem, Beth's theorem and Robinson's theorem, and will be used in Chapter 7 to derive a constructive version of Herbrand's theorem.

The main result about LK is Gentzen's cut elimination theorem. An entirely proof-theoretic argument (involving only algorithmic proof transformation steps) of the cut elimination can be given, but it is technically rather involved. Rather than giving such a proof (which can be found in Szabo, 1969, or Kleene, 1952), we will adopt the following compromise: We give a rather simple semantic proof of Gentzen's cut elimination theorem (using the completeness theorem) for LK, and we give a constructive proof of the cut elimination for a Gentzen system $G1^{nnf}$ simpler than LK (by constructive, we mean that an algorithm for converting a proof into a cut-free proof is actually given). The sequents of the system $G1^{nnf}$ are pairs of sets of formulae in negation normal form. This system is inspired from Schwichtenberg (see Barwise, 1977). The cut elimination theorem for the system $G1^{nnf}_{\underline{=}}$ which includes axioms for equality is also given.

Three applications of the cut elimination theorem will also be given:

Craig's interpolation theorem,

Beth's definability theorem and

Robinson's joint consistency theorem.

These are classical results of first-order logic, and the proofs based on cut elimination are constructive and elegant. Beth's definability theorem also illustrates the subtle interaction between syntax and semantics.

A new important theme emerges in this chapter, and will be further elaborated in Chapter 7. This is the notion of *normal form* for proofs. Gentzen's cut elimination theorem (for LK or LK_e) shows that for every provable sequent $\Gamma \to \Delta$, there is a proof in normal form, in the sense that in the system LK, the proof does not have any cuts, and for the system LK_e, it has only atomic cuts. In the next chapter, it will be shown that this normal form can be improved if the formulae in the sequent are of a certain type (prenex form or negation normal form). This normal form guaranteed by Gentzen's Sharpened Hauptsatz is of fundamental importance, since it reduces provabilily in first-order logic to provability in propositional logic.

Related to the concept of normal form is the concept of *proof transformation*. Indeed, a constructive way of obtaining proofs in normal form is to perform a sequence of proof transformations.

The concepts of normal form for proofs and of proof transformation are essential. In fact, it turns out that the completeness results obtained in the remaining chapters will be obtained via proof transformations.

First, we consider the case of a first-order language without equality.

6.2 Gentzen System LK for Languages Without Equality

The system LK is obtained from LK' by adding rules for quantified formulae.

6.2.1 Syntax of LK

The inference rules and axioms of LK are defined as follows.

Definition 6.2.1 Gentzen system LK. The system LK consists of *structural rules*, the *cut rule*, and of *logical rules*. The letters Γ, Δ, Λ, Θ stand for arbitrary (possibly empty) sequences of formulae and A, B for arbitrary formulae.

(1) Structural rules:

(i) Weakening:

$$\frac{\Gamma \rightarrow \Delta}{A, \Gamma \rightarrow \Delta} \ (left) \qquad \frac{\Gamma \rightarrow \Delta}{\Gamma \rightarrow \Delta, A} \ (right)$$

A is called the *weakening formula*

(ii) Contraction:

$$\frac{A, A, \Gamma \rightarrow \Delta}{A, \Gamma \rightarrow \Delta} \ (left) \qquad \frac{\Gamma \rightarrow \Delta, A, A}{\Gamma \rightarrow \Delta, A} \ (right)$$

(iii) Exchange:

$$\frac{\Gamma, A, B, \Delta \rightarrow \Lambda}{\Gamma, B, A, \Delta \rightarrow \Lambda} \ (left) \qquad \frac{\Gamma \rightarrow \Delta, A, B, \Lambda}{\Gamma \rightarrow \Delta, B, A, \Lambda} \ (right)$$

(2) Cut rule:

$$\frac{\Gamma \rightarrow \Delta, A \qquad A, \Lambda \rightarrow \Theta}{\Gamma, \Lambda \rightarrow \Delta, \Theta}$$

A is called the *cut formula* of this inference.

(3) Logical rules:

$$\frac{A, \Gamma \rightarrow \Delta}{A \wedge B, \Gamma \rightarrow \Delta} \ (\wedge : left) \quad \text{and} \quad \frac{B, \Gamma \rightarrow \Delta}{A \wedge B, \Gamma \rightarrow \Delta} \ (\wedge : left)$$

$$\frac{\Gamma \rightarrow \Delta, A \qquad \Gamma \rightarrow \Delta, B}{\Gamma \rightarrow \Delta, A \wedge B} \ (\wedge : right)$$

$$\frac{A, \Gamma \rightarrow \Delta \qquad B, \Gamma \rightarrow \Delta}{A \vee B, \Gamma \rightarrow \Delta} \ (\vee : left)$$

$$\frac{\Gamma \rightarrow \Delta, A}{\Gamma \rightarrow \Delta, A \vee B} \ (\vee : right) \quad \text{and} \quad \frac{\Gamma \rightarrow \Delta, B}{\Gamma \rightarrow \Delta, A \vee B} \ (\vee : right)$$

$$\frac{\Gamma \rightarrow \Delta, A \qquad B, \Lambda \rightarrow \Theta}{A \supset B, \Gamma, \Lambda \rightarrow \Delta, \Theta} \ (\supset: left) \qquad \frac{A, \Gamma \rightarrow \Delta, B}{\Gamma \rightarrow \Delta, A \supset B} \ (\supset: right)$$

$$\frac{\Gamma \to \Delta, A}{\neg A, \Gamma \to \Delta} \ (\neg : left) \qquad \frac{A, \Gamma \to \Delta}{\Gamma \to \Delta, \neg A} \ (\neg : right)$$

In the rules above, $A \vee B$, $A \wedge B$, $A \supset B$, and $\neg A$ are called the *principal formulae* and A, B the *side formulae* of the inference.

In the quantifier rules below, x is any variable and y is any variable free for x in A and not free in A, unless $y = x$ $(y \notin FV(A) - \{x\})$. The term t is any term free for x in A.

$$\frac{A[t/x], \Gamma \to \Delta}{\forall x A, \Gamma \to \Delta} \ (\forall : left) \qquad \frac{\Gamma \to \Delta, A[y/x]}{\Gamma \to \Delta, \forall x A} \ (\forall : right)$$

$$\frac{A[y/x], \Gamma \to \Delta}{\exists x A, \Gamma \to \Delta} \ (\exists : left) \qquad \frac{\Gamma \to \Delta, A[t/x]}{\Gamma \to \Delta, \exists x A} \ (\exists : right)$$

Note that in both the $(\forall : right)$-rule and the $(\exists : left)$-rule, the variable y does *not* occur free in the lower sequent. In these rules, the variable y is called the *eigenvariable* of the inference. The condition that the eigenvariable does not occur free in the conclusion of the rule is called the *eigenvariable condition*. The formula $\forall x A$ (or $\exists x A$) is called the *principal formula* of the inference, and the formula $A[t/x]$ (or $A[y/x]$) the *side formula* of the inference.

The *axioms* of the system LK are all sequents of the form $A \to A$.

Note that since the system LK contains the exchange rules, the order of the formulae in a sequent is really irrelevant. Hence, we can view a sequent as a pair of multisets (as defined in problem 2.1.8).

Proof trees and *deduction tree* are defined inductively as in definition 3.4.5, but with the rules of the system LK given in definition 6.2.1. If a sequent has a proof in the system G we say that it is *G-provable* and similarly, if it is provable in the system LK, we say that it is *LK-provable*. The system obtained from LK by removing the cut rule is denoted by $LK - \{cut\}$. We also say that a sequent is LK-provable without a cut if it has a proof tree using the rules of the system $LK - \{cut\}$.

6.2.2 The Logical Equivalence of the Systems G, LK, and $LK - \{cut\}$

We now show that the systems G and LK are logically equivalent. We will in fact prove a stronger result, namely that G, $LK - \{cut\}$ and LK are equivalent. First, we show that the system LK is sound.

Lemma 6.2.1 Every axiom of LK is valid. For every rule of LK, if the premises of the rule are valid, then the conclusion of the rule is valid. Every LK-provable sequent is valid.

Proof: The proof uses the induction principle for proofs and is straight-forward. □

lemma 6.2.1 differs from lemma 5.4.3 in the following point: It is not necessarily true that if the conclusion of a rule is valid, then the premises of that rule are valid.

Theorem 6.2.1 (Logical equivalence of G, LK, and $LK - \{cut\}$) There is an algorithm to convert any LK-proof of a sequent $\Gamma \rightarrow \Delta$ into a G-proof. There is an algorithm to convert any G-proof of a sequent $\Gamma \rightarrow \Delta$ into a proof using the rules of $LK - \{cut\}$.

Proof: If $\Gamma \rightarrow \Delta$ has an LK-proof, by lemma 6.2.1, $\Gamma \rightarrow \Delta$ is valid. By theorem 5.5.1, $\Gamma \rightarrow \Delta$ has a G-proof given by the procedure *search*. Conversely, using the induction principle for G-proofs we show that every G-proof can be converted to an $(LK - \{cut\})$-proof. The proof is similar to that of theorem 3.6.1. Every G-axiom $\Gamma \rightarrow \Delta$ contains some common formula A, and by application of the weakening and the exchange rules, an $(LK - \{cut\})$-proof of $\Gamma \rightarrow \Delta$ can be obtained from the axiom $A \rightarrow A$. Next, we have to show that every application of a G-rule can be replaced by a sequence of $(LK - \{cut\})$-rules. There are twelve cases to consider. Note that the G-rules $\wedge : right$, $\vee : left$, $\supset: right$, $\supset: left$, $\neg : right$, $\neg : left$, $\forall : right$ and $\exists : left$ can easily be simulated in $LK - \{cut\}$ using the exchange, contraction, and corresponding $(LK - \{cut\})$-rules. The rules $\vee : right$ and $\wedge : left$ are handled as in theorem 3.6.1. Finally, we show how the G-rule $\forall : left$ can be transformed into a sequence of $(LK - \{cut\})$-rules, leaving the $\exists : right$ case as an exercise.

$$\Gamma, A[t/x], \forall x A, \Delta \rightarrow \Lambda$$

(several exchanges)

$$A[t/x], \forall x A, \Gamma, \Delta \rightarrow \Lambda$$ $(\forall : left)$

$$\forall x A, \forall x A, \Gamma, \Delta \rightarrow \Lambda$$ $(contraction)$

$$\forall x A, \Gamma, \Delta \rightarrow \Lambda$$

(several exchanges)

$$\Gamma, \forall x A, \Delta \rightarrow \Lambda$$

The above $(LK - \{cut\})$-derivation simulates the G-rule $\forall : left$. This conclude the proof of the theorem. □

Corollary   (Gentzen Hauptsatz for LK) A sequent is LK-provable if and only if it is LK-provable without a cut.

Note that the *search* procedure together with the method indicated in theorem 6.2.1 actually provides an algorithm to construct a cut-free LK-proof from an LK-proof with cuts. Gentzen proved the above result by a very different method in which an LK-proof is (recursively) transformed into an LK-proof without cut. Gentzen's proof is entirely constructive since it does not use any semantic arguments and can be found either in Takeuti, 1975, Kleene, 1952, or in Gentzen's original paper in Szabo, 1969.

In the next section, we discuss the case of first-order languages with equality symbol.

PROBLEMS

6.2.1.   Give LK-proofs for the following formulae:

$$\forall x A \supset A[t/x],$$
$$A[t/x] \supset \exists x A,$$

where t is free for x in A.

6.2.2.   Let x, y be any distinct variables. Let A, B be any formulae, C, D any formulae not containing the variable x free, and let E be any formula such that x is free for y in E. Give proof trees for the following formulae:

$$\forall x C \equiv C \qquad\qquad \exists x C \equiv C$$
$$\forall x \forall y A \equiv \forall y \forall x A \qquad\qquad \exists x \exists y A \equiv \exists y \exists x A$$
$$\forall x \forall y E \supset \forall x E[x/y] \qquad\qquad \exists x E[x/y] \supset \exists x \exists y E$$
$$\forall x A \supset \exists x A$$
$$\exists x \forall y A \supset \forall y \exists x A$$

6.2.3.   First, give G-proofs for the formulae of problem 6.2.2, and then convert them into LK-proofs using the method of theorem 6.2.1.

6.2.4.   Give an LK-proof for

$$(A \supset (B \supset C)) \to (B \supset (A \supset C)).$$

Give an LK-proof for

$$(B \supset (C \supset A)), (A \supset (B \supset C)) \to (B \supset (A \equiv C))$$

using the cut rule, and another LK-proof without cut.

6.2.5. Given a set S of formulae, let $Des(S)$ be the set of immediate descendants of formulae in S as defined in problem 5.5.18, and define S^n by induction as follows:

$$S^0 = S;$$
$$S^{n+1} = Des(S^n).$$
$$\text{Let } S^* = \bigcup_{n \geq 0} S^n.$$

S^* is called the set of *descendants* of S.

(a) Every sequent $\Gamma \to \Delta$ corresponds to the set of formulae $\Gamma \cup \{\neg B \mid B \in \Delta\}$. Prove that for every deduction tree for a sequent $\Gamma_0 \to \Delta_0$, the union of the sets of formulae $\Gamma \cup \{\neg B \mid B \in \Delta\}$ for all sequents $\Gamma \to \Delta$ occurring in that tree is a subset of S^*, where $S = \Gamma_0 \cup \{\neg B \mid B \in \Delta_0\}$ (this is called the *subformula property*).

(b) Deduce from (a) that not all formulae are provable.

6.3 The Gentzen System LK_e With Equality

We now generalize the system LK to include equality.

6.3.1 Syntax of LK_e

The rules and axioms of LK_e are defined as follows.

Definition 6.3.1 (The Gentzen system LK_e) The Gentzen system LK_e is obtained from the system LK by adding the following sequents known as *equality axioms*:

Let $t, s_1, ..., s_n, t_1, ..., t_n$ be arbitrary L-terms. For every term t, the sequent

$$\to t \doteq t$$

is an axiom.

For every n-ary function symbol f,

$$s_1 \doteq t_1, s_2 \doteq t_2, ..., s_n \doteq t_n \to f s_1...s_n \doteq f t_1...t_n$$

is an axiom.

For every n-ary predicate symbol P (including $\doteq$),

$$s_1 \doteq t_1, s_2 \doteq t_2, ..., s_n \doteq t_n, P s_1...s_n \to P t_1...t_n$$

is an axiom.

It is easily shown that these axioms are valid. In order to generalize theorem 6.2.1 it is necessary to define the concept of an atomic cut.

Definition 6.3.2 If the cut formula of a cut in an LK_e-proof is an atomic formula, the cut is called *atomic*. Otherwise, it is called an *essential cut*.

theorem 6.2.1 can now be generalized, provided that we allow atomic cuts. However, some technical lemmas including an exchange lemma will be needed in the proof.

6.3.2 A Permutation Lemma for the System $G_=$

The following lemma holds in $G_=$.

Lemma 6.3.1 Given a $G_=$-proof T, there is a $G_=$-proof T' such that all leaf sequents of T' contain either atomic formulae or quantified formulae (that is, formulae of the form $\forall x B$ or $\exists x B$).

Proof: Given a nonatomic formula A not of the form $\forall x B$ nor $\exists x B$, define its *weight* as the number of logical connectives $\land$, $\lor$, $\neg$, $\supset$ in it. The weight of an atomic formula and of a quantified formula is 0. The weight of a sequent is the sum of the weights of the formulae in it. The weight of a proof tree is the maximum of the weights of its leaf sequents. We prove the lemma by induction on the weight of the proof tree T. If $weight(T) = 0$, the lemma holds trivially. Otherwise, we show how the weight of each leaf sequent of T whose weight is nonzero can be decreased. This part of the proof proceeds by cases. We cover some of the cases, leaving the others as an exercise.

Let S be any leaf sequent of T whose weight is nonzero. Then, for some formula A, S is of the form

$$\Gamma_1, A, \Gamma_2 \to \Delta_1, A, \Delta_2.$$

If A is the only formula in S whose weight is nonzero, we can extend the leaf S as follows:

Case 1: A is of the form $B \land C$.

$$\frac{\Gamma_1, B, C, \Gamma_2 \to \Delta_1, B, \Delta_2 \qquad \Gamma_1, B, C, \Gamma_2 \to \Delta_1, C, \Delta_2}{\dfrac{\Gamma_1, B, C, \Gamma_2 \to \Delta_1, B \land C, \Delta_2}{\Gamma_1, B \land C, \Gamma_2 \to \Delta_1, B \land C, \Delta_2}}$$

The weights of the new leaf sequents are strictly smaller than the weight of S.

Case 2: A is of the form $\neg B$.

$$\frac{\dfrac{B, \Gamma_1, \Gamma_2 \to B, \Delta_1, \Delta_2}{\Gamma_1, \Gamma_2 \to B, \Delta_1, \neg B, \Delta_2}}{\Gamma_1, \neg B, \Gamma_2 \to \Delta_1, \neg B, \Delta_2}$$

The weight of the new leaf sequent is strictly smaller than the weight of S.

Case 3: A is of the form $B \supset C$.

$$\frac{\dfrac{B, \Gamma_1, \Gamma_2 \to B, C, \Delta_1, \Delta_2 \qquad C, B, \Gamma_1, \Gamma_2 \to C, \Delta_1, \Delta_2}{B, \Gamma_1, B \supset C, \Gamma_2 \to C, \Delta_1, \Delta_2}}{\Gamma_1, B \supset C, \Gamma_2 \to \Delta_1, B \supset C, \Delta_2}$$

The weights of the new leaf sequents are strictly smaller than the weight of S.

We leave the case in which A is of the form $(C \vee D)$ as an exercise.

If A is not the only formula in $\Gamma_1, A, \Gamma_2 \to \Delta_1, A, \Delta_2$ whose weight is nonzero, if we apply the corresponding rule to that formula, we obtain a sequent S_1 or two sequents S_1 and S_2 still containing A on both sides of the arrow, and whose weight is strictly smaller than the weight of S.

Now, if we apply the above transformations to all leaf sequents of T whose weight is nonzero, since the weight of each new leaf sequent is strictly smaller than the weight of some leaf sequent of T, we obtain a tree T' whose weight is strictly smaller than the weight of T. We conclude by applying the induction hypothesis to T'. This completes the proof. $\square$

Lemma 6.3.2 (i) Given a $G_=$-proof tree T for a sequent $\Gamma \to A \wedge B, \Delta$ satisfying the conclusion of lemma 6.3.1, another $G_=$-proof T' can be constructed such that, $depth(T) = depth(T')$, and the rule applied at the root of T' is the $\wedge : right$ rule applied to the occurrence of $A \wedge B$ to the right of $\to$.

(ii) Given a $G_=$-proof tree T of a sequent $\Gamma, A \supset B \to \Delta$ satisfying the conclusion of lemma 6.3.1, another $G_=$-proof tree T' can be constructed such that, $depth(T) = depth(T')$, and the rule applied at the root of T' is the $\supset : left$ rule applied to the occurrence of $A \supset B$ to the left of $\to$.

Proof: (i) Let S be the initial subtree of T obtained by deleting the descendants of every node closest to the root of T, where the $\wedge : right$ rule is applied to the formula $A \wedge B$ to the right of $\to$ in the sequent labeling that node. Since the proof tree T satisfies the condition of lemma 6.3.1, the rule $\wedge : right$ is applied to each occurrence of $A \wedge B$ on the right of $\to$. Hence, the tree T has the following shape:

Tree T

$$
\frac{S_1}{\Gamma_1 \rightarrow A, \Delta_1} \qquad \frac{T_1}{\Gamma_1 \rightarrow B, \Delta_1} \qquad \frac{S_m}{\Gamma_m \rightarrow A, \Delta_m} \qquad \frac{T_m}{\Gamma_m \rightarrow B, \Delta_m}
$$

$$
\frac{}{\Gamma_1 \rightarrow A \wedge B, \Delta_1} \qquad\qquad \frac{}{\Gamma_m \rightarrow A \wedge B, \Delta_m}
$$

$$S$$

$$\Gamma \rightarrow A \wedge B, \Delta$$

where the tree with leaves $\Gamma_1 \rightarrow A \wedge B, \Delta_1, ..., \Gamma_m \rightarrow A \wedge B, \Delta_m$ is the tree S, and the subtrees S_1, T_1,..., S_m, T_m are proof trees. Let S' be the tree obtained from S by replacing every occurrence in S of the formula $A \wedge B$ to the right of $\rightarrow$ by A, and S'' the tree obtained from S by replacing every occurrence of $A \wedge B$ on the right of $\rightarrow$ by B. Since no rule is applied to an occurrence of $A \wedge B$ on the right of $\rightarrow$ in S, both S' and S'' are well defined. The following proof tree T' satisfies the conditions of the lemma:

Tree T'

$$
\frac{S_1}{\Gamma_1 \rightarrow A, \Delta_1} \qquad \frac{S_m}{\Gamma_m \rightarrow A, \Delta_m} \qquad \frac{T_1}{\Gamma_1 \rightarrow B, \Delta_1} \qquad \frac{T_m}{\Gamma_m \rightarrow B, \Delta_m}
$$

$$S' \qquad\qquad\qquad S''$$

$$
\frac{\Gamma \rightarrow A, \Delta \qquad\qquad \Gamma \rightarrow B, \Delta}{\Gamma \rightarrow A \wedge B, \Delta}
$$

It is clear that $depth(T) = depth(T')$.

(ii) The proof is similar to that of (i). The proof tree T can be converted to T' as shown:

Tree T

$$
\frac{S_1}{\Gamma_1 \rightarrow A, \Delta_1} \qquad \frac{T_1}{B, \Gamma_1 \rightarrow \Delta_1} \qquad \frac{S_m}{\Gamma_m \rightarrow A, \Delta_m} \qquad \frac{T_m}{B, \Gamma_m \rightarrow \Delta_m}
$$

$$
\frac{}{\Gamma_1, A \supset B \rightarrow \Delta_1} \qquad\qquad \frac{}{\Gamma_m, A \supset B \rightarrow \Delta_m}
$$

$$S$$

$$\Gamma, A \supset B \rightarrow \Delta$$

Tree T'

$$\frac{S_1}{\Gamma_1 \to A, \Delta_1} \qquad \frac{S_m}{\Gamma_m \to A, \Delta_m} \qquad \frac{T_1}{B, \Gamma_1 \to \Delta_1} \qquad \frac{T_m}{B, \Gamma_m \to \Delta_m}$$

$$S' \qquad\qquad\qquad S''$$

$$\frac{\Gamma \to A, \Delta \qquad B, \Gamma \to \Delta}{\Gamma, A \supset B \to \Delta}$$

Note that $depth(T) = depth(T')$. $\square$

Lemma 6.3.3 Every $G_=$-proof tree T can be converted to a proof tree T' of the same sequent such that the rule applied to every sequent $\Gamma \to A \wedge B, \Delta$ is the $\wedge : right$ rule applied to the occurrence of $A \wedge B$ to the right of $\to$, and the rule applied to every sequent $\Gamma, A \supset B \to \Delta$ is the $\supset : left$ rule applied to the occurrence of $A \supset B$ to the left of $\to$. Furthermore, if T satisfies the conditions of lemma 6.3.1, then T' has the same depth as T.

Proof: First, using lemma 6.3.1, we can assume that T has been converted to a proof tree such that in all leaf sequents, all formulae are either atomic or quantified formulae. Then, since lemma 6.3.2 preserves the depth of such proof trees, we conclude by induction on the depth of proof trees using lemma 6.3.2. Since the transformations of lemma 6.3.2 are depth preserving, the last clause of the lemma follows. $\square$

6.3.3 Logical equivalence of $G_=$, LK_e, and LK_e Without Essential Cuts: Gentzen's Hauptsatz for LK_e Without Essential Cuts

The generalization of theorem 6.2.1 is the following.

Theorem 6.3.1 (Logical equivalence of $G_=$, LK_e, and LK_e without essential cuts) There is an algorithm to convert any LK_e-proof of a sequent $\Gamma \to \Delta$ into a $G_=$-proof. There is an algorithm to convert any $G_=$-proof of a sequent $\Gamma \to \Delta$ into an LK_e-proof without essential cuts.

Proof: The proof is similar to that of theorem 6.2.1. The proof differs because atomic cuts cannot be eliminated. By lemma 6.3.1, we can assume that the axioms of proof trees are either atomic formulae or quantified formulae. We proceed by induction on $G_=$-proof trees having axioms of this form. The base case is unchanged and, for the induction step, only the equality rules of definition 6.2.1 need to be considered, since the other rules are handled as in theorem 6.2.1.

(i) The rule

$$\frac{\Gamma, t \doteq t \to \Delta}{\Gamma \to \Delta}$$

is simulated in LK_e using an atomic cut as follows:

$$\cfrac{\to t \doteq t \qquad \cfrac{\cfrac{\Gamma, t \doteq t \to \Delta}{\text{exchanges}}}{t \doteq t, \Gamma \to \Delta}}{\Gamma \to \Delta} \quad \text{atomic cut}$$

(ii) The root of the $G_=$-proof tree T is the conclusion of the rule

$$\frac{\Gamma, (s_1 \doteq t_1) \wedge \dots \wedge (s_n \doteq t_n) \supset (fs_1 \dots s_n \doteq ft_1 \dots t_n) \to \Delta}{\Gamma \to \Delta}$$

If the premise of this rule is an axiom in $G_=$, this sequent contains a same formula A on both sides, and an LK_e-proof without cuts can be obtained from $A \to A$ using the exchange and weakening rules. Otherwise, using lemma 6.3.3, an LK_e-derivation with only atomic cuts can be constructed as follows. By lemma 6.3.3, the proof tree T is equivalent to a proof tree of same depth having the following shape:

$$\cfrac{\cfrac{T_{n-1}}{\Gamma \to s_{n-1} \doteq t_{n-1}, \Delta} \qquad \cfrac{T_n}{\Gamma \to s_n \doteq t_n, \Delta}}{\Gamma \to s_{n-1} \doteq t_{n-1} \wedge s_n \doteq t_n, \Delta}$$

$$\dots$$

$$\cfrac{\cfrac{\cfrac{T_1}{\Gamma \to s_1 \doteq t_1, \Delta} \quad \Gamma \to s_2 \doteq t_2 \wedge \dots \wedge s_n \doteq t_n, \Delta}{\Gamma \to s_1 \doteq t_1 \wedge \dots \wedge s_n \doteq t_n, \Delta} \qquad \cfrac{T_0}{fs_1 \dots s_n \doteq ft_1 \dots t_n, \Gamma \to \Delta}}{\Gamma, s_1 \doteq t_1 \wedge \dots \wedge s_n \doteq t_n \supset fs_1 \dots s_n \doteq ft_1 \dots t_n \to \Delta}$$

Using the axiom

$$s_1 \doteq t_1, \dots, s_n \doteq t_n \to fs_1 \dots s_n \doteq ft_1 \dots t_n$$

and applying the induction hypothesis to the $G_=$-trees $T_0, T_1, \dots, T_n$, an LK_e-proof without essential cuts can be constructed:

$$\frac{\dfrac{T'_n}{\Gamma \to \Delta, s_n \doteq t_n} \quad \dfrac{s_1 \doteq t_1, ..., s_n \doteq t_n \to fs_1...s_n \doteq ft_1...t_n}{s_n \doteq t_n, ..., s_1 \doteq t_1 \to fs_1...s_n \doteq ft_1...t_n}}{s_{n-1} \doteq t_{n-1}, ..., s_1 \doteq t_1, \Gamma \to fs_1...s_n \doteq ft_1...t_n, \Delta}$$

$$...$$

$$\frac{\dfrac{T'_1}{\Gamma \to \Delta, s_1 \doteq t_1} \quad s_1 \doteq t_1, \Gamma \to fs_1...s_n \doteq ft_1...t_n, \Delta}{\Gamma \to fs_1...s_n \doteq ft_1...t_n, \Delta}$$

We finish the proof with one more atomic cut:

$$\frac{\dfrac{\Gamma \to fs_1..., s_n \doteq ft_1...t_n, \Delta}{\Gamma \to \Delta, fs_1..., s_n \doteq ft_1...t_n} \quad \dfrac{\dfrac{T'_0}{\Gamma, fs_1...s_n \doteq ft_1...t_n \to \Delta}}{fs_1...s_n \doteq ft_1...t_n, \Gamma \to \Delta}}{\Gamma \to \Delta}$$

The trees $T'_0, T'_1, ..., T'_n$ have been obtained using the induction hypothesis. Note that applications of exchange and contraction rules are implicit in the above proofs.

(iii) The root of the $G_=$-proof tree T is labeled with the conclusion of the rule:

$$\frac{\Gamma, ((s_1 \doteq t_1) \wedge ... \wedge (s_n \doteq t_n) \wedge Ps_1...s_n) \supset Pt_1...t_n \to \Delta}{\Gamma \to \Delta}$$

This case is handled as case (ii). This concludes the proof of the theorem. □

Corollary (A version of Gentzen Hauptsatz for LK_e) A sequent is LK_e-provable if and only if it is LK_e-provable without essential cuts.

PROBLEMS

6.3.1. The set of *closed equality axioms* is the set of closed formulae given below:

$$\forall x (x \doteq x)$$

$$\forall x_1...\forall x_n \forall y_1...\forall y_n ((x_1 \doteq y_1 \land ... \land x_n \doteq y_n) \supset$$
$$(f(x_1,...,x_n) \doteq f(y_1,...,y_n)))$$
$$\forall x_1...\forall x_n \forall y_1...\forall y_n (((x_1 \doteq y_1 \land ... \land x_n \doteq y_n) \land P(x_1,...,x_n)) \supset$$
$$P(y_1,...,y_n))$$

Prove that the closed equality axioms are LK_e-provable.

6.3.2. Prove that the axioms of definition 6.3.1 are valid.

6.3.4. Finish the proof of the cases in lemma 6.3.1.

6.3.5. Prove case (iii) in the proof of theorem 6.3.1.

6.4 Gentzen's Hauptsatz for Sequents in NNF

Combining ideas from Smullyan and Schwichtenberg (Smullyan, 1968, Barwise, 1977), we formulate a sequent calculus $G1^{nnf}$ in which sequents consist of formulae in negation normal form. Such a system shares characteristics of both G and LK but the main difference is that sequents consist of pairs of *sets* rather than pairs of *sequences*. The main reason for using sets rather than sequences is that the structural rules become unnecessary. As a consequence, an induction argument simpler than Gentzen's original argument (Szabo, 1969) can be used in the proof of the cut elimination theorem. The advantage of considering formulae in negation normal form is that fewer inference rules need to be considered. Since every formula is equivalent to a formula in negation normal form, there is actually no loss of generality.

6.4.1 Negation Normal Form

The definition of a formula in negation normal form given in definition 3.4.8 is extended to the first-order case as follows.

Definition 6.4.1 The set of formulae in *negation normal form* (for short, NNF) is the smallest set of formulae such that

(1) For every atomic formula A, A and $\neg A$ are in NNF;

(2) If A and B are in NNF, then $(A \lor B)$ and $(A \land B)$ are in NNF.

(3) If A is in NNF, then $\forall x A$ and $\exists x A$ are in NNF.

Lemma 6.4.1 Every formula is equivalent to another formula in NNF.

Proof: The proof proceeds by induction on formulae as in lemma 3.4.4. Careful checking of the proof of lemma 3.4.4 reveals that for the propositional connectives, only the case where A is of the form $\neg \forall x B$ or $\neg \exists x B$ needs to be considered. If A is of the form $\neg \forall x B$, by lemma 5.3.6(8) $\neg \forall x B$ is equivalent to $\exists x \neg B$, and since $\neg B$ has fewer connectives than $\neg \forall x B$, by the induction hypothesis $\neg B$ has a NNF B'. By lemma 5.3.7, A is equivalent to $\exists x B'$, which is in NNF. The case where A is of the form $\neg \exists x B$ is similar. Finally, if A is of the form $\forall x B$ or $\exists x B$, by the induction hypothesis B is equivalent to a formula B' is NNF, and by lemma 5.3.7 $\forall x B$ is equivalent to $\forall x B'$ and $\exists x B$ is equivalent to $\exists x B'$. $\square$

EXAMPLE 6.4.1

Let

$$A = \forall x(P(x) \vee \neg \exists y(Q(y) \wedge R(x,y))) \vee \neg(P(y) \wedge \neg \forall x P(x))$$

The NNF of $\neg \exists y(Q(y) \wedge R(x,y))$ is

$$\forall y(\neg Q(y) \vee \neg R(x,y)).$$

The NNF of $\neg(P(y) \wedge \neg \forall x P(x))$ is

$$(\neg P(y) \vee \forall x P(x)).$$

The NNF of A is

$$\forall x(P(x) \vee \forall y(\neg Q(y) \vee \neg R(x,y))) \vee (\neg P(y) \vee \forall x P(x)).$$

We now define a new Gentzen system in which sequents are pairs of sets of formulae in NNF. First, we treat the case of languages without equality.

6.4.2 The Gentzen System $G1^{nnf}$

The axioms and inference rules of $G1^{nnf}$ are defined as follows.

Definition 6.4.2 The sequents of the system $G1^{nnf}$ are pairs $\Gamma \rightarrow \Delta$, where Γ and Δ are finite *sets* of formulae in NNF. Given two sets of formuae Γ and Δ, the expression Γ, Δ denotes the *union* of the sets Γ and Δ, and similarly, if A is a formula, Γ, A denotes the set $\Gamma \cup \{A\}$. The inference rules are the rules listed below:

(1) Cut rule:

$$\frac{\Gamma \rightarrow \Delta, A \qquad A, \Lambda \rightarrow \Theta}{\Gamma, \Lambda \rightarrow \Delta, \Theta}$$

A is called the *cut formula* of this inference.

(2) Propositional logical rules:

$$\frac{A, \Gamma \rightarrow \Delta}{A \wedge B, \Gamma \rightarrow \Delta} \ (\wedge:left) \quad \text{and} \quad \frac{B, \Gamma \rightarrow \Delta}{A \wedge B, \Gamma \rightarrow \Delta} \ (\wedge:left)$$

$$\frac{\Gamma \to \Delta, A \quad \Gamma \to \Delta, B}{\Gamma \to \Delta, A \wedge B} \ (\wedge : right)$$

$$\frac{A, \Gamma \to \Delta \quad B, \Gamma \to \Delta}{A \vee B, \Gamma \to \Delta} \ (\vee : left)$$

$$\frac{\Gamma \to \Delta, A}{\Gamma \to \Delta, A \vee B} \ (\vee : right) \quad \text{and} \quad \frac{\Gamma \to \Delta, B}{\Gamma \to \Delta, A \vee B} \ (\vee : right)$$

In the rules above, $A \vee B$ and $A \wedge B$ are called the *principal formulae* and A, B the *side formulae* of the inference.

(3) Quantifier rules

In the quantifier rules below, x is any variable and y is any variable free for x in A and not free in A, unless $y = x$ ($y \notin FV(A) - \{x\}$). The term t is any term free for x in A.

$$\frac{A[t/x], \Gamma \to \Delta}{\forall x A, \Gamma \to \Delta} \ (\forall : left) \quad \frac{\Gamma \to \Delta, A[y/x]}{\Gamma \to \Delta, \forall x A} \ (\forall : right)$$

$$\frac{A[y/x], \Gamma \to \Delta}{\exists x A, \Gamma \to \Delta} \ (\exists : left) \quad \frac{\Gamma \to \Delta, A[t/x]}{\Gamma \to \Delta, \exists x A} \ (\exists : right)$$

In both the ($\forall : right$)-rule and the ($\exists : left$)-rule, the variable y does *not* occur free in the lower sequent. In these rules, the variable y is called the *eigenvariable* of the inference. The condition that the eigenvariable does not occur free in the conclusion of the rule is called the *eigenvariable condition*. The formula $\forall x A$ (or $\exists x A$) is called the *principal formula* of the inference, and the formula $A[t/x]$ (or $A[y/x]$) the *side formula* of the inference.

The *axioms* of $G1^{nnf}$ are all sequents of the form

$$\Gamma, A \to A, \Delta,$$
$$\Gamma, \neg A \to \neg A, \Delta,$$
$$\Gamma, A, \neg A \to \Delta, \text{ or}$$
$$\Gamma \to \Delta, A, \neg A,$$

with A *atomic*.

The notions of *deduction trees* and *proof trees* are defined as usual, but with the rules and axioms of $G1^{nnf}$. It is readily shown that the system $G1^{nnf}$ is sound. We can also prove that $G1^{nnf}$ is complete for sequents in NNF.

6.4.3 Completeness of $G1^{nnf}$

The following lemma is shown using theorem 5.5.1.

Lemma 6.4.2 (Completeness of $G1^{nnf}$) Every valid $G1^{nnf}$-sequent has a $G1^{nnf}$-proof.

Proof: First, we check that the proof given in theorem 5.5.1 can be adapted to hold for sequents consisting of sets rather than sequences. In order to simulate the $\forall : left$ rule and the $\exists : right$ rule of G using the quantifier rules of $G1^{nnf}$, we use the fact that in $G1^{nnf}$, sequents consist of sets. Since $\forall x B, \Gamma \rightarrow \Delta$ and $\forall x B, \forall x B, \Gamma \rightarrow \Delta$ actually denote the same sequent, we can apply the $\forall : left$ rule (of $G1^{nnf}$) to $\forall x B$, with the formulae in $\forall x B, \Gamma$ and Δ as auxiliary formulae, obtaining:

$$\frac{\forall x B, B[t/x], \Gamma \rightarrow \Delta}{\forall x B, \Gamma \rightarrow \Delta}$$

The case of $\exists : right$ is similar. We simulate the $\wedge : left$ rule of G and the $\vee : right$ of G as in the proof of theorem 3.6.1. For example, the following derivation simulates $\wedge : left$ of G:

$$\frac{\dfrac{A, B, \Gamma \rightarrow \Delta}{A \wedge B, B, \Gamma \rightarrow \Delta}}{A \wedge B, \Gamma \rightarrow \Delta}$$

$\wedge : left$ applied to A

$\wedge : left$ applied to B

We obtain proofs in which the conditions for declaring that a sequent is an axiom are as follows: A sequent $\Gamma \rightarrow \Delta$ is an axiom iff any of the following conditions holds:

(i) Γ and Δ have some formula A in common; or

(ii) Γ contains some atomic formula B and its negation $\neg B$; or

(iii) Δ contains some atomic formula B and its negation $\neg B$.

However, the formula A may not be a *literal* (that is, an atomic formula, or the negation of an atomic formula). To make sure that in (i) A is a literal, we use the method of lemma 6.3.1, complemented by the cases of quantified formulae. We consider the case in which $A = \forall x B$ is on the left of $\rightarrow$, the case $A = \exists x B$ on the right of $\rightarrow$ being similar. Assume that the axiom is $\Gamma, \forall x B \rightarrow \Delta, \forall x B$. Then we have the following proof:

$$\frac{\Gamma, B[z/x] \to \Delta, B[z/x]}{\dfrac{\Gamma, \forall x B \to \Delta, B[z/x]}{\Gamma, \forall x B \to \Delta, \forall x B}}$$

where z is a new variable.

The top sequent is an axiom, and $B[z/x]$ has fewer connectives than $\forall x B$. We conclude by induction on the number of connectives in $B[z/x]$. The details are left as an exercise. □

6.4.4 The Cut Elimination Theorem for $G1^{nnf}$

We now proceed with the proof of the cut elimination theorem for $G1^{nnf}$. The proof uses a method due to Schwichtenberg (adapted from Tait, Tait, 1968), which consists of a single induction on the cut-rank of a proof with cut. This proof is simpler than Gentzen's original, because the system $G1^{nnf}$ does not have structural rules. This is the reason a simple induction on the cut-rank works (as opposed to the the induction used in Gentzen's original proof, which uses a lexicographic ordering). Furthermore, the proof of the theorem also yields an upper bound on the size of the resulting cut-free proof. The key parameter of the proof, is the cut-rank of a $G1^{nnf}$-proof. First, we need to define the degree of a formula in NNF. Roughly speaking, the degree of a formula in NNF is the depth of the tree representing that formula, ignoring negations.

Definition 6.4.3 The *degree* $|A|$ of a formula A in NNF is defined inductively as follows:

(i) If A is an atomic formula or the negation of an atomic formula, then

$$|A| = 0;$$

(ii) If A is either of the form $(B \vee C)$ or $(B \wedge C)$, then

$$|A| = max(|B|, |C|) + 1;$$

(iii) If A is either of the form $\forall x B$ or $\exists x B$, then

$$|A| = |B| + 1.$$

The cut-rank is defined as follows.

Definition 6.4.4 Let T be a $G1^{nnf}$-proof. The *cut-rank* $c(T)$ of T is defined inductively as follows. If T is an axiom, then $c(T) = 0$. If T is not an axiom, the last inference has either one or two premises. In the first case, the premise of that inference is the root of a subtree T_1. In the second case, the left premise is the root of a subtree T_1, and the right premise is the root of a subtree T_2. If the last inference is not a cut, then if it has a single premise,

$$c(T) = c(T_1),$$

else

$$c(T) = max(c(T_1), c(T_2)).$$

If the last inference is a cut with cut formula A, then

$$c(T) = max(|A| + 1, c(T_1), c(T_2)).$$

Note that $c(T) = 0$ iff T is cut free. We will need a number of lemmas to establish the cut elimination theorem for $G1^{nnf}$. In some of these proofs, it will be necessary to replace in a proof all free occurrences of variable y by a new variable z not occurring in the proof. This substitution process is defined as follows.

Definition 6.4.5 Given a formula A, we say that the variable y is *not bound in the formula A* iff $y \notin BV(A)$. The variable y is *not bound in the sequent* $\Gamma \to \Delta$ iff y is not bound in any formula in Γ or Δ. The variable y is *not bound in the deduction tree T* iff it is not bound in any sequent occurring in T. Given a formula A and two variables y, z, the formula $A[z/y]$ is defined as in definition 5.2.6. For a sequent $\Gamma \to \Delta$, the sequent $(\Gamma \to \Delta)[z/y]$ is the sequent obtained by substituting z for y in all formulae in $\Gamma \to \Delta$. For a deduction tree T, a variable y not bound in T, and a variable z not occurring in T, the deduction tree $T[z/y]$ is the result of replacing every sequent $\Gamma \to \Delta$ in T by $(\Gamma \to \Delta)[z/y]$. This operation can be defined more precisely by induction on proof trees, the simple details being left to the reader.

A similar definition can be given for the result $T[t/y]$ of substituting a term t for a variable y not bound in T, provided that t is free for y in every formula in which it is substituted, and that y and the variables in $FV(t)$ are distinct from all eigenvariables in T. In order to justify that $T[z/y]$ are $T[t/y]$ are indeed proof trees when T is, the following technical lemma is needed.

Lemma 6.4.3 Let $\Gamma \to \Delta$ be a sequent and T an $G1^{nnf}$-proof for $\Gamma \to \Delta$. Assume that y is any variable not bound in the proof tree T.

(i) For any variable z not occurring in T, the result $T[z/y]$ of substituting z for all occurrences of y in T is a proof tree for $\Gamma[z/y] \to \Delta[z/y]$.

(ii) If t is a term free for y in every formula in which it is substituted, and y and the variables in $FV(t)$ are distinct from all eigenvariables in T, then $T[t/y]$ is a proof tree for $\Gamma[t/y] \to \Delta[t/y]$.

Proof: We proceed by induction on proof trees. We only treat some key cases, leaving the others as an exercise. We consider (i). If T is an axiom

$\Gamma \to \Delta$, it is clear that $(\Gamma \to \Delta)[z/y]$ is also an axiom. The propositional rules present no difficulty and are left to the reader. We consider two of the quantifier rules: $\forall : left$ and $\forall : right$.

Case 1: The bottom inference is $\forall : left$:

$$\frac{\dfrac{T_1}{A[t/x], \Gamma \to \Delta}}{\forall x A, \Gamma \to \Delta}$$

where t is free for x in A.

By the induction hypothesis, $T_1[z/y]$ is a proof tree of $A[t/x][z/y], \Gamma[z/y] \to \Delta[z/y]$. Since we have assumed that y is not bound in the proof T, $y \neq x$. Hence, $(\forall x A)[z/y] = \forall x A[z/y]$. By the result of problem 5.2.7,

$$A[t/x][z/y] = A[z/y][t[z/y]/x],$$

and since z does not occur in T, $t[z/y]$ is free for x in $A[z/y]$. But then, $T[z/y]$ is the proof tree:

$$\frac{\dfrac{T_1[z/y]}{A[z/y][t[z/y]/x], \Gamma[z/y] \to \Delta[z/y]}}{\forall x A[z/y], \Gamma[z/y] \to \Delta[z/y]}$$

Case 2: The bottom inference is $\forall : right$:

$$\frac{\dfrac{T_1}{\Gamma \to \Delta, A[w/x]}}{\Gamma \to \Delta, \forall x A}$$

where w is not free in $\Gamma \to \Delta, \forall x A$.

There are two subcases.

Subcase 2.1: If $y = w$, since w does not occur free in $\Gamma \to \Delta, \forall x A$, the variable w is not free in A, Γ or Δ. By the induction hypothesis, $T_1[z/y]$ is a proof tree for

$$(\Gamma \to \Delta, A[y/x])[z/y] = \Gamma \to \Delta, A[z/x].$$

Also, since z does not occur in T, z does not occur in $\Gamma \to \Delta, \forall x A$, the $\forall : right$ rule is applicable and $T[z/y]$ is a proof tree.

Subcase 2.2: $y \neq w$. By the induction hypothesis, $T_1[z/y]$ is a proof tree for $\Gamma[z/y] \to \Delta[z/y], A[w/x][z/y]$. By problem 5.2.7, we have

$$A[w/x][z/y] = A[z/y][w[z/y]/x],$$

but since $w \neq y$,

$$A[w/x][z/y] = A[z/y][w/x].$$

Also, since z does not occur in T, $z \neq w$, and so w does not occur free in $\Gamma[z/y] \rightarrow \Delta[z/y], \forall x A[z/y]$. Hence, the $\forall : right$ rule is applicable and $T[z/y]$ is a proof tree:

$$\frac{\dfrac{T_1[z/y]}{\Gamma[z/y] \rightarrow \Delta[z/y], A[z/y][w/x]}}{\Gamma[z/y] \rightarrow \Delta[z/y], \forall x A[z/y]}$$

It should be noted that in the proof of (ii), the condition that y is distinct from all eigenvariables in T rules out subcase 2.1. $\square$

Lemma 6.4.4 (Substitution lemma) Let T be a $G1^{nnf}$-proof of a sequent $\Gamma \rightarrow \Delta$ such that the variable x is not bound in T. For any term t free for x in $\Gamma \rightarrow \Delta$, a proof $T'(t)$ of $\Gamma[t/x] \rightarrow \Delta[t/x]$ can be constructed such that $T'(t)$ and T have same depth, x and the variables in $FV(t)$ are distinct from all eigenvariables in $T'(t)$ and $c(T) = c(T'(t))$.

Proof: By induction on proof trees using lemma 6.4.3. For a similar proof, see lemma 7.3.1. $\square$

Lemma 6.4.5 (Weakening lemma) Given a $G1^{nnf}$-proof T of a sequent $\Gamma \rightarrow \Delta$, for any formula A (in NNF), a proof T' of $A, \Gamma \rightarrow \Delta$ (resp. $\Gamma \rightarrow \Delta, A$) can be obtained such that T and T' have the same depth, all variables free in A are distinct from all eigenvariables in T and $c(T') = c(T)$.

Proof: Straightforward induction on proof trees, similar to that of lemma 6.4.3. $\square$

The proof T' given by the weakening lemma will also be denoted by (T, A).

Lemma 6.4.6 (Inversion lemma) (i) If a sequent $\Gamma \rightarrow \Delta, A \wedge B$ has a $G1^{nnf}$-proof T (resp. $A \vee B, \Gamma \rightarrow \Delta$ has a $G1^{nnf}$-proof T), a $G1^{nnf}$-proof T_1 of $\Gamma \rightarrow \Delta, A$ and a $G1^{nnf}$-proof T_2 of $\Gamma \rightarrow \Delta, B$ can be constructed (resp. a $G1^{nnf}$-proof T_1 of $A, \Gamma \rightarrow \Delta$ and a $G1^{nnf}$-proof T_2 of $B, \Gamma \rightarrow \Delta$ can be constructed) such that, $depth(T_1), depth(T_2) \leq depth(T)$ and $c(T_1), c(T_2) \leq c(T)$.

(ii) If $\Gamma \rightarrow \Delta, \forall x B$ has a $G1^{nnf}$-proof T (resp. $\exists x B, \Gamma \rightarrow \Delta$ has a $G1^{nnf}$-proof T), then for any variable y not bound in T and distinct from all eigenvariables in T, a $G1^{nnf}$-proof T_1 of $\Gamma \rightarrow \Delta, B[y/x]$ can be constructed (resp. a $G1^{nnf}$-proof T_1 of $B[y/x], \Gamma \rightarrow \Delta$ can be constructed), such that $depth(T_1) \leq depth(T)$ and $c(T_1) \leq c(T)$.

Proof: The proofs of (i) and (ii) are similar, both by induction on proof trees. We consider (ii), leaving (i) as an exercise.

If $\forall x B$ belongs to Δ, the result follows by the weakening lemma (lemma 6.4.5), using the proof $(T, B[y/x])$ given by that lemma. If $\forall x B$ does not belong to Δ, there are two cases.

Case 1: $\forall x B$ is not the principal formula of the last inference. There are several subcases depending on that inference. Let us consider the $\wedge : right$ rule, the other subcases being similar and left as an exercise. The proof has the form

$$\frac{\begin{array}{c} S_1 \\ \hline \Gamma \to \Delta, \forall x B, C \end{array} \qquad \begin{array}{c} S_2 \\ \hline \Gamma \to \Delta, \forall x B, D \end{array}}{\Gamma \to \Delta, \forall x B, C \wedge D}$$

By the induction hypothesis, for any variable y not bound in S_1 or S_2 and distinct from all eigenvariables in S_1 or S_2, we can find proofs T_1 for $\Gamma \to \Delta, B[y/x], C$ and T_2 for $\Gamma \to \Delta, B[y/x], D$, such that $depth(T_i) \le depth(S_i)$ and $c(T_i) \le c(S_i)$, for $i = 1, 2$. We conclude using the following proof:

$$\frac{\begin{array}{c} T_1 \\ \hline \Gamma \to \Delta, B[y/x], C \end{array} \qquad \begin{array}{c} T_2 \\ \hline \Gamma \to \Delta, B[y/x], D \end{array}}{\Gamma \to \Delta, B[y/x], C \wedge D}$$

Case 2: $\forall x B$ is the principal formula of the last inference. Using the weakening lemma, we can make sure that the last inference is of the form

$$\frac{\begin{array}{c} T_1 \\ \hline \Gamma \to \Delta, \forall x B, B[y/x] \end{array}}{\Gamma \to \Delta, \forall x B}$$

replacing the proof T by $(T, \forall x B)$ if necessary. Using lemma 6.4.3, we can also make sure that y is not bound in T_1 (or T) and is distinct from all eigenvariables in T_1 (and T). Then, the induction hypothesis applies to the variable y in the lower sequent of the proof T_1, and we can find a proof T_1' of $\Gamma \to \Delta, B[y/x]$ such that $depth(T_1) < depth(T)$ and $c(T_1) \le c(T)$. Using lemma 6.4.3 again, we actually have a proof of $\Gamma \to \Delta, B[z/x]$ for any variable z not bound in T_1' and distinct from all eigenvariables in T_1', establishing (ii). The proof of the other cases is similar. $\square$

We are now ready for the main lemma which, shows how cuts are eliminated.

Lemma 6.4.7 (Reduction lemma for $G1^{nnf}$) Let T_1 be a $G1^{nnf}$-proof of $\Gamma \to \Delta, A$, and T_2 a $G1^{nnf}$-proof of $A, \Lambda \to \Theta$, and assume that

$$c(T_1), c(T_2) \le |A|.$$

A $G1^{nnf}$-proof T of

$$\Gamma, \Lambda \to \Delta, \Theta$$

can be constructed, such that

$$depth(T) \leq depth(T_1) + depth(T_2) \quad \text{and} \quad c(T) \leq |A|.$$

Proof: We proceed by induction on $depth(T_1) + depth(T_2)$.

Case 1: Either A is not the principal formula of the last inference of T_1, or A is not the principal formula of the last inference of T_2. By symmetry, we can assume the former. There are several subcases, depending on the last inference. Let us consider the $\wedge : right$ rule, the other subcases being similar and left as an exercise. The proof has the form

$$\frac{\dfrac{S_1}{\Gamma \to \Delta', C, A} \qquad \dfrac{S_2}{\Gamma \to \Delta', D, A}}{\Gamma \to \Delta', C \wedge D, A}$$

By the induction hypothesis, we can find proofs T_1' for $\Gamma, \Lambda \to \Delta', \Theta, C$ and T_2' for $\Gamma, \Lambda \to \Delta', \Theta, D$, such that $depth(T_i') < depth(T_1) + depth(T_2)$ and $c(T_i') \leq |A|$, for i=1,2. The result follows by the inference

$$\frac{\Gamma, \Lambda \to \Delta', \Theta, C \qquad \Gamma, \Lambda \to \Delta', \Theta, D}{\Gamma, \Lambda \to \Delta', C \wedge D, \Theta}$$

Case 2: A is the principal formula of the last inference of both T_1 and T_2.

Case 2.1: A is a literal; that is, an atomic formula, or the negation of an atomic formula. In this case, both $\Gamma \to \Delta, A$ and $A, \Lambda \to \Theta$ are axioms. By considering all possible cases, it can be verified that $\Gamma, \Lambda \to \Delta, \Theta$ is an axiom. For example, if A is atomic, Δ contains $\neg A$ and Λ contains $\neg A$, $\Gamma, \Lambda \to \Delta, \Theta$ is an axiom.

Case 2.2: A is of the form $(B \vee C)$. Using the weakening lemma, we can make sure that the last inference of T_1 is of the form

$$\frac{\dfrac{S_0}{\Gamma \to \Delta, A, B}}{\Gamma \to \Delta, A}$$

or

$$\frac{\dfrac{S_0}{\Gamma \to \Delta, A, C}}{\Gamma \to \Delta, A}$$

replacing T_1 by (T_1, A) if necessary. Consider the first case, the other being similar. By the induction hypothesis, we can find a proof T_1' for $\Gamma, \Lambda \to \Delta, \Theta, B$, such that $depth(T_1') < depth(T_1) + depth(T_2)$ and $c(T_1') \leq |A|$. By the inversion lemma, we can find a proof T_2' of $B, \Lambda \to \Theta$ such that $depth(T_2') \leq depth(T_2)$ and $c(T_2') \leq |A|$. The following proof T'

$$\frac{\dfrac{T_1'}{\Gamma, \Lambda \to \Delta, \Theta, B} \qquad \dfrac{T_2'}{B, \Lambda \to \Theta}}{\Gamma, \Lambda \to \Delta, \Theta}$$

is such that $depth(T') \leq depth(T_1) + depth(T_2)$ and has cut-rank $c(T') \leq |A|$, since $|B| < |A|$.

Case 2.3: A is of the form $(B \wedge C)$. This case is symmetric to case 2.2.

Case 2.4: A is of the form $\exists x B$. As in case 2.2, we can assume that A belongs to the premise of the last inference in T_1, so that T_1 is of the form

$$\frac{\dfrac{S_0}{\Gamma \to \Delta, A, B[t/x]}}{\Gamma \to \Delta, A}$$

By the induction hypothesis, we can find a proof tree T_1' for $\Gamma, \Lambda \to \Delta, \Theta, B[t/x]$, such that $depth(T_1') < depth(T_1) + depth(T_2)$ and $c(T_1') \leq |A|$. By the inversion lemma, for any variable y not bound in T_2 and distinct from all eigenvariables in T_2, there is a proof T_2' for $B[y/x], \Lambda \to \Theta$, such that $depth(T_2') \leq depth(T_2)$ and $c(T_2') \leq |A|$. By the substitution lemma, we can construct a proof T_2'' for $B[t/x], \Lambda \to \Theta$, also such that $depth(T_2'') \leq depth(T_2)$ and $c(T_2'') \leq |A|$. Since $|B[t/x]| < |A|$, the proof obtained from T_2' and T_2'' by applying a cut to $B[t/x]$ has cut-rank $\leq |A|$.

Case 2.5: A is of the form $\forall x B$. This case is symmetric to case 2.4. This concludes all the cases. $\square$

Finally, we can prove the cut elimination theorem.

The function $exp(m, n, p)$ is defined recursively as follows:

$$exp(m, 0, p) = p;$$
$$exp(m, n + 1, p) = m^{exp(m,n,p)}.$$

This function grows extremely fast in the argument n. Indeed, $exp(m, 1, p) = m^p$, $exp(m, 2, p) = m^{m^p}$, and in general, $exp(m, n, p)$ is an iterated stack of exponentials of height n, topped with a p:

$$exp(m, n, p) = m^{m^{m^{.^{.^{.^{m^p}}}}}} \Big\} n$$

The Tait-Schwichtenberg's version of the cut elimination theorem for $G1^{nnf}$ follows.

Theorem 6.4.1 (Cut elimination theorem for $G1^{nnf}$) Let T be a $G1^{nnf}$-proof with cut-rank $c(T)$ of a sequent $\Gamma \rightarrow \Delta$. A cut-free proof T^* for $\Gamma \rightarrow \Delta$ such that $depth(T^*) \leq exp(2, c(T), depth(T))$ can be constructed.

Proof: We prove the following claim by induction on the depth of proof trees.

Claim: Let T be a $G1^{nnf}$-proof with cut-rank $c(T)$ for a sequent $\Gamma \rightarrow \Delta$. If $c(T) > 0$ then we can construct a proof T' for $\Gamma \rightarrow \Delta$, such that

$$c(T') < c(T) \quad \text{and} \quad depth(T') \leq 2^{depth(T)}.$$

Proof of Claim: If either the last inference of T is not a cut, or it is a cut and $c(T) > |A| + 1$, where A is the cut formula of the last inference, we can apply the induction hypothesis to the immediate subtrees T_1 or T_2 (or T_1) of T. We are left with the case in which the last inference is a cut and $c(T) = |A| + 1$. The proof is of the form

$$
\begin{array}{cc}
T_1 & T_2 \\
\hline
\Gamma \rightarrow \Delta, A \qquad\qquad & A, \Gamma \rightarrow \Delta \\
\end{array}
$$
$$\overline{\qquad\qquad\qquad\qquad \Gamma \rightarrow \Delta \qquad\qquad\qquad\qquad}$$

By the induction hypothesis, we can construct a proof T_1' for $\Gamma \rightarrow \Delta, A$ and a proof T_2' for $A, \Gamma \rightarrow \Delta$, such that $c(T_i') \leq |A|$ and $depth(T_i') \leq 2^{depth(T_i)}$, for $i = 1, 2$. Applying the reduction lemma, we obtain a proof T' such that, $c(T') \leq |A|$ and $depth(T') \leq depth(T_1') + depth(T_2')$. But

$$depth(T_1') + depth(T_2') \leq 2^{depth(T_1)} + 2^{depth(T_2)} \leq$$
$$2^{max(depth(T_1), depth(T_2))+1} = 2^{depth(T)}.$$

Hence, the claim holds for T'. $\square$

The proof of the theorem follows by induction on $c(T)$, and by the definition of $exp(2, m, n)$. $\square$

It is remarkable that theorem 6.4.1 provides an upper bound on the depth of cut-free proofs obtained by converting a proof with cuts. Note that the "blow up" in the size of the proof can be very large.

Cut-free proofs are "direct" (or analytic), in the sense that all inferences are purely mechanical, and thus require no ingenuity. Proofs with cuts are "indirect" (or nonanalytic), in the sense that the cut formula in a cut rule may not be a subformula of any of the formulae in the conclusion of the inference. Theorem 6.4.1 suggests that if some ingenuity is exercised in constructing proofs with cuts, the size of a proof can be reduced significantly. It gives a measure of the complexity of proofs. Without being very rigorous, we can say that theorem 6.4.1 suggests that there are theorems that have no easy proofs, in the sense that if the steps are straightforward, the proof is very long, or else if the proof is short, the cuts are very ingenious. Such an example is given in Statman, 1979.

We now add equality axioms to the system $G1^{nnf}$.

6.4.5 The System $G1_{=}^{nnf}$

The axioms and inference rules of the system $G1_{=}^{nnf}$ are defined as follows.

Definition 6.4.6 The system $G1_{=}^{nnf}$ is obtained by adding to $G1^{nnf}$ the following sequents as axioms. All sequents of the form

(i) $\Gamma \rightarrow \Delta, t \doteq t$;

(ii) For every n-ary function symbol f,

$$\Gamma, s_1 \doteq t_1, s_2 \doteq t_2, ..., s_n \doteq t_n \rightarrow \Delta, fs_1...s_n \doteq ft_1...t_n$$

(iii) For every n-ary predicate symbol P (including $\doteq$),

$$\Gamma, s_1 \doteq t_1, s_2 \doteq t_2, ..., s_n \doteq t_n, Ps_1...s_n \rightarrow \Delta, Pt_1...t_n$$

and all sequents obtained from the above sequents by applications of $\neg : left$ and $\neg : right$ rules to the atomic formulae $t \doteq t$, $s_i \doteq t_i$, $fs_1...s_n \doteq ft_1...t_n$, $Ps_1...s_n$ and $Pt_1...t_n$.

For example, $\rightarrow \neg(s \doteq t), f(s) \doteq f(t)$ is an equality axiom. It is obvious that these sequents are valid and that the system $G1_{=}^{nnf}$ is sound.

Lemma 6.4.8 (Completeness of $G1_{=}^{nnf}$) For the system $G1_{=}^{nnf}$, every valid sequent is provable.

Proof: The lemma can be proved using the completeness of LK_e (theorem 6.3.1). First, we can show that in an LK_e-proof, all weakenings can be moved above all other inferences. Then, we can show how such a proof can be simulated by a $G1_{=}^{nnf}$-proof. The details are rather straighforward and are left as an exercise. $\square$

6.4.6 The Cut Elimination Theorem for $G1_{\doteq}^{nnf}$

Gentzen's cut elimination theorem also holds for $G1_{\doteq}^{nnf}$ if an *inessential cut* is defined as a cut in which the cut formula is a *literal* B or $\neg B$, where B is *equation* $s \doteq t$ (but not an atomic formula of the form $Ps_1...s_n$, where P is a predicate symbol different from $\doteq$). This has interesting applications, such as the completeness of equational logic (see the problems). The proof requires a modification of lemma 6.4.6. The *cut-rank* of a proof is now defined by considering essential cuts only.

Lemma 6.4.9 (Reduction lemma for $G1_{\doteq}^{nnf}$) Let T_1 be a $G1_{\doteq}^{nnf}$-proof of $\Gamma \rightarrow \Delta, A$, and T_2 a $G1_{\doteq}^{nnf}$-proof of $A, \Lambda \rightarrow \Theta$, and assume that $c(T_1), c(T_2) \leq |A|$. Let m be the maximal rank of all predicate symbols P such that some literal $Pt_1...t_n$ or $\neg Pt_1...t_n$ is a cut formula in either T_1 or T_2. A $G1_{\doteq}^{nnf}$-proof T of $\Gamma, \Lambda \rightarrow \Delta, \Theta$ can be constructed such that

$$c(T) \leq |A| \quad \text{and} \quad depth(T) \leq depth(T_1) + depth(T_2) + m.$$

Proof: We proceed by induction on $depth(T_1) + depth(T_2)$. The only new case is the case in which A is a literal of the form $Pt_1...t_n$ or $\neg Pt_1...t_n$, and one of the two axioms $\Gamma \rightarrow \Delta, A$ and $A, \Lambda \rightarrow \Theta$ is an equality axiom. Assume that $A = Pt_1...t_n$, the other case being similar. If $\Gamma \rightarrow \Delta$ or $\Lambda \rightarrow \Theta$ is an axiom, then $\Gamma, \Lambda \rightarrow \Delta, \Theta$ is an axiom. Otherwise, we have three cases.

Case 1: $\Gamma \rightarrow \Delta, A$ is an equality axiom, but $A, \Lambda \rightarrow \Theta$ is not. Then, either $\neg A$ is in Λ or A is in Θ, and $\Gamma \rightarrow \Delta, A$ is obtained from some equality axiom of the form

$$\Gamma', s_1 \doteq t_1, s_2 \doteq t_2, ..., s_n \doteq t_n, Ps_1...s_n \rightarrow \Delta', Pt_1...t_n$$

by application of $\neg : rules$. Since either $Pt_1...t_n$ is in Θ or $\neg Pt_1...t_n$ is in Λ, the sequent $\Gamma, \Lambda \rightarrow \Delta, \Theta$ is an equality axiom also obtained from some sequent of the form

$$\Gamma'', s_1 \doteq t_1, s_2 \doteq t_2, ..., s_n \doteq t_n, Ps_1...s_n \rightarrow \Delta'', Pt_1...t_n.$$

Case 2: $\Gamma \rightarrow \Delta, A$ is not an equality axiom, but $A, \Lambda \rightarrow \Theta$ is. This is similar to case 1.

Case 3: Both $\Gamma \rightarrow \Delta, A$ and $A, \Lambda \rightarrow \Theta$ are equality axioms. In this case, $\Gamma \rightarrow \Delta, A$ is obtained from some equality axiom of the form

$$\Gamma', s_1 \doteq t_1, s_2 \doteq t_2, ..., s_n \doteq t_n, Ps_1...s_n \rightarrow \Delta', Pt_1...t_n$$

by application of $\neg : rules$, and $A, \Lambda \rightarrow \Theta$ is obtained from some equality axiom of the form

$$\Lambda', t_1 \doteq r_1, t_2 \doteq r_2, ..., t_n \doteq r_n, Pt_1...t_n \rightarrow \Theta', Pr_1...r_n$$

by applications of $\neg : rules$.

Then, $\Gamma, \Lambda \rightarrow \Delta, \Theta$ is obtained by applying negation rules to the sequent

$$\Gamma', \Lambda', s_1 \doteq t_1, s_2 \doteq t_2, ..., s_n \doteq t_n, t_1 \doteq r_1, t_2 \doteq r_2, ..., t_n \doteq r_n, Ps_1...s_n$$
$$\rightarrow \Delta', \Theta', Pr_1...r_n.$$

A proof with only inessential cuts can be given using axioms derived by applying $\neg : rules$ to the axioms

$$s_i \doteq t_i, t_i \doteq r_i \rightarrow s_i \doteq r_i,$$

for $i = 1, ..., n$, and the axiom

$$\Gamma', \Lambda', s_1 \doteq r_1, ..., s_n \doteq r_n, Qs_1...s_n \rightarrow \Delta', \Theta', Qr_1...r_n,$$

and n inessential cuts (the i-th cut with cut formula $s_i \doteq r_i$ or $\neg s_i \doteq r_i$). The depth of this proof is n, which is bounded by m. The rest of the proof is left as an exercise. $\square$

We obtain the following cut elimination theorem.

Theorem 6.4.2 (Cut elimination theorem for $G1_{\doteq}^{nnf}$) Let T be a $G1_{\doteq}^{nnf}$-proof with cut-rank $c(T)$ for a sequent $\Gamma \rightarrow \Delta$, and let m be defined as in lemma 6.4.9. A $G1_{\doteq}^{nnf}$-proof T^* for $\Gamma \rightarrow \Delta$ without essential cuts such that $depth(T^*) \leq exp(m+2, c(T), depth(T))$ can be constructed.

Proof: The following claim can be shown by induction on the depth of proof trees:

Claim: Let T be a $G1_{\doteq}^{nnf}$-proof with cut-rank $c(T)$ for a sequent $\Gamma \rightarrow \Delta$. If $c(T) > 0$ then we can construct a $G1_{\doteq}^{nnf}$-proof T' for $\Gamma \rightarrow \Delta$, such that

$$c(T') < c(T) \quad \text{and} \quad depth(T') \leq (m+2)^{depth(T)}.$$

The details are left as an exercise. $\square$

Note: The cut elimination theorem with inessential cuts (with atomic cut formulae of the form $s \doteq t$) also holds for LK_e. The interested reader is refered to Takeuti, 1975.

As an application of theorem 6.3.1, we shall prove Craig's interpolation theorem in the next section. The significance of a proof of Craig's theorem using theorem 6.3.1 is that an interpolant can actually be constructed. In turn, Craig's interpolation theorem implies two other important results: Beth's definability theorem, and Robinson's joint consistency theorem.

PROBLEMS

6.4.1. Convert the following formulae to NNF:

$$(\neg \forall x P(x,y) \lor \forall x R(x,y))$$
$$\forall x(P(x) \supset \neg \exists y R(x,y))$$
$$(\neg \forall x \neg \forall y \neg \forall z P(x,y) \lor \neg \exists x \neg \exists y(\neg \exists z Q(x,y,z) \supset R(x,y)))$$

6.4.2. Convert the following formulae to NNF:

$$(\exists x \forall y P(x,y) \land \forall y \exists x P(y,x))$$
$$(\neg(\forall x P(x) \lor \exists y \neg Q(y)) \lor (\forall z G(z) \lor \exists w \neg Q(w)))$$
$$(\neg \forall x(P(x) \lor \exists y \neg Q(y)) \lor (\forall z P(z) \lor \exists w \neg Q(w)))$$

6.4.3. Write a computer program for converting a formula to NNF.

6.4.4. Give the details of the proof of lemma 6.4.2.

6.4.5. Finish the proof of the cases that have been left out in the proof of lemma 6.4.3.

6.4.6. Prove lemma 6.4.4.

6.4.7. Prove lemma 6.4.5.

6.4.8. Finish the proof of the cases that have been left out in the proof of lemma 6.4.6.

6.4.9. Finish the proof of the cases that have been left out in the proof of lemma 6.4.7.

6.4.10. Prove that for any LK_e-proof, all weakenings can be moved above all other kinds of inferences. Use this result to prove lemma 6.4.8.

6.4.11. Finish the proof of the cases that have been left out in the proof of lemma 6.4.9.

6.4.12. Give the details of the proof of theorem 6.4.2.

6.4.13. Given a set S of formulae, let $Des(S)$ be the set of *immediate descendants* of formulae in S as defined in problem 5.5.18, and define S^n by induction as follows:

$$S^0 = S;$$
$$S^{n+1} = Des(S^n).$$
$$\text{Let } S^* = \bigcup_{n \geq 0} S^n.$$

S^* is called the set of descendants of S.

(a) Prove that for every deduction tree without essential cuts for a sequent $\Gamma_0 \to \Delta_0$, the formulae in the sets of formulae $\Gamma \cup \{\neg B \mid B \in \Delta\}$ for all sequents $\Gamma \to \Delta$ occurring in that tree, belong to S^* or are equations of the form $s \doteq t$, where $S = \Gamma_0 \cup \{\neg B \mid B \in \Delta_0\}$.

(b) Deduce from (a) that not all $G1^{nnf}_{\doteq}$-formulae are provable.

6.4.14. Let $\mathbf{L}$ be a first-order language with equality and with function symbols and constant symbols, but no predicate symbols. Such a language will be called *equational*.

Let $e_1, ..., e_m \to e$ be a sequent where each e_i is a closed formula of the form $\forall x_1...\forall x_n(s \doteq t)$, called a *universal equation*, where $\{x_1, ..., x_n\}$ is the set of variables free in $s \doteq t$, and e is an atomic formula of the form $s \doteq t$, called an *equation*. Using theorem 6.4.2, prove that if $e_1, ..., e_n \to e$ is $G1^{nnf}_{\doteq}$-provable, then it is provable using only axioms (including the equality axioms), the cut rule applied to equations (of the form $s \doteq t$), weakenings, and the $\forall : left$ rule.

6.4.15. Let $\mathbf{L}$ be an equational language as defined in problem 6.4.14. A *substitution function* (for short, a *substitution*) is any function $\sigma : \mathbf{V} \to TERM_{\mathbf{L}}$ assigning terms to the variables in $\mathbf{V}$. By theorem 2.4.1, there is a unique homomorphism $\hat{\sigma} : TERM_{\mathbf{L}} \to TERM_{\mathbf{L}}$ extending σ and defined recursively as follows:

For every variable $x \in \mathbf{V}$, $\hat{\sigma}(x) = s(x)$.

For every constant c, $\hat{\sigma}(c) = c$.

For every term $ft_1...t_n \in TERM_{\mathbf{L}}$,

$$\hat{\sigma}(ft_1...t_n) = f\hat{\sigma}(t_1)...\hat{\sigma}(t_n).$$

By abuse of language and notation, the function $\hat{\sigma}$ will also be called a substitution, and will often be denoted by σ.

The subset X of $\mathbf{V}$ consisting of the variables such that $s(x) \neq x$ is called the *support of the substitution*. In what follows, we will be dealing with substitutions of finite support. If a substitution σ has finite support $\{y_1, ..., y_n\}$ and $\sigma(y_i) = s_i$, for $i = 1, .., n$, for any term t, the substitution instance $\hat{\sigma}(t)$ is also denoted as $t[s_1/y_1, ..., s_n/y_n]$.

Let $E = < e_1, ..., e_n >$ be a sequence of equations, and E' the sequence of their universal closures. (Recall that for a formula A with set of free variables $FV(A) = \{x_1, ..., x_n\}$, $\forall x_1...\forall x_n A$ is the *universal closure* of A.)

We define the relation $\longrightarrow_E$ on the set $TERM_{\mathbf{L}}$ of terms as follows. For any two terms t_1, t_2,

$$t_1 \longrightarrow_E t_2$$

iff there is some term r, some equation $s \doteq t \in E$, some substitution σ with support $FV(s) \cup FV(t)$, and some tree address u in r, such that,

$$t_1 = r[u \leftarrow \hat{\sigma}(s)], \text{ and } t_2 = r[u \leftarrow \hat{\sigma}(t)].$$

When $t_1 \longrightarrow_E t_2$, we say that t_1 *rewrites* to t_2. In words, t_1 rewrites to t_2 iff t_2 is obtained from t_1 by finding a subterm $\hat{\sigma}(s)$ of t_1 (called a pattern) which is a substitution instance of the left hand side s of an equation $s \doteq t \in E$, and replacing this subterm by the subterm $\hat{\sigma}(t)$ obtained by applying the same substitution σ to the right hand side t of the equation.

Let $\longleftrightarrow_E$ be the relation defined such that

$$t_1 \longleftrightarrow_E t_2 \quad \text{iff}$$
$$\text{either } t_1 \longrightarrow_E t_2 \quad \text{or} \quad t_2 \longrightarrow_E t_1,$$

and let $\overset{*}{\longleftrightarrow}_E$ be the reflexive and transitive closure of $\longleftrightarrow_E$.

Our goal is to prove that for every sequence $E = < e_1, ..., e_n >$ of equations and for every equation $s \doteq t$, if E' is the sequence of universal closures of equations in E, then

$$E' \rightarrow s \doteq t \quad \text{is } G1\underline{\underline{}}^{nnf}\text{-provable iff} \quad s \overset{*}{\longleftrightarrow}_E t.$$

During the proof that proceeds by induction, we will have to consider formulae which are universally quantified equations of the form

$$\forall y_1 ... \forall y_m (s \doteq t),$$

where $\{y_1, ..., y_m\}$ is a subset of $FV(s) \cup FV(t)$, including the case $m = 0$ which corresponds to the quantifier-free equation $s \doteq t$. Hence, we will prove two facts.

For every sequence T consisting of (partially) universally quantified equations from E,

(1)   If $T \rightarrow s \doteq t$ is $G1\underline{\underline{}}^{nnf}$-provable then $s \overset{*}{\longleftrightarrow}_E t$.

(2)   If $s \overset{*}{\longleftrightarrow}_E t$ then $E' \rightarrow s \doteq t$ is $G1\underline{\underline{}}^{nnf}$-provable,

where E' is the universal closure of E.

This is done as follows:

(a) Prove that if $s \overset{*}{\longleftrightarrow}_E t$, then the sequent $E' \rightarrow s \doteq t$ is $G1\underline{\underline{}}^{nnf}$-provable, where E' is the universal closure of E.

Hint: Use induction on the structure of the term r in the definition of $\overset{*}{\longleftrightarrow}_E$.

(b) Given a set E of equations and any equation $v \doteq w$, let $\{E, v \doteq w\}$ denote the union of E and $\{v \doteq w\}$.

Prove that

(i) If $s_i \overset{*}{\longleftrightarrow}_E t_i$, for $1 \leq i \leq n$, then

$$f(t_1, ..., t_n) \overset{*}{\longleftrightarrow}_E f(s_1, ..., s_n).$$

(ii) If $t_1 \overset{*}{\longleftrightarrow}_E t_2$, then for every substitution σ with support $FV(t_1) \cup FV(t_2)$,

$$\widehat{\sigma}(t_1) \overset{*}{\longleftrightarrow}_E \widehat{\sigma}(t_2).$$

(iii) If $v \overset{*}{\longleftrightarrow}_E w$ and $s \overset{*}{\longleftrightarrow}_{\{E,v\doteq w\}} t$, then $s \overset{*}{\longleftrightarrow}_E t$.

Hint: Show that every rewrite step involving $v \doteq w$ as an equation can be simulated using the steps in $v \overset{*}{\longleftrightarrow}_E w$.

(c) Prove that, for every sequence T consisting of (partially) universally quantified equations from E, for any equation $s \doteq t$, if the sequent $T \rightarrow s \doteq t$ is $G1^{nnf}_{\doteq}$-provable then $s \overset{*}{\longleftrightarrow}_E t$.

Hint: Proceed by induction on $G1^{nnf}_{\doteq}$ proofs without essential cuts. One of the cases is that of an inessential cut. If the bottom inference is a cut, it must be of the form

$$\frac{\begin{array}{c} S_1 \\ \hline T_1 \rightarrow v \doteq w \end{array} \qquad \begin{array}{c} S_2 \\ \hline v \doteq w, T_2 \rightarrow s \doteq t \end{array}}{T \rightarrow s \doteq t}$$

where T_1 and T_2 are subsets of T. By the induction hypothesis, $v \overset{*}{\longleftrightarrow}_E w$ and $s \overset{*}{\longleftrightarrow}_{\{E,v\doteq w\}} t$. Conclude using (b).

(d) Prove that the sequent $E' \rightarrow s \doteq t$ is $G1^{nnf}_{\doteq}$-provable iff $s \overset{*}{\longleftrightarrow}_E t$.

The above problem shows the completeness of the rewrite rule method for (universal) *equational logic*. This is an important current area of research. For more on this approach, see the article by Huet and Oppen, in Book, 1980, and Huet, 1980.

6.5 Craig's Interpolation Theorem

First, we define the concept of an interpolant.

6.5.1 Interpolants

If a formula of the form $A \supset B$ is valid, then it is not obvious that there is
a formula I, called an *interpolant* of $A \supset B$, such that $A \supset I$ and $I \supset B$
are valid, and every predicate, constant, and free variable occurring in I also
occurs both in A and B. As a matter of fact, the existence of an interpolant
depends on the language. If we don't allow either the constant $\perp$ or equality
($\doteq$), the formula $(P \supset P) \supset (Q \supset Q)$ is valid, and yet, no formula I as above
can be found. If we allow equality and make the exception that if $\doteq$ occurs
in I, it does not necessarily occur in both A and B, then $I = \forall x(x \doteq x)$ does
the job. Alternatively, $\neg \perp$ (true!) does the job.

Craig's interpolation theorem gives a full answer to the problem of the
existence of an interpolant. An interesting application of Craig's interpolation
theorem can be found in Oppen and Nelson's method for combining decision
procedures. For details, see Nelson and Oppen, 1979, and Oppen, 1980b.

6.5.2 Craig's Interpolation Theorem Without Equality

First, we consider a first-order language without equality. For the next lemma,
we assume that the constant $\perp$ (for false) is in the language, but that $\equiv$ is
not. Let $LK_\perp$ be the extension of LK obtained by allowing the sequent $\perp\to$
as an axiom. It is easily checked that the cut elimination theorem also holds
for $LK_\perp$ (see the problems).

During the proof of the key lemma, it will be necessary to replace in a
proof all occurrences of a free variable y by a new variable z not occurring
in the proof. This substitution process is defined as in definition 6.4.5, but
for LK (and LK_e) rather than $G1^{nnf}$. Given a deduction tree T such that
y does not occur bound in T, $T[z/y]$ is the result of replacing every sequent
$\Gamma \to \Delta$ in T by $(\Gamma \to \Delta)[z/y]$. Similarly, $T[z/c]$ is the result of substituting a
new variable z for all occurrences of a constant c in a proof T. The following
technical lemma will be needed.

Lemma 6.5.1 Let $\Gamma \to \Delta$ be a sequent and T an LK-proof for $\Gamma \to \Delta$. Assume that y is a variable not occurring bound in the proof T. For any variable
z not occurring in T, the result $T[z/y]$ of substituting z for all occurrences of
y in T is a proof tree. Similarly, if c is a constant occurring in the proof tree
T, $T[z/c]$ is a proof tree. The lemma also holds for LK_e-proofs.

Proof: Similar to that of lemma 6.4.3. $\square$

The following lemma is the key to a constructive proof of the interpolation theorem.

Lemma 6.5.2 Let **L** be a first-order language without equality, without $\equiv$,
and with constant $\perp$. Given any provable sequent $\Gamma \to \Delta$, let (Γ_1, Γ_2) and
(Δ_1, Δ_2) be pairs of disjoint subsequences of Γ and Δ respectively, such that

the union of Γ_1 and Γ_2 is Γ, and the union of Δ_1 and Δ_2 is Δ. Let us call $< \Gamma_1, \Delta_1, \Gamma_2, \Delta_2 >$ a *partition* of $\Gamma \to \Delta$. Then there is a formula C of $LK_\perp$ called an *interpolant* of $< \Gamma_1, \Delta_1, \Gamma_2, \Delta_2 >$ having the following properties:

(i) $\Gamma_1 \to \Delta_1, C$ and $C, \Gamma_2 \to \Delta_2$ are $LK_\perp$-provable.

(ii) All predicate symbols (except for $\perp$), constant symbols, and variables free in C occur both in $\Gamma_1 \cup \Delta_1$ and $\Gamma_2 \cup \Delta_2$.

Proof: We proceed by induction on cut-free proof trees. We treat some typical cases, leaving the others as an exercise.

(1) $\Gamma \to \Delta$ is an axiom. Hence, it is of the form $A \to A$. There are four cases. For $< \Gamma_1, \Delta_1, \Gamma_2, \Delta_2 >=< A, A, \emptyset, \emptyset >$, $C =\perp$; for $< \Gamma_1, \Delta_1, \Gamma_2, \Delta_2 > =< \emptyset, \emptyset, A, A >$, $C = \neg \perp$; for $< \Gamma_1, \Delta_1, \Gamma_2, \Delta_2 >=< A, \emptyset, \emptyset, A >$, $C = A$; For $< \Gamma_1, \Delta_1, \Gamma_2, \Delta_2 >=< \emptyset, A, A, \emptyset >$, $C = \neg A$.

(2) The root inference is $\wedge : right$:

$$\frac{\Gamma \to \Delta, A \qquad \Gamma \to \Delta, B}{\Gamma \to \Delta, A \wedge B}$$

Assume the partition is

$$< \Gamma_1, (\Delta_1, A \wedge B), \Gamma_2, \Delta_2 >,$$

the case

$$< \Gamma_1, \Delta_1, \Gamma_2, (\Delta_2, A \wedge B) >$$

being similar. We have induced partitions $< \Gamma_1, (\Delta_1, A), \Gamma_2, \Delta_2 >$ and $< \Gamma_1, (\Delta_1, B), \Gamma_2, \Delta_2 >$. By the induction hypothesis, there are formulae C_1 and C_2 satisfying the conditions of the lemma. Since $\Gamma_1 \to \Delta_1, A, C_1$ and $\Gamma_1 \to \Delta_1, B, C_2$ are provable,

$$\Gamma_1 \to \Delta_1, A, C_1 \vee C_2$$

and

$$\Gamma_1 \to \Delta_1, B, C_1 \vee C_2$$

are provable (using $\vee : right$), and

$$\Gamma_1 \to \Delta_1, A \wedge B, C_1 \vee C_2$$

is provable, by $\wedge : right$. Since $C_1, \Gamma_2 \to \Delta_2$ and $C_2, \Gamma_2 \to \Delta_2$ are provable,

$$C_1 \vee C_2, \Gamma_2 \to \Delta_2$$

is provable by $\vee : left$. Then, we can take $C_1 \vee C_2$ as an interpolant.

(3) The root inference is $\forall : right$:

$$\frac{\Gamma \rightarrow \Delta, A[y/x]}{\Gamma \rightarrow \Delta, \forall x A}$$

where y does not occur free in the conclusion. Assume the partition is

$$< \Gamma_1, (\Delta_1, \forall x A), \Gamma_2, \Delta_2 >,$$

the case

$$< \Gamma_1, \Delta_1, \Gamma_2, (\Delta_2, \forall x A) >$$

being similar. By the induction hypothesis, there is an interpolant C such that $\Gamma_1 \rightarrow \Delta_1, A[y/x], C$ and $C, \Gamma_2 \rightarrow \Delta_2$ are provable. By condition (ii) of the lemma, C cannot contain y, since otherwise y would be in $\Gamma_1 \cup \Delta_1$, contradicting the fact that y does not occur free in $\Gamma \rightarrow \Delta, \forall x A$. Hence,

$$\Gamma_1 \rightarrow \Delta_1, \forall x A, C$$

is provable by $\forall : right$. Since by the induction hypothesis

$$C, \Gamma_2 \rightarrow \Delta_2$$

is provable, we can take C as an interpolant.

(4) The root inference is $\forall : left$:

$$\frac{A[t/x], \Gamma \rightarrow \Delta}{\forall x A, \Gamma \rightarrow \Delta}$$

where t is free for x in A.

This case is more complicated if t is not a variable. Indeed, one has to be careful to ensure condition (ii) of the lemma. Assume that the partition is $< (\forall x A, \Gamma_1), \Delta_1, \Gamma_2, \Delta_2 >$, the case $< \Gamma_1, \Delta_1, (\forall x A, \Gamma_2), \Delta_2 >$ being similar. By the induction hypothesis, there is a formula C such that

$$A[t/x], \Gamma_1 \rightarrow \Delta_1, C \quad \text{and} \quad C, \Gamma_2 \rightarrow \Delta_2$$

are provable. First, note that $\forall x A, \Gamma_1 \rightarrow \Delta_1, C$ is provable by $\forall : left$. If t is a variable z and z does not occur free in $\forall x A, \Gamma \rightarrow \Delta$, then by $\forall : right$, $\forall x A, \Gamma_1 \rightarrow \Delta_1, \forall z C$ is provable, and so is $\forall z C, \Gamma_2 \rightarrow \Delta_2$, by $\forall : left$. Since C contains predicate symbols, constants and free variables both occurring in $(A[z/x], \Gamma_1) \cup \Delta_1$ and $\Gamma_2 \cup \Delta_2$, and z does not occur free in $\forall x A, \Gamma \rightarrow \Delta$, the formula $\forall z C$ also satisfies condition (ii) of the lemma. If z occurs free in in $(\forall x A, \Gamma_1) \cup \Delta_1$, then C can serve as an interpolant, since if it contains z, z is both in $(\forall x A, \Gamma_1) \cup \Delta_1$ and also in $\Gamma_2 \cup \Delta_2$ by the induction hypothesis. If t is not a variable, we proceed as follows: Let $x_1, ...x_k$ and $c_1, ..., c_m$ be the variables and the constants in t which occur in C but do not occur in $(\forall x A, \Gamma_1) \cup \Delta_1$. In the proof of $A[t/x], \Gamma_1 \rightarrow \Delta_1, C$, replace all occurrences of $x_1, ...x_k$ and $c_1, ..., c_m$ in t and C by new variables $z_1, ..., z_{k+m}$ (not occurring in the proof). Let t' be the result of the substitution in the term t, and C' the result of the substitution in the formula C. By lemma 5.6.1, $A[t'/x], \Gamma_1 \rightarrow \Delta_1, C'$ is provable. But now, $z_1, ..., z_{k+m}$ do not occur free in $\forall x A, \Gamma_1 \rightarrow \Delta_1, \forall z_1...\forall z_{k+m} C'$, and so

$$\forall x A, \Gamma_1 \rightarrow \Delta_1, \forall z_1...\forall z_{k+m} C'$$

is provable (by applications of $\forall : right$ and $\forall : left$). But

$$\forall z_1...\forall z_{k+m} C', \Gamma_2 \rightarrow \Delta_2$$

is also provable (by applications of $\forall : left$). Hence,

$$\forall z_1...\forall z_{k+m} C'$$

can be used as an interpolant, since it also satisfies condition (ii) of the lemma. $\square$

Note that the method used in lemma 6.5.2 does not guarantee that the function symbols occurring in C also occur both in $\Gamma_1 \cup \Delta_1$ and $\Gamma_2 \cup \Delta_2$.

We are now ready to prove Craig's interpolation theorem for LK.

Theorem 6.5.1 (Craig's interpolation theorem, without equality) Let **L** be a first-order language without equality and without $\equiv$. Let A and B be two **L**-formulae such that $A \supset B$ is LK-provable.

(a) If A and B contain some common predicate symbol, then there exists a formula C called an *interpolant* of $A \supset B$, such that all predicate, constant symbols, and free variables in C occur in both A and B, and both $A \supset C$ and $C \supset B$ are LK-provable.

(b) If A and B do not have any predicate symbol in common, either $A \rightarrow$ is LK-provable or $\rightarrow B$ is LK-provable.

Proof: Using the cut rule, it is obvious that $\rightarrow A \supset B$ is LK-provable iff $A \rightarrow B$ is. Consider the partition in which $\Gamma_1 = A$, $\Delta_1 = \emptyset$, $\Gamma_2 = \emptyset$ and $\Delta_2 = B$. By lemma 6.5.2, there is a formula C of $LK_\perp$ such that $A \rightarrow C$ and $C \rightarrow B$ are $LK_\perp$-provable, and all predicate, constant symbols, and free variables in C occur in A and B. If A and B have some predicate symbol P of rank n in common, let P' be the sentence

$$\forall y_1...\forall y_n (P(y_1, ..., y_n) \wedge \neg P(y_1, ...y_n)),$$

where $y_1, ..., y_n$ are new variables. Let C' be obtained by replacing all occurrences of $\perp$ in C by P'. Since P' is logically equivalent to $\perp$, it is not difficult to obtain LK-proofs of $A \to C'$ and of $C' \to B$. But then $\to A \supset C'$ and $\to C' \supset B$ are LK-provable, and the formula C' is the desired interpolant.

If A and B have no predicate symbols in common, the formula C given by lemma 6.5.2 only contains $\perp$ as an atom. It is easily shown by induction on the structure of C that either $\to C$ is $LK_\perp$-provable or $C \to$ is $LK_\perp$-provable. But then, using the cut rule, we can show that either $A \to$ is $LK_\perp$-provable, or $\to B$ is $LK_\perp$-provable. However, since neither A nor B contains $\perp$ and the cut elimination theorem holds for $LK_\perp$, a cut-free proof in $LK_\perp$ for either $A \to$ or $\to B$ is in fact an LK-proof. $\square$

EXAMPLE 6.5.1

Given the formula

$$(P(a) \wedge \forall x Q(x)) \supset (\forall y S(y) \vee Q(b)),$$

the formula $\forall x Q(x)$ is an interpolant.

6.5.3 Craig's Interpolation Theorem With Equality

We now consider Craig's interpolation theorem for first-order languages with equality. Because cuts cannot be completely eliminated in LK_e-proofs, the technique used in lemma 6.5.2 cannot be easily extended to the system LK_e. However, there is a way around, which is to show that a sequent $S = \Gamma \to \Delta$ is LK_e-provable if and only if the sequent $S_e, \Gamma \to \Delta$ is LK-provable, where S_e is a certain sequence of closed formulae called the *closed equality axioms* for $\Gamma \to \Delta$.

Definition 6.5.1   Given a sequent $S = \Gamma \to \Delta$, the set of *closed equality axioms for S* is the set S_e of closed formulae given below, for all predicate and function symbols occuring in S:

$$\forall x (x \doteq x)$$
$$\forall x_1 ... \forall x_n \forall y_1 ... \forall y_n (x_1 \doteq y_1 \wedge ... \wedge x_n \doteq y_n \supset f(x_1, ..., x_n) \doteq f(y_1, ..., y_n))$$
$$\forall x_1 ... \forall x_n \forall y_1 ... \forall y_n (x_1 \doteq y_1 \wedge ... \wedge x_n \doteq y_n \wedge P(x_1, ..., x_n) \supset P(y_1, ..., y_n))$$

Lemma 6.5.3   A sequent $S = \Gamma \to \Delta$ is LK_e-provable iff $S_e, \Gamma \to \Delta$ is LK-provable.

Proof: We sketch the proof, leaving the details as an exercise. First, we show that if S is LK_e-provable then $S_e, \Gamma \to \Delta$ is LK-provable. For this, we prove that for every equality axiom $\Gamma' \to \Delta'$ of LK_e used in the proof of S, the sequent $S_e, \Gamma' \to \Delta'$ is LK-provable. We conclude by induction on LK_e-proof trees.

Next, assume that $S_e, \Gamma \to \Delta$ is LK-provable. First, we show that every formula in S_e is LK_e-provable. We conclude by constructing an LK_e-proof of $\Gamma \to \Delta$ using cuts on formulae in S_e. □

We can now generalize lemma 6.5.2 to LK_e as follows. Let $LK_{e,\perp}$ be the system obtained by allowing $\perp$ as a constant and the sequent $\perp \to$ as an axiom.

Lemma 6.5.4 Let **L** be a first-order language with equality and with constant $\perp$, but without $\equiv$. Given any LK_e-provable sequent $\Gamma \to \Delta$, let (Γ_1, Γ_2) and (Δ_1, Δ_2) be pairs of disjoint subsequences of Γ and Δ respectively, such that the union of Γ_1 and Γ_2 is Γ, and the union of Δ_1 and Δ_2 is Δ. $< \Gamma_1, \Delta_1, \Gamma_2, \Delta_2 >$ is called a *partition* of $\Gamma \to \Delta$. There is a formula C of $LK_{e,\perp}$ called an *interpolant* of $< \Gamma_1, \Delta_1, \Gamma_2, \Delta_2 >$ having the following properties:

(i) $\Gamma_1 \to \Delta_1, C$ and $C, \Gamma_2 \to \Delta_2$ are LK_e-provable.

(ii) All predicate symbols (except for $\doteq$), constant symbols, and variables free in C occur both in $\Gamma_1 \cup \Delta_1$ and $\Gamma_2 \cup \Delta_2$.

Proof: By lemma 6.5.3, $S = \Gamma \to \Delta$ is LK_e-provable iff $S_e, \Gamma \to \Delta$ is LK-provable. Let S_e^1 be the closed equality axioms corresponding to function and predicate symbols in $\Gamma_1 \cup \Delta_1$, and S_e^2 the closed equality axioms corresponding to function and predicate symbols in $\Gamma_2 \cup \Delta_2$. Clearly, $S_e = S_e^1 \cup S_e^2$. Consider the partition $< (S_e^1, \Gamma_1), \Delta_1, (S_e^2, \Gamma_2), \Delta_2 >$. By lemma 6.5.2, there is a formula C such that

$$S_e^1, \Gamma_1 \to \Delta_1, C \quad \text{and} \quad C, S_e^2, \Gamma_2 \to \Delta_2$$

are $LK_\perp$-provable, and the predicate symbols, constant symbols and free variables in C occur both in $S_e^1 \cup \Gamma_1 \cup \Delta_1$ and $S_e^2 \cup \Gamma_2 \cup \Delta_2$. By definition of S_e^1 and S_e^2, the predicate symbols (other than $\doteq$) in S_e^1 are those in $\Gamma_1 \cup \Delta_1$, and similarly for S_e^2 and $\Gamma_2 \cup \Delta_2$. Furthermore, all formulae in S_e^1 and S_e^2 being closed, there are no free variables occurring in them. Hence, the predicate symbols, constant and free variables in C occur both in $\Gamma_1 \cup \Delta_1$ and $\Gamma_2 \cup \Delta_2$. Using lemma 6.5.3 again, $\Gamma_1 \to \Delta_1, C$ and $C, \Gamma_2 \to \Delta_2$ are $LK_{e,\perp}$-provable. Now, we can replace all occurrence of $\perp$ in C by $\neg \forall z(z \doteq z)$, where z is a new variable, obtaining a formula C', and since $\perp$ is logically equivalent to $\neg \forall z(z \doteq z)$, it is easy to obtain LK_e-proofs for $\Gamma_1 \to \Delta_1, C'$ and $C', \Gamma_2 \to \Delta_2$. Taking C' as the interpolant this concludes the proof of the lemma. □

Using lemma 6.5.4, Craig's interpolation theorem is generalized to languages with equality.

Theorem 6.5.2 (Craig's interpolation theorem, with equality) Let **L** be a first-order language with equality and without $\equiv$. Let A and B be two **L**-formulae such that $A \supset B$ is LK_e-provable. Then there exists a formula C called an *interpolant* of $A \supset B$, such that all predicate symbols except $\doteq$, constant symbols, and free variables in C occur in both A and B, and both $A \supset C$ and $C \supset B$ are LK_e-provable.

Proof: As in theorem 6.5.1, $A \supset B$ is LK_e-provable iff $A \rightarrow B$ is LK_e-provable. We apply lemma 6.5.4 to the partition in which $\Gamma_1 = A$, $\Delta_1 = \emptyset$, $\Gamma_2 = \emptyset$ and $\Delta_2 = B$. This time, because $\doteq$ is available, the result is obtained immediately from lemma 6.5.4. $\square$

Remark: As in theorem 6.5.2, our proof does not guarantee that the function symbols occurring in the interpolant also occur in both A and B. However, this can be achieved using a different proof technique presented in problem 5.6.7, which consists in replacing function symbols by predicate symbols. Hence, in case of a first-order language with equality, we obtain a stonger version of Craig's interpolation theorem, in which all predicate symbols (different from $\doteq$), function symbols, constant symbols, and free variables occurring in the interpolant C of $A \supset B$ occur in both A and B.

In the next section, we give two applications of Craig's interpolation theorem.

PROBLEMS

6.5.1. Finish the proof of lemma 6.5.1.

6.5.2. Finish the proof of lemma 6.5.2.

6.5.3. Give a proof for lemma 6.5.3.

6.5.4. Verify that the cut elimination theorem holds for $LK_\perp$, and for $LK_{e,\perp}$ (without essential cuts).

6.5.5. Provide the details in the proof of theorem 6.5.1 regarding the replacement of $\perp$ by $\forall y_1 ... \forall y_n (P(y_1, ..., y_n) \wedge \neg P(y_1, ... y_n))$.

6.5.6. Show that $\forall x \neg P(x) \vee Q(b)$ is an interpolant for

$$[(R \supset \exists x P(x)) \supset Q(b)] \supset [\forall x((S \wedge P(x)) \supset (S \wedge Q(b)))].$$

6.5.7. Using the proof technique presented in problem 5.6.7, which consists in replacing function symbols by predicate symbols, prove the stronger version of Craig's interpolation theorem, in which all predicate symbols (different from $\doteq$), function symbols, constant symbols, and free variables occurring in the interpolant C of $A \supset B$ occur in both A and B.

6.6 Beth's Definability Theorem

First, we consider what definability means.

6.6.1 Implicit and Explicit Definability

Let **L** be a first-order language with or without equality. Let $A_1, ..., A_m$ be closed formulae containing exactly the distinct predicate symbols $P_1, ..., P_k, Q$, where Q is not $\doteq$, but some of the P_i can be $\doteq$. Assume that Q has rank $n > 0$. We can view $A_1, ..., A_m$ as the axioms of a theory. The question of interest is whether Q is definable in terms of $P_1, ..., P_k$. First, we need to make precise what definable means. Let $A(P_1, ..., P_k, Q)$ be the conjunction $A_1 \wedge ... \wedge A_m$.

A first plausible criterion for definability is that, for any two **L**-structures **A** and **B** with the same domain M, and which assign the same interpretation to the predicate symbols $P_1, ..., P_k$,

$$\text{if } \mathbf{A} \models A(P_1, ..., P_k, Q) \quad \text{and} \quad \mathbf{B} \models A(P_1, ..., P_k, Q),$$
$$\text{then for every } (a_1, ..., a_n) \in M^n,$$
$$\mathbf{A} \models Q(a_1, ..., a_n) \quad \text{and} \quad \mathbf{B} \models Q(a_1, ..., a_n).$$

The idea is that given any two interpretations of the predicate symbols Q, if these two interpretations make $A(P_1, ..., P_k, Q)$ true then they must agree on Q. This is what is called *implicit definability*.

A seemingly stronger criterion for definability is the following:

Definition 6.6.1 Assume that there is an **L**-formula $D(P_1, ..., P_k)$ whose set of free variables is a subset of $\{x_1, ..., x_n\}$ and whose predicate symbols are among $P_1, ..., P_k$, and which contains at least one of the P_i. We say that Q is *defined explicitly* from $P_1, ..., P_k$ in the theory based on $A_1, ..., A_m$ iff the following is provable (or equivalently valid, by the completeness theorem):

$$\rightarrow A(P_1, ..., P_k, Q) \supset \forall x_1...\forall x_n(Q(x_1, ..., x_n) \equiv D(P_1, ..., P_k)),$$

where $A(P_1, ..., P_k, Q)$ is the conjunction $A_1 \wedge ... \wedge A_m$.

We can modify the definition of implicit definability so that it refers to a single structure as opposed to a pair of structures, by using a new copy Q' of the predicate symbol Q, explained as follows.

Definition 6.6.2 Let $P_1, ..., P_k, Q, Q'$ be distinct predicate symbols, and let $A(P_1, ..., P_k, Q')$ be the result of substituting Q' for Q in $A(P_1, ..., P_k, Q)$. We say that Q is *defined implicitly* from $P_1, ..., P_k$ in the theory based on $A_1, ..., A_m$ iff the following formula is valid (or equivalently provable):

$$A(P_1, ..., P_k, Q) \wedge A(P_1, ..., P_k, Q') \supset$$
$$\forall x_1...\forall x_n(Q(x_1, ..., x_n) \equiv Q'(x_1, ..., x_n)),$$

where $A(P_1, ..., P_k, Q)$ is the conjunction $A_1 \wedge ... \wedge A_m$.

EXAMPLE 6.6.1

Let **L** be the language with equality having 0 as a constant, S as a unary function symbol, $+$ as a binary function symbol, and $<$ as a binary predicate symbol. Let $A(\doteq, <)$ be the conjunction of the following closed formulae:

$$\forall x \neg (S(x) \doteq 0)$$
$$\forall x \forall y (S(x) \doteq S(y) \supset x \doteq y)$$
$$\forall x (x + 0 \doteq x)$$
$$\forall x \forall y (x + S(y) \doteq S(x + y))$$
$$\forall x (0 < Sx)$$

The predicate symbol $<$ is not definable implicitly from $\doteq$ and $A(\doteq, <)$. Indeed, we can define two structures **A** and **B** with domain **N**, in which 0, S and $+$ receive the same natural interpretation, but in the first structure, $<$ is interpreted as the strict order on **N**, whereas in the second, we interpret $<$ as the predicate such that for all $x, y \in \mathbf{N}$, $x < y$ iff $x = 0$. Both **A** and **B** are models of $A(\doteq, <)$, but $<$ is interpreted differently.

On the other hand, if we add to $A(\doteq, <)$ the sentence

$$\forall x \forall y (x < y \equiv \exists z (y \doteq x + S(z))),$$

then $D = \exists z (y \doteq x + S(z))$ defines $<$ explicitly.

6.6.2 Explicit Definability Implies Implicit Definability

It is natural to ask whether the concepts of implicit and explicit definability are related. First, it is not difficult to see that explicit definability implies implicit definability. Indeed, we show that if

$$A(P_1, ..., P_k, Q) \supset \forall x_1 ... \forall x_n (Q(x_1, ..., x_n) \equiv D(P_1, ..., P_k))$$

is valid, then $A(P_1, ..., P_k, Q)$ defines Q implicitly. For if $A(P_1, ..., P_k, Q) \wedge A(P_1, ..., P_k, Q')$ is valid in any model **A**, then

$$\forall x_1 ... \forall x_n (Q(x_1, ..., x_n) \equiv D(P_1, ..., P_k)) \quad \text{and}$$
$$\forall x_1 ... \forall x_n (Q'(x_1, ..., x_n) \equiv D(P_1, ..., P_k))$$

are valid in **A**, and this implies that

$$\forall x_1 ... \forall x_n (Q(x_1, ..., x_n) \equiv Q'(x_1, ..., x_n))$$

is valid in **A**.

The fact that explicit definability implies implicit definability yields a method known as *Padoa's method*, to show that a predicate Q is not definable explicitly from $P_1, ..., P_k$ in the theory based on $A_1, ..., A_m$:

Find two structures with the same domain and assigning the same interpretation to the predicate symbols $P_1, ..., P_k$, but assigning different interpretations to Q.

However, it is not as obvious that implicit definability implies explicit definability. But this is the case, as shown by Beth's theorem. The proof given below that uses Craig's interpolation theorem even gives the defining formula $D(P_1, ..., P_k)$ constructively.

6.6.3 Beth's Definability Theorem, Without Equality

First, we consider the case of a first-order language without equality.

Theorem 6.6.1 (Beth's definability theorem, without equality) Let **L** be a first-order language without equality. Let $A(P_1, ..., P_k, Q)$ be a closed formula containing predicate symbols among the distinct predicate symbols $P_1, ..., P_k$, Q, where Q has rank $n > 0$. Assume that Q is defined implicitly from $P_1, ..., P_k$ by the sentence $A(P_1, ..., P_k, Q)$.

(a) If one of the predicate symbols P_i actually occurs in $A(P_1, ..., P_k, Q)$, then there is a formula $D(P_1, ..., P_k)$ defining Q explicitly from $P_1, ..., P_k$.

(b) If none of the predicate symbols $P_1, ..., P_k$ occur in $A(P_1, ..., P_k)$, then either

$$\models A(P_1, ..., P_k, Q) \supset \forall x_1...\forall x_n Q(x_1, ..., x_n) \quad \text{or}$$
$$\models A(P_1, ..., P_k, Q) \supset \forall x_1...\forall x_n \neg Q(x_1, ..., x_n).$$

Proof: Assume that

$$A(P_1, ..., P_k, Q) \wedge A(P_1, ..., P_k, Q') \supset$$
$$\forall x_1...\forall x_n (Q(x_1, ..., x_n) \equiv Q'(x_1, ..., x_n))$$

is valid. Since $A(P_1, ..., P_k, Q)$ and $A(P_1, ..., P_k, Q')$ are closed,

$$A(P_1, ..., P_k, Q) \wedge A(P_1, ..., P_k, Q') \supset (Q(x_1, ..., x_n) \equiv Q'(x_1, ..., x_n))$$

is also valid. This formula is of the form $(A \wedge B) \supset (C \equiv D)$. It is an easy exercise to show that if $(A \wedge B) \supset (C \equiv D)$ is valid, then $(A \wedge C) \supset (B \supset D)$ is also valid. Hence,

(1) $\quad (A(P_1, ..., P_k, Q) \wedge Q(x_1, ..., x_n)) \supset (A(P_1, ..., P_k, Q') \supset Q'(x_1, ..., x_n))$

is valid.

(a) If some of the predicate symbols P_i occurs in $A(P_1, ..., P_k, Q)$, by Craig's theorem (theorem 6.5.1) applied to (1), there is a formula C containing only predicate symbols, constant symbols, and free variables occurring in both $A(P_1, ..., P_k, Q) \wedge Q(x_1, ..., x_n)$ and $A(P_1, ..., P_k, Q') \supset Q'(x_1, ..., x_n)$, and such that the following are valid:

$$(2) \qquad\qquad (A(P_1, ..., P_k, Q) \wedge Q(x_1, ..., x_n)) \supset C,$$

and

$$(3) \qquad\qquad C \supset (A(P_1, ..., P_k, Q') \supset Q'(x_1, ..., x_n)).$$

Since $A(P_1, ..., P_k, Q) \wedge Q(x_1, ..., x_n)$ does not contain Q' and $A(P_1, ..., P_k, Q')$ $\supset Q'(x_1, ..., x_n)$ does not contain Q, the formula C only contains predicate symbols among $P_1, ..., P_k$ and free variables among $x_1, ..., x_n$.

By substituting Q for Q' in (3), we also have the valid formula

$$(4) \qquad\qquad C \supset (A(P_1, ..., P_k, Q) \supset Q(x_1, ..., x_n)),$$

which implies the valid formula

$$(5) \qquad\qquad A(P_1, ..., P_k, Q) \supset (C \supset Q(x_1, ..., x_n)).$$

But (2) implies the validity of

$$(6) \qquad\qquad A(P_1, ..., P_k, Q) \supset (Q(x_1, ..., x_n) \supset C).$$

The validity of (5) and (6) implies that

$$A(P_1, ..., P_k, Q) \supset (C \equiv Q(x_1, ..., x_n))$$

is valid, which in turns implies that C defines Q explicitly from $P_1, ..., P_k$.

(b) If none of $P_1, ..., P_k$ occurs in $A(P_1, ..., P_k, Q)$, then by Craig's theorem (theorem 6.5.1), part (ii), either

$$(7) \qquad\qquad \neg(A(P_1, ..., P_k, Q) \wedge Q(x_1, ..., x_n))$$

is valid, or

$$(8) \qquad\qquad A(P_1, ..., P_k, Q') \supset Q'(x_1, ..., x_n)$$

is valid.

Using propositional logic, either

$$A(P_1, ..., P_k, Q) \supset \neg Q(x_1, ..., x_n)$$

is valid, or

$$A(P_1, ..., P_k, Q) \supset Q(x_1, ..., x_n)$$

is valid, which implies part (b) of the theorem. $\square$

6.6.4 Beth's Definability Theorem, With Equality

We now consider Beth's definability theorem for a first-order language with equality. This time, we can define either a predicate symbol, or a function symbol, or a constant.

Theorem 6.6.2 (Beth's definability theorem, with equality) Let **L** be a first-order language with equality.

(a) Let $A(P_1, ..., P_k, Q)$ be a closed formula possibly containing equality and containing predicate symbols among the distinct predicate symbols $P_1, ..., P_k, Q$ (different from $\doteq$), where Q has rank $n > 0$. Assume that Q is defined implicitly from $P_1, ..., P_k$ by the sentence $A(P_1, ..., P_k, Q)$. Then there is a formula $D(P_1, ..., P_k)$ defining Q explicitly from $P_1, ..., P_k$.

(b) Let $A(P_1, ..., P_k, f)$ be a closed formula possibly containing equality, and containing predicate symbols among the distinct predicate symbols $P_1, ..., P_k$ (different from $\doteq$), and containing the function or constant symbol f of rank $n \geq 0$. Assume that f is defined implicitly from $P_1, ..., P_k$ by the sentence $A(P_1, ..., P_k, f)$, which means that the following formula is valid, where f' is a new copy of f:

$$A(P_1, ..., P_k, f) \wedge A(P_1, ..., P_k, f') \supset \forall x_1 ... \forall x_n (f(x_1, ..., x_n) \doteq f'(x_1, ..., x_n)).$$

Then there is a formula $D(P_1, ..., P_k)$ whose set of free variables is among $x_1, ..., x_n$ and not containing f (or f') defining f explicitly, in the sense that the following formula is valid:

$$A(P_1, ..., P_k, f) \supset \forall x_1 ... \forall x_n \forall y ((f(x_1, ..., x_n) \doteq y) \equiv D(P_1, ..., P_k)).$$

Proof: The proof of (a) is similar to that of theorem 6.6.1(a), but using theorem 6.5.2, which yields $D(P_1, ..., P_k)$ in all cases.

To prove (b), observe that $f(x_1, ..., x_n) \doteq f'(x_1, ..., x_n)$ is equivalent to

$$\forall y ((f(x_1, ..., x_n) \doteq y) \equiv (f'(x_1, ..., x_n) \doteq y)).$$

We conclude by applying the reasoning used in part (a) with $f(x_1, ..., x_n) \doteq y$ instead of $Q(x_1, ..., x_n)$. $\square$

The last application of Craig's interpolation theorem presented in the next section is Robinson's joint consistency theorem.

PROBLEMS

6.6.1. Show that in definition 6.6.2, the definability condition can be relaxed to

$$A(P_1, ..., P_k, Q) \wedge A(P_1, ..., P_k, Q') \supset$$
$$\forall x_1 ... \forall x_n (Q(x_1, ..., x_n) \supset Q'(x_1, ..., x_n)).$$

6.6.2. Give the details of the proof that explicit definability implies implicit definability.

6.7 Robinson's Joint Consistency Theorem

Let **L** be a first-order language with or without equality, and let **L**$_1$ and **L**$_2$ be two expansions of **L**. Also, let S_1 be a set of **L**$_1$-sentences, and S_2 a set of **L**$_2$-sentences. Let S be the set of **L**-sentences that are in $S_1 \cap S_2$. If S_1 and S_2 are both consistent, their union $S_1 \cup S_2$ is not necessarily consistent. Indeed, if S is *incomplete*, that is, there is some **L**-sentence C such that neither $S \to C$ nor $S \to \neg C$ is provable, S_1 could contain C and S_2 could contain $\neg C$, and $S_1 \cup S_2$ would be inconsistent. A concrete illustration of this phenomenon can be given using Gödel's incompleteness theorem for Peano's arithmetic, which states that there is a sentence C of the language of arithmetic such that neither $A_P \to C$ nor $A_P \to \neg C$ is provable, where A_P consists of the axioms of Peano's arithmetic (see example 5.6.3). (For a treatment of Gödel's incompleteness theorems, see Enderton, 1972; Kleene, 1952; Shoenfield, 1967; or Monk, 1976.) Since Peano's arithmetic is incomplete, then $\{A_P, C\}$ and $\{A_P, \neg C\}$ are both consistent, but their union is inconsistent.

A. Robinson's theorem shows that inconsistency does not arise if S is complete. Actually, one has to be a little careful about the presence in the language of function symbols or of the equality predicate $\doteq$.

Theorem 6.7.1 (Robinson's joint consistency theorem) Let **L** be a first-order language either without function symbols and without equality, or with equality (and possibly function symbols). Let **L**$_1$ and **L**$_2$ be two expansions of **L**, let S_1 be a set of **L**$_1$-sentences, and S_2 a set of **L**$_2$-sentences, and S be the set of **L**-sentences that are in $S_1 \cap S_2$. If S_1 and S_2 are consistent and S is complete, that is, for every closed **L**-formula C, either $\vdash S \to C$, or $\vdash S \to \neg C$, then the union $S_1 \cup S_2$ of S_1 and S_2 is consistent.

Proof: Assume that $S_1 \cup S_2$ is inconsistent. Then $\models S_1 \cup S_2 \to$. By the completeness theorem (theorem 5.6.1), there is a finite subsequent of $S_1 \cup S_2 \to$ that is provable. Let $A_1, ..., A_m, B_1, ..., B_n \to$ be such a sequent, where $A_1, ..., A_m \in S_1$, and $B_1, ..., B_n \in S_2$. It is immediate that

$$\vdash (A_1 \wedge ... \wedge A_m) \supset \neg(B_1 \wedge ... \wedge B_n).$$

We apply Craig's interpolation theorem (theorem 6.5.1) to this formula.

First, we consider the case where **L** does not contain function symbols and does not contain equality. Then, if $A_1 \wedge ... \wedge A_m$ and $\neg(B_1 \wedge ... \wedge B_n)$ do not have any predicate symbol in common, either

$$\vdash \neg(A_1 \wedge ... \wedge A_m), \text{ or}$$
$$\vdash \neg(B_1 \wedge ... \wedge B_n).$$

In the first case, the consistency of S_1 is contradicted, in the second case, the consistency if S_2 is contradicted.

If $(A_1 \wedge ... \wedge A_m)$ and $\neg(B_1 \wedge ... \wedge B_n)$ have some predicate in common, then there is a formula C such that

$$(1) \qquad\qquad \vdash (A_1 \wedge ... \wedge A_m) \supset C,$$

and

$$(2) \qquad\qquad \vdash C \supset \neg(B_1 \wedge ... \wedge B_n),$$

and the predicate symbols, constant symbols and variables free in C are both in $A_1 \wedge ... \wedge A_m$ and $\neg(B_1 \wedge ... \wedge B_n)$. Since these formulae are closed, C is also a closed formula, and C belongs to S. Since S is complete, either $\vdash S \rightarrow C$ or $\vdash S \rightarrow \neg C$. In the first case, by (2)

$$\vdash S \rightarrow \neg(B_1 \wedge ... \wedge B_n),$$

contradicting the consistency of S_2. In the second case, by (1)

$$\vdash S \rightarrow \neg(A_1 \wedge ... \wedge A_m),$$

contradicting the consistency of S_1.

If **L** is a language with equality, we need the strong form of Craig's interpolation theorem mentioned as a remark after theorem 6.5.2, which states that the all predicate, function and constant symbols occurring in an interpolant C of $A \supset B$ occur in both A and B. The rest of the proof is as above. $\square$

Another slightly more general version of Robinson's joint consistency theorem is given in problem 6.7.1.

PROBLEMS

6.7.1. Prove the following version of Robinson's joint consistency theorem:

Let **L** be a first-order language either without function symbols and without equality, or with equality (and possibly function symbols). Let $\mathbf{L}_1$ and $\mathbf{L}_2$ be two expansions of **L**, let S_1 be a set of $\mathbf{L}_1$-sentences, and S_2 a set of $\mathbf{L}_2$-sentences, and S be the set of **L**-sentences that are in $S_1 \cap S_2$. Assume that S_1 and S_2 are consistent. Then the union $S_1 \cup S_2$ of S_1 and S_2 is consistent iff for every closed **L**-formula C, either $S_1 \to C$ is not provable, or $S_2 \to \neg C$ is not provable.

6.7.2. Prove that the version of Robinson's joint consistency theorem given in problem 6.7.1 implies Craig's interpolation theorem.

Hint: Let $A \supset B$ be a provable formula. Let $S_1 = \{C \mid\ \vdash A \to C\}$, and $S_2 = \{C \mid\ \vdash \neg B \to C\}$. Then $S_1 \cup S_2$ is inconsistent.

Notes and Suggestions for Further Reading

Gentzen's cut elimination theorem is one of the jewels of proof theory. Originally, Gentzen's motivation was to provide constructive consistency proofs, and the cut elimination theorem is one of the main tools.

Gentzen's original proof can be found in Szabo, 1969, and other proofs are in Kleene, 1952, and Takeuti, 1975. A very elegant proof can also be found in Smullyan, 1968. The proof given in Section 6.4 is inspired by Schwichtenberg and Tait (Barwise, 1977, Tait, 1968).

Craig himself used Gentzen systems for proving his interpolation theorem. We have followed a method due to Maehara sketched in Takeuti, 1975, similar to the method used in Kleene, 1967. There are model-theoretic proofs of Craig's theorem, Beth's definability theorem, and Robinson's joint consistency theorem. The reader is referred to Chang and Keisler, 1973, or Shoenfield, 1967.

The reader interested in proof theory should also read the article by Schwichtenberg in Barwise, 1977. For an interesting analysis of analytic versus nonanalytic proofs, the reader if referred to the article by Frank Pfenning, in Shostak, 1984a.

Chapter 7

Gentzen's Sharpened Hauptsatz and Herbrand's Theorem

7.1 Introduction

We have mentioned in Chapter 6 that the cut elimination theorem shows the existence of normal forms for proofs, namely the fact that every LK-proof can be transformed to a cut-free proof (or a proof without essential cuts in LK_e).

In this chapter we shall use Gentzen's cut elimination theorem (also called Gentzen's Hauptsatz) to prove a version of Herbrand's theorem for LK and LK_e. A derivation of Herbrand's theorem from the cut elimination theorem has the advantage that it yields a constructive version of the result, in the spirit of Herbrand's original version (Herbrand, 1971). The proof given in this chapter using Gentzen's Hauptsatz is inspired from a method sketched in Kleene, 1967.

Herbrand's theorem is perhaps the most fundamental result of first-order logic because it shows how the provability of a formula of first-order logic reduces to the provability of a quantifier-free formula (obtained from the original formula by substitutions).

Before proceeding any further, we wish to emphasize that Herbrand's original theorem is concerned with *provability*, a proof-theoretic concept, and not *validity*, a semantic concept.

This is an important point because another theorem known as Skolem-Herbrand-Gödel theorem is often improperly referred to as Herbrand's theorem in the literature, thus causing a confusion. The Skolem-Herbrand-Gödel theorem is similar in form to Herbrand's theorem but deals with *unsatisfiability* (or validity), a semantic concept.

The reason for the confusion is probably that, by Gödel's completeness theorem, validity can be equated to provability. Hence, Herbrand's original theorem can also be stated for unsatisfiability (or validity).

However, Herbrand's original theorem is a deeper result, whose proof is significantly harder than the Skolem-Herbrand-Gödel theorem, and Herbrand's theorem also yields more information than the latter. More on this subject will be said at the end of Sections 7.5 and 7.6. For an illuminating discussion, the reader should consult Goldfarb's introduction to Herbrand, 1971.

In this chapter, we shall give in Section 7.5 a version of Herbrand's original theorem for prenex formulae, and in Section 7.6 a version of the Skolem-Herbrand-Gödel theorem for formulae in NNF (actually, half of this theorem is more like Herbrand's original theorem).

The Skolem-Herbrand-Gödel theorem can be viewed as the theoretical basis of most theorem proving methods, in particular the resolution method, one of the best known techniques in automatic theorem proving. In fact, the completeness of the resolution method will be shown in Chapter 8 by combining the Skolem-Herbrand-Gödel theorem and theorem 4.3.2.

A constructive version of Herbrand's theorem can be obtained relatively easily from Gentzen's sharpened Hauptsatz, which is obtained by analyzing carefully cut-free proofs of formulae of a special type.

Gentzen's sharpened Hauptsatz shows that provided the formulae in the bottom sequent have a certain form, a proof can be reorganized to yield another proof in *normal form* such that *all the quantifier inferences appear below all the propositional inferences*. The main obstacle to the permutation of inferences is the possibility of having a quantifier rule applied to the side formula arising from a propositional rule as illustrated below:

EXAMPLE 7.1.1

$$
\cfrac{P(a) \to Q(f(g(a))) \qquad \cfrac{\cfrac{Q(f(g(a))) \to Q(f(g(a)))}{\forall x Q(f(x)) \to Q(f(g(a)))}}{}}{P(a) \vee \forall x Q(f(x)) \to Q(f(g(a)))}
$$

The obvious solution that consists in permuting the $\vee : left$ rule and the $\forall : left$ rule does not work since a quantifier rule is not allowed to apply to a subformula like $\forall x Q(f(x))$ in $P(a) \vee \forall x Q(f(x))$.

There are at least two ways of resolving this difficulty:

(1) Impose restrictions on formulae so that a quantifier inference cannot be applied to the side formula of a propositional inference.

(2) Allow more general quantifier rules applying to subformulae.

The standard approach has been to enforce (1) by requiring the formulae to be *prenex formulae*. A Prenex formula is a formula consisting of a (possibly empty) string of quantified variables followed by a quantifier-free formula, and it is easily seen that (1) holds.

The second approach is perhaps not as well known, but is possible. Smullyan has given quantifier rules (page 122 in Chapter 14 of Smullyan, 1968) that allow the permutation process to be performed for unrestricted formulae, and a general version of the extended Hauptsatz is also given (theorem 2, page 123). These rules are binary branching and do not seem very convenient in practice. We will not pursue this method here and refer the interested reader to Smullyan, 1968.

We have also discovered that Andrews's version of the Skolem-Herbrand-Gödel theorem (Andrews, 1981) suggests quantifier rules such that (2) holds for formulae in negation normal form. Such rules are quite simple, and since there is a refutation method based on Andrews's version of Skolem-Herbrand-Gödel's theorem (the method of matings), we shall give a proof of the extended Hauptsatz for such a system.

Since every formula is equivalent to a prenex formula and to a formula in negation normal form, there is no loss of generality in restricting our attention to such normal forms.

7.2 Prenex Normal Form

First, we define the concept of a prenex formula.

Definition 7.2.1 (Prenex form) A formula is a *prenex formula* (or in *prenex form*) iff either it contains no quantifiers, or it is of the form $Q_1 x_1 ... Q_n x_n B$, where B is a formula with no quantifiers (quantifier-free), $x_1, ..., x_n$ are (not necessarily distinct) variables, and $Q_i \in \{\forall, \exists\}$, for $i = 1, ..., n$.

In order to show that every formula is equivalent to a prenex formula, the following lemma will be used.

Lemma 7.2.1 The following formulae are valid:

(a)
$$\forall x(A \supset B) \equiv (\exists x A \supset B)$$
$$\exists x(A \supset B) \equiv (\forall x A \supset B)$$

where x does not occur free in B.

$$\forall x(A \supset B) \equiv (A \supset \forall x B)$$
$$\exists x(A \supset B) \equiv (A \supset \exists x B)$$

where x does not occur free in A.

(b)
$$\neg \forall x A \equiv \exists x \neg A$$
$$\neg \exists x A \equiv \forall x \neg A$$

(c)
$$(\forall x A \vee B) \equiv \forall x (A \vee B)$$
$$(\forall x A \wedge B) \equiv \forall x (A \wedge B)$$
$$(\exists x A \vee B) \equiv \exists x (A \vee B)$$
$$(\exists x A \wedge B) \equiv \exists x (A \wedge B)$$

where x does not occur free in B.

(d)
$$\forall x (A \wedge B) \equiv \forall x A \wedge \forall x B$$
$$\exists x (A \vee B) \equiv \exists x A \vee \exists x B$$

(e)
$$\forall x A \equiv \forall y A[y/x]$$
$$\exists x A \equiv \exists y A[y/x]$$

where y is free for x in A, and y does not occur free in A unless $y = x$ $(y \notin FV(A) - \{x\})$.

Proof: Some of these equivalences have already been shown. We prove two new cases, leaving the others as an exercise. A convenient method is to construct a proof tree (in G).

We give proofs for $(\forall x A \wedge B) \equiv \forall x (A \wedge B)$ and $(\forall x A \vee B) \equiv \forall x (A \vee B)$, where x is not free in B. As usual, to prove $X \equiv Y$, we prove $X \supset Y$ and $Y \supset X$.

(i) Proof of $(\forall x A \wedge B) \supset \forall x (A \wedge B)$:

$$\frac{\dfrac{A[y/x], B \to A[y/x] \qquad\qquad A[y/x], B \to B}{A[y/x], B \to (A[y/x] \wedge B)}}{\dfrac{\forall x A, B \to (A \wedge B)[y/x]}{\dfrac{\forall x A, B \to \forall x (A \wedge B)}{\dfrac{(\forall x A \wedge B) \to \forall x (A \wedge B)}{\to (\forall x A \wedge B) \supset \forall x (A \wedge B)}}}} \qquad \text{for a new variable } y$$

We have used the fact that since y is not free in B, $B[y/x] = B$.

(ii) Proof of $\forall x (A \wedge B) \supset (\forall x A \wedge B)$

$$\frac{\dfrac{A[y/x], B \to A[y/x]}{\dfrac{A[y/x] \wedge B \to A[y/x]}{\dfrac{\forall x(A \wedge B) \to A[y/x]}{\forall x(A \wedge B) \to \forall x A}}} \qquad \dfrac{\dfrac{A[y/x], B \to B}{\dfrac{A[y/x] \wedge B \to B}{\forall x(A \wedge B) \to B}}}{\to \forall x(A \wedge B) \supset (\forall x A \wedge B)}$$

Again, y is a new variable and we used the fact that $B[y/x] = B$.

(iii) Proof of $(\forall x A \vee B) \supset \forall x(A \vee B)$:

$$\dfrac{\dfrac{\dfrac{A[y/x] \to A[y/x], B}{\forall x A \to A[y/x], B} \qquad B \to A[y/x], B}{\dfrac{(\forall x A \vee B) \to A[y/x], B}{\dfrac{(\forall x A \vee B) \to A[y/x] \vee B}{(\forall x A \vee B) \to \forall x(A \vee B)}}}}{\to (\forall x A \vee B) \supset \forall x(A \vee B)}$$

As in (ii) y is a new variable and we used the fact that $B[y/x] = B$.

(iv) Proof of $\forall x(A \vee B) \supset (\forall x A \vee B)$:

$$\dfrac{\dfrac{\dfrac{A[y/x] \to A[y/x], B \qquad B \to A[y/x], B}{\dfrac{A[y/x] \vee B \to A[y/x], B}{\dfrac{\forall x(A \vee B) \to A[y/x], B}{\forall x(A \vee B) \to \forall x A, B}}}}{\forall x(A \vee B) \to (\forall x A \vee B)}}{\to \forall x(A \vee B) \supset (\forall x A \vee B)}$$

As in (iii) y is a new variable and we used the fact that $B[y/x] = B$. $\square$

Theorem 7.2.1 For every formula A, a prenex formula B can be constructed such that $A \equiv B$ is valid (A and B are equivalent).

Proof: To simplify the proof, we will assume that $Q_1 x_1 ... Q_n x_n B$ denotes the quantifier-free formula B when $n = 0$. The proof proceeds using the induction principle applied to A. The base case is trivial since an atomic formula is quantifier free. The induction step requires six cases. We cover three of these cases, leaving the others as exercises.

(i) $A = (B \lor C)$. By the induction hypothesis, B is equivalent to a prenex formula $Q_1x_1...Q_mx_mB'$ and C is equivalent to a prenex formula $R_1y_1...R_ny_nC'$, where B' and C' are quantifier free and $Q_i, R_j \in \{\forall, \exists\}$. By lemma 7.2.1(e), it can be assumed that the sets of variables $\{x_1,...,x_m\}$ and $\{y_1,...,y_n\}$ are disjoint and that both sets are also disjoint from the union of the sets $FV(B')$ and $FV(C')$ of free variables in B' and C' (by renaming with new variables). Using lemma 7.2.1(c) (several times), we get the prenex formula

$$Q_1x_1...Q_mx_mR_1y_1...R_ny_n(B' \lor C'),$$

which, by lemma 5.3.7 is equivalent to A.

(ii) $A = \neg B$. By the induction hypothesis, B is equivalent to a prenex formula $Q_1x_1...Q_mx_mB'$ where B' is quantifier free. For each Q_i, let

$$R_i = \begin{cases} \forall & \text{if } Q_i = \exists; \\ \exists & \text{if } Q_i = \forall. \end{cases}$$

Using lemma 7.2.1(b) (several times), by lemma 5.3.7 the prenex formula

$$R_1x_1...R_mx_m\neg B'$$

is equivalent to A.

(iii) $A = \forall xB$. By the induction hypothesis, B is equivalent to a prenex formula $Q_1x_1...Q_mx_mB'$. Hence, by lemma 5.3.7 A is equivalent to the prenex formula $\forall xQ_1x_1...Q_mx_mB'$. $\square$

EXAMPLE 7.2.1

Let

$$A = \forall x(P(x) \lor \neg \exists y(Q(y) \land R(x,y))) \lor \neg(P(y) \land \neg \forall xP(x)).$$

The prenex form of
$$\neg \exists y(Q(y) \land R(x,y))$$

is
$$\forall y\neg(Q(y) \land R(x,y)).$$

The prenex form of

$$\forall x(P(x) \lor \neg \exists y(Q(y) \land R(x,y)))$$

is
$$\forall x \forall y(P(x) \lor \neg(Q(y) \land R(x,y))).$$

The prenex form of
$$\neg \forall xP(x)$$

is

$$\exists x \neg P(x).$$

The prenex form of

$$(P(y) \wedge \neg \forall x P(x))$$

is

$$\exists x (P(y) \wedge \neg P(x)).$$

The prenex form of

$$\neg (P(y) \wedge \neg \forall x P(x))$$

is

$$\forall x \neg (P(y) \wedge \neg P(x)).$$

The prenex form of A is

$$\forall x \forall v \forall z ((P(x) \vee \neg (Q(v) \wedge R(x,v))) \vee \neg (P(y) \wedge \neg P(z))).$$

In a prenex formula $Q_1 x_1 ... Q_n x_n B$, B is sometimes called the *matrix*. Since B is quantifier free, it is equivalent to a formula in either conjunctive or disjunctive normal form. The disjunctive normal form of a formula can be obtained by using the *search* procedure as explained in theorem 3.4.2. The following lemma, which is the dual of lemma 3.4.5, provides another method for transforming a formula into disjunctive normal form.

Lemma 7.2.2 Every quantifier-free formula A (containing the connectives $\vee$, $\wedge$, $\neg$, $\supset$) can be transformed into an equivalent formula in disjunctive normal form, by application of the following identities:

$$(A \supset B) \equiv (\neg A \vee B)$$
$$\neg \neg A \equiv A$$
$$\neg (A \wedge B) \equiv (\neg A \vee \neg B)$$
$$\neg (A \vee B) \equiv (\neg A \wedge \neg B)$$
$$A \wedge (B \vee C) \equiv (A \wedge B) \vee (A \wedge C)$$
$$(B \vee C) \wedge A \equiv (B \wedge A) \vee (C \wedge A)$$
$$(A \wedge B) \wedge C \equiv A \wedge (B \wedge C)$$
$$(A \vee B) \vee C \equiv A \vee (B \vee C)$$

Proof: The proof is dual to that of lemma 3.4.5 and is left as an exercise.

□

In the next sections, we consider versions of the sharpened Hauptsatz.

PROBLEMS

7.2.1. Convert the following formulae to prenex form:

$$(\neg\forall x P(x,y) \vee \forall x R(x,y))$$
$$\forall x(P(x) \supset \neg\exists y R(x,y))$$
$$(\neg\forall x\neg\forall y\neg\forall z P(x,y) \vee \neg\exists x\neg\exists y(\neg\exists z Q(x,y,z) \supset R(x,y)))$$

7.2.2. Convert the following formulae to prenex form:

$$(\exists x\forall y P(x,y) \wedge \forall y\exists x P(y,x))$$
$$(\neg(\forall x P(x) \vee \exists y\neg Q(y)) \vee (\forall z P(z) \vee \exists w\neg Q(w)))$$
$$(\neg\forall x(P(x) \vee \exists y\neg Q(y)) \vee (\forall z P(z) \vee \exists w\neg Q(w)))$$

7.2.3. Finish the proof of lemma 7.2.1.

7.2.4. Finish the proof of theorem 7.2.1.

7.2.5. Prove lemma 7.2.2.

7.2.6. Write a computer program for converting a formula to prenex form.

7.3 Gentzen's Sharpened Hauptsatz for Prenex Formulae

First, we will need a certain normal form for proofs.

7.3.1 Pure Variable Proofs

Gentzen's sharpened Hauptsatz for prenex formulae is obtained by observing that in a cut-free proof of a prenex formula (in LK or LK_e), the quantifier rules can be interchanged with the propositional rules, in such a way that no propositional rule appears below (closer to the root of the proof tree than) a quantifier rule. There is actually a more general permutability lemma due to Kleene, but only part of this lemma is needed to obtain the sharpened Hauptsatz.

During the course of the proof of the sharpened Hauptsatz, it will be necessary to show that every provable sequent (in LK or LK_e) has a proof in which the variables introduced by $\forall : right$ and $\exists : left$ rules satisfy certain conditions. Proofs satisfying these conditions are called "pure-variable proofs." To illustrate the necessity of these conditions, consider the following example.

EXAMPLE 7.3.1

$$\frac{\dfrac{\Gamma \to P(y), Q}{\Gamma \to \forall x P(x), Q} \; \forall : right \qquad \Gamma \to \forall x P(x), R(y)}{\Gamma \to \forall x P(x), Q \wedge R(y)} \; \wedge : right$$

(To shorten proof trees, structural rules are often not shown.) The rule $\forall : right$ occurs above the rule $\wedge : right$. We wish to permute these two rules; that is, construct a proof with the same conclusion and the same premises, but with the $\forall : right$ rule occurring below the $\wedge : right$ rule. The following "proof" almost works, except that the variable y occurs free in the conclusion of the $\forall : right$ rule, violating the eigenvariable condition.

$$\frac{\dfrac{\dfrac{\Gamma \to P(y), Q}{\Gamma \to P(y), \forall x P(x), Q}\text{weakening and exchanges} \quad \dfrac{\Gamma \to \forall x P(x), R(y)}{\Gamma \to P(y), \forall x P(x), R(y)}\text{weakening and exchanges}}{\dfrac{\Gamma \to P(y), \forall x P(x), Q \wedge R(y)}{\dfrac{\Gamma \to \forall x P(x), \forall x P(x), Q \wedge R(y)}{\Gamma \to \forall x P(x), Q \wedge R(y)}}}}{} $$

$\wedge : right$

$\forall : right$

contraction

The problem can be eliminated by making sure that for every eigen-variable y used in an application of the rule $\forall : right$ or $\exists : left$, that variable only occurs in the subtree rooted at the sequent constituting the premise of the rule (and y does not occur both free and bound, but this is already a condition required for constructing proof trees). If we rename the topmost occurrence of y in the first proof tree with the new variable z, the problem is indeed eliminated, as shown.

EXAMPLE 7.3.2

Pure-variable proof tree

$$\frac{\dfrac{\Gamma \to P(z), Q}{\Gamma \to \forall x P(x), Q} \; \forall : right \qquad \Gamma \to \forall x P(x), R(y)}{\Gamma \to \forall x P(x), Q \wedge R(y)} \; \wedge : right$$

New legal proof tree with $\wedge : right$ above $\forall : right$

$$\frac{\Gamma \rightarrow P(z), Q}{\text{weakening and exchanges}}$$
$$\frac{}{\Gamma \rightarrow P(z), \forall x P(x), Q}$$

$$\frac{\Gamma \rightarrow \forall x P(x), R(y)}{\text{weakening and exchanges}}$$
$$\frac{}{\Gamma \rightarrow P(z), \forall x P(x), R(y)} \quad \wedge : right$$

$$\frac{\Gamma \rightarrow P(z), \forall x P(x), Q \wedge R(y)}{\Gamma \rightarrow \forall x P(x), \forall x P(x), Q \wedge R(y)} \quad \forall : right$$

$$\frac{}{\Gamma \rightarrow \forall x P(x), Q \wedge R(y)} \quad \text{contraction}$$

Definition 7.3.1 A proof tree (in LK, LK_e, G or $G_=$) is a *pure-variable proof tree* iff, for every variable y occurring as the eigenvariable in an application of the rule $\forall : right$ or $\exists : left$, y does not occur both free and bound in any formula in the proof tree, and y only occurs in the subtree rooted at the sequent constituting the premise of the rule.

Lemma 7.3.1 In LK or LK_e (or G or $G_=$), every proof of a sequent $\Gamma \rightarrow \Delta$ can be converted to a pure-variable proof of the same sequent, simply by replacing occurrences of variables with new variables.

Proof: First, recall that from lemma 6.5.1, given a sequent $\Gamma \rightarrow \Delta$ with proof tree T, where the variable x does not occur bound in T, for any variable y not occurring in T, the tree $T[y/x]$ obtained by substituting y for every free occurrence of x in T is a proof tree for the sequent $\Gamma[y/x] \rightarrow \Delta[y/x]$.

The rest of the proof proceeds by induction on the number k of inferences of the form $\forall : right$ or $\exists : left$ in the proof tree T for the sequent $\Gamma \rightarrow \Delta$. If $k = 0$, the lemma is obvious. Otherwise, T is of the form

$$\frac{T_1 \quad \dots \quad T_n}{S}$$

where each tree T_i is of the form

$$\frac{Q_i}{\dfrac{\Gamma_i \rightarrow \Delta_i, A_i[y_i/x_i]}{\Gamma_i \rightarrow \Delta_i, \forall x_i A_i}}$$

or

$$\frac{Q_i}{\dfrac{A_i[y_i/x_i], \Gamma_i \rightarrow \Delta_i}{\exists x_i A_i, \Gamma_i \rightarrow \Gamma_i}}$$

where S does not contain any inferences of the form $\forall : right$ or $\exists : left$, the root of Q_i is labeled with $\Gamma_i \rightarrow \Delta_i, A_i[y_i/x_i]$ (or $A_i[y_i/x_i], \Gamma_i \rightarrow \Delta_i$), and the leaves of S are either axioms or are labeled with $\Gamma_i \rightarrow \Delta_i, \forall x_i A_i$ (or $\exists x_i A_i, \Gamma_i \rightarrow \Gamma_i$). Since each proof tree Q_i contains fewer applications of $\forall : right$ or $\exists : left$ than T, by the induction hypothesis, there is a pure-variable proof tree Q_i' for each $\Gamma_i \rightarrow \Delta_i, A_i[y_i/x_i]$ (or $A_i[y_i/x_i], \Gamma_i \rightarrow \Delta_i$). Note that no variable free in $\Gamma_i \rightarrow \Delta_i, A_i[y_i/x_i]$ (or $A_i[y_i/x_i], \Gamma_i \rightarrow \Delta_i$), including y_i, is used as an eigenvariable in Q_i'. For every $i = 1, ..., n$, let Q_i'' be the proof tree obtained from Q_i' by renaming all eigenvariables in Q_i', so that the set of eigenvariables in Q_i'' is disjoint from the set of eigenvariables in Q_j'' for $i \neq j$, and is also disjoint from the set of variables in S. Finally, for $i = 1, ..., n$, if y_i occurs in S (not below $\Gamma_i \rightarrow \Delta_i, \forall x_i A_i$ or $\exists x_i A_i, \Gamma_i \rightarrow \Delta_i$, since y_i is an eigenvariable), let z_i be a new variable (not in any of the Q_i'', and such that $z_i \neq z_j$ whenever $i \neq j$), and let T_i' be the tree

$$\frac{\dfrac{Q_i''[z_i/y_i]}{\Gamma_i \rightarrow \Delta_i, A_i[z_i/x_i]}}{\Gamma_i \rightarrow \Delta_i, \forall x_i A_i}$$

or

$$\frac{\dfrac{Q_i''[z_i/y_i]}{A_i[z_i/x_i], \Gamma_i \rightarrow \Gamma_i}}{\exists x_i A_i, \Gamma_i \rightarrow \Gamma_i}$$

obtained by substituting z_i for each occurrence of y_i in Q_i''. Then, using the above claim, one can verify that the following proof tree is a pure-variable proof tree:

$$\frac{T_1' \quad \cdots \quad T_n'}{S}$$

We will also need the following lemma, which is analogous to lemma 6.3.1.

Lemma 7.3.2 (i) Every sequent provable in LK is provable with a proof in which the axioms are sequents $A \rightarrow A$, where A is atomic.

(ii) Every sequent provable in LK_e is provable with a proof in which the axioms are either sequents $A \rightarrow A$ where A is atomic, or equality axioms.

Proof: The proof proceeds by induction on the weight of a proof as in lemma 6.3.1, and is very similar to the proof given in that lemma, except that LK (or LK_e) inferences are used. The details are left as an exercise. $\square$

7.3.2 The Permutability Lemma

The following terminology will be used in proving the permutability lemma used for establishing the sharpened Hauptsatz.

Definition 7.3.2 If in a proof tree T the conclusion of a one-premise rule R_1 is a descendant of some premise of a rule R_2, and if any intermediate rules occur between R_1 and R_2, then they are structural rules, we say that R_1 *can be permuted with* R_2, iff a proof tree T' can be constructed from T as explained below.

If T is of the form

$$
\cfrac{\cfrac{\cfrac{\cfrac{\cfrac{T_1}{S_3}}{S_1'}}{ST}}{S_1} \quad R_1 \qquad \cfrac{T_2}{S_2}}{\cfrac{S_0}{T_0}} \quad R_2
$$

or of the form

$$
\cfrac{\cfrac{T_1}{S_3} \qquad \cfrac{\cfrac{\cfrac{\cfrac{T_2}{S_2}}{S_1'}}{ST}}{S_1} \; R_1}{\cfrac{S_0}{T_0}} \quad R_2
$$

where S_0, S_1, S_2, S_3, S_1' are sequents, T_0, T_1, T_2 are trees, and the only other rules in ST between R_1 and R_2 (if any) are structural rules, then T' is of the form

$$
\cfrac{\cfrac{T_1}{S_3} \qquad \cfrac{T_2}{S_2}}{\cfrac{Q}{\cfrac{S_0}{T_0}}}
$$

and in the tree Q, the conclusion of the rule R_1 is not a descendant of any premise of the rule R_2, and the other rules used in Q besides R_1 and R_2 are structural rules.

If T is of the form

$$
\begin{array}{c}
T_1 \\
\hline
S_2 \\
\hline
S_1' \\
\hline
ST \\
\hline
S_1 \\
\hline
S_0 \\
\hline
T_0
\end{array}
\quad
\begin{array}{c}
R_1 \\
\\
\\
R_2
\end{array}
$$

and the only rules in ST (if any) between R_1 and R_2 are structural rules, then T' is of the form

$$
\begin{array}{c}
T_1 \\
\hline
S_2 \\
\hline
Q \\
\hline
S_0 \\
\hline
T_0
\end{array}
$$

where in the tree Q, the conclusion of the rule R_1 is not a descendant of the premise of the rule R_2, and the other rules used in Q besides R_1 and R_2 are structural rules.

Lemma 7.3.3 In every pure-variable, cut-free proof T in LK (proof without essential cuts in LK_e) the following properties hold:

(i) Every quantifier rule R_1 can be permuted with a rule R_2 which is either a logical rule, or a structural rule, or the cut rule, provided that the principal formula of the quantifier rule is not a side formula of the lower rule R_2.

(ii) Every quantifier rule, contraction or exchange rule R_1 can be permuted with a weakening rule R_2 whose premise is the conclusion of R_1.

Proof: There are several cases to consider. We only consider some typical cases, leaving the others as exercises.

(i) $R_1 = \forall : right$, $R_2 = \wedge : right$.

We have the following tree:

$$\frac{\Gamma' \to \Delta', A[y/x]}{\Gamma' \to \Delta', \forall x A} \quad \forall : right$$

$$\overline{\text{structural rules } ST}$$

$$\frac{\Gamma \to \Delta, \forall x A, \Lambda, B \qquad \Gamma \to \Delta, \forall x A, \Lambda, C}{\Gamma \to \Delta, \forall x A, \Lambda, B \wedge C} \quad \wedge : right$$

The permutation is achieved as follows:

$$\frac{\Gamma' \to \Delta', A[y/x]}{\text{weakening and exchanges}}$$

$$\frac{\Gamma' \to A[y/x], \Delta', \forall x A}{\text{structural rules ST}} \qquad \frac{\Gamma \to \Delta, \forall x A, \Lambda, C}{\text{weakening and exchanges}}$$

$$\frac{\Gamma \to A[y/x], \Delta, \forall x A, \Lambda, B \qquad \Gamma \to A[y/x], \Delta, \forall x A, \Lambda, C}{\Gamma \to A[y/x], \Delta, \forall x A, \Lambda, B \wedge C} \quad \wedge : right$$

$$\frac{}{\text{exchanges}}$$

$$\frac{\Gamma \to \Delta, \forall x A, \Lambda, B \wedge C, A[y/x]}{\Gamma \to \Delta, \forall x A, \Lambda, B \wedge C, \forall x A} \quad \forall : right$$

$$\frac{}{\text{exchanges, contraction, and exchanges}}$$

$$\Gamma \to \Delta, \forall x A, \Lambda, B \wedge C$$

Since the first tree is part of a pure-variable proof, y only occurs above the conclusion of the $\forall : right$ rule in it, and so, y does not occur in the conclusion of the $\forall : right$ rule in the second tree and the eigenvariable condition is satisfied.

(ii) $R_1 = \forall : left$, $R_2 = \wedge : right$.

We have the following tree:

$$\frac{A[t/x], \Gamma' \to \Delta'}{\forall x A, \Gamma' \to \Delta'} \quad \forall : left$$

$$\overline{\text{structural rules } ST}$$

$$\frac{\Gamma, \forall x A, \Delta \to \Lambda, B \qquad \Gamma, \forall x A, \Delta \to \Lambda, C}{\Gamma, \forall x A, \Delta \to \Lambda, B \wedge C} \quad \wedge : right$$

The permutation is achieved as follows:

$$
\cfrac{
  \cfrac{
    \cfrac{
      \cfrac{
        A[t/x], \Gamma' \to \Delta'
      }{\forall x A, \Gamma', A[t/x] \to \Delta'} \text{ weakening and exchanges}
    }{\Gamma, \forall x A, \Delta, A[t/x] \to \Lambda, B} \text{ structural rules ST}
    \qquad
    \cfrac{
      \cfrac{
        \Gamma, \forall x A, \Delta \to \Lambda, C
      }{\Gamma, \forall x A, \Delta, A[t/x] \to \Lambda, C} \text{ weakening and exchanges}
    }{}
  }{
    \cfrac{
      \cfrac{
        \cfrac{
          \Gamma, \forall x A, \Delta, A[t/x] \to \Lambda, B \land C
        }{A[t/x], \Gamma, \forall x A, \Delta \to \Lambda, B \land C} \text{ exchanges}
      }{\forall x A, \Gamma, \forall x A, \Delta \to \Lambda, B \land C} \ \forall : left
    }{}
  } \ \land : right
}{\Gamma, \forall x A, \Delta \to \Lambda, B \land C} \text{ exchanges, contraction, and exchanges}
$$

(iii) $R_1 = \forall : left$, $R_2 = \land : left$.

We have the following tree:

$$
\cfrac{
  \cfrac{
    \cfrac{
      A[t/x], \Gamma' \to \Delta'
    }{\forall x A, \Gamma' \to \Delta'} \ \forall : left
  }{B, \Gamma, \forall x A, \Delta \to \Lambda} \text{ structural rules ST}
}{B \land C, \Gamma, \forall x A, \Delta \to \Lambda} \ \land : left
$$

The permutation is achieved as follows:

$$\frac{A[t/x], \Gamma' \to \Delta'}{\text{weakening and exchanges}}$$

$$\frac{}{\forall x A, \Gamma', A[t/x] \to \Delta'}$$

$$\frac{}{\text{structural rules ST}}$$

$$\frac{B, \Gamma, \forall x A, \Delta, A[t/x] \to \Lambda}{B \wedge C, \Gamma, \forall x A, \Delta, A[t/x] \to \Lambda} \quad \wedge : left$$

$$\frac{}{\text{exchanges}}$$

$$\frac{A[t/x], B \wedge C, \Gamma, \forall x A, \Delta \to \Lambda}{\forall x A, B \wedge C, \Gamma, \forall x A, \Delta \to \Lambda} \quad \forall : left$$

$$\frac{}{\text{exchanges, contraction, and exchanges}}$$

$$B \wedge C, \Gamma, \forall x A, \Delta \to \Lambda$$

(iv) $R_1 = \forall : right$, $R_2 = weakening\ right$.

We have the following tree:

$$\frac{\Gamma \to \Delta, A[y/x]}{\Gamma \to \Delta, \forall x A} \quad \forall : right$$

$$\frac{}{\Gamma \to \Delta, \forall x A, B} \quad \text{weakening right}$$

The permutation is performed as follows:

$$\frac{\Gamma \to \Delta, A[y/x]}{\Gamma \to \Delta, A[y/x], B} \quad \text{weakening right}$$

$$\frac{}{\Gamma \to \Delta, B, A[y/x]} \quad \text{exchange}$$

$$\frac{}{\Gamma \to \Delta, B, \forall x A} \quad \forall : right$$

$$\frac{}{\Gamma \to \Delta, \forall x A, B} \quad \text{exchange}$$

Since the first tree is part of a pure-variable proof, y only occurs above the conclusion of the $\forall : right$ rule in it, and so, y does not occur in the conclusion of the $\forall : right$ rule in the second tree and the eigenvariable condition is satisfied.

(v) $R_1 = exchange\ right$, $R_2 = weakening\ right$.

We have the following tree:

$$\frac{\Gamma \to \Delta, A, B, \Lambda}{\frac{\Gamma \to \Delta, B, A, \Lambda}{\Gamma \to \Delta, B, A, \Lambda, C}} \quad \begin{array}{l} \text{exchange right} \\ \\ \text{weakening right} \end{array}$$

The permutation is performed as follows:

$$\frac{\Gamma \to \Delta, A, B, \Lambda}{\frac{\Gamma \to \Delta, A, B, \Lambda, C}{\Gamma \to \Delta, B, A, \Lambda, C}} \quad \begin{array}{l} \text{weakening right} \\ \\ \text{exchange right} \end{array}$$

(vi) $R_1 = contraction\ right$, $R_2 = weakening\ right$.

We have the following tree:

$$\frac{\Gamma \to \Delta, A, A}{\frac{\Gamma \to \Delta, A}{\Gamma \to \Delta, A, B}} \quad \begin{array}{l} \text{contraction right} \\ \\ \text{weakening right} \end{array}$$

The permutation is performed as follows:

$$\frac{\Gamma \to \Delta, A, A}{\Gamma \to \Delta, A, A, B} \quad \text{weakening right}$$

$$\overline{\phantom{\frac{\Gamma \to \Delta, A, A}{}}} \quad \text{exchanges}$$

$$\frac{\Gamma \to \Delta, B, A, A}{\frac{\Gamma \to \Delta, B, A}{\Gamma \to \Delta, A, B}} \quad \begin{array}{l} \text{contraction right} \\ \\ \text{exchange} \end{array}$$

(vii) $R_1 = \forall : right$, $R_2 = atomic\ cut$.

We have the following tree:

$$\frac{\Gamma' \to \Delta', A[y/x]}{\Gamma' \to \Delta', \forall x A} \quad \forall : right$$

$$\overline{\text{structural rules } ST}$$

$$\frac{\Gamma \to \Delta_1, \forall x A, \Delta_2, B \qquad B, \Lambda \to \Theta}{\Gamma, \Lambda \to \Delta_1, \forall x A, \Delta_2, \Theta} \quad \text{atomic cut } (B)$$

Since B is atomic, it is different from $\forall x A$.

The permutation is performed as follows:

$$\frac{\Gamma' \to \Delta', A[y/x]}{\text{weakening and exchanges}}$$

$$\frac{\Gamma' \to A[y/x], \Delta', \forall xA}{\text{structural rules ST}}$$

$$\frac{\Gamma \to A[y/x], \Delta_1, \forall xA, \Delta_2, B \qquad B, \Lambda \to \Theta}{\Gamma, \Lambda \to A[y/x], \Delta_1, \forall xA, \Delta_2, \Theta} \quad \text{atomic cut } (B)$$

$$\frac{}{\text{exchanges}}$$

$$\frac{\Gamma, \Lambda \to \Delta_1, \forall xA, \Delta_2, \Theta, A[y/x]}{\Gamma, \Lambda \to \Delta_1, \forall xA, \Delta_2, \Theta, \forall xA} \quad \forall : right$$

$$\frac{}{\text{exchanges, contraction, and exchanges}}$$

$$\Gamma, \Lambda \to \Delta_1, \forall xA, \Delta_2, \Theta$$

□

Remark: It is easily shown that the rules $\forall : right$ and $\exists : left$ permute with each other, permute with the rules $\forall : left$ and $\exists : right$, and the rules $\forall : left$ and $\exists : right$ permute with each other, provided that the main formula of the the upper inference is not equal to the side formula of the lower inference.

7.3.3 Gentzen's Sharpened Hauptsatz

We now prove Gentzen's sharpened Hauptsatz.

Theorem 7.3.1 (Gentzen's sharpened Hauptsatz) Given a sequent $\Gamma \to \Delta$ containing only prenex formulae, if $\Gamma \to \Delta$ is provable in LK (or LK_e), then there is a cut-free, pure-variable proof in LK (proof without essential cuts in LK_e) that contains a sequent $\Gamma' \to \Delta'$ (called the *midsequent*), which has the following properties:

(1) Every formula in $\Gamma' \to \Delta'$ is quantifier free.

(2) Every inference rule above $\Gamma' \to \Delta'$ is either structural, logical (not a quantifier rule), or the cut rule.

(3) Every inference rule below $\Gamma' \to \Delta'$ is either a quantifier rule, or a contraction, or an exchange rule (but not a weakening).

Thus, the midsequent splits the proof tree into an upper part that contains the propositional inferences, and a lower part that contains the quantifier inferences (and no weakening rules).

Proof: From theorem 6.3.1, lemma 7.3.1, and lemma 7.3.2, we can assume that $\Gamma \rightarrow \Delta$ has a cut-free proof T in LK (without essential cuts in LK_e) that is a pure-variable proof, and such that the axioms are of the form either $A \rightarrow A$ with A atomic, or an equality axiom (it is actually sufficient to assume that A is quantifier free).

For every quantifier inference R in T, let $m(R)$ be the number of propositional inferences and $n(R)$ the number of weakening inferences on the path from R to the root (we shall say that such inferences are *below* R). For a proof T, $m(T)$ is the sum of all $m(R)$ and $n(T)$ the sum of all $n(R)$, for all quantifier inferences R in T. We shall show by induction on $(m(T), n(T))$ (using the lexicographic ordering defined in Subsection 2.1.10) that the theorem holds.

(i) $m(T) = 0$, $n(T) = 0$. Then, all propositional inferences and all weakening inferences are above all quantifier inferences. Let S_0 be the premise of the highest quantifier inference (that is, such that no descendant of this inference is a quantifier inference). If S_0 only contains quantifier-free formulae, S_0 is the midsequent and we are done. Otherwise, since the proof is cut free (without essential cuts in LK_e) and since the axioms only contain quantifier-free formulae (actually, atomic formulae), prenex quantified formulae in S_0 must have been introduced by weakening rules. If T_0 is the portion of the proof with conclusion S_0, by eliminating the weakening inferences from T_0, we obtain a (quantifier-free) proof T_1 of the sequent S_1 obtained by deleting all quantifed formulae from S_0. For every quantified prenex formula $A = Q_1x_1...Q_nx_nC$ occurring in S_0, let $B = C[z_1/x_1, ..., z_n/x_n]$ be the quantifier-free formula obtained from A by substituting new variables $z_1, ..., z_n$ for $x_1, ..., x_n$ in C (we are assuming that these new variables are distinct from all the variables occurring in T_1 and distinct from the variables used for other such occurrences of quantified formulae occurring in S_0). Let $\Gamma' \rightarrow \Delta'$ be the sequent obtained from S_0 by replacing every occurrence of a quantified formula A by the corresponding quantifier-free formula B, as above. It is clear that $\Gamma' \rightarrow \Delta'$ can be obtained from S_1 by weakening inferences, and that S_0 can be obtained from $\Gamma' \rightarrow \Delta'$ by quantifier inferences. But then, $\Gamma' \rightarrow \Delta'$ is the midsequent of the proof obtained from T_1 and the intermediate inferences from S_1 to $\Gamma' \rightarrow \Delta'$ and from $\Gamma' \rightarrow \Delta'$ to S_0. Since the weakening inferences only occur above $\Gamma' \rightarrow \Delta'$, all the conditions of the theorem are satisfied.

(ii) *Case* 1: $m(T) > 0$. In this case, there is an occurrence R_1 of a quantifier inference above an occurrence R_2 of a propositional inference, and intermediate inferences (if any) are structural. Since all formulae in the root sequent are prenex, the side formula (or formulae) of R_2 are quantifier free. Hence, lemma 7.3.3(i) applies, R_1 can be permuted with R_2, yielding a new proof T' such that $m(T') = m(T) - 1$. We conclude by applying the induction hypothesis to T'.

(ii) *Case* 2: $m(T) = 0$, $n(T) > 0$. In this case, all propositional inferences are above all quantifier inferences, but there is a quantifier inference R_1 above a weakening inference R_2, and intermediate inferences (if any) are exchange or contraction rules. Using lemma 7.3.3(ii), R_2 can be moved up until R_1 and R_2 are permuted, yielding a new proof T' such that $m(T') = 0$ and $n(T') = n(T) - 1$. We conclude by applying the induction hypothesis to T'. $\square$

Remark: In the case where $m(T) > 0$, even though $m(T') = m(T) - 1$, $n(T')$ might be increased because the transformation may introduce extra weakenings. However, $n(T)$ is eventually reduced to 0, when $m(T)$ is reduced to 0. Also, note that the proof provides an algorithm for converting a pure-variable cut-free proof (proof without essential cuts in LK_e) into a proof having the properties stated in the theorem.

An example of the transformations of lemma 7.3.3 also showing the values of $m(T)$ and $n(T)$ is given below. In order to shorten proofs, some contractions and exchanges will not be shown explicitly.

EXAMPLE 7.3.3

Consider the following proof tree T in which $n(T) = 2$ and $m(T) = 0$:

$$\frac{\dfrac{\dfrac{P(f(a)) \to P(f(a))}{P(f(a)) \to P(f(a)), \exists y R(y)}}{P(f(a)) \to \exists x P(x), \exists y R(y)} \qquad \dfrac{\dfrac{R(g(a)) \to R(g(a))}{R(g(a)) \to R(g(a)), \exists x P(x)}}{R(g(a)) \to \exists x P(x), \exists y R(y)}}{P(f(a)) \vee R(g(a)) \to \exists x P(x), \exists y R(y)} \; \vee : left$$

We first exchange the $\exists : right$ inference in the right subtree with $\vee : left$ obtaining the following tree

$$\dfrac{\dfrac{T_1 \qquad\qquad T_2}{P(f(a)) \vee R(g(a)) \to \exists x P(x), \exists y R(y), R(g(a))}}{\dfrac{P(f(a)) \vee R(g(a)) \to \exists x P(x), \exists y R(y), \exists y R(y)}{P(f(a)) \vee R(g(a)) \to \exists x P(x), \exists y R(y)}} \begin{array}{l} \\ \exists : right \\ \\ \end{array}$$

where T_1 is the tree

$$\dfrac{\dfrac{\dfrac{P(f(a)) \to P(f(a))}{P(f(a)) \to P(f(a)), \exists y R(y)}}{P(f(a)) \to \exists x P(x), \exists y R(y)}}{P(f(a)) \to \exists x P(x), \exists y R(y), R(g(a))} \begin{array}{l} \\ \exists : right \\ \\ weakening \end{array}$$

and T_2 is the tree

$$R(g(a)) \rightarrow R(g(a))$$

—————————————————————

$$R(g(a)) \rightarrow R(g(a)), \exists x P(x)$$

——————— weakening, exchanges ———————

$$R(g(a)) \rightarrow \exists x P(x), \exists y R(y), R(g(a))$$

We now have $n(T) = 1$ and $m(T) = 1$, since a weakening was introduced below the $\exists : right$ inference in the tree T_1. We finally exchange the $\exists : right$ inference in T_1 with $\vee : left$. We obtain the following tree

$$\frac{\dfrac{\quad T_1 \qquad\qquad T_2 \quad}{P(f(a)) \vee R(g(a)) \rightarrow \exists x P(x), \exists y R(y), R(g(a)), P(f(a))}}{\dfrac{P(f(a)) \vee R(g(a)) \rightarrow \exists x P(x), \exists y R(y), R(g(a)), \exists x P(x)}{\dfrac{P(f(a)) \vee R(g(a)) \rightarrow \exists x P(x), \exists y R(y), R(g(a))}{\dfrac{P(f(a)) \vee R(g(a)) \rightarrow \exists x P(x), \exists y R(y), \exists y R(y)}{P(f(a)) \vee R(g(a)) \rightarrow \exists x P(x), \exists y R(y)}}}}$$

$\vee : left$

$\exists : right$

$\exists : right$

where T_1 is the tree

$$P(f(a)) \rightarrow P(f(a))$$

—————————————————————

$$P(f(a)) \rightarrow P(f(a)), \exists x P(x)$$

—————————————————————

$$P(f(a)) \rightarrow P(f(a)), \exists x P(x), \exists y R(y)$$

—————————————————————

$$P(f(a)) \rightarrow \exists x P(x), \exists y R(y), R(g(a)), P(f(a))$$

and T_2 is the tree

$$R(g(a)) \rightarrow R(g(a))$$

—————————————————————

$$R(g(a)) \rightarrow R(g(a))), \exists x P(x)$$

—————————————————————

$$R(g(a)) \rightarrow R(g(a)), \exists x P(x), \exists y R(y)$$

—————————————————————

$$R(g(a)) \rightarrow \exists x P(x), \exists y R(y), R(g(a)), P(f(a))$$

We now have $n(T) = 0$ and $m(T) = 0$ and the proof is in normal form.

Another example of a proof in the normal form given by theorem 7.3.1 is shown below.

EXAMPLE 7.3.4

Consider the sequent

$$\to \exists x \forall y \neg P(x,y), \exists y_1 \forall z \neg Q(y_1,z), \forall x_1 \exists y_2 \exists z_1 (P(x_1,y_2) \wedge Q(y_2,z_1))$$

whose validity is equivalent to the validity of the formula

$$A = \exists x \forall y \neg P(x,y) \vee \exists y_1 \forall z \neg Q(y_1,z) \vee \forall x_1 \exists y_2 \exists z_1 (P(x_1,y_2) \wedge Q(y_2,z_1))$$

The following proof satisfies the conditions of theorem 7.3.1:

$$
\dfrac{
\dfrac{
\dfrac{P(x_1,y) \to P(x_1,y)}{P(x_1,y), Q(y,z) \to P(x_1,y)} \qquad \dfrac{Q(y,z) \to Q(y,z)}{P(x_1,y), Q(y,z) \to Q(y,z)}
}{
\dfrac{P(x_1,y), Q(y,z) \to P(x_1,y) \wedge Q(y,z)}{
\dfrac{P(x_1,y) \to \neg Q(y,z), P(x_1,y) \wedge Q(y,z)}{
\dfrac{\to \neg P(x_1,y), \neg Q(y,z), P(x_1,y) \wedge Q(y,z)}{
\dfrac{\to \neg P(x_1,y), \neg Q(y,z), \exists z_1(P(x_1,y) \wedge Q(y,z_1))}{
\dfrac{\to \neg P(x_1,y), \neg Q(y,z), \exists y_2 \exists z_1(P(x_1,y_2) \wedge Q(y_2,z_1))}{
\dfrac{\to \neg P(x_1,y), \forall z \neg Q(y,z), \exists y_2 \exists z_1(P(x_1,y_2) \wedge Q(y_2,z_1))}{
\dfrac{\to \neg P(x_1,y), \exists y_1 \forall z \neg Q(y_1,z), \exists y_2 \exists z_1(P(x_1,y_2) \wedge Q(y_2,z_1))}{
\dfrac{\to \forall y \neg P(x_1,y), \exists y_1 \forall z \neg Q(y_1,z), \exists y_2 \exists z_1(P(x_1,y_2) \wedge Q(y_2,z_1))}{
\dfrac{\to \exists x \forall y \neg P(x,y), \exists y_1 \forall z \neg Q(y_1,z), \exists y_2 \exists z_1(P(x_1,y_2) \wedge Q(y_2,z_1))}{\to \exists x \forall y \neg P(x,y), \exists y_1 \forall z \neg Q(y_1,z), \forall x_1 \exists y_2 \exists z_1(P(x_1,y_2) \wedge Q(y_2,z_1))}
}}}}}}}}
}
}
$$

The midsequent is $\to \neg P(x_1,y), \neg Q(y,z), P(x_1,y) \wedge Q(y,z)$.

Notice that the proof was constructed so that the $\forall : right$ rule is always applied as soon as possible. The reason for this is that we have the freedom of choosing the terms involved in applications of the $\exists : right$ rule (as long as they are free for the substitutions), whereas the $\forall : right$ rule requires that the substituted term be a new variable. The above strategy has been chosen to allow ourselves as much freedom as possible in the substitutions.

PROBLEMS

7.3.1. Prove lemma 7.3.2.

7.3.2. Finish the proof of the cases in lemma 7.3.3.

7.3.3. Prove the following facts stated as a remark at the end of the proof of lemma 7.3.3: The rules $\forall : right$ and $\exists : left$ permute with each other, permute with the rules $\forall : left$ and $\exists : right$, and the rules $\forall : left$ and $\exists : right$ permute with each other, provided that the main formula of the the upper inference is not equal to the side formula of the lower inference.

7.3.4. Give proofs in normal form for the prenex form of the following sequents:

$$(\exists x \forall y P(x, y) \wedge \forall y \exists x P(y, x)) \rightarrow$$
$$(\neg(\forall x P(x) \vee \exists y \neg Q(y)) \vee (\forall z P(z) \vee \exists w \neg Q(w))) \rightarrow$$
$$(\neg \forall x (P(x) \vee \exists y \neg Q(y)) \vee (\forall z P(z) \vee \exists w \neg Q(w))) \rightarrow$$

7.3.5. Write a computer program converting a proof into a proof in normal form as described in theorem 7.3.1.

7.3.6. Design a *search* procedure for prenex sequents using the suggestions made at the end of example 7.3.4.

7.4 The Sharpened Hauptsatz for Sequents in NNF

In this section, it is shown that the quantifier rules of the system $G1^{nnf}$ presented in Section 6.4 can be extended to apply to certain subformulae, so that the permutation lemma holds. As a consequence, the sharpened Hauptsatz also holds for such a system. In this section, all formulae are rectified.

7.4.1 The System $G2^{nnf}$

First, we show why we chose to define such rules for formulae in NNF and not for arbitrary formulae.

EXAMPLE 7.4.1

Let
$$A = P(a) \vee \neg(Q(b) \wedge \forall x Q(x)) \rightarrow \; .$$

Since A is logically equivalent to $P(a) \vee (\neg Q(b) \vee \exists x \neg Q(x))$, the correct rule to apply to the subformula $\forall x Q(x)$ is actually the $\exists : left$ rule!

Hence, the choice of the rule applicable to a quantified subformula B depends on the parity of the number of negative subformulae that have the formula B has a subformula. This can be handled, but makes matters more complicated. However, this problem does not arise with formulae in NNF since only atoms can be negated.

Since there is no loss of generality in considering formulae in NNF and the quantifier rules for subformulae of formulae in NNF are simpler than for arbitrary formulae, the reason for considering formulae in NNF is clear.

The extended quantifier rules apply to maximal quantified subformulae of a formula. This concept is defined rigorously as follows.

Definition 7.4.1 Given a formula A in NNF, the set $QF(A)$ of *maximal quantified subformulae* of A is defined inductively as follows:

(i) If A is a literal, that is either an atomic formula B or the negation $\neg B$ of an atomic formula, then $QF(A) = \emptyset$;

(ii) If A is of the form $(B \vee C)$, then $QF(A) = QF(B) \cup QF(C)$;

(iii) If A is of the form $(B \wedge C)$, then $QF(A) = QF(B) \cup QF(C)$.

(iv) If A is of the form $\forall x B$, then $QF(A) = \{A\}$.

(v) If A is of the form $\exists x B$, then $QF(A) = \{A\}$.

EXAMPLE 7.4.2

Let

$$A = \forall u P(u) \vee (Q(b) \wedge \forall x(\neg P(x) \vee \forall y R(x,y))) \vee \exists z P(z).$$

Then

$$QF(A) = \{\forall u P(u), \forall x(\neg P(x) \vee \forall y R(x,y)), \exists z P(z)\}.$$

However, $\forall y R(x,y)$ is not a maximal quantified subformula since it is a subformula of $\forall x(\neg P(x) \vee \forall y R(x,y))$.

Since we are dealing with rectified formulae, all quantified subformulae differ by at least the outermost quantified variable, and therefore, they are all distinct. This allows us to adopt a simple notation for the result of substituting a formula for a quantified subformula.

Definition 7.4.2 Given a formula A in NNF whose set $QF(A)$ is nonempty, for any subformula B in $QF(A)$, for any formula C, the formula $A[C/B]$ is defined inductively as follows:

(i) If $A = B$, then $A[C/B] = C$;

(ii) If A is of the form $(A_1 \vee A_2)$, B belongs either to A_1 or to A_2 but not both (since subformulae in $QF(A)$ are distinct), and assume that B belongs to A_1, the other case being similar. Then, $A[C/B] = (A_1[C/B] \vee A_2)$;

(iii) If A is of the form $(A_1 \wedge A_2)$, as in (ii), assume that B belongs to A_1. Then, $A[C/B] = (A_1[C/B] \wedge A_2)$.

EXAMPLE 7.4.3

Let
$$A = \forall u P(u) \vee (Q(b) \wedge \forall x(\neg P(x) \vee \forall y R(x,y))),$$
$$B = \forall x(\neg P(x) \vee \forall y R(x,y)),$$

and
$$C = (\neg P(f(a)) \vee \forall y R(f(a), y)).$$

Then,

$$A[C/B] = \forall u P(u) \vee (Q(b) \wedge (\neg P(f(a)) \vee \forall y R(f(a), y))).$$

We now define the system $G2^{nnf}$.

Definition 7.4.3 (Gentzen system $G2^{nnf}$) The Gentzen system $G2^{nnf}$ is the system obtained from $G1^{nnf}$ by adding the weakening rules, and replacing the quantifier rules by the *quantifier rules for subformulae* listed below. Let A be any formula in NNF containing quantifiers, let C be any subformula in $QF(A)$ of the form $\forall x B$, and D any subformula in $QF(A)$ of the form $\exists x B$. Let t be any term free for x in B.

$$\frac{A[B[t/x]/C], \Gamma \to \Delta}{A, \Gamma \to \Delta} \ (\forall : left)$$

The formula $A[B[t/x]/C]$ is the *side formula* of the inference. Note that for $A = C = \forall x B$, this rule is identical to the $\forall : left$ rule of $G1^{nnf}$.

$$\frac{\Gamma \to \Delta, A[B[y/x]/C]}{\Gamma \to \Delta, A} \ (\forall : right)$$

where y is not free in the lower sequent.

The formula $A[B[y/x]/C]$ is the *side formula* of the inference. For $A = C = \forall x B$, the rule is identical to the $\forall : right$ rule of $G1^{nnf}$.

$$\frac{A[B[y/x]/D], \Gamma \to \Delta}{A, \Gamma \to \Delta} \ (\exists : left)$$

where y is not free in the lower sequent.

The formula $A[B[y/x]/D]$ is the *side formula* of the inference. For $A = D = \exists x B$, the rule is identical to the $\exists : left$ rule of $G1^{nnf}$.

$$\frac{\Gamma \to \Delta, A[B[t/x]/D]}{\Gamma \to \Delta, A} \ (\exists : right)$$

The formula $A[B[t/x]/D]$ is the *side formula* of the inference. For $A = D = \exists x B$, the rule is identical to the $\exists : right$ rule of $G1^{nnf}$.

7.4.2 Soundness of the System $G2^{nnf}$

The soundness of the quantifier rules of $G2^{nnf}$ of definition 7.4.3 is shown in the following lemma.

Lemma 7.4.1 The system $G2^{nnf}$ is sound.

Proof: We treat the case of the rules $\forall : left$ and $\forall : right$, the case of the rules $\exists : left$ and $\exists : right$ beeing similar. We need to prove that if the premise of the rule is valid, then the conclusion is valid. Equivalently, this can shown by proving that if the conclusion if falsifiable, then the premise is falsifiable.

Case 1: $\forall : left$.

To show that if the conclusion is falsifiable then the premise is falsifiable amounts to proving the following: For every structure $\mathbf{A}$, for every assignment s:

(i) $A = C$: If $\forall x B$ is satisfied in $\mathbf{A}$ by s, then $B[t/x]$ is satisfied in $\mathbf{A}$ by s. This has been shown in lemma 5.4.2.

(ii) $A \neq C$: If A is satisfied in $\mathbf{A}$ by s, then $A[B[t/x]/C]$ is satisfied in $\mathbf{A}$ by s. We proceed by induction on A.

If A is of the form $(A_1 \vee A_2)$, we can assume without loss of generality that C is in $QF(A_1)$. But $(A_1 \vee A_2)$ is satisfied in $\mathbf{A}$ by s iff either A_1 is satisfied in $\mathbf{A}$ by s or A_2 is satisfied in $\mathbf{A}$ by s. Since

$$A[B[t/x]/C] = (A_1[B[t/x]/C] \vee A_2),$$

in the second case, $(A_1[B[t/x]/C] \vee A_2)$ is also satisfied in $\mathbf{A}$ by s. In the first case, by the induction hypothesis, since A_1 is satisfied in $\mathbf{A}$ by s, $A_1[B[t/x]/C]$ is also satisfied in $\mathbf{A}$ by s, and so $(A_1[B[t/x]/C] \vee A_2)$ is satisfied in $\mathbf{A}$ by s, as desired.

If A is of the form $(A_1 \wedge A_2)$, assume as in the previous case that C occurs in A_1. If A is satisfied in $\mathbf{A}$ by s, then both A_1 and A_2 are satisfied in $\mathbf{A}$ by the same assignment s. By the induction hypothesis, $A_1[B[t/x]/C]$ is satisfied in $\mathbf{A}$ by s, and so $A[B[t/x]/C]$ is satisfied in $\mathbf{A}$ by s.

If A is equal to C, then we have to show that if $\forall x B$ is satisfied in $\mathbf{A}$ by s, then $B[t/x]$ is satisfied in $\mathbf{A}$ by s, but this reduces to Case 1(i).

Case 2: $\forall : right$.

This time, showing that if the conclusion is falsifiable then the premise is falsifiable amounts to proving the following: For every structure $\mathbf{A}$, if A is falsifiable in **A** by s, then $A[B[y/x]/D]$ is falsifiable in **A** by an assignment s' of the form $s[y := a]$, where a is some element in the domain of **A**, and y is not free in A, Γ and Δ. There are two cases:

(i) $A = D$; If $\forall x B$ is falsifiable in **A** by s, there is some element a in the domain of **A** such that $B[\mathbf{a}/x]$ is falsified in **A**, and since y does not occur free in the conclusion of the inference, by lemma 5.4.2, $B[y/x]$ is falsifiable in **A** by $s[y := a]$.

(ii) $A \neq D$; If A is falsifiable in **A** by s, then $A[B[y/x]/D]$ is falsifiable in **A** by an assignment s' of the form $s[y := a]$, where y does not occur free in A, Γ, Δ.

We prove by induction on formulae that for *every* subformula A' of A containing some formula in $QF(A)$ as a subformula, if A' is falsified in **A** by s, then $A'[B[y/x]/D]$ is falsified in **A** by an assignment s' of the form $s[y := a]$, where y is not free in A, Γ and Δ. Note that in the induction hypothesis, it is necessary to state that y is not free in A, and not just A'.

If A' is of the form $(A_1 \vee A_2)$, we can assume without loss of generality that D is in $QF(A_1)$. But $(A_1 \vee A_2)$ is falsified in **A** by s iff A_1 is falsified in **A** by s and A_2 is falsified in **A** by s. Since

$$A'[B[y/x]/D] = (A_1[B[y/x]/D] \vee A_2),$$

by the induction hypothesis $A_1[B[y/x]/D]$ is falsified in **A** by some assignment of the form $s[y := a]$ where y is not not free in A, Γ and Δ. But since A_2 is a subformula of A' and thus of A, y is not free in A_2 and by lemma 5.3.3, A_2 is also falsified in **A** by $s[y := a]$. Then $A'[B[y/x]/D]$ is falsified in **A** by the assignment $s[y := a]$.

If A' is of the form $(A_1 \wedge A_2)$, assume as in the previous case that D occurs in A_1. If A' is falsified in **A** by s, then either A_1 is falsified in **A** by s or A_2 is falsified in **A** by s. In the second case $(A_1[B[y/x]/D] \wedge A_2)$ is falsified in **A** by s. In the first case, since A_1 is falsified in **A** by s, by the induction hypothesis $A_1[B[y/x]/D]$ is falsified in **A** by some assignment s' of the form $s[y := a]$. Then $(A_1[B[y/x]/D] \wedge A_2)$ is falsified in **A** by $s[y := a]$, as desired.

If A' is equal to D, then we are back to case 1(i). $\square$

It should be emphasized that the condition that y is not free in A, Γ and Δ, and not just D, Γ, Δ plays a crucial role in the proof. Note that the system $G1^{nnf}$ is a subsystem of $G2^{nnf}$.

We now prove a version of lemma 7.3.3 and a normal form theorem analogous to theorem 7.3.1 for the system $G2^{nnf}$.

7.4.3 A Gentzen-like Sharpened Hauptsatz for $G2^{nnf}$

The definition of an inference R_1 permuting with an inference R_2 given in definition 7.3.2 extends immediately to the new quantifier rules, except that the section ST of structural rules only contains weakenings. It is also easy to see that lemmas 7.3.1 and 7.3.2 extend to the system $G2^{nnf}$. This is left as an exercise to the reader.

Consider a quantifier rule of $G2^{nnf}$, say $\forall : left$:

$$\frac{A[B[t/x]/C], \Gamma \to \Delta}{A, \Gamma \to \Delta}$$

If A belongs to Γ, letting $\Lambda = \Gamma - \{A\}$, the rule can be written as:

$$\frac{A, A[B[t/x]/C], \Lambda \to \Delta}{A, \Lambda \to \Delta}$$

where A does *not* belong to Λ. In this version, we say that the rule is the $\forall : left$ *rule with contraction*. The first version of the rule in which either $A \in \Gamma$ or A does not belong to the upper sequent is called the *rule without contraction*. The definition extends to the other quantifier rules.

Lemma 7.4.2 In every pure-variable, cut-free proof T in $G2^{nnf}$, the following holds: Every quantifier rule R_1 of $G2^{nnf}$ can be permuted with a propositional rule R_2 or a weakening.

Proof: The proof proceeds by cases. First, we prove that every quantifier rule of $G1^{nnf}$ without contraction can be permuted with a propositional rule R_2 or a weakening. It is not difficult to see that the cases treated in lemma 7.3.3 are still valid with the rules of $G1^{nnf}$. This is because sequents are pairs of sets, and contraction rules can be simulated. For example to perform the permutation involving $R_1 = \forall : right$ and $R_2 = \wedge : right$, the following deduction is used:

$$\frac{\dfrac{\dfrac{\Gamma' \to \Delta', A[y/x]}{\Gamma' \to A[y/x], \Delta', \forall x A}}{\Gamma \to A[y/x], \Delta, \forall x A, \Lambda, B} \text{ weakening rules} \qquad \dfrac{\dfrac{\Gamma \to \Delta, \forall x A, \Lambda, C}{\Gamma \to A[y/x], \Delta, \forall x A, \Lambda, C} \text{ weakening}}{} }{\dfrac{\Gamma \to \Delta, \forall x A, \Lambda, B \wedge C, A[y/x]}{\Gamma \to \Delta, \forall x A, \Lambda, B \wedge C} \; \forall : right} \; \wedge : right$$

In the lowest inference, we took advantage of the fact that

$$\Gamma \to \Delta, \forall x A, \Lambda, B \wedge C, \forall x A$$

denotes the same sequent as

$$\Gamma \to \Delta, \forall x A, \Lambda, B \wedge C,$$

since a sequent is a pair of sets. The details are left as an exercise. We still have to consider the cases of quantifier rules of $G1^{nnf}$ with contractions, quantifier rules of $G2^{nnf}$ applied to the side formula of a propositional rule, and permutation with weakenings. We treat $\wedge : left$ and $\wedge : right$, the other cases being similar.

Case 1: $\wedge : left, \forall : left$. There are two subcases:

1.1: $A \in \Gamma$ or A does not belong to the top sequent:

$$\cfrac{\cfrac{A[B[t/x]/C], \Gamma \to \Delta}{A, \Gamma \to \Delta} \quad \forall : left}{A \wedge E, \Gamma \to \Delta} \quad \wedge : left$$

The permutation is achieved as follows:

$$\cfrac{\cfrac{A[B[t/x]/C], \Gamma \to \Delta}{A[B[t/x]/C] \wedge E, \Gamma \to \Delta} \quad \wedge : left}{A \wedge E, \Gamma \to \Delta} \quad \forall : left$$

1.2: $A \notin \Gamma$ and A occurs in the top sequent.

$$\cfrac{\cfrac{A, A[B[t/x]/C], \Gamma \to \Delta}{A, \Gamma \to \Delta} \quad \forall : left}{A \wedge E, \Gamma \to \Delta} \quad \wedge : left$$

The permutation is achieved as follows:

$$\cfrac{\cfrac{\cfrac{A, A[B[t/x]/C], \Gamma \to \Delta}{A \wedge E, A[B[t/x]/C], \Gamma \to \Delta} \quad \wedge : left}{A \wedge E, A[B[t/x]/C] \wedge E, \Gamma \to \Delta} \quad \wedge : left}{A \wedge E, \Gamma \to \Delta} \quad \forall : left$$

Case 2: $\wedge : left, \exists : left$.

2.1: $A \in \Gamma$ or A does not belong to the top sequent:

$$\cfrac{\cfrac{A[B[y/x]/D], \Gamma \rightarrow \Delta}{A, \Gamma \rightarrow \Delta} \ \exists : left}{A \wedge E, \Gamma \rightarrow \Delta} \ \wedge : left$$

The variable y is a new variable occurring only above $A[B[y/x]/D], \Gamma \rightarrow \Delta$. The permutation is achieved by performing the substitution of $B[y/x]$ for D in $A \wedge E$, yielding $A[B[y/x]/D] \wedge E$. The inference is legitimate since, y being a variable that only occurs above $A[B[y/x]/D], \Gamma \rightarrow \Delta$ in the previous proof tree, y does not occur free in A, E, Γ and Δ. The permutation is achieved as follows:

$$\cfrac{\cfrac{A[B[y/x]/D], \Gamma \rightarrow \Delta}{A[B[y/x]/D] \wedge E, \Gamma \rightarrow \Delta} \ \wedge : left}{A \wedge E, \Gamma \rightarrow \Delta} \ \exists : left$$

2.2: $A \notin \Gamma$ and A occurs in the top sequent.

$$\cfrac{\cfrac{A, A[B[y/x]/D], \Gamma \rightarrow \Delta}{A, \Gamma \rightarrow \Delta} \ \exists : left}{A \wedge E, \Gamma \rightarrow \Delta} \ \wedge : left$$

The permutation is performed as follows:

$$\cfrac{\cfrac{\cfrac{A, A[B[y/x]/D], \Gamma \rightarrow \Delta}{A \wedge E, A[B[y/x]/D], \Gamma \rightarrow \Delta} \ \wedge : left}{A \wedge E, A[B[y/x]/D] \wedge E, \Gamma \rightarrow \Delta} \ \wedge : left}{A \wedge E, \Gamma \rightarrow \Delta} \ \exists : left$$

Case 3: $\wedge : right$, $\forall : right$.

3.1: $A \in \Gamma$ or A does not occur in the top sequent:

$$\cfrac{\cfrac{\Gamma \rightarrow \Delta, A[B[y/x]/C]}{\Gamma \rightarrow \Delta, A} \ \forall : right \qquad \Gamma \rightarrow \Delta, E}{\Gamma \rightarrow \Delta, A \wedge E} \ \wedge : right$$

The variable y is new and occurs only above $\Gamma \rightarrow \Delta, A[B[y/x]/C]$. The permutation is achieved as follows:

$$\frac{\Gamma \to \Delta, A[B[y/x]/C] \qquad \Gamma \to \Delta, E}{\dfrac{\Gamma \to \Delta, A[B[y/x]/C] \wedge E}{\Gamma \to \Delta, A \wedge E} \quad \forall : right} \wedge : right$$

The application of the $\forall : right$ is legal because y does not occur free in $\Gamma \to \Delta, A \wedge E$, since in the previous proof it occurs only above $\Gamma \to \Delta, A[B[y/x]/C]$.

3.2: $A \notin \Gamma$ and A occurs in the top sequent.

$$\frac{\dfrac{\Gamma \to \Delta, A, A[B[y/x]/C]}{\Gamma \to \Delta, A} \quad \forall : right \qquad \Gamma \to \Delta, E}{\Gamma \to \Delta, A \wedge E} \wedge : right$$

The permutation is performed as follows:

$$\frac{\dfrac{\Gamma \to \Delta, A, A[B[y/x]/C] \quad \dfrac{\Gamma \to \Delta, E}{\Gamma \to \Delta, A[B[y/x]/C], E}}{\Gamma \to \Delta, A[B[y/x]/C], A \wedge E} \quad \dfrac{\Gamma \to \Delta, E}{\Gamma \to \Delta, E, A \wedge E}}{\dfrac{\Gamma \to \Delta, A[B[y/x]/C] \wedge E, A \wedge E}{\Gamma \to \Delta, A \wedge E} \quad \forall : right}$$

Case 4: $\wedge : right$, $\exists right$:

4.1: $A \in \Gamma$ or A does not occur in the top sequent:

$$\frac{\dfrac{\Gamma \to \Delta, A[B[t/x]/D]}{\Gamma \to \Delta, A} \quad \exists : right \qquad \Gamma \to \Delta, E}{\Gamma \to \Delta, A \wedge E} \wedge : right$$

The permutation is achieved as follows:

$$\frac{\dfrac{\Gamma \to \Delta, A[B[t/x]/D] \qquad \Gamma \to \Delta, E}{\Gamma \to \Delta, A[B[t/x]/D] \wedge E} \quad \wedge : right}{\Gamma \to \Delta, A \wedge E} \exists : right$$

4.2: $A \notin \Gamma$ and A occurs in the top sequent.

$$\frac{\dfrac{\Gamma \to \Delta, A, A[B[t/x]/D]}{\Gamma \to \Delta, A} \;\exists : right \qquad \Gamma \to \Delta, E}{\Gamma \to \Delta, A \wedge E} \;\wedge : right$$

The permutation is achieved as follows:

$$\frac{\dfrac{\Gamma \to \Delta, A, A[B[t/x]/D] \quad \dfrac{\Gamma \to \Delta, E}{\Gamma \to \Delta, A[B[t/x]/D], E}}{\Gamma \to \Delta, A[B[t/x]/D], A \wedge E} \quad \dfrac{\Gamma \to \Delta, E}{\Gamma \to \Delta, E, A \wedge E}}{\dfrac{\Gamma \to \Delta, A[B[t/x]/D] \wedge E, A \wedge E}{\Gamma \to \Delta, A \wedge E} \;\exists : right}$$

The other cases are similar and left as an exercise. □

We can now prove the following type of normal form theorem for the systems $G1^{nnf}$ and $G2^{nnf}$.

Theorem 7.4.1 (A sharpened Hauptsatz for $G1^{nnf}$ and $G2^{nnf}$) Given a sequent $\Gamma \to \Delta$ containing only formulae in NNF, if $\Gamma \to \Delta$ is provable in $G1^{nnf}$, then there is a cut-free, pure-variable proof in $G2^{nnf}$ that contains a sequent $\Gamma' \to \Delta'$ (called the *midsequent*), which has the following properties:

(1) Every formula in $\Gamma' \to \Delta'$ is quantifier free.

(2) Every inference rule above $\Gamma' \to \Delta'$ is either a weakening or propositional (not a quantifier rule).

(3) Every inference rule below $\Gamma' \to \Delta'$ is a quantifier rule.

Thus, in such a proof in normal form, the midsequent splits the proof tree into an upper part that contains the propositional inferences (and weakenings), and a lower part that contains the quantifier inferences.

Proof: First we apply theorem 6.4.1 to obtain a cut-free $G1^{nnf}$-proof. Next, we use induction as in theorem 7.3.1. However, there is a new difficulty: Exchanges involving contractions introduce two propositional rules and sometimes a weakening above a quantifier rule, and a simple induction on $m(T)$ does not work. This difficulty can be resolved as follows. Let us assign the letter a to a quantifier rule, and the letter b to a propositional rule or a weakening. Consider the set of all strings obtained by tracing the branches of the proof tree from the root, and concatenating the letters corresponding to the quantifier and other inferences in the proof tree. The fact that a propositional inference (or a weakening) occurs below a quantifier inference is indicated by an occurrence of the substring ba. Note that every string obtained is of the form

$$a^{n_0} b^{m_1} a^{n_1} b^{m_2} a^{n_2} ... b^{m_l} a^{n_l} b^{m_{l+1}}.$$

By inspection of the exchanges given by lemma 7.4.2, note that a string ba is replaced either by ab or abb or $abbb$. From this, it is easily seen that $b^m a$ is replaced by some string ab^k, where $m \leq k \leq 3m$. Hence, $b^m a^n$ is replaced by some string $a^n b^{mk}$, where $1 \leq k \leq 3^n$.

To permute quantifier rules and propositional rules, proceed as follows: Consider all lowest occurrences of inferences of the form $b^{m_1} a^{n_1}$, and perform the exchanges using lemma 7.4.2. After this step is completed, each path

$$a^{n_0} b^{m_1} a^{n_1} b^{m_2} a^{n_2} ... b^{m_l} a^{n_l} b^{m_{l+1}}$$

is now of the form

$$a^{n_0} a^{n_1} b^{p_1} b^{m_2} a^{n_2} ... b^{m_l} a^{n_l} b^{m_{l+1}}.$$

Since the number of blocks of the form $b^{m_i} a^{n_i}$ has decreased, if we repeat the above process, we will eventually obtain a proof in which all quantifier inferences are below all propositional inferences and weakenings. When such a proof is obtained, we pick the premise of the highest quantifier inference. If this sequent M only contains quantifier-free formulae, it is the midsequent. Otherwise, since no quantifier rule is applied in the part of the proof above the sequent M, all maximal quantified subformulae in M occur in the axioms. Since a sequent is an axiom because it is either of the form $\Gamma, A, \neg A \to \Delta$, or $\Gamma, A \to A, \Delta$, or $\Gamma, \neg A \to \neg A, \Delta$ or $\Gamma \to \Gamma, A, \neg A$, where A is atomic, if we replace these quantified subformulae by quantifier-free instances obtained by applying the quantifier rules of $G2^{nnf}$, we still obtain a proof. But then, we can take the sequent obtained in this fashion as the midsequent. $\square$

Remarks: (i) The theorem deals with the two systems $G1^{nnf}$ and $G2^{nnf}$, whereas theorem 7.3.1 deals with the single system LK. Theorem 7.4.1 shows how to convert a $G1^{nnf}$-proof with cut into a cut-free normal $G2^{nnf}$-proof. This is easier to prove than the extended Hauptsatz for $G2^{nnf}$. Also, this is sufficient to derive part of Herbrand's theorem for $G1^{nnf}$.

(ii) The resulting proof may have a depth exponential in the size of the original proof.

As an illustation of theorem 7.4.1, consider the following example.

EXAMPLE 7.4.4

The following is a proof of the sequent

$$(P(a) \vee \forall x(Q(x) \vee \forall y R(y))) \to P(a), Q(a), R(f(a)) \wedge R(f(b))$$

which is not in normal form:

$$\frac{P(a) \to P(a)}{\dfrac{P(a) \to P(a), Q(a)}{\dfrac{P(a) \to P(a), Q(a), R(f(a)) \wedge R(f(b)) \qquad\qquad T_1}{(P(a) \vee \forall x(Q(x) \vee \forall y R(y))) \to P(a), Q(a), R(f(a)) \wedge R(f(b))}}}$$

where T_1 is the tree

$$
\cfrac{
\cfrac{
\cfrac{
\cfrac{Q(a) \ \to \ Q(a)}{Q(a) \ \to \ Q(a), R(f(a)) \wedge R(f(b))} \qquad T_2
}{Q(a) \vee \forall y R(y) \ \to \ Q(a), R(f(a)) \wedge R(f(b))}
}{\forall x(Q(x) \vee \forall y R(y)) \ \to \ Q(a), R(f(a)) \wedge R(f(b))}
}{\forall x(Q(x) \vee \forall y R(y)) \ \to \ P(a), Q(a), R(f(a)) \wedge R(f(b))}
$$

and T_2 is the tree

$$
\cfrac{
\cfrac{
\cfrac{R(f(a)) \ \to \ R(f(a))}{\forall y R(y) \ \to \ R(f(a))} \qquad
\cfrac{R(f(b)) \ \to \ R(f(b))}{\forall y R(y) \ \to \ R(f(b))}
}{\forall y R(y) \ \to \ R(f(a)) \wedge R(f(b))}
}{\forall y R(y) \ \to \ Q(a), R(f(a)) \wedge R(f(b))}
$$

The proof below is in normal form:

$$
\cfrac{
\cfrac{
\cfrac{
\overset{\text{propositional part}}{P(a) \vee (Q(a) \vee R(f(a))), P(a) \vee (Q(a) \vee R(f(b))) \ \to \ \Delta}
}{P(a) \vee (Q(a) \vee \forall y R(y)), P(a) \vee (Q(a) \vee R(f(a))) \ \to \ \Delta}
}{P(a) \vee (Q(a) \vee \forall y R(y)) \ \to \ P(a), Q(a), R(f(a)) \wedge R(f(b))}
}{(P(a) \vee \forall x(Q(x) \vee \forall y R(y))) \ \to \ P(a), Q(a), R(f(a)) \wedge R(f(b))}
$$

where $\Delta = P(a), Q(a), R(f(a)) \wedge R(f(b))$. The midsequent is

$$
P(a) \vee (Q(a) \vee R(f(a))), P(a) \vee (Q(a) \vee R(f(b))) \to
$$
$$
P(a), Q(a), R(f(a)) \wedge R(f(b)),
$$

which is equivalent to

$$
(P(a) \vee (Q(a)) \vee (R(f(a)) \wedge R(f(b)))) \to
$$
$$
P(a), Q(a), R(f(a)) \wedge R(f(b)).
$$

We now consider the case of languages with equality.

7.4.4 The Gentzen System $G2_{=}^{nnf}$

The system $G2_{=}^{nnf}$ has the following axioms and inference rules.

Definition 7.4.4 The system $G2^{nnf}_{\underline{\ }}$ is obtained from $G1^{nnf}_{\underline{\ }}$ by adding the quantifier rules of definition 7.4.3 and the weakening rules. It is easy to see that $G2^{nnf}_{\underline{\ }}$ is sound.

The following version of lemma 7.4.2 holds for $G2^{nnf}_{\underline{\ }}$.

Lemma 7.4.3 For every pure-variable proof T in $G2^{nnf}_{\underline{\ }}$ without essential cuts, the following holds: Every quantifier rule of $G2^{nnf}_{\underline{\ }}$ can be permuted with a propositional rule, an inessential cut, or a weakening.

Proof: The case not handled in lemma 7.4.2 is an exchange with an inessential cut. This is handled as in lemma 7.3.3. □

7.4.5 A Gentzen-like Sharpened Hauptsatz for $G2^{nnf}_{\underline{\ }}$

We also have the following sharpened Hauptsatz for $G2^{nnf}_{\underline{\ }}$.

Theorem 7.4.2 (A sharpened Hauptsatz for $G1^{nnf}_{\underline{\ }}$ and $G2^{nnf}_{\underline{\ }}$) Given a sequent $\Gamma \to \Delta$ containing only formulae in NNF, if $\Gamma \to \Delta$ is provable in $G1^{nnf}_{\underline{\ }}$, then there is a pure-variable proof in $G2^{nnf}_{\underline{\ }}$ without essential cuts that contains a sequent $\Gamma' \to \Delta'$ (called the *midsequent*), which has the following properties:

(1) Every formula in $\Gamma' \to \Delta'$ is quantifier free.

(2) Every inference rule above $\Gamma' \to \Delta'$ is either a weakening, a propositional rule, or an inessential cut (but not a quantifier rule).

(3) Every inference rule below $\Gamma' \to \Delta'$ is a quantifier rule.

Proof: The proof is essentially identical to that of theorem 7.4.1, but using lemma 7.4.3 instead of lemma 7.4.2. The details are left as an exercise. □

We shall see in Section 7.6 how theorems 7.4.1 and 7.4.2 can be used to yield Andrews's version of the Skolem-Herbrand-Gödel theorem (See Andrews, 1981), and also half of a version of Herbrand's original theorem.

PROBLEMS

7.4.1. Finish the proof of lemma 7.4.1.

7.4.2. Finish the proof of the cases left out in the proof of lemma 7.4.2

7.4.3. Prove that lemma 7.3.1 and 7.3.2 extend to the system $G2^{nnf}$.

7.4.4. Fill in the details proof of theorem 7.4.1.

7.4.5. Fill in the details proof of lemma 7.4.3.

7.4.6. Fill in the details proof of theorem 7.4.2.

7.4.7. Give a proof in normal form for the sequent

$$\forall x (P(x) \vee \forall y Q(y, f(x))) \wedge (\neg P(a) \wedge (\neg Q(a, f(a)) \vee \neg Q(b, f(a)))) \to .$$

7.4.8. It is tempting to formulate the $\forall : left$ rule so that a formula of the form $\forall x A$ can be instantiated to the formula $A[t_1/x] \wedge \ldots \wedge A[t_k/x]$, for any k terms $t_1, \ldots, t_k$ free for x in A, and never apply contractions. However, this does not work. Indeed, the resulting system is not complete. Consider the following sequent provided by Dale Miller:

$$\forall x \exists y (\neg P(x) \wedge P(y)) \to .$$

The following is a proof of the above sequent using contractions:

$$
\cfrac{
\cfrac{
\cfrac{
\cfrac{
\cfrac{
\cfrac{
\cfrac{
\cfrac{
P(u) \to \neg P(x), P(u)), P(v)
}{
\neg P(x), P(u), \neg P(u), P(v) \to
}
}{
\neg P(x), P(u), (\neg P(u) \wedge P(v)) \to
}
}{
(\neg P(x) \wedge P(u)), (\neg P(u) \wedge P(v)) \to
}
}{
(\neg P(x) \wedge P(u)), \exists y (\neg P(u) \wedge P(y)) \to
}
}{
(\neg P(x) \wedge P(u)), \forall x \exists y (\neg P(x) \wedge P(y)) \to
}
}{
\exists y (\neg P(x) \wedge P(y)), \forall x \exists y (\neg P(x) \wedge P(y)) \to
}
}{
\forall x \exists y (\neg P(x) \wedge P(y)), \forall x \exists y (\neg P(x) \wedge P(y)) \to
}
}{
\forall x \exists y (\neg P(x) \wedge P(y)) \to
}
$$

Show that a derivation involving no contractions cannot lead to a proof tree, due to the eigenvariable restriction.

7.5 Herbrand's Theorem for Prenex Formulae

In this section, we shall derive a constructive version of Herbrand's theorem from Gentzen's Hauptsatz, using a method inspired by Kleene (Kleene, 1967).

In presenting Herbrand's theorem, it is convenient to assume that we are dealing with sentences.

7.5.1 Preliminaries

The following lemma shows that there is no loss of generality in doing so.

Lemma 7.5.1 Let A be a rectified formula, and let $FV(A) = \{y_1, ..., y_n\}$ be its set of free variables. The sequent $\to A$ is (LK or LK_e) provable if and only if the sequent $\to \forall y_1 ... \forall y_n A$ is provable.

Proof: We proceed by induction on the number of free variables. The induction step consists in showing that if A is a rectified formula and has a single free variable x, then $\to A$ is provable iff $\to \forall x A$ is provable.

We can appeal to the completeness theorem and show that A is valid iff $\forall x A$ is valid, which is straightforward. We can also give a the following proof using the pure-variable lemma 7.3.1 and lemma 6.5.1.

Assume that $\to A$ is provable. Since A is rectified, the variable x is not free in $\forall x A$, and the following inference is valid.

$$\frac{\to A[x/x]}{\to \forall x A}$$

By putting the proof of $A = A[x/x]$ on top of this inference, we have a proof for $\forall x A$.

Conversely, assume that $\forall x A$ is provable. There is a minor problem, which is that the eigenvariable z used in the lowest inference in the proof of $\forall x A$ is not necessarily x, even though x is not free in $\forall x A$. However, by lemma 7.3.1, there is a pure-variable proof of A in which all eigenvariables are new and distinct. Hence, the variable x does not occur in the part of the proof above $\to A[z/x]$. Using lemma 6.5.1, we can substitute x for all occurrences of z in this part of the proof, and we get a proof of A. $\square$

We will also use the following fact which is easily shown: Given a sequent $A_1, ..., A_m \to B_1, ..., B_n$,

$$A_1, ..., A_m \to B_1, ..., B_n$$

is provable if and only if

$$\to \neg A_1 \vee ... \vee \neg A_m \vee B_1 \vee ... \vee B_n$$

is provable. Using this fact and lemma 7.5.1, a sequent consisting of formulae is provable if and only if a sentence is provable.

Hence, we will assume without loss of generality that the sequent Γ to be proved does not have sentences on the left of $\to$, that the sentences occurring on the righthand side of $\to$ are all distinct, and that they are rectified.

The main idea of Herbrand's theorem is to encode (with some inessential loss of information) the steps of the proof in which the $\forall : rule$ is applied. For this, some new function symbols called Herbrand functions or Skolem functions are introduced.

7.5.2 Skolem Function and Constant Symbols

Skolem symbols are defined as follows.

Definition 7.5.1 For every prenex formula $A = Q_n x_n ... Q_1 x_1 B$ occurring in a sequent $\to \Gamma$, for every occurrence Q_i of a universal quantifier in A then:

(i) If $n > 0$ and $m_i > 0$, where m_i is the number of existential quantifiers in the string $Q_n ... Q_{i+1}$, a new function symbol f_i^A having a number of arguments equal to m_i is created.

(ii) If $m_i = 0$ or $i = n$, the constant symbol f_i^A is created.

Such symbols are called *Skolem function symbols* and *Skolem constant symbols* (for short, *Skolem functions*).

The essence of Herbrand's theorem can be illustrated using example 7.3.4.

EXAMPLE 7.5.1

For the sequent of example 7.3.4, the unary function symbol f is associated with $\forall$ in $\exists x \forall y \neg P(x, y)$, the unary function symbol g is associated with $\forall$ in $\exists y_1 \forall z \neg Q(y_1, z)$, and the constant a with $\forall$ in $\forall x_1 \exists y_2 \exists z_1 (P(x_1, y_2) \wedge Q(y_2, z_1))$.

Now, we shall perform alterations to the proof in example 7.3.4. Moving up from the bottom sequent, instead of performing the $\forall : right$ rules, we are going to perform certain substitutions. For the instance of $\forall : right$ applied to

$$\forall x_1 \exists y_2 \exists z_1 (P(x_1, y_2) \wedge Q(y_2, z_1)),$$

substitute the Skolem constant a for all occurrences of x_1 in the proof; for the instance of $\forall : right$ applied to

$$\forall y \neg P(a, y),$$

substitute $f(a)$ for all occurrences of y in the proof; for the instance of $\forall : right$ applied to

$$\forall z \neg Q(f(a), z),$$

substitute $g(f(a))$ for all occurrences of z in the proof. Note that the resulting tree is no longer a legal proof tree because the applications of $\forall : right$ rules have been spoiled. However, the part of the proof from the new midsequent up is still a valid proof involving only propositional (and structural) rules (extending our original language with the symbols a, f, g).

The following definition will be useful for stating Herbrand's theorem.

Definition 7.5.2 Given a prenex sequent $\to \Gamma$, the *functional form* of $\to \Gamma$ is obtained as follows: For each prenex formula $A = Q_n x_n ... Q_1 x_1 B$ in

$\rightarrow \Gamma$, for each occurrence of a variable x_i bound by an occurrence Q_i of a universal quantifier, the term $f_i^A(y_1, ..., y_m)$ is substituted for x_i in B and $Q_i x_i$ is deleted from A, where f_i^A is the Skolem function symbol associated with Q_i, and $y_1, ..., y_m$ is the list of variables (in that order) bound by the existential quantifiers occurring in the string $Q_n...Q_{i+1}$ (if $m = 0$ or $i = n$, f_i^A is a constant symbol).

EXAMPLE 7.5.2

Again, referring to example 7.3.4, the sequent

$$\rightarrow \exists x \neg P(x, f(x)), \exists y_1 \neg Q(y_1, g(y_1)), \exists y_2 \exists z_1 (P(a, y_2) \wedge Q(y_2, z_1))$$

is the functional form of our original sequent. Note that the provable sequent

$$\rightarrow \neg P(a, f(a)), \neg Q(f(a), g(f(a))), P(a, f(a)) \wedge Q(f(a), g(f(a)))$$

is obtained from the functional form of the original sequent by deleting quantifiers and substituting the terms a, $f(a)$, $f(a)$ and $g(f(a))$ for x, y_1, y_2, and z_1 respectively. Example 7.5.2 illustrates part of Herbrand's theorem.

Informal statement of Herbrand's theorem: If a prenex sequent $\rightarrow \Gamma$ is provable, then a disjunction of quantifier-free formulae constructible from Γ is provable. Furthermore, this disjunction of quantifier-free formulae consists of formulae obtained by substituting ground terms (built up from the original language extended with Skolem functions) for the bound variables of the sentences occurring in the functional form of the original sequent. The converse is also true, as we shall see shortly.

The key observation used in showing the converse of the above statement is to notice that the terms a, $f(a)$, $g(f(a))$ record the order in which the $\forall$: *right* rules were applied in the original proof. The more nested the symbol, the earlier the rule was applied.

Warning: Many authors introduce Skolem function symbols to eliminate *existential quantifiers*, and not *universal quantifiers* as we do. This may seem confusing to readers who have seen this other definition of Skolem function symbols. However, there is nothing wrong with our approach. The reason the dual definition is also used (eliminating existential quantifiers using Skolem function symbols), as in the resolution method, is that the dual approach consists in showing that a formula A is valid by showing that $B = \neg A$ is *unsatisfiable*. This is equivalent to showing that the sequent $B \rightarrow$ is valid, or equivalently that $\rightarrow \neg B$ (that is $\rightarrow A$) is valid. Since in the dual approach, the *existential quantifiers* in B are Skolemized, in $\neg B = A$, the *universal quantifiers* are Skolemized. Since our approach consists in showing directly that A is valid, and not that $\neg A$ is unsatisfiable, we have defined Skolem function symbols for that purpose, and this is why they are used to eliminate *universal quantifiers*. What we have defined in definition 7.5.2 is often called in the literature the *validity functional form*, the dual form for eliminating existential quantifiers being called the *satisfiability functional form* (Herbrand, 1971). To avoid confusion, we shall call the first (the *validity functional form*) simply the *functional form*, and the second (the *satisfiability functional form*) the *Skolem normal form* (see definition 7.6.2).

7.5.3 Substitutions

Before stating and proving Herbrand's theorem, we need to define substitution functions.

Definition 7.5.3 A *substitution function* (for short, a *substitution*) is any function

$$\sigma : \mathbf{V} \to TERM_{\mathbf{L}}$$

assigning terms to the variables in $\mathbf{V}$. By theorem 2.4.1, there is a unique homomorphism

$$\hat{\sigma} : TERM_{\mathbf{L}} \to TERM_{\mathbf{L}}$$

extending σ and defined recursively as follows:

For every variable $x \in \mathbf{V}$, $\hat{\sigma}(x) = \sigma(x)$.

For every constant c, $\hat{\sigma}(c) = \sigma(c)$.

For every term $ft_1...t_n \in TERM_{\mathbf{L}}$,

$$\hat{\sigma}(ft_1...t_n) = f\hat{\sigma}(t_1)...\hat{\sigma}(t_n).$$

By abuse of language and notation, the function $\hat{\sigma}$ will also be called a substitution, and will often be denoted by σ.

The subset X of $\mathbf{V}$ consisting of the variables such that $\sigma(x) \neq x$ is called the *support* of the substitution. In what follows, we will be dealing with substitutions of finite support. If a substitution σ has finite support $\{y_1, ..., y_n\}$ and $\sigma(y_i) = s_i$, for $i = 1, .., n$, for any term t, $\hat{\sigma}(t)$ is also denoted by $t[s_1/y_1, ..., s_n/y_n]$.

Substitutions can be extended to formulae as follows. Let A be a formula, and let σ be a substitution with finite support $\{x_1, ..., x_n\}$. Assume that the variables in $\{x_1, ..., x_n\}$ are free in A and do not occur bound in A, and that each s_i ($s_i = \sigma(x_i)$) is free for x_i in A. The *substitution instance* $A[s_1/x_1, ..., s_n/x_n]$ is defined recursively as follows:

If A is an atomic formula of the form $Pt_1...t_m$, then

$$A[s_1/x_1, ..., s_n/x_n] = Pt_1[s_1/x_1, ..., s_n/x_n]...t_m[s_1/x_1, ..., s_n/x_n].$$

If A is an atomic formula of the form $(t_1 \doteq t_2)$, then

$$A[s_1/x_1, ..., s_n/x_n] = (t_1[s_1/x_1, ..., s_n/x_n] \doteq t_2[s_1/x_1, ..., s_n/x_n]).$$

If $A = \perp$, then
$$A[s_1/x_1, ..., s_n/x_n] = \perp .$$

If $A = \neg B$, then

$$A[s_1/x_1, ..., s_n/x_n] = \neg B[s_1/x_1, ..., s_n/x_n].$$

If $A = (B * C)$, where $* \in \{\vee, \wedge, \supset, \equiv\}$, then

$$A[s_1/x_1, ..., s_n/x_n] = (B[s_1/x_1, ..., s_n/x_n] * C[s_1/x_1, ..., s_n/x_n]).$$

If $A = \forall y B$ (where $y \notin \{x_1, ..., x_n\}$), then

$$A[s_1/x_1, ..., s_n/x_n] = \forall y B[s_1/x_1, ..., s_n/x_n].$$

If $A = \exists y B$ (where $y \notin \{x_1, ..., x_n\}$), then

$$A[s_1/x_1, ..., s_n/x_n] = \exists y B[s_1/x_1, ..., s_n/x_n].$$

Remark: $A[s_1/x_1][s_2/x_2]...[s_n/x_n]$, the result of substituting s_1 for x_1, ... ,s_n for x_n (as defined in definition 5.2.6) in that order, is usually different from $A[s_1/x_s, ..., s_n/x_n]$. This is because the terms $s_1,...,s_n$ may contain some of the variables in $\{x_1, ..., x_n\}$. However, if none of the variables in the support of the substitution σ occurs in the terms $s_1,...,s_n$, it is easy to see that the order in which the substitutions are performed is irrelevant, and in this case,

$$A[s_1/x_1][s_2/x_2]...[s_n/x_n] = A[s_1/x_1, ..., s_n/x_n].$$

In particular, this is the case when σ is a *ground substitution*, that is, when the terms $s_1,...,s_n$ do not contain variables.

Given a formula A and a substitution σ as above, the pair (A, σ) is called a *substitution pair*. The substitution instance defined by A and σ as above is also denoted by $\sigma(A)$. Notice that distinct substitution pairs can yield the same substitution instance.

A minor technicality has to be taken care of before proving Herbrand's theorem. If the first-order language does not have any constants, the theorem fails. For example, the sequent $\rightarrow \exists x \exists y (P(x) \vee \neg P(y))$ has a proof whose midsequent is $\rightarrow P(x) \vee \neg P(x)$, but if the language has no constants, we cannot find a ground substitution instance of $P(x) \vee \neg P(x)$ that is valid. To avoid this problem, we will assume that if any first-order language **L** does not have constants, the special constant # is added to it.

7.5.4 Herbrand's Theorem for Prenex Formulae

Before stating and proving Herbrand's theorem, recall that we can assume without loss of generality that the sequent $\rightarrow \Gamma$ to be proved consists of sentences, does not have sentences on the left of $\rightarrow$, that the sentences occurring on the righthand side of $\rightarrow$ are all distinct, and that distinct quantifiers bind occurrences of distinct variables.

Theorem 7.5.1 (Herbrand's theorem for prenex formulae) Given a sequent $\rightarrow \Gamma$ such that all sentences in Γ are prenex, $\rightarrow \Gamma$ is provable (in LK or LK_e) if and only if there is some finite sequence $< (B_1, \sigma_1), ..., (B_N, \sigma_N) >$ of substitution pairs such that the sequent $\rightarrow H$ consisting of the substitution instances $\sigma_1(B_1), ..., \sigma_N(B_N)$ is provable, where each $\sigma_i(B_i)$ is a quantifier-free substitution instance constructible from the sentences in Γ (H is called a *Herbrand disjunction*). Furthermore, each quantifier-free formula $\sigma_i(B_i)$ in the Herbrand disjunction H is a substitution instance $B_i[t_1/x_1, ..., t_k/x_k]$ of a quantifier-free formula B_i, matrix of some sentence $\exists x_1...\exists x_k B_i$ occurring in the functional form of $\rightarrow \Gamma$. The terms $t_1, ..., t_k$ are ground terms over the language consisting of the function and constant symbols in **L** occurring in $\rightarrow \Gamma$, and the Skolem function (and constant) symbols occurring in the functional form of $\rightarrow \Gamma$.

Proof: Using Gentzen's sharpened Hauptsatz (theorem 7.3.1), we can assume that we have a pure-variable cut-free proof in LK (proof without essential cuts in LK_e) with midsequent $\rightarrow \Gamma'$. We alter this proof in the following way. Starting with the bottom sequent and moving up, for every instance of the rule $\forall : right$ applied to a formula

$$\forall x_i Q_{i-1} x_{i-1} ... Q_1 x_1 B[s_1/y_1, ..., s_m/y_m]$$

which has been obtained from a prenex sentence

$$A = Q_n x_n ... Q_{i+1} x_{i+1} \forall x_i Q_{i-1} x_{i-1} ... Q_1 x_1 C$$

in Γ, substitute the term

$$f_i^A(s_1, ..., s_m)$$

for all occurrences of x_i in the proof (or Skolem constant f_i^A if $m = 0$), where f_i^A is the Skolem function symbol associated with x_i in A, $y_1, ..., y_m$ is the list of all the variables (all distinct by our hypothesis on proofs) bound by existential quantifiers in the string $Q_n x_n ... Q_{i+1} x_{i+1}$, and $s_1, ..., s_m$ is the list of terms that have been substituted for $y_1, ..., y_m$ in previous steps. Since the only inference rules used below the midsequent are quantifier, contraction or exchange rules, the resulting midsequent $\rightarrow H$ is indeed composed of substitutions instances of matrices of sentences occurring in the functional form of

$\rightarrow \Gamma$, and since the modified part of the proof above the midsequent is still a proof, the new midsequent $\rightarrow H$ is provable. If the midsequent contains variables, substitute any constant for all of these variables (by a previous assumption, the language has at least one constant, perhaps the special constant #). □

We now prove the converse of the theorem. We can assume without loss of generality that the disjuncts are distinct, since if they were not, we could suppress the duplications and still have a provable disjunct. Let $< (B_1, \sigma_1), ..., (B_N, \sigma_N) >$ be a sequence of distinct substitution pairs where each B_i is the matrix of the functional form of some sentence in $\rightarrow \Gamma$, let H be the corresponding sequence of substitution instances and assume that $\rightarrow H$ is provable. Notice that the conditions assumed before the statement of theorem 7.5.1 guarantee that every substitution pair (B, σ) corresponds to the unique pair $(Q_n x_n...Q_1 x_1 C, \sigma)$, where $\exists y_1...\exists y_m B$ is the functional form of some sentence $Q_n x_n...Q_1 x_1 C$ in $\rightarrow \Gamma$, and σ is a substitution with support $\{y_1, ..., y_m\}$.

Given the prenex sentence $A = Q_n x_n...Q_1 x_1 C$ in $\rightarrow \Gamma$, its functional form is a sentence of the form

$$\exists y_1...\exists y_m C[r_1/z_1, ..., r_p/z_p],$$

where the union of $\{y_1, ..., y_m\}$ and $\{z_1, ..., z_p\}$ is $\{x_1, ..., x_n\}$, and $\{z_1, ..., z_p\}$ is the set of variables which are universally quantified in A. Each term r_i is rooted with the Skolem function symbol f_i^A. Consider the set HT composed of all terms of the form

$$r_i[s_1/y_1, ..., s_m/y_m]$$

occurring in the Herbrand disjunction $\rightarrow H$, where r_i is a Skolem term associated with an occurrence of a universal quantifier in some prenex sentence A in Γ, and $s_1,...,s_m$ are the terms defining the substitution σ involved in the substitution pair $(C[r_1/z_1, ..., r_p/z_p], \sigma)$.

We define a partial order on the set HT as follows: For every term in HT of the form $f_i^A(t_1, ..., t_m)$, every subterm of $t_j \in HT$ (possibly t_j itself, $1 \le i \le m$) precedes $f_i^A(t_1, ..., t_m)$;

Every term in HT rooted with f_i^A precedes any term in HT rooted with f_j^A if $i > j$.

In order to reconstruct a proof, we will have to eliminate the Skolem symbols and perform $\forall : right$ rules with new variables. For this, we set up a bijection v between the set HT and a set of new variables not occurring in Γ. For every term t occurring in the Herbrand disjunction H (not only terms in HT), let $\bar{t}$ be the result of substituting the variable $v(s)$ for every maximal subterm $s \in HT$ of t (a maximal subterm in HT of t is a subterm of t in HT, which is not a proper subterm of any other subterm in HT of t). Let $< (B_1', \sigma_1'), ..., (B_N', \sigma_N') >$ be the list of substitution pairs obtained from

$< (B_1, \sigma_1), ..., (B_N, \sigma_N) >$ by replacing each term $f_j^{A_i}(y_1, ..., y_m)$ occurring in B_i (where A_i is the sentence in Γ whose functional form is $\exists y_1...\exists y_m B_i$) by $v(\sigma_i(f_j^{A_i}(y_1, ..., y_m)))$, and each term s involved in defining the substitution σ_i by $\bar{s}$ (B_i' is actually a substitution instance of the matrix of a prenex formula in Γ, where the substitution is a bijection). Let $\to H'$ be the resulting sequent of substitution instances.

We are going to show that a deduction of $\to \Gamma$ can be constructed from $\to H'$ (really $< (B_1', \sigma_1'), ..., (B_N', \sigma_N') >$) and the partially ordered set HT. First, notice that since $\to H$ is provable in the propositional part of LK (or LK_e), $\to H'$ is also provable since the above substitutions do not affect inferences used in the proof of $\to H$. The following definition will be needed.

Definition 7.5.4 Given a substitution pair (A, σ), where A is a prenex formula of the form $Q_n x_n...Q_1 x_1 C$ ($n \geq 1$) occurring in the sequent $\to \Gamma$ and σ is a substitution with support the subset of variables in $\{x_1, ..., x_n\}$ bound by existential quantifiers, the σ-*matrix of the functional form of A up to i* ($0 \leq i \leq n$) is defined inductively as follows:

For $i = 0$, the σ-matrix of the functional form of A up to 0 is the substitution instance of C obtained by substituting the new variable $v(\sigma(f_j^A(y_1, ..., y_m)))$ for x_j in C, for each occurrence of a variable x_j bound by a universal quantifier in A.

For $0 \leq i \leq n - 1$:

(i) If $Q_{i+1} = \exists$ and if the σ-matrix of the functional form of A up to i is B, the σ-matrix of the functional form of A up to $i + 1$ is $\exists x_{i+1} B$;

(ii) If $Q_{i+1} = \forall$, if the σ-matrix of the functional form of A up to i is $B[v(\sigma(f_{i+1}^A(y_1, ..., y_m)))/x_{i+1}]$, where the set of variables bound by existential quantifiers in $Q_n x_n...Q_{i+2} x_{i+2}$ is $\{y_1, ..., y_m\}$, the σ-matrix of the functional form of A up to $i + 1$ is $\forall x_{i+1} B$.

Note that the σ-matrix of the functional form of A up to n is A itself.

Next, we define inductively a sequence Π of lists of substitution pairs $< (B_1, \sigma_1), ..., (B_p, \sigma_p) >$, such that the tree of sequents $\to \sigma_1(B_1), ..., \sigma_p(B_p)$ is a deduction of $\to \Gamma$ from $\to H'$. During the construction of Π, terms will be deleted from HT, eventually emptying it. Each B_j in a pair (B_j, σ_j) is the σ-matrix of the functional form up to some i of some sentence $A = Q_n x_n...Q_i x_i...Q_1 x_1 C$ in $\to \Gamma$, where σ is one of the original substitutions in the list $< (B_1', \sigma_1'), ..., (B_N', \sigma_N') >$. Each σ_j is a substitution whose support is the set of variables in $\{x_n, ..., x_{i+1}\}$ which are bound by existential quantifiers in A.

The first element of Π is

$$< (B_1', \sigma_1'), ..., (B_N', \sigma_N') > .$$

If the last element of the list Π constructed so far is $< (B_1, \sigma_1), ..., (B_p, \sigma_p) >$ and the corresponding sequent of substitution instances is $\to \Delta$, the next list of substitution pairs is determined as follows:

If Δ differs from Γ and no formula B_j for some pair (B_j, σ_j) in the list $< (B_1, \sigma_1), ..., (B_p, \sigma_p) >$ is the σ-matrix of the functional form up to i of some sentence

$$A = Q_n x_n ... \exists x_{i+1} Q_i x_i ... Q_1 x_1 C$$

occurring in $\to \Gamma$, select the leftmost formula

$$B_j = Q_i x_i ... Q_1 x_1 B[v(f_{i+1}^A(s_1, ..., s_m))/x_{i+1}]$$

which is the σ-matrix of the functional form up to i of some sentence

$$A = Q_n x_n ... \forall x_{i+1} Q_i x_i ... Q_1 x_1 C$$

in $\to \Gamma$, and for which the variable $v(f_{i+1}^A(s_1, ..., s_m))$ corresponds to a maximal term in (the current) HT. Then, apply the $\forall : right$ rule to $v(f_{i+1}^A(s_1, ..., s_m))$, obtaining the sequent $\to \Delta'$ in which

$$\sigma_j(\forall x_{i+1} Q_i x_i ... Q_1 x_1 B)$$

replaces

$$\sigma_j(Q_i x_i ... Q_1 x_1 B[v(f_{i+1}^A(s_1, ..., s_m))/x_{i+1}]).$$

At the end of this step, delete $f_{i+1}^A(s_1, ..., s_m)$ from HT, and perform contractions (and exchanges) if possible. The new list of substitution pairs is obtained by first replacing

$$(B_j, \sigma_j)$$

by

$$(\forall x_{i+1} Q_i x_i ... Q_1 x_1 B, \sigma_j),$$

and performing the contractions and exchanges specified above.

Otherwise, there is a formula

$$B_j = Q_i x_i ... Q_1 x_1 B$$

which is the σ-matrix of the functional form up to i of some sentence

$$A = Q_n x_n ... \exists x_{i+1} Q_i x_i ... Q_1 x_1 C$$

in $\to \Gamma$. Then, apply the $\exists : right$ rule to the leftmost substitution instance

$$\sigma_j(B_j)$$

in $\to \Delta$ of such a formula. The conclusion of this inference is the sequent $\to \Delta'$ in which

$$\sigma_j'(\exists x_{i+1} Q_i x_i ... Q_1 x_1 B)$$

replaces

$$\sigma_j(Q_i x_i ... Q_1 x_1 B),$$

where σ'_j is the restriction of the substitution σ_j obtained by eliminating x_{i+1} from the support of σ_j (Hence, $\sigma'_j(x_{i+1}) = x_{i+1}$). The pair

$$(\exists x_{i+1} Q_i x_i ... Q_1 x_1, \sigma'_j)$$

replaces the pair

$$(B_j, \sigma_j)$$

in the list of substitution pairs. After this step, perform contractions (and exchanges) if possible. Note that in this step, no term is deleted from HT.

Repeat this process until a list of substitution pairs $< (B_1, \sigma_1), ..., (B_p, \sigma_p) >$ is obtained, such that every substitution σ_j has empty support and $\rightarrow B_1, ..., B_p$ is the sequent $\rightarrow \Gamma$.

We claim that Π defines a deduction of $\rightarrow \Gamma$ from $\rightarrow H'$. First, it is easy to see that the sequence Π ends with the sequent $\rightarrow \Gamma$, since we started with substitution instances of matrices of functional forms of sentences in $\rightarrow \Gamma$, and since every step brings some formula in Δ "closer" to the corresponding formula in Γ. We leave the details as an exercise.

To show that the eigenvariable condition is satisfied for every application of the $\forall : right$ rule, we show the following claim by induction on the number of $\forall : right$ steps in Π.

Claim: Just before any application of a $\forall : right$ rule, the set of terms of the form $f_i^A(s_1, ..., s_m)$ such that $v(f_i^A(s_1, ..., s_m))$ occurs (free) in Δ is the current set HT, and for every maximal term $f_i^A(s_1, ..., s_m) \in HT$, the variable $v(f_i^A(s_1, ..., s_m))$ occurs free in at most one formula in Δ of the form

$$Q_{i-1} x_{i-1} ... Q_1 x_1 B[v(f_i^A(s_1, ..., s_m))/x_i].$$

Proof of claim: Just before the first $\forall : right$ step, since all the formulae in $\rightarrow \Delta$ are of the form $\exists x_{i-1} ... \exists x_1 B$, and since the formulae in $\rightarrow H$ are substitution instances of matrices of sentences occurring in the functional form of $\rightarrow \Gamma$, it is clear that the set of terms of the form $f_i^A(s_1, ..., s_m)$ such that $v(f_i^A(s_1, ..., s_m))$ occurs in $\rightarrow \Delta$ is the initial set HT. Since a term $f_i^A(s_1, ..., s_m)$ is maximal in HT if and only if it corresponds to the rightmost occurrence of a universal quantifier in A, $v(f_i^A(s_1, ..., s_m))$ occurs free at most in a single formula

$$\exists x_{i-1} ... \exists x_1 B[v(f_i^A(s_1, ..., s_m))/x_i]$$

(substitution instance of the σ-matrix of the functional form up to $i - 1$ of the sentence $A = Q_n x_n ... \forall x_i \exists x_{i-1} ... \exists x_1 C$ in $\rightarrow \Gamma$). Next, assuming the induction hypothesis, let $\rightarrow \Delta_1$ be the sequent and HT_1 the set of terms just before an application of a $\forall : right$ step, and $\to \Delta_2$ be the sequent and HT_2 the set of terms just before the next $\forall : right$ step. Since the maximal term $f_i^A(s_1, ..., s_m)$ is deleted from HT_1 during the $\forall : right$ step applied to $\to \Delta_1$, and since the following steps until the next $\forall : right$ step are $\exists : right$ rules which do not affect $HT_1 - \{f_i^A(s_1, ..., s_m)\}$,

$$HT_2 = HT_1 - \{f_i^A(s_1, ..., s_m)\}.$$

Since a term $f_i^A(s_1, ..., s_m)$ is maximal in HT_2 if and only if it corresponds to the rightmost occurrence of a universal quantifier in the prefix $Q_n x_n ... Q_{i+1} x_{i+1} \forall x_i$ of the formula

$$A = Q_n x_n ... Q_{i+1} x_{i+1} \forall x_i Q_{i-1} x_{i-1} ... Q_1 x_1 C$$

in $\to \Gamma$, it must correspond to $\forall x_i$. If the variable $v(f_i^A(s_1, ..., s_m))$ occurs free in some other formula $R_{j-1} x_{j-1} ... R_1 x_1 B'$ in $\to \Delta_2$, since $R_j = \forall$, $v(f_i^A(s_1, ..., s_m))$ occurs within a term of the form $f_j^{A'}(s_1', ..., s_q')$, contradicting the maximality of $f_i^A(s_1, ..., s_m)$, since $f_j^{A'}(s_1', ..., s_q')$ is also in HT_2 (as a result of the induction hypothesis). Therefore, $v(f_i^A(s_1, ..., s_m))$ may only occur free in the formula

$$Q_{i-1} x_{i-1} ... Q_1 x_1 B[v(f_i^A(s_1, ..., s_m))/x_i]$$

in $\to \Delta_2$, substitution instance of the σ-matrix of the functional form up to $i - 1$ of the formula

$$A = Q_n x_n ... \forall x_i Q_{i-1} x_{i-1} ... Q_1 x_1 C$$

in $\to \Gamma$. Hence, the eigenvariable condition is satisfied. In a $\exists : step$, since the variables occurring in the term s are distinct from the variables occurring bound in the formulae in Γ, the term s is free for x_i in the substitution, and the inference is valid. Hence, Π yields a deduction of $\to \Gamma$ from $\to H'$, which can be extended to a proof of $\to \Gamma$ from axioms, since $\to H'$ is provable. This concludes the proof of Herbrand's theorem. $\square$

The method for reconstructing a proof from a list of substitution pairs is illustrated in the following example.

EXAMPLE 7.5.3

Consider the sequent $\to \Gamma$ given by:

$$\to \exists x \forall y \neg P(x, y), \exists y_1 \forall z \neg Q(y_1, z), \forall x_1 \exists y_2 \exists z_1 (P(x_1, y_2) \wedge Q(y_2, z_1)),$$

whose functional form is:

$$\to \exists x \neg P(x, f(x)), \exists y_1 \neg Q(y_1, g(y_1)), \exists y_2 \exists z_1 (P(a, y_2) \wedge Q(y_2, z_1)).$$

The provable sequent $\rightarrow H$ given by

$$\rightarrow \neg P(a, f(a)), \neg Q(f(a), g(f(a))), P(a, f(a)) \land Q(f(a), g(f(a)))$$

is obtained from the functional form of the original sequent by deleting quantifiers and substituting the terms a, $f(a)$, $f(a)$ and $g(f(a))$ for x, y_1, y_2, and z_1 respectively. We have $HT = \{a, f(a), g(f(a))\}$, with the ordering $a < f(a)$, $a < g(f(a))$, $f(a) < g(f(a))$.

Define the bijection v' such that $v'(g(f(a))) = u$, $v'(f(a)) = v$ and $v'(a) = w$. The result of replacing in $\rightarrow H$ the maximal terms in HT by the variables given by v' is the sequent $\rightarrow H'$ given by

$$\rightarrow \neg P(w, v), \neg Q(v, u), P(w, v) \land Q(v, u).$$

The formula $P(w, v) \land Q(v, u)$ is the σ-matrix of the functional form up to 0 of the formula $\forall w \exists v \exists u (P(w, v) \land Q(v, u))$. Hence we have a $\exists : right$ step.

$$\frac{\rightarrow \neg P(w, v), \neg Q(v, u), P(w, v) \land Q(v, u)}{\rightarrow \neg P(w, v), \neg Q(v, u), \exists z_1 (P(w, v) \land Q(v, z_1))}$$

Similarly, $\exists z_1 (P(w, v) \land Q(v, z_1))$ is the σ-matrix of the functional form up to 1 of $\forall w \exists v \exists z_1 (P(w, v) \land Q(v, z_1))$. Hence, we have another $\exists : right$ step.

$$\frac{\rightarrow \neg P(w, v), \neg Q(v, u), P(w, v) \land Q(v, u)}{\frac{\rightarrow \neg P(w, v), \neg Q(v, u), \exists z_1 (P(w, y_2) \land Q(y_2, z_1))}{\rightarrow \neg P(w, v), \neg Q(v, u), \exists y_2 \exists z_1 (P(w, y_2) \land Q(y_2, z_1))}}$$

Now, only a $\forall : right$ step can be applied. According to the algorithm, we apply it to the leftmost formula for which the variable $v'(t)$ corresponds to a maximal term $t \in HT$. This must be $\neg Q(v, u)$, since $v'(g(f(a))) = u$ and $g(f(a))$ is the largest element of HT. We also delete $g(f(a))$ from HT.

Note that it would be wrong to apply the $\forall : right$ rule to any of the other formulae, since both w and v would occur free in the conclusion of that inference.

$$\frac{\rightarrow \neg P(w, v), \neg Q(v, u), P(w, v) \land Q(v, u)}{\frac{\rightarrow \neg P(w, v), \neg Q(v, u), \exists z_1 (P(w, y_2) \land Q(y_2, z_1))}{\frac{\rightarrow \neg P(w, v), \neg Q(v, u), \exists y_2 \exists z_1 (P(w, y_2) \land Q(y_2, z_1))}{\rightarrow \neg P(w, v), \forall z \neg Q(v, z), \exists y_2 \exists z_1 (P(w, y_2) \land Q(y_2, z_1))}}}$$

Now, we can apply a $\exists : right$ step to $\forall z \neg Q(v, z)$.

$$\frac{\rightarrow \neg P(w, v), \neg Q(v, u), P(w, v) \wedge Q(v, u)}{\dfrac{\rightarrow \neg P(w, v), \neg Q(v, u), \exists z_1 (P(w, y_2) \wedge Q(y_2, z_1))}{\dfrac{\rightarrow \neg P(w, v), \neg Q(v, u), \exists y_2 \exists z_1 (P(w, y_2) \wedge Q(y_2, z_1))}{\dfrac{\rightarrow \neg P(w, v), \forall z \neg Q(v, z), \exists y_2 \exists z_1 (P(w, y_2) \wedge Q(y_2, z_1))}{\rightarrow \neg P(w, v), \exists y_1 \forall z \neg Q(y_1, z), \exists y_2 \exists z_1 (P(w, y_2) \wedge Q(y_2, z_1))}}}}$$

At this point, a $\forall : right$ step is the only possibility. Since the next largest term in $HT = \{a, f(a)\}$ is $f(a)$, we apply it to $\neg P(w, v)$.

$$\frac{\rightarrow \neg P(w, v), \neg Q(v, u), P(w, v) \wedge Q(v, u)}{\dfrac{\rightarrow \neg P(w, v), \neg Q(v, u), \exists z_1 (P(w, y_2) \wedge Q(y_2, z_1))}{\dfrac{\rightarrow \neg P(w, v), \neg Q(v, u), \exists y_2 \exists z_1 (P(w, y_2) \wedge Q(y_2, z_1))}{\dfrac{\rightarrow \neg P(w, v), \forall z \neg Q(v, z), \exists y_2 \exists z_1 (P(w, y_2) \wedge Q(y_2, z_1))}{\dfrac{\rightarrow \neg P(w, v), \exists y_1 \forall z \neg Q(y_1, z), \exists y_2 \exists z_1 (P(w, y_2) \wedge Q(y_2, z_1))}{\rightarrow \forall y \neg P(w, y), \exists y_1 \forall z \neg Q(y_1, z), \exists y_2 \exists z_1 (P(w, y_2) \wedge Q(y_2, z_1))}}}}}$$

We can now apply a $\exists : right$ step to $\forall y \neg P(w, y)$.

$$\frac{\rightarrow \neg P(w, v), \neg Q(v, u), P(w, v) \wedge Q(v, u)}{\dfrac{\rightarrow \neg P(w, v), \neg Q(v, u), \exists z_1 (P(w, y_2) \wedge Q(y_2, z_1))}{\dfrac{\rightarrow \neg P(w, v), \neg Q(v, u), \exists y_2 \exists z_1 (P(w, y_2) \wedge Q(y_2, z_1))}{\dfrac{\rightarrow \neg P(w, v), \forall z \neg Q(v, z), \exists y_2 \exists z_1 (P(w, y_2) \wedge Q(y_2, z_1))}{\dfrac{\rightarrow \neg P(w, v), \exists y_1 \forall z \neg Q(y_1, z), \exists y_2 \exists z_1 (P(w, y_2) \wedge Q(y_2, z_1))}{\dfrac{\rightarrow \forall y \neg P(w, y), \exists y_1 \forall z \neg Q(y_1, z), \exists y_2 \exists z_1 (P(w, y_2) \wedge Q(y_2, z_1))}{\rightarrow \exists x \forall y \neg P(x, y), \exists y_1 \forall z \neg Q(y_1, z), \exists y_2 \exists z_1 (P(w, y_2) \wedge Q(y_2, z_1))}}}}}}$$

Finally, since $HT = \{a\}$ and only a $\forall : right$ step is possible, a $\forall : right$ step is applied to $\exists y_2 \exists z_1 (P(w, y_2) \wedge Q(y_2, z_1))$.

$$\frac{\rightarrow \neg P(w,v), \neg Q(v,u), P(w,v) \wedge Q(v,u)}{}$$

$$\frac{\rightarrow \neg P(w,v), \neg Q(v,u), \exists z_1(P(w,y_2) \wedge Q(y_2, z_1))}{}$$

$$\frac{\rightarrow \neg P(w,v), \neg Q(v,u), \exists y_2 \exists z_1(P(w,y_2) \wedge Q(y_2, z_1))}{}$$

$$\frac{\rightarrow \neg P(w,v), \forall z \neg Q(v,z), \exists y_2 \exists z_1(P(w,y_2) \wedge Q(y_2, z_1))}{}$$

$$\frac{\rightarrow \neg P(w,v), \exists y_1 \forall z \neg Q(y_1,z), \exists y_2 \exists z_1(P(w,y_2) \wedge Q(y_2, z_1))}{}$$

$$\frac{\rightarrow \forall y \neg P(w,y), \exists y_1 \forall z \neg Q(y_1,z), \exists y_2 \exists z_1(P(w,y_2) \wedge Q(y_2, z_1))}{}$$

$$\frac{\rightarrow \exists x \forall y \neg P(x,y), \exists y_1 \forall z \neg Q(y_1,z), \exists y_2 \exists z_1(P(w,y_2) \wedge Q(y_2, z_1))}{}$$

$$\rightarrow \exists x \forall y \neg P(x,y), \exists y_1 \forall z \neg Q(y_1,z), \forall x_1 \exists y_2 \exists z_1(P(x_1,y_2) \wedge Q(y_2, z_1))$$

This last derivation is a deduction of $\rightarrow \Gamma$ from $\rightarrow H'$. Observe that this proof is identical to the proof of example 7.3.4.

Remarks: (1) The difference between first-order logic without equality and first-order logic with equality noted in the paragraph following the proof of theorem 5.6.1 shows up again in Herbrand's theorem. For a language without equality, in view of the second corollary to theorem 5.5.1, the hard part in finding a proof is to find appropriate substitutions yielding a valid Herbrand disjunction. Indeed, as soon as such a quantifier-free formula is obtained, there is an algorithm for deciding whether it is provable (or valid). However, in view of the remark following the corollary, Church's theorem implies that there is no algorithm for finding these appropriate substitutions.

For languages with equality, the situation is worse! Indeed, even if we can find appropriate substitutions yielding a quantifier-free formula, we are still facing the problem of finding an algorithm for deciding the provability (or validity) of quantifier-free formulae if equality is present. As we mentioned in Chapter 5, there is such an algorithm presented in Chapter 10, but it is nontrivial. Hence, it appears that automatic theorem proving in the presence of equality is harder than automatic theorem proving without equality. This phenomenon will show up again in the resolution method.

(2) Note that the last part of the proof of theorem 7.5.1 provides an algorithm for constructing a proof of $\rightarrow \Gamma$ from the Herbrand disjunction H (really, the list of substitution pairs) and its proof. Similarly, the first part of the proof provides an algorithm for constructing an Herbrand disjunction and its proof, from a proof satisfying the conditions of Gentzen's sharpened Hauptsatz. Actually, since the proof of the sharpened Hauptsatz from Gentzen's cut elimination theorem is entirely constructive, a Herbrand disjunction and its proof can be constructed from a pure-variable, cut-free proof. The only step that has not been justified constructively in our presentation is the fact that a provable sequent has a cut-free proof. This is because even though the *search* procedure yields a cut-free proof of a provable sequent, the correctness and termination of the *search* procedure for provable sequents is established by

semantic means involving a nonconstructive step: the existence of the possibly infinite counter-example tree (considering the case where the sequent is falsifiable). However, Gentzen gave a completely constructive (syntactic) proof of the cut elimination theorem, and so, the version of Herbrand's theorem given in this section is actually entirely constructive, as is Herbrand's original version (Herbrand, 1971). See also lemma 7.6.2.

As mentioned at the beginning of this chapter, there is a theorem similar in form to Herbrand's theorem and known as the Skolem-Herbrand-Gödel theorem. Since a version of that theorem will be proved in Section 7.6, we postpone a discussion of the relationship between the two theorems to the end of Section 7.6.

PROBLEMS

7.5.1. Prove the following fact: Given a sequent $A_1, ..., A_m \rightarrow B_1, ..., B_n$, $A_1, ..., A_m \rightarrow B_1, ..., B_n$ is provable (in LK) if and only if $\rightarrow \neg A_1 \vee ... \vee \neg A_m \vee B_1 \vee ... \vee B_n$ is provable (in LK).

7.5.2. The method given in Section 7.2 for converting a formula to prenex form used in conjunction with the Skolemization method of Section 7.5 tends to create Skolem functions with more arguments than necessary.

(a) Prove that the following method for Skolemizing is correct:

Step 1: Eliminate redundant quantifiers; that is, quantifiers $\forall x$ or $\exists x$ such that the input formula contains a subformula of the form $\forall x B$ or $\exists x B$ in which x does not occur in B.

Step 2: Rectify the formula.

Step 3: Eliminate the connectives $\supset$ and $\equiv$.

Step 4: Convert to NNF.

Step 5: Push quantifiers to the right. By this, we mean: Replace

$$\exists x(A \vee B) \text{ by } \begin{cases} A \vee \exists x B & \text{if } x \text{ is not free in } A, \\ \exists x A \vee B & \text{if } x \text{ is not free in } B. \end{cases}$$

$$\forall x(A \vee B) \text{ by } \begin{cases} A \vee \forall x B & \text{if } x \text{ is not free in } A, \\ \forall x A \vee B & \text{if } x \text{ is not free in } B. \end{cases}$$

$$\exists x(A \wedge B) \text{ by } \begin{cases} A \wedge \exists x B & \text{if } x \text{ is not free in } A, \\ \exists x A \wedge B & \text{if } x \text{ is not free in } B. \end{cases}$$

$$\forall x(A \wedge B) \text{ by } \begin{cases} A \wedge \forall x B & \text{if } x \text{ is not free in } A, \\ \forall x A \wedge B & \text{if } x \text{ is not free in } B. \end{cases}$$

Step 6: Eliminate universal quantifiers using Skolem function and constant symbols.

Step 7: Move existential quantifiers to the left, using the inverse of the transformation of step 5.

(b) Compare the first method and the method of this problem for the formula

$$\forall x_2 \exists y_1 \forall x_1 \exists y_2 (P(x_1, y_1) \wedge Q(x_2, y_2)).$$

Note: Step 5 is the step that reduces the number of arguments of Skolem functions.

7.5.3. Prove that the following formulae are valid using Herbrand's theorem:

$$\neg(\exists x \forall y P(x, y) \wedge \forall y \exists x P(y, x))$$
$$\neg(\neg(\forall x P(x) \vee \exists y \neg Q(y)) \vee (\forall z P(z) \vee \exists w \neg Q(w)))$$
$$\neg(\neg \forall x (P(x) \vee \exists y \neg Q(y)) \vee (\forall z P(z) \vee \exists w \neg Q(w)))$$

7.5.4. Give an example in which $A[s_1/x_1][s_2/x_2]...[s_n/x_n]$, the result of substituting s_1 for x_1, ... ,s_n for x_n (as defined in definition 5.2.6) in that order, is different from $A[s_1/x_s, ..., s_n/x_n]$.

Show that if none of the variables in the support of the substitution σ occurs in the terms $s_1,...,s_n$, the order in which the substitutions are performed is irrelevant, and in this case,

$$A[s_1/x_1][s_2/x_2]...[s_n/x_n] = A[s_1/x_1, ..., s_n/x_n].$$

7.5.5. Fill in the missing details in the proof of theorem 7.5.1.

7.5.6. Consider the following formula given by

$$\neg \exists y \forall z (P(z, y) \equiv \neg \exists x (P(z, x) \wedge P(x, z))).$$

(a) Prove that the above formula is equivalent to the following prenex formula A:

$$\forall y \exists z \forall u \exists x [(P(z, y) \wedge P(z, x) \wedge P(x, z)) \vee$$
$$(\neg P(z, y) \wedge (\neg P(z, u) \vee \neg P(u, z)))].$$

(b) Show that A can be Skolemized to the formula

$$B = \exists z \exists x [(P(z, a) \wedge P(z, x) \wedge P(x, z)) \vee$$
$$(\neg P(z, a) \wedge (\neg P(z, f(z)) \vee \neg P(f(z), z)))],$$

and that the formula C given by

$$[(P(a, a) \wedge P(a, f(a)) \wedge P(f(a), a)) \vee$$
$$(\neg P(a, a) \wedge (\neg P(a, f(a)) \vee \neg P(f(a), a)))]$$

is valid.

(c) Using the method of theorem 7.5.1, reconstruct a proof of A from the valid Herbrand disjunction C.

7.5.7. Consider a first-order language without equality. Show that Herbrand's theorem provides an algorithm for deciding the validity of prenex sentences of the form

$$\forall x_1...\forall x_m \exists y_1...\exists y_n B.$$

7.5.8. Write a computer program implementing the method given in the proof of theorem 7.5.1 for reconstructing a proof from a Herbrand's disjunction.

7.6 Skolem-Herbrand-Gödel's Theorem for Formulae in NNF

In this section, we shall state a version of the Skolem-Herbrand-Gödel theorem for unsatisfiability as opposed to validity.

7.6.1 Skolem-Herbrand-Gödel's Theorem in Unsatisfiability Form

Using the results of Section 7.4, we shall derive a version of the Herbrand-Skolem-Gödel theorem for formulae in NNF due to Andrews (Andrews, 1981). Actually, we shall prove more. We shall also give half of a version of Herbrand's theorem for sentences in NNF, the part which states that if a sequent $A \rightarrow$ is provable, then a quantifier-free formula C whose negation $\neg C$ is provable can be effectively constructed.

We believe that it is possible to give a constructive Herbrand-like version of this theorem similar to theorem 7.5.1, but the technical details of the proof of the converse of the theorem appear to be very involved. Hence we shall use a mixed strategy: Part of the proof will be obtained constructively from theorem 7.4.1, the other part by a semantic argument showing that a sentence is satisfiable iff its Skolem form is satisfiable. This last result is also interesting in its own right, and can be used to prove other results, such as the compactness theorem, and the Löwenheim-Skolem theorem (see the problems).

Since a formula A is valid iff $\neg A$ is unsatisfiable, any unsatisfiability version of the Skolem-Herbrand-Gödel theorem yields a validity version of the theorem, and vice versa. Since one of the most important applications of the Skolem-Herbrand-Gödel theorem is the completeness of refutation-oriented procedures such as the resolution method (to be presented in Chapter 8) and the method of matings (Andrews, 1981), it will be useful for the reader to see a treatment of this theorem for unsatisfiablity.

As discussed in Section 7.5, since our goal is now to prove that a formula A is valid by showing that $\neg A$ is unsatisfiable, we are going to use the dual of the method used in Section 7.5, that is, eliminate *existential quantifiers* using Skolem functions. However, it is not quite as simple to define the conversion of a formula in NNF to Skolem normal form (satisfiability functional form) as it is to convert a formula in prenex form into (validity) functional form. We present an example first.

EXAMPLE 7.6.1

Consider the formula

$$A = \exists x((P(x) \vee \exists y R(x,y)) \supset (\exists z R(x,z) \vee P(a))).$$

The NNF of its negation is

$$B = \forall x((P(x) \vee \exists y R(x,y)) \wedge (\forall z \neg R(x,z) \wedge \neg P(a))).$$

The following is a $(G2^{nnf})$ proof in normal form of the sequent $B \rightarrow$:

$$
\cfrac{
  \cfrac{
    \cfrac{P(a), \neg P(a) \rightarrow}{P(a), \neg R(a,y), \neg P(a) \rightarrow}
    \qquad
    \cfrac{R(a,y), \neg R(a,y) \rightarrow}{R(a,y), \neg R(a,y), \neg P(a) \rightarrow}
  }{
    \cfrac{
      \cfrac{(P(a) \vee R(a,y)), \neg R(a,y), \neg P(a) \rightarrow}{(P(a) \vee R(a,y)), (\neg R(a,y) \wedge \neg P(a)) \rightarrow}
    }{
      \cfrac{
        \cfrac{(P(a) \vee R(a,y)) \wedge (\neg R(a,y) \wedge \neg P(a)) \rightarrow}{(P(a) \vee R(a,y)) \wedge (\forall z \neg R(a,z) \wedge \neg P(a)) \rightarrow}
      }{
        \cfrac{(P(a) \vee \exists y R(a,y)) \wedge (\forall z \neg R(a,z) \wedge \neg P(a)) \rightarrow}{\forall x((P(x) \vee \exists y R(x,y)) \wedge (\forall z \neg R(x,z) \wedge \neg P(a))) \rightarrow}
      }
    }
  }
}{}
$$

(In order to shorten the proof, the $\wedge : left$ rule of G was used rather than the $\wedge : left$ rule of $G2^{nnf}$. We leave it as an exercise to make the necessary alterations to obtain a pure $G2^{nnf}$-proof.) The midsequent is

$$(P(a) \vee R(a,y)) \wedge (\neg R(a,y) \wedge \neg P(a)) \quad \rightarrow .$$

The existential quantifier can be eliminated by introducing the unary Skolem function symbol f, and we have the following sequent:

$$(*) \qquad \forall x((P(a) \vee R(x, f(x))) \wedge (\forall z \neg R(x,z) \wedge \neg P(a))) \rightarrow$$

If in the above proof we replace all occurrences of the eigenvariable y by $f(a)$, we obtain a proof of the sequent $(*)$ whose midsequent is:

$$(P(a) \vee R(a, f(a))) \wedge (\neg R(a, f(a)) \wedge \neg P(a)) \rightarrow \; .$$

This illustates the Skolem-Herbrand-Gödel's theorem stated in unsatisfiability form: A formula B is unsatisfiable iff some special kind of quantifier-free substitution instance of B is unsatisfiable.

Such instances are called *compound instances* by Andrews (Andrews, 1981). We shall now define precisely all the concepts mentioned in the above example.

7.6.2 Skolem Normal Form

We begin with the notion of universal scope of a subformula.

Definition 7.6.1 Given a (rectified) formula A in NNF, the set $US(A)$ of pairs $< B, L >$ where B is a subformula of A and L is a sequence of variables is defined inductively as follows:

$$US_0 = \{< A, <>>\};$$
$$US_{k+1} = US_k \cup \{< C, L >, < D, L > \; | \; < B, L >\in US_k,$$
$$B \text{ is of the form } (C \wedge D) \text{ or } (C \vee D)\}$$
$$\cup \{< C, L > \; | \; < \exists x C, L >\in US_k\}$$
$$\cup \{< C, < y_1, ..., y_m, x >> \; | \; < \forall x C, < y_1, ..., y_m >>\in US_k\}.$$

For every subformula B of A, the sequence L of variables such that $< B, L >$ belongs to $US(A) = \bigcup US_k$ is the *universal scope* of B.

In the process of introducing Skolem symbols to eliminate existential quantifiers, we shall consider the subset of US consisting of the pairs $< \exists x B, L >$, where $\exists x B$ is a subformula of A.

EXAMPLE 7.6.2

Let

$$A = \forall x (P(a) \vee \exists y (Q(y) \wedge \forall z (P(y, z) \vee \exists u Q(x, u)))) \vee \exists w Q(a, w).$$

Then,

$$< \exists y (Q(y) \wedge \forall z (P(y, z) \vee \exists u Q(x, u))), < x >>,$$
$$< \exists u Q(x, u), < x, z >> \quad \text{and}$$
$$< \exists w Q(a, w), <>>$$

define the universal scope of the subformulae of A of the form $\exists x B$. We now define the process of Skolemization.

Definition 7.6.2 Given a rectified sentence A, the *Skolem form* $SK(A)$ of A (or *Skolem normal form*) is defined recursively as follows using the set $US(A)$. Let A' be any subformula of A:

(i) If A' is either an atomic formula B or the negation $\neg B$ of an atomic formula B, then

$$SK(A') = A'.$$

(ii) If A' is of the $(B * C)$, where $* \in \{\vee, \wedge\}$, then

$$SK(A') = (SK(B) * SK(C)).$$

(iii) If A' is of the form $\forall x B$, then

$$SK(A') = \forall x SK(B).$$

(iv) If A' is of the form $\exists x B$, then if $< y_1, ..., y_m >$ is the universal scope of $\exists x B$ (that is, the sequence of variables such that $< \exists x B, < y_1, ..., y_m >>\in US(A)$) then

(a) If $m > 0$, create a new *Skolem function symbol* $f_{A'}$ of rank m and let

$$SK(A') = SK(B[f_{A'}(y_1, ..., y_m)/x]).$$

(b) If $m = 0$, create a new *Skolem constant symbol* $f_{A'}$ and let

$$SK(A') = SK(B[f_{A'}/x]).$$

Observe that since the sentence A is rectified, all subformulae A' are distinct, and since the Skolem symbols are indexed by the subformulae A', they are also distinct.

EXAMPLE 7.6.3

Let

$$A = \forall x(P(a) \vee \exists y(Q(y) \wedge \forall z(P(y, z) \vee \exists u Q(x, u)))) \vee \exists w Q(a, w),$$

as in example 7.6.2.

$SK(\exists w Q(a, w)) = Q(a, c),$
$SK(\exists u Q(x, u)) = Q(x, f(x, z)),$
$SK(\exists y(Q(y) \wedge \forall z(P(y, z) \vee \exists u Q(x, u)))) =$
$$(Q(g(x)) \wedge \forall z(P(g(x), z) \vee Q(x, f(x, z)))), \quad \text{and}$$
$SK(A) =$
$$\forall x(P(a) \vee (Q(g(x)) \wedge \forall z(P(g(x), z) \vee Q(x, f(x, z))))) \vee Q(a, c).$$

The symbol c is a Skolem constant symbol, g is a unary Skolem function symbol, and f is a binary Skolem function symbol.

7.6.3 Compound Instances

At the end of example 7.6.1, we mentioned that the midsequent of a proof in normal form of a sequent $B \to$ where B is in Skolem normal form consists of certain formulae called compound instances. The formal definition is given below.

Definition 7.6.3 Let A be a rectified sentence and let B its Skolem form. The set of *compound instances* (for short, *c-instances*) of B is defined inductively as follows:

(i) If B is either an atomic formula C or the negation $\neg C$ of an atomic formula, then B is its only c-instance;

(ii) If B is of the form $(C * D)$, where $* \in \{\vee, \wedge\}$, for any c-instance H of C and c-instance K of D, $(H * K)$ is a c-instance of B;

(iii) If B is of the form $\forall x C$, for any k closed terms $t_1,...,t_k$, if H_i is a c-instance of $C[t_i/x]$ for $i = 1, ..., k$, then $H_1 \wedge ... \wedge H_k$ is a c-instance of B.

EXAMPLE 7.6.4

Let

$$B = \forall x(P(x) \vee \forall y Q(y, f(x))) \wedge (\neg P(a) \wedge (\neg Q(a, f(a)) \vee \neg Q(b, f(a)))).$$

Then,

$$(P(a) \vee (Q(a, f(a)) \wedge Q(b, f(a)))) \wedge (\neg P(a) \wedge (\neg Q(a, f(a)) \vee \neg Q(b, f(a))))$$

is a c-instance of B.

Note that c-instances are quantifier free. The following lemma shows that in a certain sense, c-instances are closed under conjunctions.

Lemma 7.6.1 Let A be a sentence in NNF and in Skolem normal form. For any two c-instances K and L of A, a c-instance D such that $D \supset (K \wedge L)$ is provable can be constructed.

Proof: We proceed by induction on A.

(i) If A is either of the form B or $\neg B$ for an atomic formula B, then $K = L = A$ and we let $D = A$.

(ii) If A is of the form $(B * C)$, where $* \in \{\vee, \wedge\}$, then K is of the form $(B_1 * C_1)$ and L is of the form $(B_2 * C_2)$, where B_1 and B_2 are c-instances of B, and C_1 and C_2 are c-instances of C. By the induction hypothesis, there are c-instances D_1 of B and D_2 of C such that $D_1 \supset (B_1 * B_2)$ and

$D_2 \supset (C_1 * C_2)$ are provable. But then, $D = (D_1 * D_2)$ is a c-instance of A such that $D \supset (K * L)$ is provable.

(iii) If A is of the form $\forall x C$, then K has the form $H_1 \wedge ... \wedge H_m$, where H_i is a c-instance of $C[s_i/x]$ for $i = 1, ..., m$, and L has the form $K_1 \wedge ... \wedge K_n$, where K_j is a c-instance of $C[t_j/x]$ for $j = 1, ..., n$. It is clear that $D = K \wedge L$ satisfies the lemma. $\square$

We are now ready for the constructive part of Herbrand's theorem

7.6.4 Half of a Herbrand-like Theorem for Sentences in NNF

In the rest of this section, it is assumed that first-order languages have at least one constant symbol. This can always be achieved by adjoining the special symbol # as a constant. First, we prove the constructive half of Herbrand's theorem announced in the introduction to this section. This part asserts a kind of completeness result: If a sequent $A \to$ is provable, then this can be demonstrated constructively by providing a (propositional) proof of the negation $\neg C$ of a quantifier-free formula C obtained from A.

Lemma 7.6.2 Let **L** be a first-order language with or without equality. Let A be an **L**-sentence in NNF, and let B be its Skolem normal form. If $A \to$ is provable (in $G1^{nnf}$ or $G1^{nnf}_=$), then a (quantifier-free) c-instance C of B can be constructed such that $\neg C$ is provable.

Proof: From theorems 7.4.1 and 7.4.2, if $A \to$ is provable, it has a proof in normal form, in which all quantifier rules are below all propositional rules. Let us now perform the following alteration to the proof:

In a bottom-up fashion, starting from $B \to$, whenever the $\exists : left$ rule is applied to a subformula A' of the form $\exists x B$ with eigenvariable z, if the universal scope of $\exists x B$ is $< y_1, ..., y_m >$ and the terms $t_1, ..., t_m$ have been substituted for $y_1, ..., y_m$ in previous $\forall : right$ steps, substitute $f_{A'}(t_1, ..., t_m)$ for all occurrence of z in the proof. If the midsequent of the resulting tree still has variables, substitute any constant for all occurrences of these variables.

It is not difficult to prove by induction on proof trees that the resulting deduction is a proof of the sequent $B \to$, where B is the Skolem form of A. This is left as an (easy) exercise.

We can also prove that the midsequent consists of c-instances of B. For this, we prove the following claim by induction on proof trees:

Claim: For every proof in normal form of a sequent $\Gamma \to$ consisting of Skolem forms of sentences, the formulae in the midsequent are c-instances of the sentences in Γ.

Proof of claim: We have three cases depending on the sentence B to which the quantifier rule is applied.

The result is trivial if B is either an atomic formula or the negation of an atomic formula.

If B is of the form $(C * D)$, where $* \in \{\wedge, \vee\}$, assume without loss of generality that the $\forall : left$ rule is applied to C. Hence, in the premise of the rule, $(C * D)$ is replaced by $(C[F[t/x]])/E] * D)$, where E is a maximal sub-formula of C of the form $\forall x F$. By the induction hypothesis, the midsequent consists of c-instances of $(C[F[t/x]])/E] * D)$ and of the other formulae in Γ. However, one can easily show by induction on the formula C that a c-instance of $(C[F[t/x]]/E] * D)$ is also a c-instance of $(C * D)$. This is left as an exercise to the reader. Hence, the result holds.

If B is of the form $\forall x C$, then in the premise of the rule, B is replaced by $C[t/x]$ for some closed term t. By the induction hypothesis, the midsequent consists of c-instances of $C[t/x]$ and of the other formulae in Γ. By definition, a c-instance of $C[t/x]$ is a c-instance of B. Hence, the midsequent consists of c-instances of the sentences in Γ. $\square$

If $C_1,...,C_m$ are the c-instances of B occurring in the midsequent, since $C_1, ..., C_m \rightarrow$ is provable, $\neg(C_1 \wedge ... \wedge C_m)$ is provable. Applying lemma 7.6.1 $m - 1$ times, a c-instance C of B can be constructed such that $C \supset (C_1 \wedge ... \wedge C_m)$ is provable. But then, $\neg C$ is provable since $\neg(C_1 \wedge ... \wedge C_m)$ is provable. $\square$

Remark: In Andrews, 1981, a semantic proof of lemma 7.6.2 is given for languages *without equality*. Our result applies to languages with equality as well, and is constructive, because we are using the full strength of the normal form theorems (theorems 7.4.1 and 7.4.2).

7.6.5 Skolem-Herbrand-Gödel's Theorem (Sentences in NNF)

In order to prove the converse of the above lemma, we could as in the proof of theorem 7.5.1 try to reconstruct a proof from an unsatisfiable c-instance C of B. This is a rather delicate process whose justification is very tedious, and we will follow a less constructive but simpler approach involving semantic arguments. Hence, instead of proving the converse of the half of Herbrand's theorem shown in lemma 7.6.2, we shall prove a version of the Skolem-Herbrand-Gödel theorem.

The following fact will be used:

$(*)$ If the Skolem form B of a sentence A is satisfiable then A is satisfiable.

Since it is easy to prove that if C is a c-instance of B then $(B \supset C)$ is valid, if C is unsatisfiable, then B must be unsatisfiable, and by $(*)$ A is also unsatisfiable.

We shall actually prove not only $(*)$ but also its converse. This is more than we need for the part of the proof of the Skolem-Herbrand-Gödel theorem, but since this result can be used to give a semantic proof of the Skolem-Herbrand-Gödel theorem (see the problems), it is interesting to prove it in full.

Lemma 7.6.3 Let **L** be a first-order language with or without equality. Let A be a rectified **L**-sentence in NNF, and let B be its Skolem normal form. The sentence A is satisfiable iff its Skolem form B is satisfiable.

Proof: Let C be any subformula of A. We will show that the following properties hold:

(a) For every structure **A** such that all function, predicate, and constant symbols in the Skolem form $SK(C)$ of C receive an interpretation, for every assignment s, if $\mathbf{A} \models SK(C)[s]$ then $\mathbf{A} \models C[s]$.

(b) For every structure **A** such that exactly all function, predicate, and constant symbols in C receive an interpretation, for every assignment s (with range A), if $\mathbf{A} \models C[s]$ then there is an expansion **B** of **A** such that $\mathbf{B} \models SK(C)[s]$.

Recall from definition 5.4.6 that if **B** is an expansion of **A**, then **A** and **B** have the same domain, and the interpretation function of **A** is a restriction of the interpretation function of **B**.

The proof proceeds by induction on subformulae of A.

(i) If C is either of the form D or $\neg D$ where D is atomic, $SK(C) = C$ and both (a) and (b) are trivial.

(ii) If C is of the form $(D \wedge E)$, then $SK(C) = (SK(D) \wedge SK(E))$.

(a) If $\mathbf{A} \models SK(C)[s]$ then $\mathbf{A} \models SK(D)[s]$ and $\mathbf{A} \models SK(E)[s]$. By the induction hypothesis, $\mathbf{A} \models D[s]$ and $\mathbf{A} \models E[s]$. But then, $\mathbf{A} \models C[s]$.

(b) Let **A** be a structure such that exactly all function, predicate, and constant symbols in C receive an interpretation. Since $\mathbf{A} \models C[s]$, we have $\mathbf{A} \models D[s]$ and $\mathbf{A} \models E[s]$. By the induction hypothesis, there are expansions $\mathbf{B}_1$ and $\mathbf{B}_2$ of **A** such that $\mathbf{B}_1 \models SK(D)[s]$ and $\mathbf{B}_2 \models SK(E)[s]$. Since $\mathbf{B}_1$ and $\mathbf{B}_2$ are both expansions of **A**, their interpretation functions agree on all the predicate, function, and constant symbols occurring in both D and E, and since the sets of Skolem symbols in D and E are disjoint (because A is rectified), we can take the union of the two interpretation functions to obtain an expansion **B** of **A** such that $\mathbf{B} \models (SK(D) \wedge SK(E))[s]$, that is, $\mathbf{B} \models SK(C)[s]$.

(iii) C is of the form $(D \vee E)$. This case is similar to case (ii) and is left as an exercise.

(iv) C is of the form $\forall x D$. Then $SK(C) = \forall x SK(D)$.

(a) If $\mathbf{A} \models SK(C)[s]$, then $\mathbf{A} \models SK(D)[s[x := a]]$ for all $a \in A$. By the induction hypothesis, $\mathbf{A} \models D[s[x := a]]$ for all $a \in A$, that is, $\mathbf{A} \models C[s]$.

(b) If $\mathbf{A} \models C[s]$, then $\mathbf{A} \models D[s[x := a]]$ for all $a \in A$. By the induction hypothesis, there is an expansion $\mathbf{B}$ of $\mathbf{A}$ such that for any assignment s', if $\mathbf{A} \models D[s']$ then $\mathbf{B} \models SK(D)[s']$. But then, for all $a \in A$, we have $\mathbf{B} \models SK(D)[s[x := a]]$, that is, $\mathbf{B} \models SK(C)[s]$.

(v) C is of the form $\exists x D$. Then $SK(C) = SK(D)[f_C(y_1, ..., y_m)/x]$, where $< y_1, ..., y_m >$ is the universal scope of C in A, and f_C is the Skolem symbol associated with C.

(a) If $\mathbf{A} \models SK(C)$, then by lemma 5.4.1, letting

$$a = (f_C(y_1, ..., y_m))_\mathbf{A}[s],$$

we have $\mathbf{A} \models (SK(D)[\mathbf{a}/x])[s]$, which by lemma 5.3.1 is equivalent to $\mathbf{A} \models (SK(D))[s[y := a]]$. By the induction hypothesis, $\mathbf{A} \models D[s[y := a]]$, that is, $\mathbf{A} \models (\exists x D)[s]$.

(b) Assume that $\mathbf{A} \models C[s]$ for every s. Then, for some $a \in A$, $\mathbf{A} \models D[s[x := a]]$. By the induction hypothesis, there is an expansion $\mathbf{B}$ of $\mathbf{A}$ such that $\mathbf{B} \models (SK(D))[s[x := a]]$. Observe that by the recursive definition of the Skolem form of a formula, $FV(SK(D)) = \{x, y_1, ..., y_m\}$, where $< y_1, ..., y_m >$ is the universal scope of C in A. In order to expand $\mathbf{B}$ to a structure for $SK(C)$, we need to interpret f_C in $\mathbf{B}$. For any $(a_1, ..., a_m) \in A^m$, let s be any assignment such that $s(y_i) = a_i$, for $i = 1, .., m$. By the induction hypothesis, there is some $a \in A$ such that

$$\mathbf{B} \models (SK(D))[s[x := a]].$$

Define the value of $f_C(a_1, ..., a_m)$ in $\mathbf{B}$ as any chosen $a \in A$ such that

$$(*) \qquad \mathbf{B} \models (SK(D))[s[x := a]].$$

Since the only free variables in $SK(D)$ are $\{x, y_1, ..., y_m\}$, by lemma 5.3.3, this definition only depends on the values of $y_1, ..., y_m$. Given the interpretation for f_C given in $(*)$, for any assignment s, for $a = (f_C(y_1, ..., y_m))_\mathbf{B}[s]$, we have $\mathbf{B} \models (SK(D))[s[x := a]]$. By lemma 5.3.1 and lemma 5.4.1, we have

$$\mathbf{B} \models (SK(D))[s[x := a]] \quad \text{iff}$$
$$\mathbf{B} \models (SK(D)[\mathbf{a}/x])[s] \quad \text{iff}$$
$$\mathbf{B} \models (SK(D)[f_C(y_1, ..., y_m)/x])[s].$$

Hence, $\mathbf{B} \models SK(C)[s]$, as desired. $\square$

We are now ready to prove the following version of the Skolem-Herbrand-Gödel theorem extending Andrews's theorem (Andrews, 1981) to first-order languages with equality.

Theorem 7.6.1 (Skolem-Herbrand-Gödel theorem, after Andrews) Let **L** be a first-order language with or without equality. Given any rectified sentence A, if B is the Skolem form of A, then the following holds:

(a) A is unsatisfiable if and only if some compound instance C of B is unsatisfiable.

(b) Given a $(G1^{nnf}$ or $G1_{=}^{nnf})$ proof of the sequent $A \rightarrow$, a compound instance C of B such that $\neg C$ is provable can be effectively constructed from the proof of $A \rightarrow$.

Proof: Part (b) is lemma 7.6.2. Part (a) is proved as follows.

If A is unsatisfiable, then $\neg A$ is valid and by the completeness theorem, $\neg A$ is provable. Hence, $A \rightarrow$ is provable, and by lemma 7.6.2, a compound instance C of B such that $\neg C$ is provable can be effectively constructed. By soundness, $\neg C$ is valid, and so C is unsatisfiable.

Conversely, assume that some compound instance C of B is unsatisfiable. We prove that for any compound instance C of B, $(B \supset C)$ is valid. This is shown by induction on B.

If B is either an atomic formula or the negation of an atomic formula, $C = B$ and $(B \supset C)$ is valid.

If B is of the form $(D * E)$, where $* \in \{\lor, \land\}$, then C is of the form $(K * L)$ where K is a compound instance of D and L is a compound instance of D. By the induction hypothesis, both $(D \supset K)$ and $(E \supset L)$ are valid. But then, $(D * E) \supset (K * L)$ is valid.

If B is of the form $\forall x D$, then C is of the form $H_1 \land ... \land H_k$, where H_i is a compound instance of $D[t_i/x]$, for some closed terms t_i, $i = 1, ..., k$. By the induction hypothesis, $(D[t_i/x] \supset H_i)$ is valid for $i = 1, ..., k$. But $(\forall x D \supset D[t/x])$ is valid for every closed term t (in fact for every term t free for x in D), as shown in the proof of lemma 5.4.2. Therefore, $(B \supset H_i)$ is valid for $i = 1, ..., k$, which implies that $(B \supset (H_1 \land ... \land H_k))$ is valid.

This concludes the proof that $(B \supset C)$ is valid. $\square$

Now, since C is unsatisfiable and $(B \supset C)$ is valid, B is also unsatisfiable. By lemma 7.6.3 (part (a)), if A is satisfiable then B is satisfiable. Since B is unsatisfiable, A must be unsatisfiable. $\square$

If the sentence A in NNF is also prenex, observe that a compound instance C of the Skolem form B of A is in fact a conjunction of ground substitution instances of the matrix of B. Hence, we ob·ain the following useful corollary.

Corollary If A is a prenex sentence in NNF, A is unsatisfiable if and only if a finite conjunction of ground substitution instances of the matrix of the Skolem form B of A is unsatisfiable. $\square$

Remarks: (1) The first remark given at the end of the proof of theorem 7.5.1 regarding the difference between logic without equality and logic with equality also applies here. There is an algorithm for deciding whether a *c*-instance is unsatisfiable if equality is absent, but in case equality is present, such an algorithm is much less trivial. For details, see Chapter 10.

(2) The version of theorem 7.6.1(a) for languages without equality is due to Andrews (Andrews, 1981). Andrews's proof uses semantic arguments and assumes the result of lemma 7.6.3. Instead of using lemma 7.6.2, assuming that every compound instance of B is satisfiable, Andrews constructs a model of B using the compactness theorem. His proof is more concise than ours, but there is more to the theorem, as revealed by part (b). Indeed, there is a constructive aspect to this theorem reflected in the part of our proof using lemma 7.6.2, which is not brought to light by Andrews's semantic method.

Actually, we have not pushed the constructive approach as far as we could, since we did not show how a proof of $A \rightarrow$ can be reconstructed from a proof of a compound instance $C \rightarrow$. This last construction appears to be feasible, but we have not worked out the technical details, which seem very tedious. Instead, we have proved a version of the Skolem-Herbrand-Gödel theorem using the easy semantic argument that consists of showing that $(B \supset C)$ is valid for every compound instance of B, and part (a) of lemma 7.6.3.

7.6.6 Comparison of Herbrand and Skolem-Herbrand-Gödel Theorems

We now wish to discuss briefly the differences between Herbrand's original theorem and the Skolem-Herbrand-Gödel theorem.

First, Herbrand's theorem deals with provability whereas the Skolem-Herbrand-Gödel deals with unsatisfiability (or validity). Herbrand's theorem is also a deeper result, whose proof is harder, but it yields more information. Roughly speaking, Herbrand's theorem asserts the following:

Herbrand's original theorem:

(1) If a formula A is provable in a formal system Q_H defined by Herbrand, then a Herbrand disjunction H and its proof can be obtained constructively via primitive recursive functions from the proof of A.

(2) From a Herbrand disjunction H and its proof, a proof of A in Q_H can be obtained constructively via a primitive recursive function.

The concept of a primitive recursive function is covered in some detail in Section 7.7, and in the following discussion, we shall content ourselves with an informal definition. Roughly speaking, a primitive recursive function is a function over the natural numbers whose rate of growth is reasonably well behaved. The rate of growth of each primitive recursive function is uniformly bounded, in the sense that no primitive recursive function can grow faster than a certain given function (which itself is not primitive recursive). The class of primitive recursive function contains some simple functions, called the base functions, and is closed under two operations: composition and primitive recursion (see Section 7.7). Primitive recursion is a certain constrained type of recursion, and this is why the rate of growth of the primitive recursive functions is not arbitrary.

The fact that in Herbrand's theorem proofs can be obtained constructively and with reasonable complexity via primitive recursive functions, is a very essential part of the theorem. This last point is well illustrated by the history of the theorem, discussed extensively by Goldfarb, in Herbrand, 1971.

Herbrand's original version of the theorem was sometimes difficult to follow and its proof contained errors. As a matter of fact, Herbrand's original statement of the theorem did not refer to the concept of a primitive recursive function. Denton and Dreben were able to repair the defective proofs, and they realized the fact that the constructions are primitive recursive (see Note G and and Note H, in Herbrand, 1971). In his thesis, Herbrand mistakingly claimed simpler functions. It is also interesting to know that Herbrand did not accept the concept of validity because it is an infinitistic concept, and that this is the reason he gave an argument that is entirely proof-theoretic. However, Herbrand had an intuitive sense of the semantic contents of his theorem. As mentioned by Goldfarb in his introduction to Herbrand, 1971:

"Herbrand intends the notion of expansion to furnish more, namely a finitistic surrogate for the model-theoretic notion of infinite satisfiability."

We close this discussion with a final remark showing the central role occupied by Herbrand's theorem. First, observe that it is possible to prove the Skolem-Herbrand-Gödel theorem without appealing to the completeness theorem (see problem 7.6.11). Then, the following hold:

(1) Herbrand's theorem together with the Skolem-Herbrand-Gödel theorem implies the completeness theorem.

(2) Herbrand's theorem together with the completeness theorem implies the Skolem-Herbrand-Gödel theorem (see problem 7.6.15).

Of course, such proofs are a bit of an overkill, but we are merely illustrating the depth of Herbrand's theorem.

The version of Herbrand's theorem that we have presented in theorem 7.5.1 (and in the part in lemma 7.6.2) has the constructive nature of Herbrand's original theorem. What has not been shown is the primitive recursive nature of the functions yielding on the one hand the Herbrand disjunction H and its proof from the prenex sequent $\rightarrow \Gamma$, and on the other hand the proof of $\rightarrow \Gamma$ from a proof of the Herbrand disjunction H. However, we have shown the recursive nature of these functions. In view of the above discussion regarding Herbrand's original version of the theorem, it would be surprising if these functions were not primitive recursive.

For details on Herbrand's original theorem, the interested reader is referred to Herbrand, 1971; Van Heijenoort, 1967; and Joyner's Ph.D thesis (*Automatic theorem Proving and The Decision Problem*, Ph.D thesis, W. H. Joyner, Harvard University, 1974), which contains a Herbrand-like theorem for the resolution method.

PROBLEMS

7.6.1. Show that the Skolem form of the negation of the formula

$$A = \exists x \forall y [(P(x) \equiv P(y)) \supset (\exists x P(x) \equiv \forall y P(y))]$$

is the formula

$$C = \forall y [(\neg P(c) \vee P(y)) \wedge (\neg P(y) \vee P(c))] \wedge$$
$$[(P(d) \wedge \neg P(e)) \vee (\forall z P(z) \wedge \forall x \neg P(x))].$$

Using Skolem-Herbrand-Gödel's theorem, prove that A is valid.

7.6.2. Convert the negation of the following formula to Skolem form:

$$\neg \exists y \forall z (P(z, y) \equiv \neg \exists x (P(z, x) \wedge P(x, z))).$$

7.6.3. Convert the negation of the following formula to Skolem form:

$$\exists x \exists y \forall z ([P(x, y) \supset (P(y, z) \wedge P(z, z))] \wedge$$
$$[(P(x, y) \wedge Q(x, y)) \supset (Q(x, z) \wedge Q(z, z))]).$$

7.6.4. Write a computer program for converting a formulae in NNF to SKolem normal form, incorporating the optimization suggested in problem 7.5.2.

7.6.5. Fill in the missing details in the proof of lemma 7.6.2.

7.6.6. Prove the following fact: A c-instance of $(C[F[t/x]]/E] * D)$ is a c-instance of $(C * D)$.

7.6.7. Use the Skolem-Herbrand-Gödel theorem to show that the formula of problem 7.6.2 is valid.

7.6.8. Use the Skolem-Herbrand-Gödel theorem to show that the formula of problem 7.6.3 is valid.

7.6.9. Use lemma 7.6.3 to prove that a set of sentences is satisfiable iff the set of their Skolem forms is satisfiable. (Assume that for any two distinct sentences, the sets of Skolem symbols are disjoint.)

7.6.10. Let **L** be a first-order language without equality. A *free structure* **H** is an **L**-structure with domain the set $H_{\mathbf{L}}$ of all closed **L**-terms, and whose interpretation function satisfies the following property:

(i) For every function symbol f of rank n, for all $t_1,...,t_n \in H$,

$$f_{\mathbf{H}}(t_1, ..., t_n) = ft_1...t_n, \text{ and}$$

(ii) For every constant symbol c,

$$c_{\mathbf{H}} = c.$$

(a) Prove that the Skolem form B of a sentence A in NNF is satisfiable iff B is satisfiable in a free structure.

(b) Prove that a set of sentences is satisfiable iff it is satisfiable in a free structure. (Use problem 7.6.9.)

(c) Prove that (b) and (c) are false for languages with equality, but that they are true if we replace free structure by quotient of a free structure.

* **7.6.11.** Let **L** be a first-order language without equality. Given a set S of **L**-sentences in NNF, let H be the free structure built up from the set of function and constant symbols occurring in the Skolem forms of the sentences in S. The *Herbrand expansion* $E(C, H)$ of a quantifier-free formula C in NNF over the free universe H, is the set of all formulae of the form $C[t_1/x_1, ..., t_m/x_m]$, where $\{x_1, ..., x_m\}$ is the set of free variables in C, and $t_1,...,t_m \in H$. For each sentence $A \in S$, let $E(B^*, H)$ be the Herbrand expansion of the quantifier-free formula B^* obtained by deleting the universal quantifiers in the Skolem form B of A. The Herbrand expansion $E(A, H)$ of the sentence A is equal to $E(B^*, H)$.

(a) Prove that S is satisfiable iff the union of all the expansions $E(B^*, H)$ defined above is satisfiable. (Use problem 7.6.10.)

(b) Using the compactness theorem for propositional logic, prove the following version of the Skolem-Herbrand-Gödel theorem:

A sentence A is unsatisfiable iff some finite conjunction of quantifier-free formulae in the Herbrand Expansion $E(A, H)$ is unsatisfiable.

(c) Use (a) to prove the Löwenheim-Skolem theorem.

* **7.6.12.** Let **L** be a first-order language without equality. Let **L'** be an expansion of **L** obtained by adding function and constant symbols. Let H be the set of all closed **L**-terms, and H' the set of all closed **L'**-terms.

Prove that for any sentence A, if the Herbrand expansion $E(A, H)$ is satisfiable, then $E(B^*, H')$ is also satisfiable, where B^* is the quantifier-free formula obtained by deleting the universal quantifiers in the Skolem form B of A.

Hint: Define a function $h : H' \to H$ as follows: Let t_0 be any fixed term in H.

(i) $h(t) = t_0$ if t is either a constant in H' not occurring in B^* or a term of the form $f(t_1, ..., t_k)$ such that f does not occur in B^*.

(ii) $h(t) = t$ if t is a constant occurring in B^*, or $f(h(t_1), .., h(t_n))$ if t is of the form $f(t_1, ..., t_n)$ and f occurs in B^*.

Assume that $E(A, H)$ is satisfiable in a free structure $\mathbf{A}$. Expand $\mathbf{A}$ to an $\mathbf{L}'$-structure $\mathbf{B}$ using the following definition: For every predicate symbol of rank n, for all $t_1, ..., t_n \in H'$,

$$\mathbf{B} \models Pt_1...t_n \quad \text{iff} \quad \mathbf{A} \models Ph(t_1)...h(t_n).$$

Prove that $E(B^*, H')$ is satisfied in $\mathbf{B}$.

7.6.13. Let $\mathbf{L}$ be a first-order language without equality. Prove the compactness theorem using problems 7.6.11, 7.6.12, and the compactness theorem for propositional logic.

7.6.14. State and prove a validity version of theorem 7.6.1.

7.6.15. Consider first-order languages without equality. In this problem, assume that Gentzen's original proof of the cut elimination theorem is used to avoid the completeness theorem in proving theorem 7.5.1.

(a) Prove that Herbrand's theorem (theorem 7.5.1) and the Skolem-Herbrand-Gödel theorem proved in problem 7.6.11 (without the completeness theorem) yield the completeness theorem (for prenex formulae).

(b) Prove that Herbrand's theorem (theorem 7.5.1) and the completeness theorem yield the Skolem-Herbrand-Gödel theorem.

∗ 7.7 The Primitive Recursive Functions

First, we discuss informally the notion of computability.

7.7.1 The Concept of Computability

In the discussion at the end of Section 7.6, the concept of a primitive recursive function was mentioned. In this section, we discuss this concept very briefly. For more details on recursive function theory and complexity theory, the reader is referred to Lewis and Papadimitriou, 1981; Davis and Weyuker,

1983; Machtey and Young, 1978; or Rogers, 1967. For excellent surveys, we recommend Enderton and Smorynski's articles in Barwise, 1977.

At the end of the nineteenth century, classical mathematics was shaken by paradoxes and inconsistencies. The famous mathematician Hilbert proposed the following program in order to put mathematics on solid foundations: Formalize mathematics completely, and exploit the finitist nature of proofs to prove the consistency of the formalized theory (that is, the absence of a contradiction) in the theory itself. In 1930, the famous logician Kurt Gödel made a major announcement; Hilbert's consistency program could not be carried out. Indeed, Gödel had proved two incompleteness theorems that showed the impossibility of Hilbert's program. The second theorem roughly states that in any consistent formal theory T containing arithmetic, the sentence asserting the consistency of T is not provable in T.

To prove his theorems, Gödel invented a technique now known as *Gödel-numbering*, in which syntactic objects such as formulae and proofs are encoded as natural numbers. The functions used to perform such encodings are definable in arithmetic, and are in some intuitive sense computable. These functions are the primitive recursive functions. To carry out Hilbert's program, it was also important to understand what is a computable function, since one of the objectives of the program was to check proofs mechanically.

The concern for providing logical foundations for mathematics prompted important and extensive research (initiated in the early thirties) on the topic of computability and undecidability, by Herbrand, Gödel, Church, Rosser, Kleene, Turing, and Post, to name only the pioneers in the field. Summarizing more than 60 years of research in a few lines, the following important and surprising facts (at least at the time of their finding) were discovered:

Several *models of computations* were proposed by different researchers, and were shown to be equivalent, in the sense that they all define the same class of functions called the *partial recursive functions*.

Among these models are the Turing machine already discussed in Subsection 3.3.5, and the class of partial recursive functions (due to Herbrand, Kleene, Gödel).

The above led to what is usually known as the *Church-Turing thesis*, which states that any "reasonable" definition of the concept of an effectively (or algorithmically) computable function is equivalent to the concept of a partial recursive function.

The Church-Turing thesis cannot be proved because the notion of a reasonable definition of computability is not clearly defined, but most researchers in the field believe it.

The other important theme relevant to our considerations is that of the complexity of computing a function. Indeed, there are computable functions (such as Ackermann's function, see Davis and Weyuker, 1983) that require so much time and space to be computed that they are computable only for very small arguments (may be $n = 0, 1, 2$). For some of these functions, the rate of growth is so "wild" that it is actually beyond imagination. For an entertaining article on this topic, consult Smorynski, 1983.

The class of *primitive recursive functions* is a subclass of the computable functions that, for all practical purposes, contains all the computable functions that one would ever want to compute. It is generally agreed that if an algorithm corresponds to a computable function that is not primitive recursive, it is not a simple algorithm. Herbrand's theorem says that the algorithms that yields a Herbrand's disjunction from a proof of the input formula and its converse are primitive recursive functions. Hence, from a computational point of view, the transformation given by Herbrand's theorem are reasonably simple, or at least not too bad.

The primitive recursive functions also play an important role in Gödel's incompleteness results (see Enderton, 1972, or Monk, 1976).

7.7.2 Definition of the Primitive Recursive Functions

In the rest of this section, we are considering functions of the natural numbers. In order to define the class of primitive recursive functions we need to define two operations on functions:

(1) Composition;

(2) Primitive Recursion.

Definition 7.7.1 Given a function f of $m > 0$ arguments and m functions $g_1,...,g_m$ each of $n > 0$ arguments, the *composition*

$$f \circ (g_1, .., g_m)$$

of f and $g_1,...,g_m$ is the function h of n arguments such that, for all $x_1, .., x_n \in$ **N**,

$$h(x_1, ..., x_n) = f(g_1(x_1, ..., x_n), ..., g_m(x_1, ..., x_n)).$$

Primitive recusion is defined as follows.

Definition 7.7.2 Given a function f of n arguments $(n > 0)$, and a function g of $n+1$ arguments, the function h of $n+1$ arguments is defined by *primitive recursion* from f and g iff the following holds: For all $x_1, ..., x_n, y \in$ **N**,

$$h(x_1, ..., x_n, 0) = f(x_1, ..., x_n);$$
$$h(x_1, ..., x_n, y+1) = g(y, h(x_1, ..., x_n, y), x_1, ..., x_n).$$

In the special case $n = 0$, let m be any given integer. Then

$$h(0) = m;$$
$$h(y+1) = g(y, h(y)).$$

We also define the base functions.

Definition 7.7.3 The *base functions* are the following functions:

(i) The *successor function* S, such that for all $x \in \mathbf{N}$,

$$S(x) = x + 1;$$

(ii) The *zero function* Z, such that for all $x \in \mathbf{N}$,

$$Z(x) = 0;$$

(iii) The *projections functions*. For every $n > 0$, for every i, $1 \leq i \leq n$, for all $x_1, ..., x_n \in \mathbf{N}$,

$$P_i^n(x_1, ..., x_n) = x_i.$$

The class of primitive recursive functions is defined inductively as follows.

Definition 7.7.4 The class of *primitive recursive functions* is the least class of total functions over $\mathbf{N}$ containing the base functions and closed under composition and primitive recursion.

It can be shown that composition and primitive recursion preserve totality, so the definition makes sense.

7.7.3 The Partial Recursive Functions

In order to define the partial recursive functions, we need one more operation, the operation of minimization.

Definition 7.7.5 Given a function g of $n + 1$ arguments, the function f of $n > 0$ arguments is defined by *minimization* from g iff the following holds: For all $x_1, ..., x_n \in \mathbf{N}$:

(i) $f(x_1, ..., x_n)$ is defined iff there is some $y \in \mathbf{N}$ such that $g(x_1, ..., x_n, z)$ is defined for all $z \leq y$ and $g(x_1, ..., x_n, y) = 0$;

(ii) If $f(x_1, ..., x_n)$ is defined, then $f(x_1, ..., x_n)$ is equal to the least y satisfying (i). In other words,

$$f(x_1, ..., x_n) = y \quad \text{iff} \quad g(x_1, ..., x_n, y) = 0 \quad \text{and}$$
$$\text{for all } z < y, \ g(x_1, ..., x_n, z) \text{ is defined and nonzero.}$$

The condition that $g(x_1, ..., x_n, z)$ is defined for all $z \leq y$ is essential to the definition, since otherwise one could define noncomputable functions. The function f is also denoted by

$$min_y(g(x_1, ..., x_n, y) = 0)$$

(with a small abuse of notation, since the variables $x_1, ..., x_n$ should not be present).

Definition 7.7.6 The class of *partial recursive functions* is the least class of partial functions over **N** containing the base functions and closed under composition, primitive recursion, and minimization.

A function is *recursive* iff it is a total partial recursive function.

Obviously, the class of primitive recursive functions is a subclass of the class of recursive functions. Contrary to the other closure operations, if f is obtained by minimization from a total function g, f is not necessarily a total function. For example, if g is the function such that $g(x, y) = x + y + 1$, $min_y(g(x, y) = 0)$ is the partial function undefined everywhere.

It can be shown that there are (total) recursive functions that are not primitive recursive. The following function kown as *Ackermann's function* is such as example: (See example 2.1.1, in Chapter 2.)

EXAMPLE 7.7.1

$A(x, y) = if\ x = 0\ then\ y + 1$
$$else\ if\ y = 0\ then\ A(x - 1, 1)$$
$$else\ A(x - 1, A(x, y - 1))$$

A problem A (encoded as a set of natural numbers) is said to be *decidable* iff there is a (total) recursive function h_A such that for all $n \in \mathbf{N}$,

$$n \in A \quad iff \quad h_A(n) = 1,\ \text{otherwise}\ h_A(n) = 0.$$

A problem A is *partially decidable* iff there is a partial recursive function h_A such that for all $n \in \mathbf{N}$,

$$n \in A \quad iff \quad h_A(n) = 1,\ \text{otherwise either}\ h_A\ \text{is undefined or}\ h_A(n) = 0.$$

Church's theorem states that the problem of deciding whether a first-order formula is valid is partially decidable, but is not decidable (see Enderton, 1972; Monk, 1976; Lewis and Papadimitriou, 1981).

We conclude with a short list of examples of primitive recursive functions. One of the unpleasant properties of definition 7.7.4 is the rigid format of primitive recursion, which forces one to use projections and composition to permute or drop arguments. Since the purpose of this section is only to give a superficial idea of what the primitive recursive functions are, we will ignore these details in the definitions given below. We leave as an (tedious) exercise to the reader the task to rewrite the definitions below so that they fit definition 7.7.4.

7.7.4 Some Primitive Recursive Functions

EXAMPLE 7.7.2

(a) Addition:
$$x + 0 = P_1^1(x)$$
$$x + (y + 1) = S(x + y)$$

(b) Multiplication:

$$x * 0 = Z(x)$$
$$x * (y + 1) = (x * y) + x$$

(c) Exponentiation:

$$exp(x, 0) = 1$$
$$exp(x, y + 1) = exp(x, y) * x$$

(d) Factorial:
$$fact(0) = 1$$
$$fact(y + 1) = fact(y) * S(y)$$

(e) Iterated exponentiation:

$$ex(0) = 0$$
$$ex(y + 1) = exp(2, ex(y))$$

(f) N-th prime number:

$$pr(x) = \text{ the } x\text{-th prime number.}$$

PROBLEMS

7.7.1. Prove that composition and primitive recursion applied to total functions yield total functions.

7.7.2. Prove that the functions given in example 7.7.2 are primitive recursive.

Notes and Suggestions for Further Reading

Gentzen's cut elimination theorem, Gentzen's sharpened Hauptsatz, and Herbrand's theorem are perhaps the most fundamental proof-theoretic results of first-order logic.

Gentzen's theorems show that there are normal forms for proofs, and reduce the provability of a first-order sentence to the provability of a quantifier-free formula. Similarly, Herbrand's theorem provides a deep characterization of the notion of provability, and a reduction to the quantifier-free case.

Interestingly, Herbrand's proof (Herbrand, 1971) and Gentzen's proofs (Szabo, 1969) are significantly different. It is often felt that Herbrand's arguments are difficult to follow, whereas Gentzen's arguments are crystal clear.

Since Gentzen's Hauptsatz requires formulae to be prenex, but Herbrand's original theorem holds for arbitrary formulae, it is often said that Herbrand's theorem is more general than Gentzen's sharpened Hauptsatz. However, using Kleene's method (presented in Section 7.5) and the method partially developed in Section 7.4, it appears that this is not the case.

For more on Gentzen systems, the reader is referred to Szabo, 1969; Takeuti, 1975; Kleene, 1952; Smullyan, 1968; Prawitz, 1965; and Schwichtenberg's article in Barwise, 1977. Another interesting application of Herbrand's theorem is its use to find decidable classes of formulae. The reader is referred to Dreben and Goldfarb, 1979. A companion book by Lewis (Lewis, 1979) deals with unsolvable classes of formulae.

We have not explored complexity issues related to Herbrand's theorem, or Gentzen's theorems, but these are interesting. It is known that a proof of a sentence can be transformed into a proof of a disjunction of quantifier-free (ground) instances of that sentence, and that the transformation is primitive recursive, but how complex is the resulting proof?

A result of Statman (Statman, 1979) shows that a significant increase in the length of the proof can occur. It is shown in Statman, 1979, that for a certain set X of universally quantified equations, for each n, there is a closed equation E_n such that E_n is provable from X in a proof of size linear in n, but that for any set Y of ground instances of equations in X such that $Y \to E_n$ is provable, Y has cardinality at least $ex(n)/2$, where $ex(n)$ is the iterated exponential function defined at the end of Section 7.7. For related considerations, the reader is referred to Statman's article in Barwise, 1977.

Another interesting topic that we have not discussed is the possibility of extending Herbrand's theorem to higher-order logic. Such a generalization is investigated in Miller, 1984.

Chapter 8

Resolution in
First-Order Logic

8.1 Introduction

In this chapter, the resolution method presented in Chapter 4 for propositional logic is extended to first-order logic without equality. The point of departure is the Skolem-Herbrand-Gödel theorem (theorem 7.6.1). Recall that this theorem says that a sentence A is unsatisfiable iff some compound instance C of the Skolem form B of A is unsatisfiable. This suggests the following procedure for checking unsatisfiability:

Enumerate the compound instances of B systematically one by one, testing each time a new compound instance C is generated, whether C is unsatisfiable.

If we are considering a first-order language without equality, there are algorithms for testing whether a quantifier-free formula is valid (for example, the *search* procedure) and, if B is unsatisfiable, this will be eventually discovered. Indeed, the *search* procedure halts for every compound instance, and for some compound instance C, $\neg C$ will be found valid.

If the logic contains equality, the situation is more complex. This is because the *search* procedure does not necessarily halt for quantifier-free formulae that are not valid. Hence, it is possible that the procedure for checking unsatisfiability will run forever even if B is unsatisfiable, because the *search* procedure can run forever for some compound instance that is not unsatisfiable. We can fix the problem as follows:

Interleave the generation of compound instances with the process of checking whether a compound instance is unsatisfiable, proceeding by rounds. A round consists in running the *search* procedure a fixed number of steps for each compound instance being tested, and then generating a new compound instance. The process is repeated with the new set of compound instances. In this fashion, at the end of each round, we have made progress in checking the unsatisfiability of all the activated compound instances, but we have also made progress in the number of compound instances being considered.

Needless to say, such a method is horribly inefficient. Actually, it is possible to design an algorithm for testing the unsatisfiability of a quantifier-free formula with equality by extending the congruence closure method of Oppen and Nelson (Nelson and Oppen, 1980). This extension is presented in Chapter 10.

In the case of a language without equality, any algorithm for deciding the unsatisfiability of a quantifier-free formula can be used. However, the choice of such an algorithm is constrained by the need for efficiency. Several methods have been proposed. The *search* procedure can be used, but this is probably the least efficient choice. If the compound instances C are in CNF, the resolution method of Chapter 4 is a possible candidate. Another method called the method of *matings* has also been proposed by Andrews (Andrews, 1981).

In this chapter, we are going to explore the method using resolution. Such a method is called *ground resolution*, because it is applied to quantifier-free clauses with no variables.

From the point of view of efficiency, there is an undesirable feature, which is the need for systematically generating compound instances. Unfortunately, there is no hope that the process of finding a refutation can be purely mechanical. Indeed, by Church's theorem (mentioned in the remark after the proof of theorem 5.5.1), there is no algorithm for deciding the unsatisfiability (validity) of a formula.

There is a way of avoiding the systematic generation of compound instances due to J. A. Robinson (Robinson, 1965). The idea is not to generate compound instances at all, but instead to generalize the resolution method so that it applies directly to the clauses in B, as opposed to the (ground) clauses in the compound instance C. The completeness of this method was shown by Robinson. The method is to show that every ground refutation can be lifted to a refutation operating on the original clauses, as opposed to the closed (or ground) substitution instances. In order to perform this lifting operation the process of *unification* must be introduced. We shall define these concepts in the following sections.

It is also possible to extend the resolution method to first-order languages with equality using the *paramodulation* method due to Robinson and Wos (Robinson and Wos, 1969, Loveland, 1978), but the completeness proof is rather delicate. Hence, we will restrict our attention to first-order languages without equality, and refer the interested reader to Loveland, 1978, for an exposition of paramodulation.

As in Chapter 4, the resolution method for first-order logic (without equality) is applied to special conjunctions of formulae called clauses. Hence, it is necessary to convert a sentence A into a sentence A' in clause form, such that A is unsatisfiable iff A' is unsatisfiable. The conversion process is defined below.

8.2 Formulae in Clause Form

First, we define the notion of a formula in clause form.

Definition 8.2.1 As in the propositional case, a *literal* is either an atomic formula B, or the negation $\neg B$ of an atomic formula. Given a literal L, its *conjugate* $\overline{L}$ is defined such that, if $L = B$ then $\overline{L} = \neg B$, else if $L = \neg B$ then $\overline{L} = B$. A sentence A is in *clause form* iff it is a conjunction of (prenex) sentences of the form $\forall x_1...\forall x_m C$, where C is a disjunction of literals, and the sets of bound variables $\{x_1, ..., x_m\}$ are disjoint for any two distinct clauses. Each sentence $\forall x_1...\forall x_m C$ is called a *clause*. If a clause in A has no quantifiers and does not contain any variables, we say that it is a *ground clause*.

For simplicity of notation, the universal quantifiers are usually omitted in writing clauses.

Lemma 8.2.1 For every (rectified) sentence A, a sentence B' in clause form such that A is valid iff B' is unsatisfiable can be constructed.

Proof: Given a sentence A, first $B = \neg A$ is converted to B_1 in NNF using lemma 6.4.1. Then B_1 is converted to B_2 in Skolem normal form using the method of definition 7.6.2. Next, by lemma 7.2.1, B_2 is converted to B_3 in prenex form. Next, the matrix of B_3 is converted to conjunctive normal form using theorem 3.4.2, yielding B_4. In this step, theorem 3.4.2 is applicable because the matrix is quantifier free. Finally, the quantifiers are distributed over each conjunct using the valid formula $\forall x(A \wedge B) \equiv \forall x A \wedge \forall x B$, and renamed apart using lemma 5.3.4.

Let the resulting sentence be called B'. The resulting formula B' is a conjunction of clauses.

By lemma 6.4.1, B is unsatisfiable iff B_1 is. By lemma 7.6.3, B_1 is unsatisfiable iff B_2 is. By lemma 7.2.1, B_2 is unsatisfiable iff B_3 is. By theorem 3.4.2 and lemma 5.3.7, B_3 is unsatisfiable iff B_4 is. Finally, by lemma 5.3.4 and lemma 5.3.7, B_4 is unsatisfiable iff B' is. Hence, B is unsatisfiable iff B' is. Since A is valid iff $B = \neg A$ is unsatisfiable, then A is valid iff B' is unsatisfiable. $\square$

EXAMPLE 8.2.1

Let

$$A = \neg\exists y \forall z (P(z,y) \equiv \neg\exists x (P(z,x) \wedge P(x,z))).$$

First, we negate A and eliminate $\equiv$. We obtain the sentence

$$\exists y \forall z [(\neg P(z,y) \vee \neg\exists x (P(z,x) \wedge P(x,z))) \wedge$$
$$(\exists x (P(z,x) \wedge P(x,z)) \vee P(z,y))].$$

Next, we put in this formula in NNF:

$$\exists y \forall z [(\neg P(z,y) \vee \forall x (\neg P(z,x) \vee \neg P(x,z))) \wedge$$
$$(\exists x (P(z,x) \wedge P(x,z)) \vee P(z,y))].$$

Next, we eliminate existential quantifiers, by the introduction of Skolem symbols:

$$\forall z [(\neg P(z,a) \vee \forall x (\neg P(z,x) \vee \neg P(x,z))) \wedge$$
$$((P(z,f(z)) \wedge P(f(z),z)) \vee P(z,a))].$$

We now put in prenex form:

$$\forall z \forall x [(\neg P(z,a) \vee (\neg P(z,x) \vee \neg P(x,z))) \wedge$$
$$((P(z,f(z)) \wedge P(f(z),z)) \vee P(z,a))].$$

We put in CNF by distributing $\wedge$ over $\vee$:

$$\forall z \forall x [(\neg P(z,a) \vee \neg P(z,x) \vee \neg P(x,z)) \wedge$$
$$(P(z,f(z)) \vee P(z,a)) \wedge (P(f(z),z) \vee P(z,a))].$$

Omitting universal quantifiers, we have the following three clauses:

$$C_1 = (\neg P(z_1,a) \vee \neg P(z_1,x) \vee \neg P(x,z_1)),$$
$$C_2 = (P(z_2,f(z_2)) \vee P(z_2,a)) \text{ and}$$
$$C_3 = (P(f(z_3),z_3)) \vee P(z_3,a)).$$

We will now show that we can prove that $B = \neg A$ is unsatisfiable, by instantiating C_1, C_2, C_3 to ground clauses and use the resolution method of Chapter 4.

8.3 Ground Resolution

The *ground resolution method* is the resolution method applied to sets of ground clauses.

EXAMPLE 8.3.1

Consider the following ground clauses obtained by substitution from C_1, C_2 and C_3:

$$G_1 = (\neg P(a,a)) \text{ (from } C_1, \text{ substituting } a \text{ for } x \text{ and } z_1)$$
$$G_2 = (P(a,f(a)) \vee P(a,a)) \text{ (from } C_2, \text{ substituting } a \text{ for } z_2)$$
$$G_3 = (P(f(a),a)) \vee P(a,a)) \text{ (from } C_3, \text{ substituting } a \text{ for } z_3).$$
$$G_4 = (\neg P(f(a),a) \vee \neg P(a,f(a))) \text{ (from } C_1, \text{ substituting } f(a)$$
$$\text{for } z_1 \text{ and } a \text{ for } x).$$

The following is a refutation by (ground) resolution of the set of ground clauses G_1, G_2, G_3, G_4.

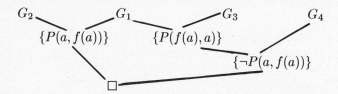

We have the following useful result.

Lemma 8.3.1 (Completeness of ground resolution) The ground resolution method is complete for ground clauses.

Proof: Observe that the systems G' and $GCNF'$ are complete for quantifier-free formulae of a first-order language without equality. Hence, by theorem 4.3.1, the resolution method is also complete for sets of ground clauses. □

However, note that this is not the case for quantifier-free formulae with equality, due to the need for equality axioms and for inessential cuts, in order to retain completeness.

Since we have shown that a conjunction of ground instances of the clauses C_1, C_2, C_3 of example 8.2.1 is unsatisfiable, by the Skolem-Herbrand-Gödel theorem, the sentence A of example 8.2.1 is valid.

Summarizing the above, we have a method for finding whether a sentence B is unsatisfiable known as *ground resolution*. This method consists in converting the sentence B into a set of clauses B', instantiating these clauses to ground clauses, and applying the ground resolution method.

By the completeness of resolution for propositional logic (theorem 4.3.1), and the Skolem-Herbrand-Gödel theorem (actually the corollary to theorem 7.6.1 suffices, since the clauses are in CNF, and so in NNF), this method is complete.

However, we were lucky to find so easily the ground clauses G_1, G_2, G_3 and G_4. In general, all one can do is enumerate ground instances one by one, testing for the unsatisfiabiliy of the current set of ground clauses each time. This can be a very costly process, both in terms of time and space.

8.4 Unification and the Unification Algorithm

The fundamental concept that allows the lifting of the ground resolution method to the first-order case is that of a most general unifier.

8.4.1 Unifiers and Most General Unifiers

We have already mentioned that Robinson has generalized ground resolution to arbitrary clauses, so that the systematic generation of ground clauses is unnecessary.

The new ingredient in this new form of resolution is that in forming the resolvent, one is allowed to apply substitutions to the parent clauses.

For example, to obtain $\{P(a, f(a))\}$ from

$$C_1 = (\neg P(z_1, a) \vee \neg P(z_1, x) \vee \neg P(x, z_1)) \quad \text{and}$$
$$C_2 = (P(z_2, f(z_2)) \vee P(z_2, a)),$$

first we substitute a for z_1, a for x, and a for z_2, obtaining

$$G_1 = (\neg P(a, a)) \quad \text{and} \quad G_2 = (P(a, f(a)) \vee P(a, a)),$$

and then we resolve on the literal $P(a, a)$.

Note that the two sets of literals $\{P(z_1, a), P(z_1, x), P(x, z_1)\}$ and $\{P(z_2, a)\}$ obtained by dropping the negation sign in G_1 have been "unified" by the substitution $(a/x, a/z_1, a/z_2)$.

In general, given two clauses B and C whose variables are disjoint, given a substitution σ with support the union of the set of variables in B and C, if $\sigma(B)$ and $\sigma(C)$ contain a literal Q and its conjugate, there must be a subset $\{B_1, ..., B_m\}$ of the sets of literals of B, and a subset $\{\overline{C_1}, ..., \overline{C_n}\}$ of the set of literals in C such that

$$\sigma(B_1) = ... = \sigma(B_m) = \sigma(C_1) = ... = \sigma(C_n).$$

We say that σ is a *unifier* for the set of literals $\{B_1, ..., B_m, C_1, ..., C_n\}$. Robinson showed that there is an algorithm called the *unification algorithm*, for deciding whether a set of literals is unifiable, and if so, the algorithm yields what is called a *most general unifier* (Robinson, 1965). We will now explain these concepts in detail.

First, we adopt the following notation for substitutions.

Definition 8.4.1 Given a substitution σ with support $\{x_1, ..., x_m\}$, if $\sigma(x_i)$ $= s_i$ for $i = 1, ..., m$, we also denote the substitution σ by $(s_1/x_1, ..., s_m/x_m)$.

The notions of a unifier and a most general unifier are defined for arbitrary trees over a ranked alphabet (see Subsection 2.2.6). Since terms and atomic formulae have an obvious representation as trees (rigorously, since they are freely generated, we could define a bijection recursively), it is perfectly suitable to deal with trees, and in fact, this is intuitively more appealing due to the graphical nature of trees.

Definition 8.4.2 Given a ranked alphabet Σ, given any set $S = \{t_1, ..., t_n\}$ of finite Σ-trees, we say that a substitution σ with support the set of variables occurring in the set S is a *unifier of S* iff

$$\sigma(t_1) = ... = \sigma(t_n).$$

We say that a substitution σ is a *most general unifier* of S iff it is a unifier of S, and for any other unifier σ' of S, there is a substitution θ such that

$$\sigma' = \sigma \circ \theta,$$

the result of composing σ with θ (in this order). The tree $t = \sigma(t_1) = ... = \sigma(t_n)$ is called a *most common instance* of $t_1, ..., t_n$.

EXAMPLE 8.4.1

(i) Let $t_1 = f(x, g(y))$ and $t_2 = f(g(u), g(z))$. The substitution $(g(u)/x, y/z)$ is a most general unifier yielding the most common instance $f(g(u), g(y))$.

(ii) However, $t_1 = f(x, g(y))$ and $t_2 = f(g(u), h(z))$ are not unifiable since this requires $g = h$.

(iii) A slightly more devious case of non unifiability is the following:

Let $t_1 = f(x, g(x), x)$ and $t_2 = f(g(u), g(g(z)), z)$. To unify these two trees, we must have $x = g(u) = z$. But we also need $g(x) = g(g(z))$, that is, $x = g(z)$. This implies $z = g(z)$, which is impossible for finite trees.

This last example suggest that unifying trees is similar to solving systems of equations by variable elimination, and there is indeed such an analogy. This analogy is explicated in Gorn, 1984.

First, we show that we can reduce the problem of unifying any set of trees to the problem of unifying two trees.

Lemma 8.4.1 Let $t_1, ..., t_m$ be any m trees, and let $\#$ be a symbol of rank m not occurring in any of these trees. A substitution σ is a unifier for the set $\{t_1, ..., t_m\}$ iff σ is a unifier for the set $\{\#(t_1, ..., t_m), \#(t_1, ..., t_1)\}$.

Proof: Since a substitution σ is a homomorphism (see definition 7.5.3),

$$\sigma(\#(t_1,...,t_m)) = \#(\sigma(t_1),...,\sigma(t_m)) \quad \text{and}$$
$$\sigma(\#(t_1,...,t_1)) = \#(\sigma(t_1),...,\sigma(t_1)).$$

Hence,

$$\sigma(\#(t_1,...,t_m)) = \sigma(\#(t_1,...,t_1)) \quad \text{iff}$$
$$\#(\sigma(t_1),...,\sigma(t_m)) = \#(\sigma(t_1),...,\sigma(t_1)) \quad \text{iff}$$
$$\sigma(t_1) = \sigma(t_1),\ \sigma(t_2) = \sigma(t_1),...,\ \sigma(t_m) = \sigma(t_1) \quad \text{iff}$$
$$\sigma(t_1) = ... = \sigma(t_m).$$

$\square$

Before showing that if a set of trees is unifiable then it has a most general unifier, we note that most general unifiers are essentially unique when they exist.

Lemma 8.4.2 If a set of trees S is unifiable and σ and θ are any two most general unifiers for S, then there exists a bijective substitution ρ such that $\theta = \sigma \circ \rho$, where ρ is a substitution such that for every variable x in its support, $\rho(x)$ is a variable.

Proof: First, note that a bijective substitution must be a bijective renaming of variables. Indeed, if ρ is bijective, there is a substitution ρ' such that $\rho \circ \rho' = Id$ and $\rho' \circ \rho = Id$. But then, if $\rho(x)$ is not a variable for some x in the support of ρ, $\rho(x)$ is a constant or a tree t of depth ≥ 1. Since $\rho \circ \rho' = Id$, we have $\rho'(t) = x$. Since a substitution is a homomorphism, if t is a constant c, $\rho'(c) = c \neq x$, and otherwise $\rho'(t)$ has depth at least 1, and so $\rho'(t) \neq x$. Hence, $\rho(x)$ must be a variable for every x (and similarly for ρ'). A reasoning similar to the above also shows that for any two substitutions σ and ρ, if $\sigma = \sigma \circ \rho$, then ρ is the identity. But then, if both σ and θ are most general unifiers, there exist σ' and θ' such that $\theta = \sigma \circ \theta'$ and $\sigma = \theta \circ \sigma'$. Hence,

$$\theta = \theta \circ (\sigma' \circ \theta') \quad \text{and} \quad \sigma = \sigma \circ (\theta' \circ \sigma').$$

This implies that $\sigma' \circ \theta' = Id$ and $\theta' \circ \sigma' = Id$, which proves that both σ' and θ' are bijective renamings of variables. $\square$

We shall now present a version of Robinson's unification algorithm.

8.4.2 The Unification Algorithm

In view of lemma 8.4.1, we restrict our attention to pairs of trees. The main idea of the unification algorithm is to find how two trees "disagree," and try to force them to agree by substituting trees for variables, if possible. There are two types of disagreements:

(1) Fatal disagreements, which are of two kinds:

    (i) For some tree address u both in $dom(t_1)$ and $dom(t_2)$, the labels $t_1(u)$ and $t_2(u)$ are not variables and $t_1(u) \neq t_2(u)$. This is illustrated by case (ii) in example 8.4.1;

    (ii) For some tree address u in both $dom(t_1)$ and $dom(t_2)$, $t_1(u)$ is a variable say x, and the subtree t_2/u rooted at u in t_2 is not a variable and x occurs in t_2/u (or the symmetric case in which $t_2(u)$ is a variable and t_1/u isn't). This is illustrated in case (iii) of example 8.4.1.

(2) Repairable disagreements: For some tree address u both in $dom(t_1)$ and $dom(t_2)$, $t_1(u)$ is a variable and the subtree t_2/u rooted at u in t_2 does not contain the variable $t_1(u)$.

In case (1), unification is impossible (although if we allowed infinite trees, disagreements of type (1)(ii) could be fixed; see Gorn, 1984). In case (2), we force "local agreement" by substituting the subtree t_2/u for all occurrences of the variable x in both t_1 and t_2.

It is rather clear that we need a systematic method for finding disagreements in trees. Depending on the representation chosen for trees, the method will vary. In most presentations of unification, it is usually assumed that trees are represented as parenthesized expressions, and that the two strings are scanned from left to right until a disagreement is found. However, an actual method for doing so is usually not given explicitly. We believe that in order to give a clearer description of the unification algorithm, it is better to be more explicit about the method for finding disagreements, and that it is also better not to be tied to any string representation of trees. Hence, we will give a recursive algorithm inspired from J. A. Robinson's original algorithm, in which trees are defined in terms of tree domains (as in Section 2.2), and the disagreements are discovered by performing two parallel top-down traversals of the trees t_1 and t_2.

The type of traversal that we shall be using is a recursive traversal in which the root is visited first, and then, from left to right, the subtrees of the root are recursively visited (this kind of traversal is called a *preorder traversal*, see Knuth, 1968, Vol. 1).

We define some useful functions on trees. (The reader is advised to review the definitions concerning trees given in Section 2.2.)

Definition 8.4.3 For any tree t, for any tree address $u \in dom(t)$:

$leaf(u) = true$ iff u is a leaf;

$variable(t(u)) = true$ iff $t(u)$ is a variable;

$left(u) = if \ leaf(u) \ then \ nil \ else \ u1$;

$right(ui) = if \ u(i+1) \in dom(t) \ then \ u(i+1) \ else \ nil$.

We also assume that we have a function $dosubstitution(t, \sigma)$, where t is a tree and σ is a substitution.

Definition 8.4.4 (A unification algorithm) The formal parameters of the algorithm *unification* are the two input trees t_1 and t_2, an output flag indicating whether the two trees are unifiable or not (*unifiable*), and a most general unifier (*unifier*) (if it exists).

The main program *unification* calls the recursive procedure *unify*, which performs the unification recursively and needs procedure *test-and-substitute* to repair disagreements found, as in case (2) discussed above. The variables *tree1* and *tree2* denote trees (of type *tree*), and the variables *node*, *newnode* are tree addresses (of type *treereference*). The variable *unifier* is used to build a most general unifier (if any), and the variable *newpair* is used to form a new substitution component (of the form (t/x), where t is a tree and x is a variable). The function *compose* is simply function composition, where *compose(unifier, newpair)* is the result of composing *unifier* and *newpair*, in this order. The variables *tree1*, *tree2*, and *node* are global variables to the procedure *unification*. Whenever a new disagreement is resolved in *test-and-substitute*, we also apply the substitution *newpair* to *tree1* and *tree2* to remove the disagreement. This step is not really necessary, since at any time, $dosubstitution(t_1, unifier) = tree1$ and $dosubstitution(t_2, unifier) = tree2$, but it simplifies the algorithm.

Procedure to Unify Two Trees t_1 and t_2

procedure $unification(t_1, t_2 : tree;$ **var** $unifiable :$ **boolean**;
$\qquad\qquad\qquad\qquad\qquad$ **var** $unifier : substitution);$
$\quad$ **var** $node : treereference;$ $tree1, tree2 : tree;$

$\quad$ **procedure** $test\text{-}and\text{-}substitute($**var** $node : treereference;$
$\qquad$ **var** $tree1, tree2 : tree;$
$\qquad$ **var** $unifier : substitution;$ **var** $unifiable :$ **boolean**);
$\quad$ **var** $newpair : substitution;$

$\quad${This procedure tests whether the variable $tree1(node)$ belongs to the subtree of $tree2$ rooted at $node$. If it does, the unification fails. Otherwise, a new substitution $newpair$ consisting of the subtree $tree2/node$ and the variable $tree1(node)$ is formed, the current $unifier$ is composed with $newpair$, and the new pair is added to the $unifier$. To simplify the algorithm, we also apply $newpair$ to $tree1$ and $tree2$ to remove the disagreement}

$\quad$ **begin**
$\quad${test whether the variable $tree1(node)$ belongs to the subtree $tree2/node$, known in the literature as "occur check"}

```
        if tree1(node) ∈ tree2/node then
          unifiable := false
        else

            {create a new substitution pair consisting of the
            subtree tree2/node at address node, and the
            variable tree1(node) at node in tree1}

            newpair := ((tree2/node)/tree1(node));

            {compose the current partial unifier with
            the new pair newpair}

            unifier := compose(unifier, newpair);

            {updates the two trees so that they now agree on
            the subtrees at node}

            tree1 := dosubstitution(tree1, newpair);
            tree2 := dosubstitution(tree2, newpair)
        endif
      end test-and-substitute;


procedure unify(var node : treereference;
    var unifiable : boolean; var unifier : substitution);
var newnode : treereference;

{Procedure unify recursively unifies the subtree
of tree1 at node and the subtree of tree2 at node}

  begin
    if tree1(node) <> tree2(node) then
      {the labels of tree1(node) and tree2(node) disagree}
      if variable(tree1(node)) or variable(tree2(node)) then
        {one of the two labels is a variable}
        if variable(tree1(node)) then
          test-and-substitute(node, tree1, tree2, unifier, unifiable)
        else
          test-and-substitute(node, tree2, tree1, unifier, unifiable)
        endif
      else
        {the labels of tree1(node) and tree2(node)
        disagree and are not variables}
        unifiable := false
      endif
    endif;
```

{At this point, if *unifiable* = **true**, the labels at *node* agree. We recursively unify the immediate subtrees of *node* in *tree*1 and *tree*2 from left to right, if *node* is not a leaf}

```
if (left(node) <> nil) and unifiable then
    newnode := left(node);
    while (newnode <> nil) and unifiable do
        unify(newnode, unifiable, unifier);
        if unifiable then
            newnode := right(newnode)
        endif
    endwhile
endif
end unify;
```

Body of Procedure Unification

```
begin
    tree1 := t₁;
    tree2 := t₂;
    unifiable := true;
    unifier := nil;      {empty unification}
    node := e;           {start from the root}
    unify(node, unifiable, unifier)
end unification
```

Note that if successful, the algorithm could also return the tree *tree*1 (or *tree*2), which is a most common form of t_1 and t_2. As presented, the algorithm performs a single parallel traversal, but we also have the cost of the *occur check* in *test-and-substitute*, and the cost of the substitutions. Let us illustrate how the algorithm works.

EXAMPLE 8.4.2

Let $t_1 = f(x, f(x, y))$ and $t_2 = f(g(y), f(g(a), z))$, which are represented as trees as follows:

Tree t_1

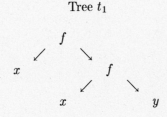

Tree t_2

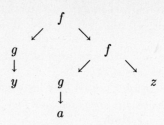

Initially, $tree1 = t_1$, $tree2 = t_2$ and $node = e$. The first disagreement is found for $node = 1$. We form $newpair = (g(y)/x)$, and $unifier = newpair$. After applying $newpair$ to $tree1$ and $tree2$, we have:

Tree $tree1$

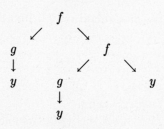

Tree $tree2$

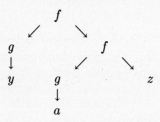

The next disagreement is found for $node = 211$. We find that $newpair = (a/y)$, and compose $unifier = (g(y)/x)$ with $newpair$, obtaining $(g(a)/x, a/y)$. After applying $newpair$ to $tree1$ and $tree2$, we have:

Tree $tree1$

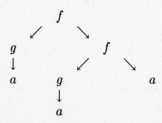

Tree *tree2*

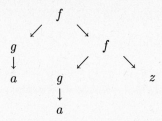

The last disagreement occurs for *node* = 22. We form *newpair* = (a/z), and compose *unifier* with *newpair*, obtaining

$$unifier = (g(a)/x, a/y, a/z).$$

The algorithm stops successfully with the most general unifier $(g(a)/x, a/y, a/z)$, and the trees are unified to the last value of *tree1*.

In order to prove the correctness of the unification algorithm, the following lemma will be needed.

Lemma 8.4.3 Let # be any constant. Given any two trees $f(s_1, ..., s_n)$ and $f(t_1, ..., t_n)$ the following properties hold:

(a) For any i, $1 \leq i \leq n$, if σ is a most general unifier for the trees

$$f(s_1, ..., s_{i-1}, \#, ..., \#) \quad \text{and} \quad f(t_1, ..., t_{i-1}, \#, ..., \#), \quad \text{then}$$
$$f(s_1, ..., s_i, \#, ..., \#) \quad \text{and} \quad f(t_1, ..., t_i, \#, ..., \#) \quad \text{are unifiable iff}$$
$$\sigma(f(s_1, ..., s_i, \#, ..., \#)) \quad \text{and} \quad \sigma(f(t_1, ..., t_i, \#, ..., \#)) \quad \text{are unifiable.}$$

(b) For any i, $1 \leq i \leq n$, if σ is a most general unifier for the trees $f(s_1, ..., s_{i-1}, \#, ..., \#)$ and $f(t_1, ..., t_{i-1}, \#, ..., \#)$, and θ is a most general unifier for the trees $\sigma(s_i)$ and $\sigma(t_i)$, then $\sigma \circ \theta$ is a most general unifier for the trees $f(s_1, ..., s_i, \#, ..., \#)$ and $f(t_1, ..., t_i, \#, ..., \#)$.

Proof: (a) The case $i = 1$ is trivial. Clearly, if σ is a most general unifier for the trees $f(s_1, ..., s_{i-1}, \#, ..., \#)$ and $f(t_1, ..., t_{i-1}, \#, ..., \#)$ and if $\sigma(f(s_1, ..., s_i, \#, ..., \#))$ and $\sigma(f(t_1, ..., t_i, \#, ..., \#))$ are unifiable, then $f(s_1, ..., s_i, \#, ..., \#)$ and $f(t_1, ..., t_i, \#, ..., \#)$ are unifiable.

We now prove the other direction. Let θ be a unifier for

$$f(s_1, ..., s_i, \#, ..., \#) \quad \text{and} \quad f(t_1, ..., t_i, \#, ..., \#).$$

Then,

$$\theta(s_1) = \theta(t_1), ..., \theta(s_i) = \theta(t_i).$$

Hence, θ is a unifier for

$$f(s_1, ..., s_{i-1}, \#, ..., \#) \quad \text{and} \quad f(t_1, ..., t_{i-1}, \#, ..., \#).$$

Since σ is a most general unifier, there is some θ' such that $\theta = \sigma \circ \theta'$. Then,

$$\theta'(\sigma(f(s_1, ..., s_i, \#, ..., \#))) = \theta(f(s_1, ..., s_i, \#, ..., \#))$$
$$= \theta(f(t_1, ..., t_i, \#, ..., \#)) = \theta'(\sigma(f(t_1, ..., t_i, \#, ..., \#))),$$

which shows that θ' unifies

$$\sigma(f(s_1, ..., s_i, \#, ..., \#)) \quad \text{and} \quad \sigma(f(t_1, ..., t_i, \#, ..., \#)).$$

(b) Again, the case $i = 1$ is trivial. Otherwise, clearly,

$$\sigma(s_1) = \sigma(t_1), ..., \sigma(s_{i-1}) = \sigma(t_{i-1}) \text{ and } \theta(\sigma(s_i)) = \theta(\sigma(t_i))$$

implies that $\sigma \circ \theta$ is a unifier of

$$f(s_1, ..., s_i, \#, ..., \#) \quad \text{and} \quad f(t_1, ..., t_i, \#, ..., \#).$$

If λ unifies $f(s_1, ..., s_i, \#, ..., \#)$ and $f(t_1, ..., t_i, \#, ..., \#)$, then

$$\lambda(s_1) = \lambda(t_1), ..., \lambda(s_i) = \lambda(t_i).$$

Hence, λ unifies

$$f(s_1, ..., s_{i-1}, \#, ..., \#) \quad \text{and} \quad f(t_1, ..., t_{i-1}, \#, ..., \#).$$

Since σ is a most general unifier of these two trees, there is some σ' such that $\lambda = \sigma \circ \sigma'$. But then, since $\lambda(s_i) = \lambda(t_i)$, we have $\sigma'(\sigma(s_i)) = \sigma'(\sigma(t_i))$, and since θ is a most general unifier of $\sigma(s_i)$ and $\sigma(t_i)$, there is some θ' such that $\sigma' = \theta \circ \theta'$. Hence,

$$\lambda = \sigma \circ (\theta \circ \theta') = (\sigma \circ \theta) \circ \theta',$$

which proves that $\sigma \circ \theta$ is a most general unifier of $f(s_1, ..., s_i, \#, ..., \#)$ and $f(t_1, ..., t_i, \#, ..., \#)$. $\square$

We will now prove the correctness of the unification algorithm.

Theorem 8.4.1 (Correctness of the unification algorithm) (i) Given any two finite trees t_1 and t_2, the unification algorithm always halts. It halts with output *unifiable* = **true** iff t_1 and t_2 are unifiable.

(ii) If t_1 and t_2 are unifiable, then they have a most general unifier and the output of procedure *unify* is a most general unifier.

Proof: Clearly, the procedure *test-and-substitute* always terminates, and we only have to prove the termination of the *unify* procedure. The difficulty in proving termination is that the trees $tree1$ and $tree2$ may grow. However, this can only happen if *test-and-substitute* is called, and in that case, since unifiable is not **false** iff the variable $x = tree1(node)$ does not belong to $t = tree2/node$, after the substitution of t for all occurrences of x in both $tree1$ and $tree2$, the variable x has been completely eliminated from both $tree1$ and $tree2$. This suggests to try a proof by induction over the well-founded lexicographic ordering $<<$ defined such that, for all pairs (m,t) and (m',t'), where m, m' are natural numbers and t, t' are finite trees,

$$(m,t) << (m',t') \quad \text{iff either } m < m',$$
$$\text{or } m = m' \text{ and } t \text{ is a proper subtree of } t'.$$

We shall actually prove the input-output correctness assertion stated below for the procedure $unify$.

Let s_0 and t_0 be two given finite trees, σ a substitution such that none of the variables in the support of σ is in s_0 or t_0, u any tree address in both $dom(\sigma(s_0))$ and $dom(\sigma(t_0))$, and let $s = \sigma(s_0)/u$ and $t = \sigma(t_0)/u$. Let $tree1_0$, $tree2_0$, $node_0$, $unifiable_0$ and $unifier_0$ be the input values of the variables $tree1$, $tree2$, $unifiable$, and $unifier$, and $tree1'$, $tree2'$, $node'$ $unifiable'$ and $unifier'$ be their output value (if any). Also, let m_0 be the sum of the number of variables in $\sigma(s_0)$ and $\sigma(t_0)$, and m' the sum of the number of variables in $tree1'$ and $tree2'$.

Correctness assertion:

$$\text{If } tree1_0 = \sigma(s_0), \quad tree2_0 = \sigma(t_0), \quad node_0 = u,$$
$$unifiable_0 = \textbf{true} \quad \text{and} \quad unifier_0 = \sigma, \text{ then}$$

the following holds:

(1) The procedure $unify$ always terminates;

(2) $unifiable' = \textbf{true}$ iff s and t are unifiable and, if $unifiable' = \textbf{true}$, then $unifier' = \sigma \circ \theta$, where θ is a most general unifier of s and t, $tree1' = unifier'(s_0)$, $tree2' = unifier'(t_0)$, and no variable in the support of $unifier'$ occurs in $tree1'$ or $tree2'$.

(3) If $tree1' \neq \sigma(s_0)$ or $tree2' \neq \sigma(t_0)$ then $m' < m_0$, else $m' = m_0$.

Proof of assertion: We proceed by complete induction on (m,s), where m is the sum of the number of variables in $tree1$ and $tree2$ and s is the subtree $tree1/node$.

(i) Assume that s is a constant, the case in which t is a constant being similar. Then u is a leaf node in $\sigma(s_0)$. If $t \neq s$, the comparison of $tree1(node)$ and $tree2(node)$ fails, and $unifiable$ is set to **false**. The procedure terminates with failure. If $s = t$, since u is a leaf node in $\sigma(s_0)$ and $\sigma(t_0)$, the procedure terminates with success, $tree1' = \sigma(s_0)$, $tree2' = \sigma(t_0)$, and $unifier' = \sigma$. Hence the assertion holds with the identity substitution for θ.

(ii) Assume that s is a variable say x, the case in which t is a variable being similar. Then u is a leaf node in $\sigma(s_0)$. If $t = s$, this case reduces to case (i). Otherwise, $t \neq x$ and the occur check is performed in *test-and-substitute*. If x occurs in t, then *unifiable* is set to **false**, and the procedure terminates. In this case, it is clear that x and t are not unifiable, and the assertion holds. Otherwise, the substitution $\theta = (t/x)$ is created, $unifier' = \sigma \circ \theta$, and $tree1' = \theta(\sigma(s_0)) = unifier'(s_0)$, $tree2' = \theta(\sigma(t_0)) = unifier'(t_0)$. Clearly, θ is a most general unifier of x and t, and since x does not occur in t, since no variable in the support of σ occurs in $\sigma(s_0)$ or $\sigma(t_0)$, no variable in the support of $unifier'$ occurs in $tree1' = \theta(\sigma(s_0))$ or $tree2' = \theta(\sigma(s_0))$. Since the variable x does not occur in $tree1'$ and $tree2'$, (3) also holds. Hence, the assertion holds.

(iii) Both s and t have $depth \geq 1$. Assume that $s = f(s_1, ..., s_m)$ and $t = f'(t_1, ..., t_n)$. If $f \neq f'$, the test $tree1(node) = tree2(node)$ fails, and *unify* halts with failure. Clearly, s and t are not unifiable, and the claim holds. Otherwise, $s = f(s_1, ..., s_n)$ and $t = f(t_1, ..., t_n)$.

We shall prove the following claim by induction:

Claim: (1) For every i, $1 \leq i \leq n + 1$, the first $i - 1$ recursive calls in the while loop in *unify* halt with success iff $f(s_1, ..., s_i, \#, ..., \#)$ and $f(t_1, ..., t_i, \#, ..., \#)$ are unifiable, and otherwise one of the calls halts with failure;

(2) If the first $i - 1$ recursive calls halt with success, the input values at the end of the i-th iteration are:

$$node_i = ui, \quad unifiable_i = \textbf{true}, \quad unifier_i = \sigma \circ \theta_{i-1},$$

where θ_{i-1} is a most general unifier for the trees $f(s_1, ..., s_{i-1}, \#, ..., \#)$ and $f(t_1, ..., t_{i-1}, \#, ..., \#)$, (with $\theta_0 = Id$, the identity substitution),

$$tree1_i = unifier_i(s_0), \quad tree2_i = unifier_i(t_0),$$

and no variable in the support of $unifier_i$ occurs in $tree1_i$ or $tree2_i$.

(3) If $tree1_i \neq tree1_0$ or $tree2_i \neq tree2_0$, if m_i is the sum of the number of variables in $tree1_i$ and $tree2_i$, then $m_i < m_0$.

Proof of claim: For $i = 1$, the claim holds because before entering the while loop for the first time,

$$tree1_1 = s_0, \quad tree2_1 = t_0, \quad node_1 = u1,$$
$$unifier_1 = \sigma, \quad unifiable_1 = \textbf{true}.$$

Now, for the induction step. We only need to consider the case where the first $i - 1$ recursive calls were successful. If we have $tree1_i = tree1_0$ and

$tree2_i = tree2_0$, then we can apply the induction hypothesis for the assertion to the address ui, since $tree1_0/ui$ is a proper subtree of $tree1_0/u$. Otherwise, $m_i < m_0$, and we can also also apply the induction hypothesis for the assertion to address ui. Note that

$$tree1_i/u = \theta_{i-1}(f(s_1, ..., s_i, ..., s_n)) \quad \text{and}$$
$$tree2_i/u = \theta_{i-1}(f(t_1, ..., t_i, ..., s_n)), \quad \text{since}$$
$$unifier_i = \sigma \circ \theta_{i-1}.$$

By lemma 8.4.3(a), since θ_{i-1} is a most general unifier for the trees

$$f(s_1, ..., s_{i-1}, \#, ..., \#) \quad \text{and} \quad f(t_1, ..., t_{i-1}, \#, ..., \#), \quad \text{then}$$
$$f(s_1, ..., s_i, \#, ..., \#) \quad \text{and} \quad f(t_1, ..., t_i, \#, ..., \#) \quad \text{are unifiable, iff}$$
$$\theta_{i-1}(f(s_1, ..., s_i, \#, ..., \#)) \quad \text{and} \quad \theta_{i-1}f((t_1, ..., t_i, \#, ..., \#)) \quad \text{are unifiable.}$$

Hence, $unify$ halts with success for this call for address ui, iff

$$f(s_1, ..., s_i, \#, ..., \#) \quad \text{and} \quad f(t_1, ..., t_i, \#, ..., \#) \quad \text{are unifiable.}$$

Otherwise, $unify$ halts with failure. This proves part (1) of the claim.

By part (2) of the assertion, the output value of the variable $unifier$ is of the form $unifier_i \circ \lambda_i$, where λ_i is a most general unifier for $\theta_{i-1}(s_i)$ and $\theta_{i-1}(t_i)$ (the subtrees at ui), and since θ_{i-1} is a most general unifier for

$$f(s_1, ..., s_{i-1}, \#, ..., \#) \quad \text{and} \quad f(t_1, ..., t_{i-1}, \#, ..., \#),$$

λ_i is a most general unifier for

$$\theta_{i-1}(f(s_1, ..., s_i, \#, ..., \#)) \quad \text{and} \quad \theta_{i-1}f((t_1, ..., t_i, \#, ..., \#)).$$

By lemma 8.4.3(b), $\theta_{i-1} \circ \lambda_i$ is a most general unifier for

$$f(s_1, ..., s_i, \#, ..., \#) \quad \text{and} \quad f(t_1, ..., t_i, \#, ..., \#).$$

Letting
$$\theta_i = \theta_{i-1} \circ \lambda_i,$$

it is easily seen that part (2) of the claim is satisfied. By part (3) of the assertion, part (3) of the claim also holds.

This concludes the proof of the claim. $\square$

For $i = n + 1$, we see that all the recursive calls in the while loop halt successfully iff s and t are unifiable, and if s and t are unifiable, when the loop is exited, we have

$$unifier_{n+1} = \sigma \circ \theta_n,$$

where θ_n is a most general unifier of s and t,

$$treel_{n+1} = unifier_{n+1}(s_0), \quad tree2_{n+1} = unifier_{n+1}(t_0),$$

and part (3) of the assertion also holds. This concludes the proof of the assertion. $\square$

But now, we can apply the assertion to the input trees t_1 and t_2, with $u = e$, and σ the identity substitution. The correctness assertion says that *unify* always halts, and if it halts with success, the output variable *unifier* is a most general unifier for t_1 and t_2. This concludes the correctness proof. $\square$

The subject of unification is the object of current research because fast unification is crucial for the efficiency of programming logic systems such as PROLOG. Some fast unification algorithms have been published such as Paterson and Wegman, 1978; Kapur, Krishnamoorthy, and Narendran, 1982; and Huet, 1976. For a survey on unification, see the article by Siekman in Shostak, 1984a. Huet, 1976, also contains a thorough study of unification, including higher-order unification.

PROBLEMS

8.4.1. Convert the following formulae to clause form:

$$\forall y(\exists x(P(y,x) \vee \neg Q(y,x)) \wedge \exists x(\neg P(x,y) \vee Q(x,y)))$$
$$\forall x(\exists y P(x,y) \wedge \neg Q(y,x)) \vee (\forall y \exists z(R(x,y,z) \wedge \neg Q(y,z)))$$
$$\neg(\forall x \exists y P(x,y) \supset (\forall y \exists z \neg Q(x,z) \wedge \forall y \neg \forall z R(y,z)))$$
$$\forall x \exists y \forall z(\exists w(Q(x,w) \vee R(x,y)) \equiv \neg \exists w \neg \exists u(Q(x,w) \wedge \neg R(x,u)))$$

8.4.2. Apply the unification algorithm to the following clauses:

$$\{P(x,y), P(y,f(z))\}$$
$$\{P(a,y,f(y)), P(z,z,u)\}$$
$$\{P(x,g(x)), P(y,y)\}$$
$$\{P(x,g(x),y), P(z,u,g(u))\}$$
$$\{P(g(x),y), P(y,y), P(u,f(w))\}$$

8.4.3. Let S and T be two finite sets of terms such that the set of variables occurring in S is disjoint from the set of variables occurring in T. Prove that if $S \cup T$ is unifiable, σ_S is a most general unifier of S, σ_T is a most general unifier of T, and $\sigma_{S,T}$ is a most general unifier of $\sigma_S(S) \cup \sigma_T(T)$, then

$$\sigma_S \circ \sigma_T \circ \sigma_{S,T}$$

is a most general unifier of $S \cup T$.

8.4.4. Show that the most general unifier of the following two trees contains a tree with 2^{n-1} occurrences of the variable x_1:

$$f(g(x_1, x_1), g(x_2, x_2), ..., g(x_{n-1}, x_{n-1})) \quad \text{and}$$
$$f(x_2, x_3, ..., x_n)$$

$*$ **8.4.5.** Define the relation $\leq$ on terms as follows: Given any two terms t_1, t_2,

$$t_1 \leq t_2 \quad \text{iff} \quad \text{there is a substitution } \sigma \text{ such that } t_2 = \sigma(t_1).$$

Define the relation $\cong$ such that

$$t_1 \cong t_2 \quad \text{iff} \quad t_1 \leq t_2 \text{ and } t_2 \leq t_1.$$

(a) Prove that $\leq$ is reflexive and transitive and that $\cong$ is an equivalence relation.

(b) Prove that $t_1 \cong t_2$ iff there is a bijective renaming of variables ρ such that $t_1 = \rho(t_2)$. Show that the relation $\leq$ induces a partial ordering on the set of equivalence classes of terms modulo the equivalence relation $\cong$.

(c) Prove that two terms have a least upper bound iff they have a most general unifier (use a separating substitution, see Section 8.5).

(d) Prove that any two terms always have a greatest lower bound.

Remark: The structure of the set of equivalence classes of terms modulo $\cong$ under the partial ordering $\leq$ has been studied extensively in Huet, 1976. Huet has shown that this set is well founded, that every subset has a greatest lower bound, and that every bounded subset has a least upper bound.

8.5 The Resolution Method for First-Order Logic

Recall that we are considering first-order languages without equality. Also, recall that even though we usually omit quantifiers, clauses are universally quantified sentences. We extend the definition of a resolvent given in definition 4.3.2 to arbitrary clauses using the notion of a most general unifier.

8.5.1 Definition of the Method

First, we define the concept of a separating pair of substitutions.

Definition 8.5.1 Given two clauses A and A', a *separating pair of substitutions* is a pair of substitutions ρ and ρ' such that:

ρ has support $FV(A)$, ρ' has support $FV(A')$, for every variable x in A, $\rho(x)$ is a variable, for every variable y in A', $\rho'(y)$ is a variable, ρ and ρ' are bijections, and the range of ρ and the range of ρ' are disjoint.

Given a set S of literals, we say that S is *positive* if all literals in S are atomic formulae, and we say that S is *negative* if all literals in S are negations of atomic formulae. If a set S is positive or negative, we say that the literals in S are of the *same sign*. Given a set of literals $S = \{A_1, ..., A_m\}$, the *conjugate* of S is defined as the set

$$\overline{S} = \{\overline{A_1}, ..., \overline{A_m}\}$$

of conjugates of literals in S. If S is a positive set of literals we let $|S| = S$, and if S is a negative set of literals, we let $|S| = \overline{S}$.

Definition 8.5.2 Given two clauses A and B, a clause C is a *resolvent* of A and B iff the following holds:

(i) There is a subset $A' = \{A_1, ..., A_m\} \subseteq A$ of literals all of the same sign, a subset $B' = \{B_1, ..., B_n\} \subseteq B$ of literals all of the opposite sign of the set A', and a separating pair of substitutions (ρ, ρ') such that the set

$$|\rho(A') \cup \rho'(\overline{B'})|$$

is unifiable;

(ii) For some most general unifier σ of the set

$$|\rho(A') \cup \rho'(\overline{B'})|,$$

we have

$$C = \sigma(\rho(A - A') \cup \rho'(B - B')).$$

EXAMPLE 8.5.1

Let
$$A = \{\neg P(z, a), \neg P(z, x), \neg P(x, z)\} \quad \text{and}$$
$$B = \{P(z, f(z)), P(z, a)\}.$$

Let
$$A' = \{\neg P(z, a), \neg P(z, x)\} \quad \text{and} \quad B' = \{P(z, a)\},$$
$$\rho = (z_1/z), \quad \rho' = (z_2/z).$$

Then,
$$|\rho(A') \cup \rho'(\overline{B'})| = \{P(z_1, a), P(z_1, x), P(z_2, a)\}$$

is unifiable,

$$\sigma = (z_1/z_2, a/x)$$

is a most general unifier, and

$$C = \{\neg P(a, z_1), P(z_1, f(z_1))\}$$

is a resolvent of A and B.

If we take $A' = A$, $B' = \{P(z, a)\}$,

$$|\rho(A') \cup \rho'(\overline{B'})| = \{P(z_1, a), P(z_1, x), P(x, z_1), P(z_2, a)\}$$

is also unifiable,

$$\sigma(a/z_1, a/z_2, a/x)$$

is the most general unifier, and

$$C = \{P(a, f(a))\}$$

is a resolvent.

Hence, two clauses may have several resolvents.

The generalization of definition 4.3.3 of a resolution DAG to the first-order case is now obvious.

Definition 8.5.3 Given a set $S = \{C_1, ..., C_n\}$ of first-order clauses, a *resolution DAG* for S is any finite set

$$G = \{(t_1, R_1), ..., (t_m, R_m)\}$$

of distinct DAGs labeled in the following way:

(1) The leaf nodes of each underlying tree t_i are labeled with clauses in S.

(2) For every DAG (t_i, R_i), every nonleaf node u in t_i is labeled with some triple $(C, (\rho, \rho'), \sigma)$, where C is a clause, (ρ, ρ') is a separating pair of substitutions, σ is a substitution and the following holds:

For every nonleaf node u in t_i, u has exactly two successors $u1$ and $u2$, and if $u1$ is labeled with a clause C_1 and u_2 is labeled with a clause C_2 (not necessarily distinct from C_1), then u is labeled with the triple $(C, (\rho, \rho'), \sigma)$, where (ρ, ρ') is a separating pair of substitutions for C_1 and C_2 and C is resolvent of C_1 and C_2 obtained with the most general unifier σ.

A resolution DAG is a *resolution refutation* iff it consists of a single DAG (t, R) whose root is labeled with the empty clause. The nodes of a DAG that are not leaves are also called *resolution steps*.

We will often use a simplified form of the above definition by dropping (ρ, ρ') and σ from the interior nodes, and consider that nodes are labeled with clauses. This has the effect that it is not always obvious how a resolvent is obtained.

EXAMPLE 8.5.2

Consider the following clauses:

$$C_1 = \{\neg P(z_1, a), \neg P(z_1, x), \neg P(x, z_1)\},$$
$$C_2 = \{P(z_2, f(z_2)), P(z_2, a)\} \quad \text{and}$$
$$C_3 = \{P(f(z_3), z_3)), P(z_3, a)\}.$$

The following is a resolution refutation:

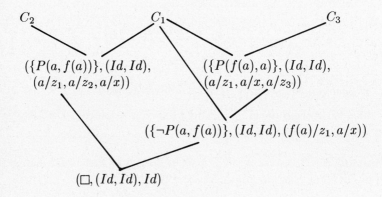

8.5.2 Soundness of the Resolution Method

In order to prove the soundness of the resolution method, we prove the following lemma, analogous to lemma 4.3.1.

Lemma 8.5.1 Given two clauses A and B, let $C = \sigma(\rho(A - A') \cup \rho'(B - B'))$ be any resolvent of A and B, for some subset $A' \subseteq A$ of literals of A, subset $B' \subseteq B$ of literals of B, separating pair of substitutions (ρ, ρ'), with $\rho = (z_1/x_1, ..., z_m/x_m)$, $\rho' = (z_{m+1}/y_1, ..., z_{m+n}/y_n)$ and most general unifier $\sigma = (t_1/u_1, ..., t_k/u_k)$, where $\{u_1, ..., u_k\}$ is a subset of $\{z_1, ..., z_{m+n}\}$. Also, let $\{v_1, ..., v_p\} = FV(C)$. Then,

$$\models (\forall x_1...\forall x_m A \wedge \forall y_1...\forall y_n B) \supset \forall v_1...\forall v_p C.$$

Proof: We show that we can constuct a G-proof for

$$(\forall x_1...\forall x_m A \wedge \forall y_1...\forall y_n B) \rightarrow \forall v_1...\forall v_p C.$$

Note that $\{z_1, ..., z_{m+n}\} - \{u_1, ..., u_k\}$ is a subset of $\{v_1, ..., v_p\}$. First, we perform $p \; \forall : right$ steps using p entirely new variables $w_1, ..., w_p$. Let

$$\sigma' = \sigma \circ (w_1/v_1, ..., w_p/v_p) = (t_1'/z_1, ..., t_{m+n}'/z_{m+n}),$$

be the substitution obtained by composing σ and the substitution replacing each occurrence of the variable v_i by the variable w_i. Then, note that the support of σ' is disjoint from the set $\{w_1, ..., w_p\}$, which means that for every tree t,

$$\sigma'(t) = t[t_1'/z_1]...[t_{m+n}'/z_{m+n}]$$

(the order being irrelevant). At this point, we have the sequent

$$(\forall x_1...\forall x_m A \wedge \forall y_1...\forall y_n B) \rightarrow \sigma'(\rho(A - A')), \sigma'(\rho'(B - B')).$$

Then apply the $\wedge : left$ rule, obtaining

$$\forall x_1...\forall x_m A, \forall y_1...\forall y_n B \rightarrow \sigma'(\rho(A - A')), \sigma'(\rho'(B - B')).$$

At this point, we apply $m+n \; \forall : left$ rules as follows: If $\rho(x_i)$ is some variable u_j, do the substitution t_j'/x_i, else $\rho(x_i)$ is some variable v_j not in $\{u_1, .., u_k\}$, do the substitution w_j/v_j.

If $\rho'(y_i)$ is some variable u_j, do the substitution t_j'/y_i, else $\rho'(y_j)$ is some variable v_j not in $\{u_1, ..., u_k\}$, do the substitution w_j/v_j.

It is easy to verify that at the end of these steps, we have the sequent

$$(\sigma'(\rho(A - A')), Q), (\sigma'(\rho'(B - B')), \overline{Q}) \rightarrow \sigma'(\rho(A - A')), \sigma'(\rho'(B - B'))$$

where $Q = \sigma'(\rho(A'))$ and $\overline{Q} = \sigma'(\rho'(B'))$ are conjugate literals, because σ is a most general unifier of the set $|\rho(A') \cup \rho'(\overline{B'})|$.

Hence, we have a quantifier-free sequent of the form

$$(A_1 \vee Q), (A_2 \vee \neg Q) \rightarrow A_1, A_2,$$

and we conclude that this sequent is valid using the proof of lemma 4.3.1. $\square$

As a consequence, we obtain the soundness of the resolution method.

Lemma 8.5.2 (Soundness of resolution without equality) If a set of clauses has a resolution refutation DAG, then S is unsatisfiable.

Proof: The proof is identical to the proof of lemma 4.3.2, but using lemma 8.5.1, as opposed to lemma 4.3.1. $\square$

8.5.3 Completeness of the Resolution Method

In order to prove that completeness of the resolution method for first-order languages without equality, we shall prove the following lifting lemma.

Lemma 8.5.3 (Lifting lemma) Let A and B be two clauses, σ_1 and σ_2 two ground substitutions, and assume that D is a resolvent of the ground clauses $\sigma_1(A)$ and $\sigma_2(B)$. Then, there is a resolvent C of A and B and a substitution θ such that $D = \theta(C)$.

Proof: First, let (ρ, ρ') be a separating pair of substitutions for A and B. Since ρ and ρ' are bijections they have inverses ρ^{-1} and ρ'^{-1}. Let σ be the substitution formed by the union of $\rho^{-1} \circ \sigma_1$ and $\rho'^{-1} \circ \sigma_2$, which is well defined, since the supports of ρ^{-1} and ρ'^{-1} are disjoint. It is clear that

$$\sigma(\rho(A)) = \sigma_1(A) \quad \text{and} \quad \sigma(\rho'(B)) = \sigma_2(B).$$

Hence, we can work with $\rho(A)$ and $\rho'(B)$, whose sets of variables are disjoint. If D is a resolvent of the clauses $\sigma_1(A)$ and $\sigma_2(B)$, there is a ground literal Q such that $\sigma(\rho(A))$ contains Q and $\sigma(\rho'(B))$ contains its conjugate. Assume that Q is positive, the case in which Q is negative being similar. Then, there must exist subsets $A' = \{A_1, ..., A_m\}$ of A and $B' = \{\neg B_1, ..., \neg B_n\}$ of B, such that

$$\sigma(\rho(A_1)) = ... = \sigma(\rho(A_m)) = \sigma(\rho'(B_1)) = ..., \sigma(\rho'(B_n)) = Q,$$

and σ is a unifier of $\rho(A') \cup \rho'(\overline{B'})$. By theorem 8.4.1, there is a most general unifier λ and a substitution θ such that

$$\sigma = \lambda \circ \theta.$$

Let C be the resolvent

$$C = \lambda(\rho(A - A') \cup \rho'(B - B')).$$

Clearly,

$$
\begin{aligned}
D &= (\sigma(\rho(A)) - \{Q\}) \cup (\sigma(\rho'(B)) - \{\neg Q\}) \\
&= (\sigma(\rho(A - A')) \cup \sigma(\rho'(B - B'))) \\
&= \theta(\lambda(\rho(A - A') \cup \rho'(B - B'))) = \theta(C).
\end{aligned}
$$

$\square$

Using the above lemma, we can now prove the following lemma which shows that resolution DAGs of ground instances of clauses can be lifted to resolution DAGs using the original clauses.

Lemma 8.5.4 (Lifting lemma for resolution refutations) Let S be a finite set of clauses, and S_g be a set of ground instances of S, so that every clause in

S_g is of the form $\sigma_i(C_i)$ for some clause C_i in S and some ground substitution σ_i.

For any resolution DAG H_g for S_g, there is a resolution DAG H for S, such that the DAG H_g is a homomorphic image of the DAG H in the following sense:

There is a function $F : H \rightarrow H_g$ from the set of nodes of H to the set of nodes of H_g, such that, for every node u in H, if $u1$ and $u2$ are the immediate descendants of u, then $h(u1)$ and $h(u2)$ are the immediate descendants of $h(u)$, and if the clause C_i is the label of u, then $h(u)$ is labeled by the clause $\theta_i(C_i)$, where θ_i is some ground substitution.

Proof: We prove the lemma by induction on the underlying tree of H_g.

(i) If H_g has a single resolution step, we have clauses $\sigma_1(A)$, $\sigma_2(B)$ and their resolvent D. By lemma 8.5.3, there exists a resolvent C of A and B and a substitution θ such that $\theta(C) = D$. Note that it is possible that A and B are distinct, but $\sigma_1(A)$ and $\sigma_2(B)$ are not. In the first case, we have the following DAGs:

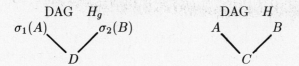

The homomorphism F is such that $F(e) = e$, $F(1) = 1$ and $F(2) = 1$.

In the second case, $\sigma_1(A) \neq \sigma_2(B)$, but we could have $A = B$. Whether or not $A = B$, we create the following DAG H with three distinct nodes, so that the homomorphism is well defined:

DAG H_g DAG H

$\sigma_1(A)$ $\sigma_2(B)$ A B

D C

The homomorphism F is the identity on nodes.

(ii) If H_g has more than one resolution step, it is of the form either

(ii)(a)

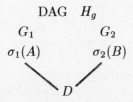

where $\sigma_1(A)$ and $\sigma_2(B)$ are distinct, or of the form

(ii)(b)

$$\text{DAG} \quad H_g$$
$$G_1$$
$$\sigma_1(A) = \sigma_2(B)$$
$$()$$
$$D$$

if $\sigma_1(A) = \sigma_2(B)$.

   (a) In the first case, by the induction hypothesis, there are DAGs H_1 and H_2 and homomorphisms $F_1 : H_1 \to G_1$ and $F_2 : H_2 \to G_2$, where H_1 is rooted with A and H_2 is rooted with B. By lemma 8.5.3, there is a resolvent C of A and B and a substitution θ such that $\theta(C) = D$. We can construct H as the DAG obtained by making C as the root, and even if $A = B$, by creating two distinct nodes 1 and 2, with 1 labeled A and 2 labeled B:

$$\text{DAG} \quad H$$
$$H_1 \qquad\qquad H_2$$
$$A \qquad\qquad\quad B$$

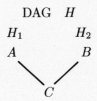

$$C$$

   The homomorphism $F : H \to H_g$ is defined such that $F(e) = e$, $F(1) = 1$, $F(2) = 2$, and it behaves like F_1 on H_1 and like F_2 on H_2. The root clause C is mapped to $\theta(C) = D$.

   (b) In the second case, either $A = B$ or $A \neq B$. If $A = B$, by the induction hypothesis, there is a DAG H_1 rooted with $A = B$ and a homomorphism $F_1 : H_1 \to G_1$. It is clear that we can form H so that C is a root node with two edges connected to A, and F is the homomorphism such that $F(e) = e$, $F(1) = 1$, and F behaves like F_1 on H_1.

$$\text{DAG} \quad H$$
$$H_1$$
$$A = B$$
$$()$$
$$C$$

   The clause C is mapped onto $D = \theta(C)$.

   If $A \neq B$ and $\sigma_1(A) = \sigma_2(B)$, by the induction hypothesis, there are two DAGs G_1 and G_2 and two homomorphisms $F_1 : H_1 \to G_1$ and $F_2 : H_2 \to G_1$.

As in case (ii)(a), there is a resolvent C of A and B and a substitution θ such that $\theta(C) = D$. The DAG H is formed as in case (ii)(a). The homomorphism F is defined so that $F(e) = e$, $F(1) = 1$, $F(2) = 1$, and F behaves like F_1 on H_1 and like F_2 on H_2. Since both F_1 and F_2 have range G_1, F is well defined. This concludes the proof. $\square$

EXAMPLE 8.5.3

The following shows a lifting of the ground resolution of example 8.3.1 for the clauses:

$$C_1 = \{\neg P(z_1, a), \neg P(z_1, x), \neg P(x, z_1)\}$$
$$C_2 = \{P(z_2, f(z_2)), P(z_2, a)\}$$
$$C_3 = \{P(f(z_3), z_3)), P(z_3, a)\}.$$

Recall that the ground instances are

$$G_1 = \{\neg P(a, a)\}$$
$$G_2 = \{P(a, f(a)), P(a, a)\}$$
$$G_3 = \{P(f(a), a)), P(a, a)\}$$
$$G_4 = \{\neg P(f(a), a), \neg P(a, f(a))\}.$$

Ground resolution-refutation H_g
for the set of ground clauses G_1, G_2, G_3, G_4

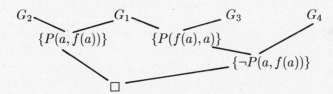

Lifting H of the above resolution refutation
for the clauses C_1, C_2, C_3

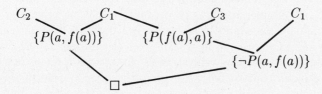

The homomorphism is the identity on the nodes, and the substitutions are, (a/z_2) for node 11 labeled C_2, $(a/z_1, a/x)$ for node 12 labeled C_1,

(a/z_3) for node 212 labeled C_3, and $(f(a)/z_1, a/x)$ for node 22 labeled C_1.

Note that this DAG is not as concise as the DAG of example 8.5.1. This is because is has been designed so that there is a homomorphism from H to H_g.

As a consequence of the lifting theorem, we obtain the completeness of resolution.

Theorem 8.5.1 (Completeness of resolution, without equality) If a finite set S of clauses is unsatisfiable, then there is a resolution refutation for S.

Proof: By the Skolem-Herbrand-Gödel theorem (theorem 7.6.1, or its corollary), S is unsatisfiable iff a conjunction S_g of ground substitution instances of clauses in S is unsatisfiable. By the completeness of ground resolution (lemma 8.3.1), there is a ground resolution refutation H_g for S_g. By lemma 8.5.4, this resolution refutation can be lifted to a resolution refutation H for S. This concludes the proof. $\square$

Actually, we can also prove the following type of Herbrand theorem for the resolution method, using the constructive nature of lemma 7.6.2.

Theorem 8.5.2 (A Herbrand-like theorem for resolution) Consider a first-order language without equality. Given any prenex sentence A whose matrix is in CNF, if $A \rightarrow$ is LK-provable, then a resolution refutation of the clause form of A can be obtained constructively.

Proof: By lemma 7.6.2, a compound instance C of the Skolem form B of A can be obtained constructively. Observe that the Skolem form B of A is in fact a clause form of A, since A is in CNF. But C is in fact a conjunction of ground instances of the clauses in the clause form of A. Since $\neg C$ is provable, the *search* procedure will give a proof that can be converted to a $GCNF'$-proof. Since theorem 4.3.1 is constructive, we obtain a ground resolution refutation H_g. By the lifting lemma 8.5.4, a resolution refutation H can be constructively obtained for S_g. Hence, we have shown that a resolution refutation for the clause form of A can be constructively obtained from an LK-proof of $A \rightarrow$. $\square$

It is likely that theorem 8.5.2 has a converse, but we do not have a proof of such a result. A simpler result is to prove the converse of lemma 8.5.4, the lifting theorem. This would provide another proof of the soundness of resolution. It is indeed possible to show that given any resolution refutation H of a set S of clauses, a resolution refutation H_g for a certain set S_g of ground instances of S can be constructed. However, the homomorphism property does not hold directly, and one has to exercise care in the construction. The interested reader should consult the problems.

It should be noted that a Herbrand-like theorem for the resolution method and a certain Hilbert system has been proved by Joyner in his Ph.D thesis (Joyner, 1974). However, these considerations are somewhat beyond the scope of this text, and we will not pursue this matter any further.

PROBLEMS

8.5.1. Give separating pairs of substitutions for the following clauses:

$$\{P(x, yf(z))\}, \ \{P(y, z, f(z))\}$$
$$\{P(x, y), P(y, z)\}, \ \{Q(y, z), P(z, f(y))\}$$
$$\{P(x, g(x))\}, \ \{P(x, g(x))\}$$

8.5.2. Find all resolvents of the following pairs of clauses:

$$\{P(x, y), P(y, z)\}, \ \{\neg P(u, f(u))\}$$
$$\{P(x, x), \neg R(x, f(x))\}, \ \{R(x, y), Q(y, z)\}$$
$$\{P(x, y), \neg P(x, x), Q(x, f(x), z)\}, \ \{\neg Q(f(x), x, z), P(x, z)\}$$
$$\{P(x, f(x), z), P(u, w, w)\}, \ \{\neg P(x, y, z), \neg P(z, z, z)\}$$

8.5.3. Establish the unsatisfiability of each of the following formulae using the resolution method.

$$(\forall x \exists y P(x, y) \wedge \exists x \forall y \neg P(x, y))$$

$$(\forall x \exists y \exists z (L(x, y) \wedge L(y, z) \wedge Q(y) \wedge R(z) \wedge (P(z) \equiv R(x)))) \wedge$$
$$\forall x \forall y \forall z ((L(x, y) \wedge L(y, z)) \supset L(x, z)) \wedge \exists x \forall y \neg (P(y) \wedge L(x, y)))$$

8.5.4. Consider the following formulae asserting that a binary relation is symmetric, transitive, and total:

$$S_1 : \forall x \forall y (P(x, y) \supset P(y, x))$$
$$S_2 : \forall x \forall y \forall z ((P(x, y) \wedge P(y, z)) \supset P(x, z))$$
$$S_3 : \forall x \exists y P(x, y)$$

Prove by resolution that

$$S_1 \wedge S_2 \wedge S_3 \supset \forall x P(x, x).$$

In other words, if P is symmetric, transitive and total, then P is reflexive.

8.5.5. Complete the details of the proof of lemma 8.5.1.

* **8.5.6.** (a) Prove that given a resolution refutation H of a set S of clauses, a resolution refutation H_g for a certain set S_g of ground instances of S can be constructed.

Apply the above construction to the following refutation:

$$\{\neg P(a), Q(a)\} \qquad \{P(x)\} \qquad \{\neg P(f(a)), \neg Q(a)\}$$
$$\{Q(a)\} \qquad \{\neg Q(a)\}$$
$$\Box$$

(b) Using (a), give another proof of the soundness of the resolution method.

* **8.5.7.** As in the propositional case, another way of presenting the resolution method is as follows. Given a (finite) set S of clauses, let

$$R(S) = S \cup \{C \mid C \text{ is a resolvent of two clauses in } S\}.$$

Also, let
$$R^0(S) = S,$$
$$R^{n+1}(S) = R(R^n(S)), (n \geq 0), \text{ and let}$$
$$R^*(S) = \bigcup_{n \geq 0} R^n(S).$$

(a) Prove that S is unsatisfiable if and only if $R^*(S)$ is unsatisfiable.

(b) Prove that if S is finite, there is some $n \geq 0$ such that

$$R^*(S) = R^n(S).$$

(c) Prove that there is a resolution refutation for S if and only if the empty clause $\Box$ is in $R^*(S)$.

(d) Prove that S is unsatisfiable if and only if $\Box$ belongs to $R^*(S)$.

* **8.5.8.** Prove that the resolution method is still complete if the resolution rule is restricted to clauses that are not tautologies (that is, clauses not containing both A and $\neg A$ for some atomic formula A).

* **8.5.9.** We say that a clause C_1 *subsumes* a clause C_2 if there is a substitution σ such that $\sigma(C_1)$ is a subset of C_2. In the version of the resolution method described in problem 8.5.7, let

$$R_1(S) = R(S) - \{C \mid C \text{ is subsumed by some clause in } R(S)\}.$$
$$\text{Let } R_1^0 = S,$$
$$R_1^{n+1}(S) = R_1(R_1^n(S)) \text{ and}$$
$$R_1^*(S) = \bigcup_{n \geq 0} R_1^n(S).$$

Prove that S is unsatisfiable if and only if $\square$ belongs to $R_1^*(S)$.

8.5.10. The resolution method described in problem 8.5.7 can be modified by introducing the concept of *factoring*. Given a clause C, if C' is any subset of C and C' is unifiable, the clause $\sigma(C)$ where σ is a most general unifier of C' is a *factor* of C. The *factoring rule* is the rule that allows any factor of a clause to be added to $R(S)$. Consider the simplification of the resolution rule in which a resolvent of two clauses A and B is obtained by resolving sets A' and B' consisting of a single literal. This restricted version of the resolution rule is sometimes called *binary resolution*.

(a) Show that binary resolution together with the factoring rule is complete.

(b) Show that the factoring rule can be restricted to sets C' consisting of a pair of literals.

(c) Show that binary resolution alone is not complete.

8.5.11. Prove that the resolution method is also complete for infinite sets of clauses.

8.5.12. Write a computer program implementing the resolution method.

8.6 A Glimpse at Paramodulation

As we have noted earlier, equality causes complications in automatic theorem proving. Several methods for handling equality with the resolution method have been proposed, including the *paramodulation method* (Robinson and Wos, 1969), and the *E-resolution method* (Morris, 1969; Anderson, 1970). Due to the lack of space, we will only define the *paramodulation rule*, but we will not give a full treatment of this method.

In order to define the paramodulation rule, it is convenient to assume that the *factoring rule* is added to the resolution method. Given a clause A, if A' is any subset of A and A' is unifiable, the clause $\sigma(A)$ where σ is a most general unifier of A' is a *factor* of A. Using the factoring rule, it is easy to see that the resolution rule can be simplified, so that a resolvent of two clauses A and B is obtained by resolving sets A' and B' consisting of a single literal. This restricted version of the resolution rule is sometimes called *binary resolution* (this is a poor choice of terminology since both this restricted rule and the general resolution rule take two clauses as arguments, but yet, it is used in the literature!). It can be shown that binary resolution alone is not complete, but it is easy to show that it is complete together with the factoring rule (see problem 8.5.10).

The *paramodulation rule* is a rule that treats an equation $s \doteq t$ as a (two way) *rewrite rule*, and allows the replacement of a subterm r unifiable with

s (or t) in an atomic formula Q, by the other side of the equation, modulo substitution by a most general unifier.

More precisely, let

$$A = ((s \doteq t) \vee C)$$

be a clause containing the equation $s \doteq t$, and

$$B = (Q \vee D)$$

be another clause containing some literal Q (of the form $Pt_1...t_n$ or $\neg Pt_1...t_n$, for some predicate symbol P of rank n, possibly the equality symbol $\doteq$, in which case $n = 2$), and assume that for some tree address u in Q, the subterm $r = Q/u$ is unifiable with s (or that r is unifiable with t). If σ is a most general unifier of s and r, then the clause

$$\sigma(C \vee Q[u \leftarrow t] \vee D)$$

or $\sigma(C \vee Q[u \leftarrow s] \vee D)$, if r and t are unifiable) is a *paramodulant* of A and B. (Recall from Subsection 2.2.5, that $Q[u \leftarrow t]$ (or $Q[u \leftarrow s]$) is the result of replacing the subtree at address u in Q by t (or s)).

EXAMPLE 8.6.1

Let

$$A = \{f(x, h(y)) \doteq g(x, y), P(x)\}, \quad B = \{Q(h(f(h(x), h(a))))\}.$$

Then

$$\{Q(h(g(h(z), h(a)))), P(h(z))\}$$

is a paramodulant of A and B, in which the replacement is performed in B at address 11.

EXAMPLE 8.6.2

Let

$$A = \{f(g(x), x) \doteq h(a)\}, \quad B = \{f(x, y) \doteq h(y)\}.$$

Then,

$$\{h(z) \doteq h(a)\}$$

is a paramodulant of A and B, in which the replacement is performed in A at address e.

It can be shown that the resolution method using the (binary) resolution rule, the factoring rule, and the paramodulation rule, is complete for any finite set S of clauses, provided that the reflexity axiom and the functional reflexivity axioms are added to S. The *reflexivity axiom* is the clause

$$\{x \doteq x\},$$

and the *functional reflexivity axioms* are the clauses

$$\{f(x_1, ..., x_n) \doteq f(x_1, ..., x_n)\},$$

for each function symbol f occurring in S, of any rank $n > 0$.

The proof that this method is complete is more involved than the proof for the case of a first-language without equality, partly because the lifting lemma does not extend directly. For details, the reader is referred to Loveland, 1978.

Notes and Suggestions for Further Reading

The resolution method has been studied extensively, and there are many refinements of this method. Some of the refinements are still complete for all clauses (linear resolution, model elimination), others are more efficient but only complete for special kinds of clauses (unit or input resolution). For a detailed exposition of these methods, the reader is referred to Loveland, 1978; Robinson, 1979, and to the collection of original papers compiled in Siekman and Wrightson, 1983. One should also consult Boyer and Moore, 1979, for advanced techniques in automatic theorem proving, induction in particular. For a more introductory approach, the reader may consult Bundy, 1983, and Kowalski, 1979.

The resolution method has also been extended to higher-order logic by Andrews. The interested reader should consult Andrews, 1970; Pietrzykowski, 1973; and Huet, 1973.

Chapter 9

SLD-Resolution and Logic Programming (PROLOG)

9.1 Introduction

We have seen in Chapter 8 that the resolution method is a complete procedure for showing unsatisfiability. However, finding refutations by resolution can be a very expensive process in the general case. If subclasses of formulae are considered, more efficient procedures for producing resolution refutations can be found. This is the case for the class of Horn clauses. A Horn clause is a disjunction of literals containing at most one positive literal. For sets of Horn clauses, there is a variant of resolution called SLD-resolution, which enjoys many nice properties. SLD-resolution is a special case of a refinement of the resolution method due to Kowalski and Kuehner known as SL-resolution (Kowalski and Kuehner, 1970), and applies to special kinds of Horn clauses called definite clauses. We shall present SLD-resolution and show its completeness for Horn clauses.

SLD-resolution is also interesting because it is the main computation procedure used in PROLOG. PROLOG is a programming language based on logic, in which a computation is in fact a refutation. The idea to define a program as a logic formula and view a refutation as a computation is a very fruitful one, because it reduces the complexity of proving the correctness of programs. In fact, it is often claimed that logic programs are obviously correct, because these programs "express" the assertions that they should satisfy. However, this is not quite so, because the notion of correctness is relative, and one still needs to define the semantics of logic programs in some independent fashion. This will be done in Subsection 9.5.4, using a model-theoretic semantics. Then, the correctness of SLD-resolution (as a computation procedure) with respect to the model-theoretic semantics will be proved.

In this chapter, as in Chapter 8, we begin by studying SLD-resolution in the propositional case, and then use the lifting technique to extend the results obtained in the propositional case to the first-order case. Fortunately, the lifting process goes very smoothly.

As in Chapter 4, in order to prove the completeness of SLD-resolution for propositional Horn clauses, we first show that Horn clauses have *GCNF'*-proofs of a certain kind, that we shall call *GCNF'*-proofs in SLD-form. Then, we show that every *GCNF'*-proof in SLD-form can be mapped into a linear SLD-refutation. Hence, the completeness proof for SLD-resolution is constructive.

The arguments used for showing that every unsatisfiable Horn clause has a *GCNF'*-proof in SLD-form are quite basic and combinatorial in nature. Once again, the central concept is that of *proof transformation*.

We conclude this chapter by discussing the notion of logic program and the idea of viewing SLD-resolution as a computation procedure. We provide a rigorous semantics for logic programs, and show the correctness and completeness of SLD-resolution with respect to this semantics.

The contents of Section 9.5 can be viewed as the theoretical foundations of logic programming, and PROLOG in particular.

9.2 *GCNF'*-Proofs in SLD-Form

First, we shall prove that every unsatisfiable propositional Horn clause has a *GCNF'*-proof of a certain type, called a proof in SLD-form. In order to explain the method for converting a *GCNF'*-proof into an SLD-resolution proof, it is convenient to consider the special case of sets of Horn clauses, containing exactly one clause containing no positive literals (clause of the form $\{\neg P_1, ..., \neg P_m\}$). Other Horn clauses will be called *definite clauses*.

9.2.1 The Case of Definite Clauses

These concepts are defined as follows.

Definition 9.2.1 A *Horn clause* is a disjunction of literals containing at most one positive literal. A Horn clause is a *definite clause* iff it contains a (single) positive literal. Hence, a definite clause is either of the form

$$\{Q\}, \quad \text{or} \quad \{\neg P_1, ..., \neg P_m, Q\}.$$

A Horn clause of the form

$$\{\neg P_1, ..., \neg P_m\}$$

is called a *negative clause* or *goal clause*.

For simplicity of notation, a clause $\{Q\}$ will also be denoted by Q. In the rest of this section, we restrict our attention to sets S of clauses consisting of definite clauses except for one goal clause. Our goal is to show that for a set S consisting of definite clauses and of a single goal B, if S is $GCNF'$-provable, then there is a proof having the property that whenever a $\vee : left$ rule is applied to a definite clause $\{\neg P_1, ..., \neg P_m, Q\}$, the rule splits it into $\{\neg P_1, ..., \neg P_m\}$ and $\{Q\}$, the sequent containing $\{Q\}$ is an axiom, and the sequent containing $\{\neg P_1, ..., \neg P_m\}$ does not contain $\neg Q$.

EXAMPLE 9.2.1

Consider the set S of Horn clauses with goal $\{\neg P_1, \neg P_2\}$ given by:

$$S = \{\{P_3\}, \{P_4\}, \{\neg P_1, \neg P_2\}, \{\neg P_3, \neg P_4, P_1\}, \{\neg P_3, P_2\}\}.$$

The following is a $GCNF'$-proof:

$$
\cfrac{
  \cfrac{
    P_3, \neg P_3 \to \quad P_4, \neg P_4 \to
  }{
    P_3, P_4, \{\neg P_3, \neg P_4\} \to
  }
  \qquad
  \cfrac{
    \neg P_1, P_1 \to \quad
    \cfrac{
      \neg P_2, P_2 \to \quad P_3, \neg P_3 \to
    }{
      P_3, \neg P_2, \{\neg P_3, P_2\} \to
    }
  }{
    P_3, \{\neg P_1, \neg P_2\}, P_1, \{\neg P_3, P_2\} \to
  }
}{
  P_3, P_4, \{\neg P_1, \neg P_2\}, \{\neg P_3, \neg P_4, P_1\}, \{\neg P_3, P_2\} \to
}
$$

Another proof having the properties mentioned above is

$$
\cfrac{
  \cfrac{
    \neg P_1, P_1 \to \quad
    \cfrac{
      P_3, \neg P_3 \to \quad P_4, \neg P_4 \to
    }{
      P_3, P_4, \{\neg P_3, \neg P_4\} \to
    }
  }{
    P_3, P_4, \neg P_1, \{\neg P_3, \neg P_4, P_1\} \to
  }
  \qquad
  \cfrac{
    \neg P_2, P_2 \to \quad P_3, \neg P_3 \to
  }{
    P_3, \neg P_2, \{\neg P_3, P_2\} \to
  }
}{
  P_3, P_4, \{\neg P_1, \neg P_2\}, \{\neg P_3, \neg P_4, P_1\}, \{\neg P_3, P_2\} \to
}
$$

Observe that in the above proof, the $\vee : left$ rule is first applied to the goal clause $\{\neg P_1, \neg P_2\}$, and then it is applied to split each definite clause $\{\neg Q_1, ..., \neg Q_m, Q\}$ into $\{\neg Q_1, ..., \neg Q_m\}$ and $\{Q\}$, in such a way that the sequent containing $\{Q\}$ is the axiom $Q, \neg Q \to$. Note also that each clause $\{\neg Q_1, ..., \neg Q_m\}$ resulting from splitting a definite clause as indicated above is the only goal clause in the sequent containing it.

9.2.2 *GCNF'*-Proofs in SLD-Form

The above example suggests that if a set of definite clauses with goal B is *GCNF'*-provable, it has a proof obtained by following rules described below, starting with a one-node tree containing the goal $B = \{\neg P_1, ..., \neg P_m\}$:

(1) If no leaf of the tree obtained so far contains a clause consisting of a single negative literal $\neg Q$ then,

As long as the tree is not a *GCNF'*-proof tree, apply the $\vee : left$ rule to each goal clause B of the form $\{\neg Q_1, ..., \neg Q_m\}$ $(m > 1)$ in order to form m immediate descendants of B, else

(2) For every goal clause consisting of a single negative literal $\neg Q$, find a definite clause $\{\neg P_1, ..., \neg P_k, Q\}$ (or Q when $k = 0$), and split $\{\neg P_1, ..., \neg P_k, Q\}$ using the $\vee : left$ rule in order to get the axiom $\neg Q, Q \rightarrow$ in one node, $\{\neg P_1, ..., \neg P_m\}$ in the other, and drop $\neg Q$ from that second node.

Go back to (1).

In is not clear that such a method works, and that in step (2), the existence of a definite clause $\{\neg P_1, ..., \neg P_k, Q\}$ such that Q cancels $\neg Q$ is guaranteed. However, we are going to prove that this is always the case. First, we define the type of proofs arising in the procedure described above.

Definition 9.2.2 Given a set S of clauses consisting of definite clauses and of a single goal B, a *GCNF'*-proof is in *SLD-form* iff the conditions below are satisfied:

For every node B in the tree that is not an axiom:

(1) If the set of clauses labeling that node does not contain any clause consisting of a single negative literal $\neg Q$, then it contains a single goal clause of the form $\{\neg Q_1, ..., \neg Q_m\}$ $(m > 1)$, and the $\vee : left$ rule is applied to this goal clause in order to form m immediate descendants of B.

(2) If the set of clauses labeling that node contains some single negative literal, for such a clause $\neg Q$, there is some definite clause

$$\{\neg P_1, ..., \neg P_k, Q\},$$

$(k > 0)$, such that the $\vee : left$ rule is applied to

$$\{\neg P_1, ..., \neg P_k, Q\}$$

in order to get the axiom $\neg Q, Q \rightarrow$ and a sequent containing the single goal clause $\{\neg P_1, ..., \neg P_k\}$.

We are now going to prove that if a set of clauses consisting of definite clauses and of a single goal clause if provable in *GCNF'*, then it has a proof in SLD-form. For this, we are going to perform proof transformations, and use simple combinatorial properties.

9.2.3 Completeness of Proofs in SLD-Form

First, we need to show that every $GCNF'$-provable set of clauses has a proof in which no weakenings takes place. This is defined as follows.

Definition 9.2.3 A $GCNF'$-proof is *without weakenings* iff every application of the $\vee : left$ rule is of the form:

$$\frac{\Gamma, A_1, ..., A_m \to \qquad \Gamma, B \to}{\Gamma, (A_1 \vee B), ..., (A_m \vee B) \to}$$

We have the following normal form lemma.

Lemma 9.2.1 If a set S of clauses is $GCNF'$-provable, then a $GCNF'$-proof without weakenings and in which all the axioms contain only literals can be constructed.

Proof: Since G' is complete, $S \to$ has a G'-proof T. By lemma 6.3.1 restricted to propositions, $S \to$ has a G'-proof T' in which all axioms are atomic. Using lemma 4.2.2, $S \to$ has a G'-proof T'' in which all axioms are atomic, and in which all applications of the $\vee : left$ rule precede all applications of the $\neg : left$ rule. The tree obtained from T'' by retaining the portion of the proof tree that does not contain $\neg : left$ inferences is the desired $GCNF'$-proof. $\square$

The following permutation lemma is the key to the conversion to SLD-form.

Lemma 9.2.2 Let S be a set of clauses that has a $GCNF'$-proof T. Then, for any clause C in S having more than one literal, for any partition of the literals in C into two disjunctions A and B such $C = (A \vee B)$, there is a $GCNF'$-proof T' in which the $\vee : left$ rule is applied to $(A \vee B)$ at the root. Furthermore, if the proof T of S is without weakenings and all axioms contain only literals, the proof T' has the same depth as T.

Proof: Observe that representing disjunctions of literals as unordered sets of literals is really a convenience afforded by the associativity, commutativity and idempotence of $\vee$, but that this convenience does not affect the completeness of G'. Hence, no matter how C is split into a disjunction $(A \vee B)$, the sequent $\Gamma, (A \vee B) \to$ is G'-provable. By converting a G'-proof of $\Gamma, (A \vee B) \to$ given by lemma 9.2.1 into a $GCNF'$-proof, we obtain a $GCNF'$-proof without weakenings, and in which the $\vee : left$ rule is applied to A and B only after it is applied to $(A \vee B)$. If the $\vee : left$ rule applied at the root does not apply to $(A \vee B)$, it must apply to some other disjunction $(C \vee D)$. Such a proof T must be of the following form:

Tree T

$$\frac{\Pi_1 \qquad\qquad \Pi_2}{\Gamma, (A \vee B), (C \vee D) \to}$$

where Π_1 is the tree

$$\frac{\dfrac{T_1}{\Gamma_1, A \to} \qquad \dfrac{S_1}{\Gamma_1, B \to}}{\Gamma_1, (A \vee B) \to} \qquad\qquad \frac{\dfrac{T_m}{\Gamma_m, A \to} \qquad \dfrac{S_m}{\Gamma_m, B \to}}{\Gamma_m, (A \vee B) \to}$$

R

$$\Gamma, (A \vee B), C \to$$

and where Π_2 is the tree

$$\frac{\dfrac{T'_1}{\Delta_1, A \to} \qquad \dfrac{S'_1}{\Delta_1, B \to}}{\Delta_1, (A \vee B) \to} \qquad\qquad \frac{\dfrac{T'_n}{\Delta_n, A \to} \qquad \dfrac{S'_n}{\Delta_n, B \to}}{\Delta_n, (A \vee B) \to}$$

S

$$\Gamma, (A \vee B), D \to$$

In the above proof, we have indicated the nodes to which the $\vee : left$ rule is applied, nodes that must exist since all axioms consist of literals. The inferences above $\Gamma, (A \vee B), C$ and below applications of the $\vee : left$ rule to $(A \vee B)$ are denoted by R, and the similar inferences above $\Gamma, (A \vee B), D$ are denoted by S. We can transform T into T' by applying the $\vee : left$ rule at the root as shown below:

Tree T'

$$\frac{\Pi'_1 \qquad\qquad \Pi'_2}{\Gamma, (A \vee B), (C \vee D) \to}$$

where Π_1' is the tree

$$\frac{T_1}{\Gamma_1, A \to} \qquad \frac{T_m}{\Gamma_m, A \to} \qquad \frac{T_1'}{\Delta_1, A \to} \qquad \frac{T_n'}{\Delta_n, A \to}$$

$$R \qquad\qquad\qquad S$$

$$\frac{\Gamma, A, C \to \qquad \Gamma, A, D \to}{\Gamma, A, (C \vee D) \to}$$

and where Π_2' is the tree

$$\frac{S_1}{\Gamma_1, B \to} \qquad \frac{S_m}{\Gamma_m, B \to} \qquad \frac{S_1'}{\Delta_1, B \to} \qquad \frac{S_n'}{\Delta_n, B \to}$$

$$R \qquad\qquad\qquad S$$

$$\frac{\Gamma, B, C \to \qquad \Gamma, B, D \to}{\Gamma, B, (C \vee D) \to}$$

Clearly, $depth(T') = depth(T)$. $\square$

Note that T' is obtained from T by permutation of inferences. We need another crucial combinatorial property shown in the following lemma.

Lemma 9.2.3 Let S be an arbitrary set of clauses such that the subset of clauses containing more than one literal is the nonempty set $\{C_1, ..., C_n\}$ and the subset consisting of the one-literal clauses is J. Assume that S is $GCNF'$-provable, and that we have a proof T without weakenings such that all axioms consist of literals. Then, every axiom is labeled with a set of literals of the form $\{L_1, ..., L_n\} \cup J$, where each literal L_i is in C_i, $i = 1, ..., n$.

Proof: We proceed by induction on proof trees. Since S contains at least one clause with at least two literals and the axioms only contain literals, $depth(T) \geq 1$. If T has depth 1, then there is exactly one application of the $\vee : rule$ and the proof is of the following form:

$$\frac{J, L_1 \to \qquad J, L_2 \to}{J, (L_1 \vee L_2) \to}$$

Clearly, the lemma holds.

If T is a tree of depth $k + 1$, it is of the following form,

$$\frac{\dfrac{T_1}{\Gamma, A \to} \qquad \qquad \dfrac{T_2}{\Gamma, B \to}}{\Gamma, (A \vee B) \to}$$

where we can assume without loss of generality that $C_n = (A \vee B)$. By the induction hypothesis, each axiom of T_1 is labeled with a set of clauses of the form $\{L_1, ..., L_n\} \cup J$, where each literal L_i is in C_i for $i = 1, ..., n - 1$, and either $L_n = A$ if A consists of a single literal, or L_n belongs to A. Similarly, each axiom of T_2 is labeled with a set of clauses of the form $\{L_1, ..., L_n\} \cup J$, where each literal L_i is in C_i for $i = 1, ..., n-1$, and either $L_n = B$ if B consists of a single literal, or L_n belongs to B. Since the union of A and B is C_n, every axiom of T is labeled with a set of clauses of the form $\{L_1, ..., L_n\} \cup J$, where each literal L_i is in C_i, $i = 1, ..., n$. Hence, the lemma holds. $\square$

As a consequence, we obtain the following useful corollary.

Lemma 9.2.4 Let S be a set of Horn clauses. If S is $GCNF'$-provable, then S contains at least one clause consisting of a single positive literal, and at least one goal (negative) clause.

Proof: It S is an axiom, this is obvious. Otherwise, by lemma 9.2.3, if S is $GCNF'$-provable, then it has a proof T without weakenings such that every axiom is labeled with a set of literals of the form $\{L_1, ..., L_n\} \cup J$, where each literal L_i is in C_i, $i = 1, ..., n$, and J is the set of clauses in S consisting of a single literal. If J does not contain any positive literals, since every Horn clause C_i contains a negative literal say $\neg A_i$, the set $\{\neg A_1, ..., \neg A_n\} \cup J$ contains only negative literals, and so cannot be an axiom. If every clause in J is positive and every clause C_i contains some positive literal say A_i, then $\{A_1, ..., A_n\} \cup J$ contains only positive literals and cannot be an axiom. $\square$

In order to prove the main theorem of this section, we will need to show that the provability of a set of Horn clauses with several goals (negative clauses) reduces to the case of a set of Horn clauses with a single goal.

Lemma 9.2.5 Let S be a set of Horn clauses consisting of a set J of single positive literals, goal clauses $N_1, ..., N_k$, and definite clauses $C_1, ..., C_m$ containing at least two literals.

If S is $GCNF'$-provable, then there is some i, $1 \leq i \leq k$, such that

$$J \cup \{C_1, ..., C_m\} \cup \{N_i\}$$

is $GCNF'$-provable. Furthermore, if T is a $GCNF'$-proof of S without weakenings and such that the axioms contain only literals, $J \cup \{C_1, ..., C_m\} \cup \{N_i\}$ has a proof of depth less than or equal to the depth of T.

Proof: We proceed by induction on proof trees. Let T be a $GCNF'$-proof of S without weakenings and such that all axioms contain only literals.

Case 1: $J \cup \{C_1, ..., C_m\} \cup \{N_1, ..., N_k\}$ is an axiom. Then, one of the positive literals in J must be the conjugate of some negative clause N_i, and the lemma holds.

Case 2: The bottom $\lor : left$ rule is applied to one of the N_i. Without loss of generality, we can assume that it is $N_1 = \{\neg Q_1, ..., \neg Q_j, \neg P\}$.

Letting $\mathcal{C} = C_1, ..., C_m$, the proof is of the form

$$\frac{\dfrac{T_1}{J, \mathcal{C}, N_2, ..., N_k, \{\neg Q_1, ..., \neg Q_j\} \to} \qquad \dfrac{T_2}{J, \mathcal{C}, N_2, ..., N_k, \neg P \to}}{J, \mathcal{C}, N_1, ..., N_k \to}$$

Observe that the bottom sequents of T_1 and T_2 satisfy the conditions of the induction hypothesis. There are two subcases. If both

$$J, C_1, ..., C_m, \{\neg Q_1, ..., \neg Q_j\} \to \quad \text{and}$$
$$J, C_1, ..., C_m, \neg P \to$$

are provable, then

$$J, C_1, ..., C_m, \{\neg Q_1, ..., \neg Q_j, \neg P\} \to$$

is provable by application of the $\lor : rule$, and the lemma holds. If

$$J, C_1, ..., C_m, N_i \to$$

is provable for some i, $2 \leq i \leq k$, then the lemma also holds.

Case 3: The bottom $\lor : rule$ is applied to one of the C_i. Without loss of generality, we can assume that it is $C_1 = \{\neg Q_1, ..., \neg Q_j, P\}$. There are two subcases:

Case 3.1: Letting $\mathcal{N} = N_1, ..., N_k$, the proof is of the form

$$\frac{\dfrac{T_1}{J, C_2, ..., C_m, \mathcal{N}, \{\neg Q_1, ..., \neg Q_j\} \to} \qquad \dfrac{T_2}{J, P, C_2, ..., C_m, \mathcal{N} \to}}{J, C_1, ..., C_m, \mathcal{N} \to}$$

Again the induction hypothesis applies to both T_1 and T_2. If

$$J, C_2, ..., C_m, \{\neg Q_1, ..., \neg Q_j\} \to \quad \text{is provable and}$$
$$J, P, C_2, ..., C_m, N_i \to \quad \text{is provable}$$

for some i, $1 \leq i \leq k$, then by the $\vee : rule$,

$$J, C_1, ..., C_m, N_i \rightarrow$$

is also provable, and the lemma holds. If

$$J, C_2, ..., C_m, N_i \rightarrow$$

is provable for some i, $1 \leq i \leq k$, then

$$J, C_1, ..., C_m, N_i \rightarrow$$

is also provable (using weakening in the last $\vee : rule$).

Case 3.2: Letting $\mathcal{N} = N_1, ..., N_k$, the proof is of the form

$$\frac{\begin{array}{c} T_1 \\ \hline J, C_2, ..., C_m, \mathcal{N}, \{\neg Q_2, ..., \neg Q_j, P\} \rightarrow \end{array} \qquad \begin{array}{c} T_2 \\ \hline J, C_2, ..., C_m, \neg Q_1, \mathcal{N} \rightarrow \end{array}}{J, C_1, ..., C_m, \mathcal{N} \rightarrow}$$

Applying the induction hypothesis, either

$$J, C_2, ..., C_m, N_i, \{\neg Q_2, ..., \neg Q_j, P\}$$

is provable for some i, $1 \leq i \leq k$, and

$$J, C_2, ..., C_m, \neg Q_1 \rightarrow$$

is provable, and by the $\vee : rule$, $J, C_1, ..., C_m, N_i$ is provable and the lemma holds. Otherwise,

$$J, C_2, ..., C_m, N_i$$

is provable for some i, $1 \leq i \leq k$, and so $J, C_1, ..., C_m, N_i$ is also provable using weakening in the last $\vee : rule$. This concludes the proof. $\square$

We are now ready to prove the main theorem of this section.

Theorem 9.2.1 (Completeness of proofs in SLD-form) If a set S consisting of definite clauses and of a single goal $B = \{\neg P_1, ..., \neg P_n\}$ is $GCNF'$-provable, then it has a $GCNF'$-proof in SLD-form.

Proof: Assume that S is not an axiom. By lemma 9.2.2, there is a $GCNF'$-proof T without weakenings, and such that all axioms consist of literals. We proceed by induction on the depth of proof trees. If $depth(T) = 1$, the proof is already in SLD-form (this is the base case of lemma 9.2.3). If $depth(T) > 1$, by n applications of lemma 9.2.2, we obtain a proof tree T' having the same depth as T, such that the i-th inference using the $\vee : left$ rule is applied to $\{\neg P_i, ..., \neg P_n\}$. Hence, letting $\mathcal{C} = C_1, ..., C_m$, the tree T' is of the form:

$$\frac{\dfrac{T_{n-1}}{J, C, \neg P_{n-1} \rightarrow} \qquad \dfrac{T_n}{J, C, \neg P_n \rightarrow}}{J, C, \{\neg P_{n-1}, \neg P_n\} \rightarrow}$$

$$\ldots$$

$$\frac{T_1}{J, C, \neg P_1 \rightarrow} \qquad \frac{\dfrac{T_2}{J, C, \neg P_2 \rightarrow} \qquad J, C, \{\neg P_3, ..., \neg P_n\} \rightarrow}{J, C, \{\neg P_2, ..., \neg P_n\} \rightarrow}$$
$$\overline{\qquad\qquad J, C, \{\neg P_1, ..., \neg P_n\} \rightarrow \qquad\qquad}$$

where J is the set of clauses consisting of a single positive literal, and each clause C_i has more than one literal. For every subproof rooted with $J, C_1, ...,$ $C_m, \neg P_i \rightarrow$, by lemma 9.2.3, each axiom is labeled with a set of literals

$$\{L_1, ..., L_m\} \cup \{\neg P_i\} \cup J,$$

where each L_j is in C_j, $1 \leq j \leq m$. In particular, since each clause C_j contains a single positive literal A_j, for every i, $1 \leq i \leq n$, $\{A_1, ..., A_m\} \cup \{\neg P_i\} \cup J$ must be an axiom. Clearly, either some literal in J is of the form P_i, or there is some definite clause $C = \{\neg Q_1, ..., \neg Q_p, A_j\}$ among $C_1, ..., C_m$, with positive literal $A_j = P_i$. In the first case, $J, C_1, ..., C_m, \neg P_i \rightarrow$ is an axiom and the tree T_i is not present. Otherwise, let $C' = \{C_1, ..., C_m\} - \{C\}$. Using lemma 9.2.2 again, we obtain a proof R_i of

$$J, C_1, ..., C_m, \neg P_i \rightarrow$$

(of depth equal to the previous one) such that the the $\vee : left$ rule is applied to C:

$$\frac{P_i, \neg P_i \rightarrow \qquad \dfrac{T_i'}{J, \{\neg Q_1, ..., \neg Q_p\}, C', \neg P_i \rightarrow}}{J, \{\neg Q_1, ..., \neg Q_p, P_i\}, C', \neg P_i \rightarrow}$$

Note that
$$J, \{\neg Q_1, ..., \neg Q_p\}, C', \neg P_i \rightarrow$$

has two goal clauses. By lemma 9.2.5, either

$$J, \{\neg Q_1, ..., \neg Q_p\}, C' \rightarrow$$

has a proof U_i, or

$$J, C', \neg P_i \to$$

has a proof V_i, and the depth of each proof is no greater than the depth of the proof R_i of $J, \{\neg Q_1, ..., \neg Q_p, P_i\}, C', \neg P_i \to$. In the second case, by performing a weakening in the last inference of V_i, we obtain a proof for $J, C_1 ..., C_m, \neg P_i \to$ of smaller depth than the original, and the induction hypothesis applies, yielding a proof in SLD-form for $J, C_1, ..., C_m, \neg P_i \to$. In the first case, $\neg P_i$ is dropped and, by the induction hypothesis, we also have a proof in SLD-form of the form:

$$\cfrac{P_i, \neg P_i \to \qquad \cfrac{T_i''}{J, \{\neg Q_1, ..., \neg Q_p\}, C' \to}}{J, \{\neg Q_1, ..., \neg Q_p, P_i\}, C', \neg P_i \to}$$

Hence, by combining these proofs in SLD-form, we obtain a proof in SLD-form for S. $\square$

Combining theorem 9.2.1 and lemma 9.2.5, we also have the following theorem.

Theorem 9.2.2 Let S be a set of Horn clauses, consisting of a set J of single positive literals, goal clauses $N_1, ..., N_k$, and definite clauses $C_1, ..., C_m$ containing at least two literals. If S is $GCNF'$-provable, then there is some i, $1 \leq i \leq k$, such that

$$J \cup \{C_1, ..., C_m\} \cup \{N_i\}$$

has a $GCNF'$-proof in SLD-form. $\square$

Proof: Obvious by theorem 9.2.1 and lemma 9.2.5.

In the next section, we shall show how proofs in SLD-form can be converted into resolution refutations of a certain type.

PROBLEMS

9.2.1. Give a $GCNF'$-proof in SLD-form for each of the following sequents:

$$\{\neg P_3, \neg P_4, P_5\}, \{\neg P_1, P_2\}, \{\neg P_2, P_1\}, \{\neg P_3, P_4\}, \{P_3\},$$
$$\{\neg P_1, \neg P_2\}, \{\neg P_5, P_2\} \to$$

$$\{P_1\}, \{P_2\}, \{P_3\}, \{P_4\}, \{\neg P_1, \neg P_2, P_6\}, \{\neg P_3, \neg P_4, P_7\},$$
$$\{\neg P_6, \neg P_7, P_8\}, \{\neg P_8\} \to$$

$$\{\neg P_2, P_3\}, \{\neg P_3, P_4\}, \{\neg P_4, P_5\}, \{P_3\}, \{P_1\}, \{P_2\}, \{\neg P_1\},$$
$$\{\neg P_3, P_6\}, \{\neg P_3, P_7\}, \{\neg P_3, P_8\} \rightarrow$$

9.2.2. Complete the missing details in the proof of lemma 9.2.5.

9.2.3. Write a computer program for building proof trees in SLD-form for Horn clauses.

* **9.2.4.** Given a set S of Horn clauses, we define an H-tree for S as a tree labeled with propositional letters and satisfying the following properties:

(i) The root of T is labeled with **F** (false);

(ii) The immediate descendants of **F** are nodes labeled with propositional letters $P_1,...,P_n$ such that $\{\neg P_1, ..., \neg P_n\}$ is some goal clause in S;

(iii) For every nonroot node in the tree labeled with some letter Q, either the immediate descendants of that node are nodes labeled with letters $P_1,...,P_k$ such that $\{\neg P_1, ..., \neg P_k, Q\}$ is some clause in S, or this node is a leaf if $\{Q\}$ is a clause in S.

Prove that S is unsatisfiable iff it has an H-tree.

9.3 SLD-Resolution in Propositional Logic

SLD-refutations for sets of Horn clauses can be viewed as linearizations of $GCNF'$-proofs in SLD-form.

9.3.1 SLD-Derivations and SLD-Refutations

First, we show how to linearize SLD-proofs.

Definition 9.3.1 The *linearization procedure* is a recursive algorithm that converts a $GCNF'$-proof in SLD-form into a sequence of negative clauses according to the following rules:

(1) Every axiom $\neg P, P \rightarrow$ is converted to the sequence $< \{\neg P\}, \square >$.

(2) For a sequent $R \rightarrow$ containing a goal clause $N = \{\neg P_1, ..., \neg P_n\}$, with $n > 1$, if C_i is the sequence of clauses that is the linearization of the subtree with root the i-th descendant of the sequent $R \rightarrow$, construct the sequence obtained as follows:

Concatenate the sequences $C'_1,...,C'_{n-1}, C_n$, where, for each i, $1 \leq i \leq n - 1$, letting n_i be the number of clauses in the sequence C_i, the sequence C'_i has $n_i - 1$ clauses such that, for every j, $1 \leq j \leq n_i - 1$, if the j-th clause of C_i is

$$\{B_1, ..., B_m\},$$

then the j-th clause of C_i' is

$$\{B_1, ..., B_m, \neg P_{i+1}, ..., \neg P_n\}.$$

(3) For every nonaxiom sequent $\Gamma, \neg P \rightarrow$ containing some negative literal $\neg P$, if the definite clause used in the inference is $\{\neg P_1, ..., \neg P_m, P\}$, letting $\Delta = \Gamma - \{\neg P_1, ..., \neg P_m, P\}$, then if the sequence of clauses for the sequent $\Delta, \{\neg P_1, ..., \neg P_m\} \rightarrow$ is C, form the sequence obtained by concatenating $\neg P$ and the sequence C.

Note that by (1), (2), and (3), in (2), the first clause of each C_i', $(1 \leq i \leq n - 1)$, is

$$\{\neg P_i, \neg P_{i+1}, ..., \neg P_n\},$$

and the first clause of C_n is $\{\neg P_n\}$.

The following example shows how such a linearization is done.

EXAMPLE 9.3.1

Recall the proof tree in SLD-form given in example 9.2.1:

$$\cfrac{\cfrac{P_3, \neg P_3 \rightarrow \quad P_4, \neg P_4 \rightarrow}{\neg P_1, P_1 \rightarrow \quad P_3, P_4, \{\neg P_3, \neg P_4\} \rightarrow} \qquad \cfrac{}{\neg P_2, P_2 \rightarrow \quad P_3, \neg P_3 \rightarrow}}{}$$

$$\cfrac{P_3, P_4, \neg P_1, \{\neg P_3, \neg P_4, P_1\} \rightarrow \qquad P_3, \neg P_2, \{\neg P_3, P_2\} \rightarrow}{P_3, P_4, \{\neg P_1, \neg P_2\}, \{\neg P_3, \neg P_4, P_1\}, \{\neg P_3, P_2\} \rightarrow}$$

The sequence corresponding to the left subtree is

$$< \{\neg P_1\}, \{\neg P_3, \neg P_4\}, \{\neg P_4\}, \Box >$$

and the sequence corresponding to the right subtree is

$$< \{\neg P_2\}, \{\neg P_3\}, \Box >$$

Hence, the sequence corresponding to the proof tree is

$$< \{\neg P_1, \neg P_2\}, \{\neg P_3, \neg P_4, \neg P_2\}, \\ \{\neg P_4, \neg P_2\}, \{\neg P_2\}, \{\neg P_3\}, \Box > .$$

This last sequence is an SLD-refutation, as defined below.

Definition 9.3.2 Let S be a set of Horn clauses consisting of a set D of definite clauses and a set $\{G_1, ..., G_q\}$ of goals. An *SLD-derivation* for S is a sequence $< N_0, N_1, ..., N_p >$ of negative clauses satisfying the following properties:

(1) $N_0 = G_j$, where G_j is one of the goals;

(2) For every N_i in the sequence, $0 < i < p$, if

$$N_i = \{\neg A_1, ..., \neg A_{k-1}, \neg A_k, \neg A_{k+1}, ..., \neg A_n\},$$

then there is some definite clause

$$C_i = \{\neg B_1, ..., \neg B_m, A_k\}$$

in D such that, if $m > 0$, then

$$N_{i+1} = \{\neg A_1, ..., \neg A_{k-1}, \neg B_1, ..., \neg B_m, \neg A_{k+1}, ..., \neg A_n\}$$

else if $m = 0$ then

$$N_{i+1} = \{\neg A_1, ..., \neg A_{k-1}, \neg A_{k+1}, ..., \neg A_n\}.$$

An SLD-derivation is an *SLD-refutation* iff $N_p = \square$. The *SLD-resolution method* is the method in which a set of of Horn clauses is shown to be unsatisfiable by finding an SLD-refutation.

Note that an SLD-derivation is a linear representation of a resolution DAG of the following special form:

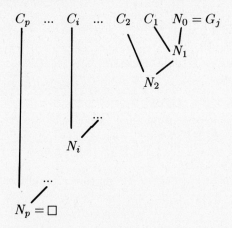

At each step, the clauses

$$\{\neg A_1, ..., \neg A_{k-1}, \neg A_k, \neg A_{k+1}, ..., \neg A_n\} \quad \text{and}$$
$$\{\neg B_1, ..., \neg B_m, A_k\}$$

are resolved, the literals A_k and $\neg A_k$ being canceled. The literal A_k is called the *selected atom* of N_i, and the clauses $N_0, C_1, ..., C_p$ are the *input clauses*.

Such a resolution method is a form of linear input resolution, because it resolves the current clause N_k with some clause in the input set D.

By the soundness of the resolution method (lemma 4.3.2), the SLD-resolution method is sound.

EXAMPLE 9.3.2

The sequence

$$< \{\neg P_1, \neg P_2\}, \{\neg P_3, \neg P_4, \neg P_2\},$$
$$\{\neg P_4, \neg P_2\}, \{\neg P_2\}, \{\neg P_3\}, \square >$$

of example 9.3.1 is an SLD-refutation.

9.3.2 Completeness of SLD-Resolution for Horn Clauses

In order to show that SLD-resolution is complete for Horn clauses, since by theorem 9.2.2 every set of Horn clauses has a $GCNF'$-proof in SLD-form, it is sufficient to prove that the linearization algorithm of definition 9.3.1 converts a proof in SLD-form to an SLD-refutation.

Lemma 9.3.1 (Correctness of the linearization process) Given any $GCNF'$-proof T in SLD-form, the linearization procedure outputs an SLD-refutation.

Proof: We proceed by induction on proofs. If T consists of an axiom, then the set S of Horn clauses contains a goal $\neg Q$ and a positive literal Q, and we have the SLD-refutation $< \{\neg Q\}, \square >$.

Otherwise, because it is in SLD-form, letting $\mathcal{C} = C_1, ..., C_m$, the tree T has the following structure:

$$\cfrac{\cfrac{T_{n-1}}{J, \mathcal{C}, \neg P_{n-1} \rightarrow} \qquad \cfrac{T_n}{J, \mathcal{C}, \neg P_n \rightarrow}}{J, \mathcal{C}, \{\neg P_{n-1}, \neg P_n\} \rightarrow}$$

$$...$$

$$\cfrac{T_1}{J, \mathcal{C}, \neg P_1 \rightarrow} \qquad \cfrac{\cfrac{T_2}{J, \mathcal{C}, \neg P_2 \rightarrow} \qquad J, \mathcal{C}, \{\neg P_3, ..., \neg P_n\} \rightarrow}{J, \mathcal{C}, \{\neg P_2, ..., \neg P_n\} \rightarrow}}{J, \mathcal{C}, \{\neg P_1, ..., \neg P_n\} \rightarrow}$$

Each tree T_i that is not an axiom is also in SLD-form and has the following shape:

$$\frac{P_i, \neg P_i \to \qquad \dfrac{T_i'}{J, \{\neg Q_1, ..., \neg Q_p\}, C' \to}}{J, \{\neg Q_1, ..., \neg Q_p, P_i\}, C', \neg P_i \to}$$

where $C' = \{C_1, ..., C_m\} - \{C\}$, for some definite clause $C = \{\neg Q_1, ..., \neg Q_p, P_i\}$.

By the induction hypothesis, each tree T_i' is converted to an SLD-refutation

$$Y_i = < \{\neg Q_1, ..., \neg Q_p\}, N_2, ..., N_q > .$$

By rule (3), the proof tree T_i is converted to the SLD-refutation X_i obtained by concatenating $\{\neg P_i\}$ and Y_i. But then,

$$X_i = < \{\neg P_i\}, \{\neg Q_1, ..., \neg Q_p\}, N_2, ..., N_q >$$

is an SLD-refutation obtained by resolving $\{\neg P_i\}$ with $\{\neg Q_1, ..., \neg Q_p, P_i\}$.

If T_i is an axiom then by rule (1) it is converted to $< \{\neg P_i\}, \square >$, which is an SLD-refutation.

Finally, rule (2) combines the SLD-refutations $X_1, ..., X_n$ in such a way that the resulting sequence is an SLD-refutation. Indeed, for every i, $1 \le i \le n - 1$, X_i becomes the SLD-derivation X_i', where

$$X_i' = < \{\neg P_i, \neg P_{i+1}..., \neg P_n\}, \{\neg Q_1, ..., \neg Q_p, \neg P_{i+1}..., \neg P_n\},$$
$$N_2 \cup \{\neg P_{i+1}..., \neg P_n\}, ..., N_{q-1} \cup \{\neg P_{i+1}..., \neg P_n\} >,$$

and so the entire sequence $X_1', ..., X_{n-1}', X_n$ is an SLD-refutation starting from the goal $\{\neg P_1, ..., \neg P_n\}$. $\square$

As a corollary, we have the completeness of SLD-resolution for Horn clauses.

Theorem 9.3.1 (Completeness of SLD-resolution for Horn clauses) The SLD-resolution method is complete for Horn clauses. Furthermore, if the first negative clause is $\{\neg P_1, ..., \neg P_n\}$, for every literal $\neg P_i$ in this goal, there is an SLD-resolution whose first selected atom is P_i.

Proof: Completeness is a consequence of lemma 9.3.1 and theorem 9.2.2. It is easy to see that in the linearization procedure, the order in which the subsequences are concatenated does not matter. This implies the second part of the lemma. $\square$

Actually, since SLD-refutations are the result of linearizing proof trees in SLD-form, it is easy to show that any atom P_i such that $\neg P_i$ belongs to a negative clause N_k in an SLD-refutation can be chosen as the selected atom.

By theorem 9.2.2, if a set S of Horn clauses with several goals $N_1, ..., N_k$ is $GCNF'$-provable, then there is some goal N_i such that $S - \{N_1, ..., N_{i-1}, N_{i+1}, ..., N_k\}$ is $GCNF'$-provable. This does not mean that there is a unique such N_i, as shown by the following example.

EXAMPLE 9.3.2

Consider the set S of clauses:

$$\{P\}, \{Q\}, \{\neg S, R\}, \{\neg R, \neg P\}, \{\neg R, \neg Q\}, \{S\}.$$

We have two SLD-refutations:

$$< \{\neg R, \neg P\}, \{\neg R\}, \{\neg S\}, \square >$$

and

$$< \{\neg R, \neg Q\}, \{\neg R\}, \{\neg S\}, \square > .$$

In the next section, we generalize SLD-resolution to first-order languages without equality, using the lifting technique of Section 8.5.

PROBLEMS

9.3.1. Apply the linearization procedure to the proof trees in SLD-form obtained in problem 9.2.1.

9.3.2. Give different SLD-resolution refutations for the following sets of clauses:

$$\{P_1\}, \{P_2\}, \{P_3\}, \{P_4\}, \{\neg P_1, \neg P_2, P_6\}, \{\neg P_3, \neg P_4, P_7\},$$
$$\{\neg P_6, \neg P_7, P_8\}, \{\neg P_8\}.$$

$$\{\neg P_2, P_3\}, \{\neg P_3, P_4\}, \{\neg P_4, P_5\}, \{P_3\}, \{P_1\}, \{P_2\}, \{\neg P_1\},$$
$$\{\neg P_3, P_6\}, \{\neg P_3, P_7\}, \{\neg P_3, P_8\}.$$

9.3.3. Write a computer program implementing the linearization procedure.

9.4 SLD-Resolution in First-Order Logic

In this section we shall generalize SLD-resolution to first-order languages without equality. Fortunately, it is relatively painless to generalize results about propositional SLD-resolution to the first-order case, using the lifting technique of Section 8.5.

9.4.1 Definition of SLD-Refutations

Since the main application of SLD-resolution is to PROLOG, we shall also revise our notation to conform to the PROLOG notation.

Definition 9.4.1 A *Horn clause* (in PROLOG notation) is one of the following expressions:

(i) $B : -A_1, ..., A_m$

(ii) B

(iii) $: -A_1, ..., A_m$

In the above, B, A_1,...,A_m are atomic formulae of the form $Pt_1...t_k$, where P is a predicate symbol of rank k, and t_1,...,t_k are terms.

A clause of the form (i) or (ii) is called a *definite clause*, and a clause of the form (iii) is called a *goal clause* (or *negative clause*).

The translation into the standard logic notation is the following:

The clause $B : -A_1, ..., A_m$ corresponds to the formula

$$(\neg A_1 \vee ... \vee \neg A_m \vee B);$$

The clause B corresponds to the atomic formula B;

The clause $: -A_1, ..., A_m$ corresponds to the formula

$$(\neg A_1 \vee ... \vee \neg A_m).$$

Actually, as in definition 8.2.1, it is assumed that a Horn clause is the universal closure of a formula as above (that is, of the form $\forall x_1...\forall x_n C$, where $FV(C) = \{x_1, ..., x_n\}$). The universal quantifiers are dropped for simplicity of notation, but it is important to remember that they are implicitly present.

The definition of SLD-derivations and SLD-refutations is extended by combining definition 9.3.2 and the definition of a resolvent given in definition 8.5.2.

Definition 9.4.2 Let S be a set of Horn clauses consisting of a set D of definite clauses and a set $\{G_1, ..., G_q\}$ of goals. An *SLD-derivation* for S is a sequence $< N_0, N_1, ..., N_p >$ of negative clauses satisfying the following properties:

(1) $N_0 = G_j$, where G_j is one of the goals;

(2) For every N_i in the sequence, $0 < i < p$, if

$$N_i =: -A_1, ..., A_{k-1}, A_k, A_{k+1}, ..., A_n,$$

then there is some definite clause $C_i = A : -B_1, ..., B_m$ in D such that A_k and A are unifiable, and for some most general unifier σ_i of A_k and $\rho_i(A)$, where (Id, ρ_i) is a separating substitution pair, if $m > 0$, then

$$N_{i+1} =: -\sigma_i(A_1, ..., A_{k-1}, \rho_i(B_1), ..., \rho_i(B_m), A_{k+1}, ..., A_n)$$

else if $m = 0$ then

$$N_{i+1} =: -\sigma_i(A_1, ..., A_{k-1}, A_{k+1}, ..., A_n).$$

An SLD-derivation is an *SLD-refutation* iff $N_p = \square$.

Note that an SLD-derivation is a linear representation of a resolution DAG of the following special form:

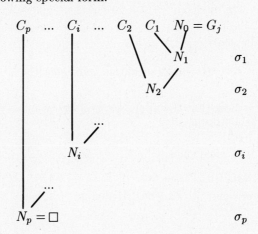

At each step, the clauses

$$: -A_1, ..., A_{k-1}, A_k, A_{k+1}, ..., A_n$$

and

$$A : -B_1, ..., B_m$$

are resolved, the atoms A_k and $\rho_i(A)$ being canceled, since they are unified by the most general unifier σ_i. The literal A_k is called the *selected atom* of N_i, and the clauses $N_0, C_1, ..., C_p$ are the *input clauses*.

When the derivation is a refutation, the substitution

$$\sigma = (\rho_1 \circ \sigma_1) \circ ... \circ (\rho_p \circ \sigma_p)$$

obtained by composing the substitutions occurring in the refutation is called the *result substitution* or *answer substitution*. It is used in PROLOG to extract the output of an SLD-computation.

Since an SLD-derivation is a special kind of resolution DAG, (a linear input resolution), its soundness is a consequence of lemma 8.5.2.

Lemma 9.4.1 (Soundness of SLD-resolution) If a set of Horn clauses has an SLD-refutation, then it is unsatisfiable.

Proof: Immediate from lemma 8.5.2. $\square$

Let us give an example of an SLD-refutation in the first-order case.

EXAMPLE 9.4.1

Consider the following set of definite clauses, axiomatizing addition of natural numbers:

$$C_1 : add(X, 0, X).$$
$$C_2 : add(X, succ(Y), succ(Z)) : -add(X, Y, Z).$$

Consider the goal

$$B : -add(succ(0), V, succ(succ(0))).$$

We wish to show that the above set is unsatisfiable. We have the following SLD-refutation:

Goal clause	Input clause	Substitution
$: -add(succ(0), V, succ(succ(0)))$	C_2	
$: -add(succ(0), Y_2, succ(0))$	C_1	σ_1
$\square$		σ_2

where
$$\sigma_1 = (succ(0)/X_1, succ(0)/Z_1, succ(Y_2)/V),$$
$$\sigma_2 = (succ(0)/X_2, 0/Y_2)$$

The variables X_1, Z_1, Y_2, X_2 were introduced by separating substitutions in computing resolvents. The result substitution is

$$(succ(0)/V, succ(0)/X_1, succ(0)/Z_1, succ(0)/X_2).$$

The interesting component is $succ(0)/V$. Indeed, there is a computational interpretation of the unsatisfiability of the set $\{C_1, C_2, B\}$. For this, it is necessary to write quantifiers explicitly and remember that goal clauses are negative. Observe that

$$\forall X C_1 \wedge \forall X \forall Y \forall Z C_2 \wedge \forall V B$$

is unsatisfiable, iff

$$\neg(\forall X C_1 \wedge \forall X \forall Y \forall Z C_2 \wedge \forall V\, B)$$

is valid, iff

$$(\forall X C_1 \wedge \forall X \forall Y \forall Z C_2) \supset \exists V \neg B$$

is valid. But $\exists V \neg B$ is actually

$$\exists V\, add(succ(0), V, succ(0)).$$

Since $(\forall X C_1 \wedge \forall X \forall Y \forall Z C_2)$ defines addition in the intuitive sense that any X, Y, Z satisfying the above sentence are such that $Z = X + Y$, we are trying to find some V such that $succ(0) + V = succ(succ(0))$, or in other words, compute the difference of $succ(succ(0))$ and $succ(0)$, which is indeed $succ(0)$!

This interpretation of a refutation showing that a set of Horn clauses is unsatisfiable as a computation of the answer to a query, such as

$$(\forall X C_1 \wedge \forall X \forall Y \forall Z C_2) \supset \exists V \neg B,$$

"find some V satisfying $\neg B$ and such that some conditional
axioms $\forall X C_1$ and $\forall X \forall Y \forall Z C_2$ hold,"

is the essense of PROLOG. The set of clauses $\{C_1, C_2\}$ can be viewed as a *logic program*.

We will come back to the idea of refutations as computations in the next section.

9.4.2 Completeness of SLD-Resolution for Horn Clauses

The completeness of SLD-resolution for Horn clauses is shown in the following theorem.

Theorem 9.4.1 (Completeness of SLD-Resolution for Horn Clauses) Let **L** be any first-order language without equality. Given any finite set S of Horn clauses, if S is unsatisfiable, then there is an SLD-refutation with first clause some negative clause $: -B_1, ..., B_n$ in S.

Proof: We shall use the lifting technique provided by lemma 8.5.4. First, by the Skolem-Herbrand-Gödel theorem, if S is unsatisfiable, there is a set S_g of ground instances of clauses in S which is unsatisfiable. Since substitution instances of Horn clauses are Horn clauses, by theorem 9.3.1, there is an SLD-refutation for S_g, starting from some negative clause in S_g. Finally, we conclude by observing that if we apply the lifting technique of lemma 8.5.4, we obtain an SLD-refutation. This is because we always resolve a negative clause (N_i) against an input clause (C_i). Hence, the result is proved. $\square$

From theorem 9.3.1, it is also true that if the first negative clause is : $-B_1, ..., B_n$, for every atom B_i in this goal, there is an SLD-resolution whose first selected atom is B_i. As a matter of fact, this property holds for any clause N_i in the refutation.

Even though SLD-resolution is complete for Horn clauses, there is still the problem of choosing among many possible SLD-derivations. The above shows that the choice of the selected atom is irrelevant. However, we still have the problem of choosing a definite clause $A : -B_1, ..., B_m$ such that A unifies with one of the atoms in the current goal clause : $-A_1, ..., A_{k-1}, A_k, A_{k+1}, ...,$ A_n.

Such problems are important and are the object of current research in programming logic, but we do not have the space to address them here. The interested reader is referred to Kowalski, 1979, or Campbell, 1983, for an introduction to the methods and problems in programming logic.

In the next section, we discuss the use of SLD-resolution as a computation procedure for PROLOG.

PROBLEMS

9.4.1. Prove using SLD-resolution that the following set of clauses is unsatisfiable:
$$add(X, 0, X)$$
$$add(X, succ(Y), succ(Z)) : -add(X, Y, Z)$$
$$: -add(succ(succ(0)), succ(succ(0)), U).$$

9.4.2. Prove using SLD-resolution that the following set of clauses is unsatisfiable:
$$add(X, 0, X)$$
$$add(X, succ(Y), succ(Z)) : -add(X, Y, Z)$$
$$: -add(U, V, succ(succ(succ(0)))).$$

Find all possible SLD-refutations.

9.4.3. Using SLD-resolution, show that the following set of Horn clauses is unsatisfiable:
$$hanoi(N, Output) : -move(a, b, c, N, Output).$$
$$move(A, B, C, succ(M), Output) : -move(A, C, B, M, Out1),$$
$$move(C, B, A, M, Out2),$$
$$append(Out1, cons(to(A, B), Out2), Output).$$
$$move(A, B, C, 0, nil).$$
$$append(cons(A, L1), L2, cons(A, L3)) : -append(L1, L2, L3).$$
$$append(nil, L1, L1).$$
$$: -hanoi(succ(succ(0)), Z)$$

9.5 SLD-Resolution, Logic Programming (PROLOG)

We have seen in example 9.4.1 that an SLD-refutation for a set of Horn clauses can be viewed as a computation. This illustrates an extremely interesting use of logic as a *programming language*.

9.5.1 Refutations as Computations

In the past few years, Horn logic has been the basis of a new type of programming language due to Colmerauer named PROLOG. It is not the purpose of this book to give a complete treatment of PROLOG, and we refer the interested reader to Kowalski, 1979, or Clocksin and Mellish, 1981, for details. In this section, we shall lay the foundations of the programming logic PROLOG. It will be shown how SLD-resolution can be used as a computational procedure to solve certain problems, and the correctness and completeness of this approach will be proved.

In a logic programming language like PROLOG, one writes programs as sets of assertions in the form of Horn clauses, or more accurately, definite clauses, except for the goal. A set P of definite clauses is a *logic program*. As we said in Section 9.4, it is assumed that distinct Horn clauses are universally quantified.

Roughly speaking, a logic program consists of facts and assertions. Given such a logic program, one is usually interested in extracting facts that are consequences of the logic program P. Typically, one has a certain "query" (or goal) G containing some free variables $z_1,...,z_q$, and one wants to find term instances t_1, ..., t_q for the variables z_1, ..., z_q, such that the formula

$$P \supset G[t_1/z_1, ..., t_q/z_q]$$

is valid.

For simplicity, it will be assumed that the query is a positive atomic formula G. More complicated formulae can be handled (anti-Horn clauses), but we will consider this case later. In PROLOG, a goal statement G is denoted by $? - G$.

From a logical point of view, the problem is to determine whether the sentence

$$P \supset (\exists z_1...\exists z_q G)$$

is valid.

From a computational point of view, the problem is to find term values $t_1,...,t_q$ for the variables $z_1,...,z_q$ that make the formula

$$P \supset G[t_1/z_1, ..., t_q/z_q]$$

valid, and perhaps all such assignments.

Remarkably, SLD-resolution can be used not only as a proof procedure, but also a a computational procedure, because it returns a result substitution. The reason is as follows:

$$\text{The formula } P \supset (\exists z_1 ... \exists z_q G) \text{ is valid iff}$$
$$\neg (P \supset (\exists z_1 ... \exists z_q G)) \text{ is unsatisfiable iff}$$
$$P \wedge (\forall z_1 ... \forall z_q \neg G) \text{ is unsatisfiable.}$$

But since G is an atomic formula, $\neg G$ is a goal clause $: -G$, and $P \wedge (\forall z_1 ... \forall z_q \neg G)$ is a conjuction of Horn clauses!

Hence, SLD-resolution can be used to test for unsatisfiability, and if it succeeds, it returns a result substitution σ. The crucial fact is that the components of the substitution σ corresponding to the variables $z_1,...,z_q$ are answers to the query G. However, this fact is not obvious. A proof will be given in the next section. As a preliminary task, we give a rigorous definition of the semantics of a logic program.

9.5.2 Model-Theoretic Semantics of Logic Programs

We begin by defining what kind of formula can appear as a goal.

Definition 9.5.1 An *anti-Horn clause* is a formula of the form

$$\exists x_1 ... \exists x_m B,$$

where B is a conjunctions of literals $L_1 \wedge ... \wedge L_p$, with at most one negative literal and $FV(B) = \{x_1, ..., x_m\}$.

A *logic program* is a pair (P, G), where the *program* P is a set of (universal) Horn clauses, and the *query* G is a disjunction

$$(G_1 \vee ... \vee G_n)$$

of anti-Horn clauses $G_i = \exists y_1 ... \exists z_{m_i} B_i$.

It is also assumed that for all $i \neq j$, $1 \leq i, j \leq n$, the sets of variables $FV(B_i)$ and $FV(B_j)$ are disjoint. The union $\{z_1, ..., z_q\}$ of the sets of free variables occurring in each B_i is called the set of *output variables associated with* G.

Note that an anti-Horn clause is not a clause. However, the terminology is justified by the fact that the negation of an anti-Horn clause is a (universal) Horn clause, and that $\neg G$ is equivalent to a conjunction of universal Horn clauses.

Remark: This definition is more general than the usual definition used in PROLOG. In (standard) PROLOG, P is a set of definite clauses (that is,

P does not contain negative clauses), and G is a formula that is a conjunction of atomic formulae. It is shown in the sequel that more general queries can be handled, but that the semantics is a bit more subtle. Indeed, indefinite answers may arise.

EXAMPLE 9.5.1

The following is a logic program, where P consists of the following clauses:

$$rocksinger(jackson).$$
$$teacher(jean).$$
$$teacher(susan).$$
$$rich(X) : -rocksinger(X).$$
$$: -teacher(X), rich(X).$$

The query is the following disjunction:

$$? - \neg rocksinger(Y) \vee rich(Z)$$

EXAMPLE 9.5.2

The following is the program of a logic program:

$$hanoi(N, Output) : -move(a, b, c, N, Output).$$
$$move(A, B, C, succ(M), Output) : -move(A, C, B, M, Out1),$$
$$move(C, B, A, M, Out2),$$
$$append(Out1, cons(to(A, B), Out2), Output).$$
$$move(A, B, C, 0, nil).$$
$$append(cons(A, L1), L2, cons(A, L3)) : -append(L1, L2, L3).$$
$$append(nil, L1, L1).$$

The query is:

$$? - hanoi(succ(succ(succ(0))), Output).$$

The above program is a logical version of the well known problem known as the tower of Hanoi (see Clocksin and Mellish, 1981).

In order to give a rigorous definition of the semantics of a logic program, it is convenient to define the concept of a free structure. Recall that we are only dealing with first-order languages without equality, and that if the language has no constants, the special constant $\#$ is added to it.

Definition 9.5.2 Given a first-order language **L** without equality and with at least one constant, a *free structure* (or *Herbrand structure*) **H** is an **L**-structure with domain the set $H_{\mathbf{L}}$ of all closed **L**-terms, and whose interpretation function satisfies the following property:

(i) For every function symbol f of rank n, for all $t_1,...,t_n \in H_\mathbf{L}$,

$$f_\mathbf{H}(t_1,...,t_n) = ft_1...t_n \quad \text{and}$$

(ii) For every constant symbol c,

$$c_\mathbf{H} = c.$$

The set of terms $H_\mathbf{L}$ is called the *Herbrand universe* of $\mathbf{L}$. For simplicity of notation, the set $H_\mathbf{L}$ is denoted as H when $\mathbf{L}$ is understood. The following lemma shows that free structures are universal. This lemma is actually not necessary for giving the semantics of Horn clauses, but it is of independent interest.

Lemma 9.5.1 A sentence X in NNF containing only universal quantifiers is satisfiable in some model iff it is satisfiable in some free structure.

Proof: Clearly, if X is satisfied in a free structure, it is satisfiable in some model. For the converse, assume that X has some model $\mathbf{A}$. We show how a free structure can be constructed from $\mathbf{A}$. We define the function $h : H \to A$ as follows:

For every constant c, $h(c) = c_\mathbf{A}$;

For every function symbol f of rank $n > 0$, for any n terms $t_1,...,t_n \in H$,

$$h(ft_1...t_n) = f_\mathbf{A}(h(t_1),...,h(t_n)).$$

Define the interpretation of the free structure H such that, for any predicate symbol P of rank n, for any n terms $t_1,...,t_n \in H$,

$$\mathbf{H} \models P(t_1,...,t_n) \quad \text{iff} \quad \mathbf{A} \models P(h(t_1),...,h(t_n)). \tag{$*$}$$

We now prove by induction on formulae that, for every assignment $s : \mathbf{V} \to H$, if $\mathbf{A} \models X[s \circ h]$, then $\mathbf{H} \models X[s]$.

(i) If X is a literal, this amounts to the definition $(*)$.

(ii) If X is of the form $(B \wedge C)$, then $\mathbf{A} \models X[s \circ h]$ implies that

$$\mathbf{A} \models B[s \circ h] \quad \text{and} \quad \mathbf{A} \models C[s \circ h].$$

By the induction hypothesis,

$$\mathbf{H} \models B[s] \quad \text{and} \quad \mathbf{H} \models C[s],$$

that is, $\mathbf{H} \models X[s]$.

(iii) If X is of the form $(B \vee C)$, then $\mathbf{A} \models X[s \circ h]$ implies that

$$\mathbf{A} \models B[s \circ h] \quad \text{or} \quad \mathbf{A} \models C[s \circ h].$$

By the induction hypothesis,

$$\mathbf{H} \models B[s] \quad \text{or} \quad \mathbf{H} \models C[s],$$

that is, $\mathbf{H} \models X[s]$.

(iv) X is of the form $\exists x B$. This case is not possible since X does not contain existential quantifiers.

(v) X is of the form $\forall x B$. If $\mathbf{A} \models X[s \circ h]$, then for every $a \in A$,

$$\mathbf{A} \models B[(s \circ h)[x := a]].$$

Now, since $h : H \to A$, for every $t \in H$, $h(t) = a$ for some $a \in A$, and so, for every t in H,

$$\mathbf{A} \models B[(s \circ h)[x := h(t)]], \text{ that is, } \mathbf{A} \models B[(s[x := t]) \circ h].$$

By the induction hypothesis, $\mathbf{H} \models B[s[x := t]]$ for all $t \in H$, that is, $\mathbf{H} \models X[s]$.
$\square$

It is obvious that lemma 9.5.1 also applies to sets of sentences. Also, since a formula is unsatisfiable iff it has no model, we have the following corollary:

Corollary   Given a first-order language without equality and with some constant, a set of sentences in NNF and only containing universal quantifiers is unsatisfiable iff it is unsatisfiable in every free (Herbrand) structure. $\square$

We now provide a rigorous semantics of logic programs.

Given a logic program (P, G), the question of interest is to determine whether the formula $P \supset G$ is valid. Actually, we really want more. If $\{z_1, ..., z_q\}$ is the the set of output variables occurring in G, we would like to find some (or all) tuple(s) $(t_1, ..., t_q)$ of ground terms such that

$$\models P \supset (B_1 \vee ... \vee B_n)[t_1/z_1, ..., t_q/z_q].$$

As we shall see, such tuples do not always exist. However, indefinite (or disjunctive) answers always exist, and if some conditions are imposed on P and G, definite answers (tuples of ground terms) exist.

Assume that $P \supset G$ is valid. This is equivalent to $\neg(P \supset G)$ being unsatisfiable. But $\neg(P \supset G)$ is equivalent to $P \wedge \neg G$, which is equivalent to a conjunction of universal Horn clauses. By the Skolem-Herbrand-Gödel theorem (theorem 7.6.1), if $\{x_1, ..., x_m\}$ is the set of all universally quantified variables in $P \wedge \neg G$, there is *some set*

$$\{(t_1^1, ..., t_m^1), ..., (t_1^k, ..., t_m^k)\}$$

of m-tuples of ground terms such that the conjunction

$$(P \wedge \neg G)[t_1^1/x_1, ..., t_m^1/x_m] \wedge ... \wedge (P \wedge \neg G)[t_1^k/x_1, ..., t_m^k/x_m]$$

is unsatisfiable (for some $k \geq 1$). From this, it is not difficult to prove that

$$\models P \supset G[t_1^1/x_1, ..., t_m^1/x_m] \vee ... \vee G[t_1^k/x_1, ..., t_m^k/x_m].$$

However, we cannot claim that $k = 1$, as shown by the following example.

EXAMPLE 9.5.3

Let $P = \neg Q(a) \vee \neg Q(b)$, and $G = \exists x \neg Q(x)$. $P \supset G$ is valid, but there is no term t such that

$$\neg Q(a) \vee \neg Q(b) \supset \neg Q(t)$$

is valid.

As a consequence, the answer to a query may be indefinite, in the sense that it is a disjunction of substitution instances of the goal. However, definite answers can be ensured if certain restrictions are met.

Lemma 9.5.2 (Definite answer lemma) If P is a (finite) set of definite clauses and G is a query of the form

$$\exists z_1 ... \exists z_q (B_1 \wedge ... \wedge B_l),$$

where each B_i is an atomic formula, if

$$\models P \supset \exists z_1 ... \exists z_q (B_1 \wedge ... \wedge B_l),$$

then there is some tuple $(t_1, ..., t_q)$ of ground terms such that

$$\models P \supset (B_1 \wedge ... \wedge B_l)[t_1/z_1, ..., t_q/z_q].$$

Proof:

$$\models P \supset \exists z_1 ... z_q (B_1 \wedge ... \wedge B_l) \quad \text{iff}$$
$$P \wedge \forall z_1 ... \forall z_q (\neg B_1 \vee ... \vee \neg B_l) \quad \text{is unsatisfiable.}$$

By the Skolem-Herbrand-Gödel theorem, there is a set C of ground substitution instances of the clauses in $P \cup \{\neg B_1, ..., \neg B_l\}$ that is unsatisfiable. Since the only negative clauses in C come from $\{\neg B_1, ..., \neg B_l\}$, by lemma 9.2.5, there is some substitution instance

$$(\neg B_1 \vee ... \vee \neg B_l)[t_1/z_1, ..., t_q/z_q]$$

such that
$$P' \cup \{(\neg B_1 \vee ... \vee \neg B_l)[t_1/z_1, ..., t_q/z_q]\}$$

is unsatisfiable, where P' is the subset of C consisting of substitution instances of clauses in P. But then, it is not difficult to show that

$$\models P \supset (B_1 \wedge ... \wedge B_l)[t_1/z_1, ..., t_q/z_q].$$

$\square$

The result of lemma 9.5.2 justifies the reason that in PROLOG only programs consisting of definite clauses and queries consisting of conjunctions of atomic formulae are considered. With such restrictions, definite answers are guaranteed. The above discussion leads to the following definition.

Definition 9.5.3 Given a logic program (P, G) with query $G = \exists z_1 ... \exists z_q B$ and with $B = (B_1 \vee ... \vee B_n)$, the *semantics* (or *meaning*) of (P, G) is the set

$$M(P, G) = \bigcup \{\{(t_1^1, ..., t_q^1), ..., (t_1^k, ..., t_q^k)\}, \ k \geq 1, \ (t_1^k, ..., t_q^k) \in H^q \mid$$
$$\models P \supset B[t_1^1/z_1, ..., t_q^1/z_q] \vee ... \vee B[t_1^k/z_1, ..., t_q^k/z_q]\}$$

of sets q-tuples of terms in the Herbrand universe H that make the formula

$$P \supset B[t_1^1/z_1, ..., t_q^1/z_q] \vee ... \vee B[t_1^k/z_1, ..., t_q^k/z_q]$$

valid (in every free structure).

If P is a set of definite clauses and B is a conjunction of atomic formulae, $k = 1$.

9.5.3 Correctness of SLD-Resolution as a Computation Procedure

We now prove that for every SLD-refutation of the conjunction of clauses in $P \wedge \neg G$, the components of the result substitution σ restricted to the output variables belong to the semantics $M(P, G)$ of (P, G). We prove the following slightly more general lemma, which implies the fact mentioned above.

Lemma 9.5.3 Given a set P of Horn clauses, let $\mathcal{R}$ be an SLD-refutation

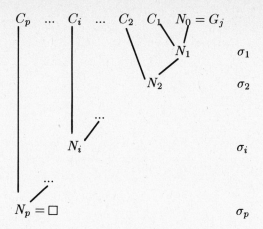

with result substitution σ (not necessarily ground). Let $\theta_p = \rho_p \circ \sigma_p$, and for every i, $1 \le i \le p-1$, let

$$\theta_i = (\rho_i \circ \sigma_i) \circ \theta_{i+1}.$$

(Note that $\sigma = \theta_1$, the result substitution.) The substitutions θ_i are also called *result substitutions*.) Then the set of quantifier-free clauses

$$\{\theta_1(N_0), \theta_1(C_1), ..., \theta_p(C_p)\}$$

is unsatisfiable (using the slight abuse of notation in which the matrix D of a clause $C = \forall x_1 ... \forall x_k D$ is also denoted by C).

Proof: We proceed by induction on the length of the derivation.

(i) If $p = 1$, N_0 must be a negative formula : $-B$ and C_1 a positive literal A such that A and B are unifiable, and it is clear that $\{\neg\theta_1(B), \theta_1(C_1)\}$ is unsatisfiable.

(ii) If $p > 1$, then by the induction hypothesis, taking N_1 as the goal of an SLD-refutation of length $p-1$ the set

$$\{\theta_2(N_1), \theta_2(C_2), ..., \theta_p(C_p)\}$$

is unsatisfiable. But N_0 is some goal clause

$$: -A_1, ..., A_{k-1}, A_k, A_{k+1}, ..., A_n,$$

and C_1 is some definite clause

$$A : -B_1, ..., B_m,$$

such that A and A_k are unifiable. Furthermore, the resolvent is given by

$$N_1 =: -\sigma_1(A_1, ..., A_{k-1}, \rho_1(B_1), ..., \rho_1(B_m), A_{k+1}, ..., A_n),$$

where σ_1 is a most general unifier, and we know that

$$\sigma_1(N_0) \wedge (\rho_1 \circ \sigma_1)(C_1) \supset N_1$$

is valid (by lemma 8.5.1). Since ρ_1 is a renaming substitution, it is the identity on N_0, and by the definition of θ_1, we have

$$\{\theta_2(\sigma_1(N_0)), \theta_2(\rho_1 \circ \sigma_1(C_1)), \theta_2(C_2), ..., \theta_p(C_p)\}$$
$$= \{\theta_1(N_0), \theta_1(C_1), \theta_2(C_2), ..., \theta_p(C_p)\}.$$

If $\{\theta_1(N_0), \theta_1(C_1), ..., \theta_p(C_p)\}$ was satisfiable, since

$$\sigma_1(N_0) \wedge (\rho_1 \circ \sigma_1)(C_1) \supset N_1$$

is valid,

$$\{\theta_2(N_1), \theta_2(C_2), ..., \theta_p(C_p)\}$$

would also be satisfiable, a contradiction. Hence,

$$\{\theta_1(N_0), \theta_1(C_1), ..., \theta_p(C_p)\}$$

is unsatisfiable. $\square$

Theorem 9.5.1 (Correctness of SLD-resolution as a computational procedure) Let (P, G) be a logic program with query $G = \exists z_1 ... \exists z_q B$, with $B = (B_1 \vee ... \vee B_n)$. For every SLD-refutation $\mathcal{R} =< N_0, N_1, ..., N_p >$ for the set of Horn clauses in $P \wedge \neg G$, if $\mathcal{R}$ uses (as in lemma 9.5.3) the list of definite clauses $< C_1, ..., C_p >$, the list of result substitutions (not necessarily ground) $< \theta_1, ..., \theta_p >$, and if $< \neg B_{i_1}, ..., \neg B_{i_k} >$ is the subsequence of $< N_0, C_1, ..., C_p >$ consisting of the clauses in $\{\neg B_1, ..., \neg B_n\}$, then

$$\models P \supset \theta_{i_1}(B_{i_1}) \vee ... \vee \theta_{i_k}(B_{i_k}).$$

Proof: Let P' be the set of formulae obtained by deleting the universal quantifiers from the clauses in P. By lemma 9.5.3, there is a sequence of clauses $< N_0, C_1, ..., C_p >$ from the set $P' \cup \{\neg B_1, ..., \neg B_n\}$ such that

$$\{\theta_1(N_0), \theta_1(C_1), ..., \theta_p(C_p)\}$$

is unsatisfiable. But then, it is easy to construct a proof of

$$P \supset \theta_{i_1}(B_{i_1}) \vee ... \vee \theta_{i_k}(B_{i_k})$$

(using $\forall : right$ rules as in lemma 8.5.4), and this yields the result. $\square$

Note: The formulae $B_{i_1}, ..., B_{i_k}$ are not necessarily distinct, but the substitutions $\theta_{i_1}, ..., \theta_{i_k}$ might be.

Corollary   Let (P, G) be a logic program such that P is a set of definite clauses and G is a formula of the form $\exists z_1 ... \exists z_q B$, where B is a conjunction of atomic formulae. For every SLD-refutation of the set of Horn clauses $P \wedge \neg G$, if σ is the result substitution and every term in the substitution belongs to H, if $(t_1/z_1, ..., t_q/z_q)$ is any substitution such that for every variable z_i in the support of σ, $t_i = \sigma(z_i)$, and otherwise t_i is any arbitrary term in H, then

$$\models P \supset B[t_1/z_1, ..., t_q/z_q].$$

Proof: First, observe that $\neg B$ must be the goal clause N_0. Also, if some output variable z_i does not occur in the support of the output substitution σ, this means that $\sigma(z_i) = z_i$. But then, it is immediate by lemma 9.5.3 that the result of substituting arbitrary terms in H for these variables in

$$\{\theta_1(N_0), \theta_1(C_1), ..., \theta_p(C_p)\}$$

is also unsatisfiable. $\square$

Theorem 9.5.1 shows that SLD-resolution is a correct method for computing elements of $M(P, G)$, since every set $\{(t_1^1, ..., t_q^1), ..., (t_1^k, ..., t_q^k)\}$ of tuples of terms in H returned by an SLD-refutation (corresponding to the output variables) makes

$$P \supset B[t_1^1/z_1, ..., t_q^1/z_q] \vee ... \vee B[t_1^k/z_1, ..., t_q^k/z_q]$$

valid.

Remark: Normally, we are interested in tuples of terms in H, because we want the answers to be interpretable as *definite* elements of the Herbrand universe. However, by lemma 9.5.3, *indefinite answers* (sets of tuples of terms containing variables) have to be considered. This is illustrated in the next example.

EXAMPLE 9.5.4

Consider the logic program of example 9.5.1. The set of clauses corresponding to $P \wedge \neg G$ is the following:

$$rocksinger(jackson).$$
$$teacher(jean).$$
$$teacher(susan).$$
$$rich(X) : -rocksinger(X).$$
$$: -teacher(X), rich(X).$$
$$rocksinger(Y).$$
$$: -rich(Z)$$

Note the two negative clauses. There are four SLD-refutations, two with goal : $-teacher(X), rich(X)$, and two with goal : $-rich(Z)$.

(i) SLD-refutation with output $(jean/Y)$:

Goal clause	Input clause	Substitution
$: -teacher(X), rich(X)$	$teacher(jean)$	
$: -rich(jean)$	$rich(X) : -rocksinger(X)$	$(jean/X)$
$: -rocksinger(jean)$	$rocksinger(Y)$	$(jean/X_1)$
$\square$		$(jean/Y_1)$

The result substitution is $(jean/Y, jean/X)$. Also, Z is any element of the Herbrand universe.

(ii) SLD-refutation with output $(susan/Y)$: Similar to the above.

(iii) SLD-refutation with output $(jackson/Z)$:

Goal clause	Input clause	Substitution
$: -rich(Z)$	$rich(X) : -rocksinger(X)$	
$: -rocksinger(X_1)$	$rocksinger(jackson)$	(X_1/Z)
$\square$		$(jackson/X_1)$

Y is any element of the Herbrand universe.

(iv) SLD-refutation with output $(Y_1/Y, Y_1/Z)$:

Goal clause	Input clause	Substitution
$: -rich(Z)$	$rich(X) : -rocksinger(X)$	
$: -rocksinger(X_1)$	$rocksinger(Y)$	(X_1/Z)
$\square$		(Y_1/X_1)

In this last refutation, we have an indefinite answer that says that for any Y_1 in the Herbrand universe, $Y = Y_1$, $Z = Y_1$ is an answer. This is indeed correct, since the clause $rich(X) : -rocksinger(X)$ is equivalent to $\neg rocksinger(X) \lor rich(X)$, and so

$$\models P \supset (\neg rocksinger(Y_1) \lor rich(Y_1)).$$

We now turn to the completeness of SLD-resolution as a computation procedure.

9.5.4 Completeness of SLD-Resolution as a Computational Procedure

The correctness of SLD-resolution as a computational procedure brings up immediately the question of its completeness. For any set of tuples in $M(P,G)$, is there an SLD-refutation with that answer? This is indeed the case, as shown below. We state and prove the following theorem for the special case of definite clauses, leaving the general case as an exercise.

Theorem 9.5.2 Let (P,G) be a logic program such that P is a set of definite clauses and G is a goal of the form $\exists z_1...\exists z_q B$, where B is a conjunction $B_1 \wedge ... \wedge B_n$ of atomic formulae. For every tuple $(t_1, ..., t_q) \in M(P,G)$, there is an SLD-refutation with result substitution σ and a (ground) substitution η such that the restriction of $\sigma \circ \eta$ to $z_1,...,z_q$ is $(t_1/z_1, ..., t_q/z_q)$.

Proof: By definition, $(t_1, ..., t_q) \in M(P,G)$ iff

$$\models P \supset (B_1 \wedge ... \wedge B_n)[t_1/z_1, ..., t_q/z_q] \quad \text{iff}$$
$$P \wedge (\neg B_1 \vee ... \vee \neg B_n)[t_1/z_1, ..., t_q/z_q] \quad \text{is unsatisfiable.}$$

By theorem 9.5.1, there is an SLD-refutation with output substitution θ_1. Since

$$(\neg B_1 \vee ... \vee \neg B_n)[t_1/z_1, ..., t_q/z_q]$$

is the only negative clause, by lemma 9.5.3, for some sequence of clauses $< C_1, ..., C_p >$ such that the universal closure of each clause C_i is in P,

$$\{\theta_1((\neg B_1 \vee ... \vee \neg B_n)[t_1/z_1, ..., t_q/z_q]), \theta_1(C_1), ..., \theta_p(C_p)\}$$

is also unsatisfiable. If θ_1 is not a ground substitution, we can substitute ground terms for the variables and form other ground substitutions $\theta'_1,...,\theta'_p$ such that,

$$\{\theta'_1((\neg B_1 \vee ... \vee \neg B_n)[t_1/z_1, ..., t_q/z_q]), \theta'_1(C_1), ..., \theta'_p(C_p)\}$$

is still unsatisfiable. Since the terms $t_1,...,t_q$ are ground terms,

$$\theta'_1((\neg B_1 \vee ... \vee \neg B_n)[t_1/z_1, ..., t_q/z_q]) = (\neg B_1 \vee ... \vee \neg B_n)[t_1/z_1, ..., t_q/z_q].$$

By theorem 9.3.1, there is a ground SLD-refutation $\mathcal{R}_g$ with sequence of input clauses

$$< \{(\neg B_1 \vee ... \vee \neg B_n)[t_1/z_1, ..., t_q/z_q], C'_1, ..., C'_r > \quad \text{for}$$
$$\{(\neg B_1 \vee ... \vee \neg B_n)[t_1/z_1, ..., t_q/z_q], \theta'_1(C_1), ..., \theta'_p(C_p)\}.$$

By the lifting lemma (lemma 8.5.4), there is an SLD-refutation $\mathcal{R}$ with sequence of input clauses

$$< \{\neg B_1, ..., \neg B_n\}, C''_1, ..., C''_r > \quad \text{for}$$
$$\{\{\neg B_1, ..., \neg B_n\}, C_1, ..., C_p\},$$

such that for every clause C_i'' in $\mathcal{R}$, $C_i'' = \eta_i(C_i')$, for some ground substitution η_i. Let $\eta = \eta_1 \circ \, ... \, \circ \eta_r$ and let σ be the result substitution of the SLD-refutation $\mathcal{R}$. It is easily shown that

$$(\neg B_1 \vee ... \vee \neg B_n)[t_1/z_1, ..., t_q/z_q] = (\sigma \circ \eta)(\neg B_1 \vee ... \vee \neg B_n),$$

which shows that $(t_1/z_1, ..., t_q/z_q)$ is equal to the restriction of $\sigma \circ \eta$ to $z_1, ..., z_q$.
$\square$

9.5.5 Limitations of PROLOG

Theorem 9.5.1 and theorem 9.5.2 show that SLD-Resolution is a correct and complete procedure for computing the sets of tuples belonging to the meaning of a logic program. From a theoretical point of view, this is very satisfactory. However, from a practical point of view, there is still something missing. Indeed, we still need a procedure for producing SLD-refutations, and if possible, efficiently. It is possible to organize the set all SLD-refutations into a kind of tree (the search space), and the problem is then reduced to a tree traversal. If one wants to retain completeness, the kind of tree traversal chosen must be a breadth-first search, which can be very inefficient. Most implementations of PROLOG sacrifice completeness for efficiency, and adopt a depth-first traversal strategy.

Unfortunately, we do not have the space to consider these interesting issues, but we refer the interested reader to Kowalski, 1979, and to Apt and Van Emden, 1982, where the semantics of logic programming is investigated in terms of fixedpoints.

Another point worth noting is that not all first-order formulae (in Skolem form) can be expressed as Horn clauses. The main limitation is that negative premises are not allowed, in the sense that a formula of the form

$$B : -A_1, ..., A_{i-1}, \neg A, A_{i+1}, ..., A_n.$$

is not equivalent to any Horn clause (see problem 3.5.9).

This restriction can be somewhat overcome by the *negation by failure* strategy, but one has to be careful in defining the semantics of such programs (see Kowalski, 1979, or Apt and Van Emden , 1982).

PROBLEMS

9.5.1.   (a) Give an SLD-resolution and the result substitution for the following set of clauses:

$French(Jean)$.

$French(Jacques)$.

$British(Peter)$.

$likewine(X, Y) : -French(X), wine(Y)$.

$likewine(X, Bordeaux) : -British(X)$.

$wine(Burgundy)$.

$wine(Bordeaux)$.

$: -likewine(U, V)$.

(b) Derive all possible answers to the query $likewine(U, V)$.

9.5.2. Give an SLD-resolution and the result substitution for the following set of clauses:

$append(cons(A, L1), L2, cons(A, L3)) : -append(L1, L2, L3)$.

$append(nil, L1, L1)$.

$: -append(cons(a, cons(b, nil)), cons(b, cons(c, nil)), Z)$

9.5.3. Give an SLD-resolution and the result substitution for the following set of clauses:

$hanoi(N, Output) : -move(a, b, c, N, Output)$.

$move(A, B, C, succ(M), Output) : -move(A, C, B, M, Out1),$
$\qquad move(C, B, A, M, Out2),$
$\qquad\qquad append(Out1, cons(to(A, B), Out2), Output)$.

$move(A, B, C, 0, nil)$.

$append(cons(A, L1), L2, cons(A, L3)) : -append(L1, L2, L3)$.

$append(nil, L1, L1)$.

$: -hanoi(succ(succ(succ(0))), Z)$

9.5.4. Complete the proof of theorem 9.5.1 by filling in the missing details.

9.5.5. State and prove a generalization of theorem 9.5.2 for the case of arbitrary logic programs.

* **9.5.6** Given a set S of Horn clauses, an H-tree for S is a tree labeled with substitution instances of atomic formulae in S defined inductively as follows:

(i) A tree whose root is labeled with **F** (false), and having n immediate successors labeled with atomic formulae $B_1, ..., B_n$, where $: -B_1, ..., B_n$ is some goal clause in S, is an H-tree.

(ii) If T is an H-tree, for every leaf node u labeled with some atomic formulae X that is not a substitution instance of some atomic formula

B in S (a definite clause consisting of a single atomic formula), if X is unifiable with the lefthand side of any clause $A : -B_1, ..., B_k$ in S, if σ is a most general unifier of X and A, the tree T' obtained by applying the substitution σ to all nodes in T and adding the k ($k > 0$) immediate successors $\sigma(B_1), ..., \sigma(B_k)$ to the node u labeled with $\sigma(X) = \sigma(A)$ is an H-tree (if $k = 0$, the tree T becomes the tree T' obtained by applying the substitution σ to all nodes in T. In this case, $\sigma(X)$ is a substitution instance of an axiom.)

An H-tree for S is a proof tree iff all its leaves are labeled with substitution instances of axioms in S (definite clauses consisting of a single atomic formula).

Prove that S is unsatisfiable iff there is some H-tree for S which is a proof tree.

Hint: Use problem 9.2.4 and adapt the lifting lemma.

Notes and Suggestions for Further Reading

The method of SLD-resolution is a special case of the SL-resolution of Kowalski and Kuehner (see Siekman and Wrightson, 1983).

To the best of our knowledge, the method used in Sections 9.2 and 9.3 for proving the completeness of SLD-resolution for (propositional) Horn clauses by linearizing a Gentzen proof in SLD-form to an SLD-refutation is original.

For an introduction to logic as a problem-solving tool, the reader is referred to Kowalski, 1979, or Bundy, 1983. Issues about the implementation of PROLOG are discussed in Campbell, 1983. So far, there are only a few articles and texts on the semantic foundations of PROLOG, including Kowalski and Van Emden , 1976; Apt and Van Emden , 1982; and Lloyd, 1984. The results of Section 9.5 for disjunctive goals appear to be original.

Chapter 10

Many-Sorted
First-Order Logic

10.1 Introduction

There are situtations in which it is desirable to express properties of structures of different types (or sorts). For instance, this is the case if one is interested in axiomatizing data strutures found in computer science (integers, reals, characters, stacks, queues, lists, etc). By adding to the formalism of first-order logic the notion of *type* (also called *sort*), one obtains a flexible and convenient logic called *many-sorted first-order logic*, which enjoys the same properties as first-order logic.

In this chapter, we shall define and give the basic properties of many-sorted first-order logic. It turns out that the semantics of first-order logic can be given conveniently using the notion of a many-sorted algebra defined in Section 2.5, given in the appendix. Hence, the reader is advised to review the appendix before reading this chapter.

At the end of this chapter, we give an algorithm for deciding whether a quantifier-free formula is valid, using the method of congruence closure due to Nelson and Oppen (Nelson and Oppen, 1980).

Due to the lack of space, we can only give a brief presentation of many-sorted first-order logic, and most proofs are left as exercises. Fortunately, they are all simple variations of proofs given in the first-order case.

10.2 Syntax

First, we define alphabets for many-sorted first-order languages.

10.2.1 Many-Sorted First-Order Languages

In contrast to standard first-order languages, in many-sorted first-order languages, the arguments of function and predicate symbols may have different types (or sorts), and constant and function symbols also have some type (or sort). Technically, this means that a many-sorted alphabet is a many-sorted ranked alphabet as defined in Subsection 2.5.1.

Definition 10.2.1 The *alphabet of a many-sorted first-order language* consists of the following sets of symbols:

A countable set $S \cup \{bool\}$ of *sorts* (or *types*) containing the special sort *bool*, and such that S is nonempty and does not contain *bool*.

Logical connectives: $\wedge$ (and), $\vee$ (or), $\supset$ (implication), $\equiv$ (equivalence), all of rank $(bool.bool, bool)$, $\neg$ (not) of rank $(bool, bool)$, $\bot$ (of rank $(e, bool)$);

Quantifiers: For every sort $s \in S$, $\forall_s$ (for all), $\exists_s$ (there exists), each of rank $(s, bool)$;

For every sort s in S, the *equality symbol* $\doteq_s$, of rank $(ss, bool)$.

Variables: For every sort $s \in S$, a countably infinite set $\mathbf{V}_s = \{x_0, x_1, x_2, ...\}$, each variable x_i being of rank (e, s). The family of sets $\mathbf{V}_s$ is denoted by $\mathbf{V}$.

Auxiliary symbols: "(" and ")".

An $(S \cup \{bool\})$-ranked alphabet $\mathbf{L}$ of *nonlogical symbols* consisting of:

(i) *Function symbols*: A (countable, possibly empty) set $\mathbf{FS}$ of symbols $f_0, f_1,...$, and a *rank function* $r : \mathbf{FS} \to S^+ \times S$, assigning a pair $r(f) = (u, s)$ called *rank* to every function symbol f. The string u is called the *arity* of f, and the symbol s the *sort* (or *type*) of f.

(ii) *Constants*: For every sort $s \in S$, a (countable, possibly empty) set $\mathbf{CS}_s$ of symbols $c_0, c_1,...$, each of rank (e, s). The family of sets $\mathbf{CS}_s$ is denoted by $\mathbf{CS}$.

(iii) *Predicate symbols*: A (countable, possibly empty) set $\mathbf{PS}$ of symbols $P_0, P_1,...$, and a *rank function* $r : \mathbf{PS} \to S^* \times \{bool\}$, assigning a pair $r(P) = (u, bool)$ called *rank* to each predicate symbol P. The string u is called the *arity* of P. If $u = e$, P is a propositional letter.

It is assumed that the sets $\mathbf{V}_s$, $\mathbf{FS}$, $\mathbf{CS}_s$, and $\mathbf{PS}$ are disjoint for all $s \in S$. We will refer to a many-sorted first-order language with set of nonlogical symbols $\mathbf{L}$ as the language $\mathbf{L}$. Many-sorted first-order languages obtained by omitting the equality symbol are referred to as *many-sorted first-order languages without equality.*

Observe that a standard (one sorted) first-order language corresponds to the special case of a many-sorted first-order language for which the set S of sorts contains a single element.

We now give inductive definitions for the sets of many-sorted terms and formulae.

Definition 10.2.2 Given a many-sorted first-order language **L**, let Γ be the union of the sets **V**, **CS**, **FS**, **PS**, and $\{\perp\}$, and let Γ_s be the subset of Γ^+ consisting of all strings whose leftmost symbol is of sort $s \in S \cup \{bool\}$.

For every function symbol f of rank $(u_1...u_n, s)$, let C_f be the function $C_f : \Gamma_{u_1} \times ... \times \Gamma_{u_n} \to \Gamma_s$ such that, for all strings $t_1, ..., t_n$, with each t_i of sort u_i,

$$C_f(t_1, ..., t_n) = ft_1...t_n.$$

For every predicate symbol P of arity $u_1...u_n$, let C_P be the function $C_P : \Gamma_{u_1} \times ... \times \Gamma_{u_n} \to \Gamma_{bool}$ such that, for all strings $t_1, ..., t_n$, each t_i of sort u_i,

$$C_P(t_1, ..., t_n) = Pt_1...t_n.$$

Also, let $C_{\doteq}^s$ be the function $C_{\doteq}^s : (\Gamma_s)^2 \to \Gamma_{bool}$ (of rank $(ss, bool)$) such that, for all strings t_1, t_2 of sort s,

$$C_{\doteq}^s(t_1, t_2) = \doteq_s t_1 t_2.$$

Finally, the functions $C_\wedge$, $C_\vee$, $C_\supset$, $C_\equiv$, $C_\neg$ are defined as in definition 3.2.2, with $C_\wedge$, $C_\vee$, $C_\supset$, $C_\equiv$ of rank $(bool.bool, bool)$, $C_\neg$ of rank $(bool, bool)$, and the functions $A_i^s, E_i^s : \Gamma_s \to \Gamma_{bool}$ (of rank $(s, bool)$) are defined such that, for any string A in Γ_s,

$$A_i^s(A) = \forall_s x_i A, \text{ and } E_i^s(A) = \exists_s x_i A.$$

The $(S \cup \{bool\})$-indexed family $(\Gamma_s)_{s \in (S \cup \{bool\})}$ is made into a many-sorted algebra also denoted by Γ as follows:

Each carrier of sort $s \in (S \cup \{bool\})$ is the set of strings Γ_s;

Each constant c of sort s in **CS** is interpreted as the string c;

Each predicate symbol P of rank $(e, bool)$ in **PS** (propositional symbol) is interpreted as the string P;

The constant $\perp$ is interpreted as the string $\perp$.

The operations are the functions C_f, C_P, $C_\wedge$, $C_\vee$, $C_\supset$, $C_\equiv$, $C_\neg$, $C_{\doteq}^s$, A_i^s and E_i^s.

From Subsection 2.5.5, we know that there is a least subalgebra $T(\mathbf{L}, \mathbf{V})$ containing the $(S \cup \{bool\})$-indexed family of sets **V** (with the component of sort *bool* being empty).

The set of terms $TERM^s_\mathbf{L}$ of **L**-*terms of sort s* (for short, terms of sort s) is the carrier of sort s of $T(\mathbf{L}, \mathbf{V})$, and the set $FORM_\mathbf{L}$ of **L**-*formulae* (for short, formulae) is the carrier of sort *bool* of $T(\mathbf{L}, \mathbf{V})$.

A less formal way of stating definition 10.2.2 is the following. Terms and atomic formulae are defined as follows:

(i) Every constant and every variable of sort s is a term of sort s.

(ii) If $t_1, ..., t_n$ are terms, each t_i of sort u_i, and f is a function symbol of rank $(u_1...u_n, s)$, then $ft_1...t_n$ is a term of sort s.

(iii) Every propositional letter is an atomic formula, and so is $\bot$.

(iv) If $t_1, ..., t_n$ are terms, each t_i of sort u_i, and P is a predicate symbol of arity $u_1...u_n$, then $Pt_1...t_n$ is an atomic formula; If t_1 and t_2 are terms of sort s, then $\doteq_s t_1 t_2$ is an atomic formula;

Formulae are defined as follows:

(i) Every atomic formula is a formula.

(ii) For any two formulae A and B, $(A \wedge B)$, $(A \vee B)$, $(A \supset B)$, $(A \equiv B)$ and $\neg A$ are also formulae.

(iii) For any variable x_i of sort s and any formula A, $\forall_s x_i A$ and $\exists_s x_i A$ are also formulae.

EXAMPLE 10.2.1

Let **L** be following many-sorted first-order language for stacks, where $S = \{stack, integer\}$, $\mathbf{CS}_{integer} = \{0\}$, $\mathbf{CS}_{stack} = \{\Lambda\}$, $\mathbf{FS} = \{Succ, +, *, push, pop, top\}$, and $\mathbf{PS} = \{<\}$.

The rank functions are given by:

$r(Succ) = (integer, integer)$;

$r(+) = r(*) = (integer.integer, integer)$;

$r(push) = (stack.integer, stack)$;

$r(pop) = (stack, stack)$;

$r(top) = (stack, integer)$;

$r(<) = (integer.integer, bool)$.

Then, the following are terms:

$Succ\ 0$

$top\ push\ \Lambda\ Succ\ 0$

EXAMPLE 10.2.2

Using the first-order language of example 10.2.1, the following are formulae:

$< \ 0 \ Succ \ 0$

$\forall_{integer} x \forall_{stack} y \ \dot{=}_{stack} \ pop \ push \ y \ x \ y$.

10.2.2 Free Generation of Terms and Formulae

As in Subsections 5.2.2 and 5.2.3, it is possible to show that terms and formulae are freely generated. This is left as an exercise for the reader.

Theorem 10.2.1 The many-sorted algebra $T(\mathbf{L}, \mathbf{V})$ is freely generated by the family $\mathbf{V}$. □

As a consequence, the family $TERM_\mathbf{L}$ of many-sorted terms is freely generated by the set of constants and variables as atoms and the functions C_f, and the set of $\mathbf{L}$-formulae is freely generated by the atomic formulae as atoms and the functions C_X (X a logical connective), A_i^s and E_i^s.

Remarks:

(1) Instead of defining terms and atomic formulae in prefix notation, one can define them as follows (using parentheses):

The second clause of definition 10.2.2 is changed to: For every function symbol f of rank $(u_1...u_n, s)$ and any terms $t_1, ..., t_n$, with each t_i of sort u_i, $f(t_1, ..., t_n)$ is a term of sort s. Also, atomic formulae are defined as follows:

For every predicate symbol P of arity $u_1...u_n$, for any terms $t_1, ..., t_n$, with each t_i of sort u_i, $P(t_1, ..., t_n)$ is an atomic formula; For any terms t_1 and t_2 of sort s, $(t_1 \dot{=}_s t_2)$ is an atomic formula.

We will also use the notation $\forall x : sA$ and $\exists x : sA$, instead of $\forall_s x A$ and $\exists_s x A$.

One can still show that the terms and formulae are freely generated. In the sequel, we shall use either notation. For simplicity, we shall also frequently use $=$ instead of $\dot{=}_s$ and omit parentheses whenever possible, using the conventions adopted in Chapter 5.

(2) The many-sorted algebra $T(\mathbf{L}, \mathbf{V})$ is free on the set of variables $\mathbf{V}$ (as defined in Section 2.5), and is isomorphic to the tree algebra $T_\mathbf{L}(\mathbf{V})$ (in $T_\mathbf{L}(\mathbf{V})$, the term $A_i^s(A)$ is used instead of $\forall x_i : sA$, and $E_i^s(A)$ instead of $\exists x_i : sA$).

10.2.3 Free and Bound Variables, Substitutions

The definitions of *free and bound variables*, *substitution*, and *term free for a variable in a formula* given in Chapter 5 extend immediately to the many-sorted case. The only difference is that in a substitution $s[t/x]$ or $A[t/x]$ of a term t for a variable x, the sort of the term t must be equal to the sort of the variable x.

PROBLEMS

10.2.1. Prove theorem 10.2.1.

10.2.2. Examine carefully how the definitions of Section 5.2 generalize to the many-sorted case (free and bound variables, substitutions, etc.).

10.2.3. Give a context-free grammar describing many-sorted terms and many-sorted formulae, as in problem 5.2.8.

10.2.4. Generalize the results of problems 5.2.3 to 5.2.7 to the many-sorted case.

10.3 Semantics of Many-Sorted First-Order Languages

First, we need to define many-sorted first-order structures.

10.3.1 Many-Sorted First-Order Structures

Given a many-sorted first-order language **L**, the semantics of formulae is defined as in Section 5.3, but using the concept of a many-sorted algebra rather than the concept of a (one-sorted) structure.

Recall from definition 10.2.1 that the nonlogical part **L** of a many-sorted first-order language is an $(S \cup \{bool\})$-sorted ranked alphabet.

Definition 10.3.1 Given a many-sorted first-order language **L**, a *many-sorted* **L**-*structure* **M**, (for short, a *structure*) is a many-sorted **L**-algebra as defined in Section 2.5.2, such that the carrier of sort *bool* is the set $BOOL = \{\mathbf{T}, \mathbf{F}\}$.

Recall that the carrier of sort s is nonempty and is denoted by M_s. Every function symbol f of rank $(u_1...u_n, s)$ is interpreted as a function $f_{\mathbf{M}} : M^u \to M_s$, with $M^u = M^{u_1} \times ... \times M^{u_n}$, $u = u_1...u_n$, each constant c of sort s is interpreted as an element $c_{\mathbf{M}}$ in M_s, and each predicate P of arity $u_1...u_n$ is interpreted as a function $P_{\mathbf{M}} : M^u \to BOOL$.

10.3.2 Semantics of Formulae

We now wish to define the semantics of formulae by generalizing the definition given in Section 5.3.

Definition 10.3.2 Given a first-order many-sorted language **L** and an **L**-structure **M**, an *assignment of sort s* is any function $v_s : \mathbf{V}_s \to M_s$ from the set of variables $\mathbf{V}_s$ to the domain M_s. The family of all such functions is denoted as $[\mathbf{V}_s \to M_s]$. An assignment v is any S-indexed family of assignments of sort s. The set of all assignments is denoted by $[\mathbf{V} \to M]$.

To define the meaning of a formula, we shall construct the function $A_\mathbf{M}$ recursively, using the fact that formulae are freely generated from the atomic formulae, and using theorem 2.5.1. As in Chapter 5, the meaning of a formula is a function from $[\mathbf{V} \to M]$ to $BOOL$. In order to apply theorem 2.5.1, it is necessary to extend the connectives to the set $[[\mathbf{V} \to M] \to BOOL]$ of all functions from $[\mathbf{V} \to M]$ to $BOOL$, to make this set into a many-sorted algebra. For this, the next two definitions are needed.

Definition 10.3.3 For all $i \geq 0$, for every sort s, define the functions $(A_i^s)_\mathbf{M}$ and $(E_i^s)_\mathbf{M}$ (of rank $(s, bool)$) from $[[\mathbf{V} \to M] \to BOOL]$ to $[[\mathbf{V} \to M] \to BOOL]$ as follows: For every function f in $[[\mathbf{V} \to M] \to BOOL]$, $(A_i^s)_\mathbf{M}(f)$ is the function such that: For every assignment v in $[\mathbf{V} \to M]$,

$$(A_i^s)_\mathbf{M}(f)(v) = \mathbf{F} \quad \text{iff} \quad f(v[x_i := a]) = \mathbf{F} \text{ for some } a \in M_s;$$

The function $(E_i^s)_\mathbf{M}(f)$ is the function such that: For every assignment v in $[\mathbf{V} \to M]$,

$$(E_i^s)_\mathbf{M}(f)(v) = \mathbf{T} \quad \text{iff} \quad f(v[x_i := a]) = \mathbf{T} \text{ for some } a \in M_s.$$

Note that $(A_i^s)_\mathbf{M}(f)(v) = \mathbf{T}$ iff the function $g_v : M_s \to BOOL$ such that $g_v(a) = f(v[x_i := a])$ for all $a \in M_s$, is the constant function whose value is $\mathbf{T}$, and that $(E_i^s)_\mathbf{M}(f)(v) = \mathbf{F}$ iff the function $g_v : M_s \to BOOL$ defined above is the constant function whose value is $\mathbf{F}$.

Definition 10.3.4 Given an S-sorted family M of nonempty domain M_s, the functions $\wedge_\mathbf{M}$, $\vee_\mathbf{M}$, $\supset_\mathbf{M}$ and $\equiv_\mathbf{M}$ of rank $(bool.bool, bool)$ from $[[\mathbf{V} \to M] \to BOOL] \times [[\mathbf{V} \to M] \to BOOL]$ to $[[\mathbf{V} \to M] \to BOOL]$, and the function $\neg_\mathbf{M}$ of rank $(bool, bool)$ from $[[\mathbf{V} \to M] \to BOOL]$ to $[[\mathbf{V} \to M] \to BOOL]$ are defined as follows: For every two functions f, g in $[[\mathbf{V} \to M] \to BOOL]$, for every assignment v in $[\mathbf{V} \to M]$,

$$\wedge_\mathbf{M}(f, g)(v) = H_\wedge(f(v), g(v));$$

$$\vee_\mathbf{M}(f, g)(v) = H_\vee(f(v), g(v));$$

$$\supset_\mathbf{M}(f, g)(v) = H_\supset(f(v), g(v));$$

$$\equiv_\mathbf{M}(f, g)(v) = H_\equiv(f(v), g(v));$$

$$\neg_\mathbf{M}(f)(v) = H_\neg(f(v)).$$

With the above definitions, we obtain the many-sorted algebra in which every carrier of sort $s \in S$ and the carrier of sort $bool$ is the set of functions $[[\mathbf{V} \to M] \to BOOL]$. Using theorem 2.5.1, we can now define the meaning $t_\mathbf{M}$ of a term and the meaning $A_\mathbf{M}$ of a formula. We begin with terms.

Definition 10.3.5 Given a many-sorted L-structure $\mathbf{M}$ and an assignment $v : \mathbf{V} \to M$, the function $t_\mathbf{M} : [\mathbf{V} \to M] \to M$ defined by a term t is the function such that for every assignments v in $[\mathbf{V} \to M]$, the value $t_\mathbf{M}[v]$ is defined recursively as follows:

(i) $x_{\mathbf{M}}[v] = v_s(x)$, for a variable x of sort s;

(ii) $c_{\mathbf{M}}[v] = c_{\mathbf{M}}$, for a constant c;

(iii) $(ft_1...t_n)_{\mathbf{M}}[v] = f_{\mathbf{M}}((t_1)_{\mathbf{M}}[v], ..., (t_n)_{\mathbf{M}}[v])$.

The recursive definition of the function $A_{\mathbf{M}} : [\mathbf{V} \to M] \to BOOL$ is now given.

Definition 10.3.6 The function $A_{\mathbf{M}} : [\mathbf{V} \to M] \to BOOL$ is defined recursively by the following clauses:

(1) For atomic formulae: $A_{\mathbf{M}}$ is the function such that, for every assignment v,

$(Pt_1...t_n)_{\mathbf{M}}[v] = P_{\mathbf{M}}((t_1)_{\mathbf{M}}[v], ..., (t_n)_{\mathbf{M}}[v])$;

$(\doteq_s t_1 t_2)_{\mathbf{M}}[v] = if \ (t_1)_{\mathbf{M}}[v] \ = \ (t_2)_{\mathbf{M}}[v] \ then \ \mathbf{T} \ else \ \mathbf{F}$;

$(\perp)_{\mathbf{M}}[v] = \mathbf{F}$;

(2) For nonatomic formulae:

$(A * B)_{\mathbf{M}} = *_{\mathbf{M}}(A_{\mathbf{M}}, B_{\mathbf{M}})$, where $*$ is in $\{\wedge, \vee, \supset, \equiv\}$, and $*_{\mathbf{M}}$ is the corresponding function defined in definition 10.3.4;

$(\neg A)_{\mathbf{M}} = \neg_{\mathbf{M}}(A_{\mathbf{M}})$;

$(\forall x_i : sA)_{\mathbf{M}} = (A_i^s)_{\mathbf{M}}(A_{\mathbf{M}})$;

$(\exists x_i : sA)_{\mathbf{M}} = (E_i^s)_{\mathbf{M}}(A_{\mathbf{M}})$.

Note that by definitions 10.3.3, 10.3.4, 10.3.5, and 10.3.6, for every assignment v,

$(\forall x_i : sA)_{\mathbf{M}}[v] = \mathbf{T} \quad iff \quad A_{\mathbf{M}}[v[x_i := m]] = \mathbf{T}$, for all $m \in M_s$, and

$(\exists x_i : sA)_{\mathbf{M}}[v] = \mathbf{T} \quad iff \quad A_{\mathbf{M}}[v[x_i := m]] = \mathbf{T}$, for some $m \in M_s$.

The notions of *satisfaction*, *validity*, and *model* are defined as in Subsection 5.3.3. As in Subsection 5.3.4, it is also possible to define the semantics of formulae using modified formulae obtained by substitution.

10.3.3 An Alternate Semantics

The following result analogous to lemma 5.3.2 can be shown.

Lemma 10.3.1 For any formula B, for any assignment v in $[\mathbf{V} \to M]$, and any variable x_i of sort s, the following hold:

(1) $(\forall x_i : sB)_{\mathbf{M}}[v] = \mathbf{T}$ iff $(B[\mathbf{m}/x_i])_{\mathbf{M}}[v] = \mathbf{T}$ for all $m \in M_s$;

(2) $(\exists x_i : sB)_{\mathbf{M}}[v] = \mathbf{T}$ iff $(B[\mathbf{m}/x_i])_{\mathbf{M}}[v] = \mathbf{T}$ for some $m \in M_s$. $\square$

In view of lemma 10.3.1, the recursive clauses of the definition of satisfaction can also be stated more informally as follows:

$$\mathbf{M} \models (\neg A)[v] \text{ iff } not \ \mathbf{M} \models A[v],$$
$$\mathbf{M} \models (A \wedge B)[v] \text{ iff } \mathbf{M} \models A[v] \ and \ \mathbf{M} \models B[v],$$
$$\mathbf{M} \models (A \vee B)[v] \text{ iff } \mathbf{M} \models A[v] \ or \ \mathbf{M} \models B[v],$$
$$\mathbf{M} \models (A \supset B)[v] \text{ iff } not \ \mathbf{M} \models A[v] \ or \ \mathbf{M} \models B[v],$$
$$\mathbf{M} \models (A \equiv B)[v] \text{ iff } (\mathbf{M} \models A[v] \text{ iff } \mathbf{M} \models B[v]),$$
$$\mathbf{M} \models (\forall x_i : sA)[v] \text{ iff } \mathbf{M} \models (A[\mathbf{a}/x_i])[v] \text{ for every } a \in M_s,$$
$$\mathbf{M} \models (\exists x_i : sA)[v] \text{ iff } \mathbf{M} \models (A[\mathbf{a}/x_i])[v] \text{ for some } a \in M_s.$$

It is also easy to show that the semantics of a formula A only depends on the set $FV(A)$ of variables free in A.

10.3.4 Semantics and Free Variables

The following lemma holds.

Lemma 10.3.2 Given a formula A with set of free variables $\{y_1, ..., y_n\}$, for any two assignments s_1, s_2 such that $s_1(y_i) = s_2(y_i)$, for $1 \leq i \leq n$, $A_{\mathbf{M}}[s_1] = A_{\mathbf{M}}[s_2]$. $\square$

10.3.5 Subformulae and Rectified Formulae

Subformulae and *rectified formulae* are defined as in Subsection 5.3.6, and lemma 5.3.4 can be generalized easily. Similarly, the results of the rest of Section 5.3 can be generalized to many-sorted logic. The details are left as exercises.

PROBLEMS

10.3.1. Prove lemma 10.3.1.

10.3.2. Prove lemma 10.3.2.

10.3.3. Generalize lemma 5.3.4 to the many-sorted case.

10.3.4. Examine the generalization of the other results of Section 5.3 to the many-sorted case.

10.4 Proof Theory of Many-Sorted Languages

The system G defined in Section 5.4 is extended to the many-sorted case as follows.

10.4.1 Gentzen System G for Many-Sorted Languages Without Equality

We first consider the case of a many-sorted first-order language **L** *without equality*.

Definition 10.4.1 (Gentzen system G for languages without equality) The symbols Γ, Δ, Λ will be used to denote arbitrary sequences of formulae and A, B to denote formulae. The rules of the sequent calculus G are the following:

$$\frac{\Gamma, A, B, \Delta \to \Lambda}{\Gamma, A \wedge B, \Delta \to \Lambda} \ (\wedge : left) \qquad \frac{\Gamma \to \Delta, A, \Lambda \quad \Gamma \to \Delta, B, \Lambda}{\Gamma \to \Delta, A \wedge B, \Lambda} \ (\wedge : right)$$

$$\frac{\Gamma, A, \Delta \to \Lambda \quad \Gamma, B, \Delta \to \Lambda}{\Gamma, A \vee B, \Delta \to \Lambda} \ (\vee : left) \qquad \frac{\Gamma \to \Delta, A, B, \Lambda}{\Gamma \to \Delta, A \vee B, \Lambda} \ (\vee : right)$$

$$\frac{\Gamma, \Delta \to A, \Lambda \quad B, \Gamma, \Delta \to \Lambda}{\Gamma, A \supset B, \Delta \to \Lambda} \ (\supset : left) \qquad \frac{A, \Gamma \to B, \Delta, \Lambda}{\Gamma \to \Delta, A \supset B, \Lambda} \ (\supset : right)$$

$$\frac{\Gamma, \Delta \to A, \Lambda}{\Gamma, \neg A, \Delta \to \Lambda} \ (\neg : left) \qquad \frac{A, \Gamma \to \Delta, \Lambda}{\Gamma \to \Delta, \neg A, \Lambda} \ (\neg : right)$$

In the quantifier rules below, x is any variable of sort s and y is any variable of sort s free for x in A and not free in A, unless $y = x$ ($y \notin FV(A) - \{x\}$). The term t is any term of sort s free for x in A.

$$\frac{\Gamma, A[t/x], \forall x : sA, \Delta \to \Lambda}{\Gamma, \forall x : sA, \Delta \to \Lambda} \ (\forall : left) \qquad \frac{\Gamma \to \Delta, A[y/x], \Lambda}{\Gamma \to \Delta, \forall x : sA, \Lambda} \ (\forall : right)$$

$$\frac{\Gamma, A[y/x], \Delta \to \Lambda}{\Gamma, \exists x : sA, \Delta \to \Lambda} \ (\exists : left) \qquad \frac{\Gamma \to \Delta, A[t/x], \exists x : sA, \Lambda}{\Gamma \to \Delta, \exists x : sA, \Lambda} \ (\exists : right)$$

Note that in both the ($\forall : right$)-rule and the ($\exists : left$)-rule, the variable y does *not* occur free in the lower sequent. In these rules, the variable y is called the *eigenvariable* of the inference. The condition that the eigenvariable does not occur free in the conclusion of the rule is called the *eigenvariable condition*. The formula $\forall x : sA$ (or $\exists x : sA$) is called the *principal formula* of the inference, and the formula $A[t/x]$ (or $A[y/x]$) the *side formula* of the inference.

The *axioms* of G are all sequents $\Gamma \to \Delta$ such that Γ and Δ contain a common formula.

10.4.2 Deduction Trees for the System G

Deduction trees and *proof trees* are defined as in Subsection 5.4.2.

10.4.3 Soundness of the System G

The soundness of the system G is obtained easily from the proofs given in Subsection 5.4.3.

Lemma 10.4.1 (Soundness of G, many-sorted case) Every sequent provable in G is valid. $\square$

10.4.4 Completeness of G

It is also possible to prove the completeness of the system G by adapting the definitions and proofs given in Sections 5.4 and 5.5 to the many-sorted case. For this it is necessary to modify the definition of a Hintikka set so that it applies to a many-sorted algebra. We only state the result, leaving the proof as a sequence of problems.

Theorem 10.4.1 (Gödel's extended completeness theorem for G) A sequent (even infinite) is valid iff it is G-provable. $\square$

10.5 Many-Sorted First-Order Logic With Equality

The equality rules for many-sorted languages with equality are defined as follows.

10.5.1 Gentzen System $G_=$ for Languages With Equality

Let $G_=$ be the Gentzen system obtained from the Gentzen system G (defined in definition 10.4.1) by adding the following rules.

Definition 10.5.1 (Equality rules for $G_=$) Let Γ, Δ, Λ denote arbitrary sequences of formulae (possibly empty) and let $t, s_1, ..., s_n, t_1, ..., t_n$ denote arbitrary **L**-terms. For every sort s, for every term t of sort s:

$$\frac{\Gamma, t \doteq_s t \to \Delta}{\Gamma \to \Delta}$$

For each function symbol f of rank $(u_1...u_n, s)$ and any terms $s_1, ..., s_n, t_1, ..., t_n$ such that s_i and t_i are of sort u_i:

$$\frac{\Gamma, (s_1 \doteq_{u_1} t_1) \wedge ... \wedge (s_n \doteq_{u_n} t_n) \supset (fs_1...s_n \doteq_s ft_1...t_n) \to \Delta}{\Gamma \to \Delta}$$

For each predicate symbol P (including $\doteq_s$) of arity $u_1...u_n$ and any terms $s_1, ..., s_n, t_1, ..., t_n$ such that s_i and t_i are of sort u_i:

$$\frac{\Gamma, ((s_1 \doteq_{u_1} t_1) \wedge ... \wedge (s_n \doteq_{u_n} t_n) \wedge Ps_1...s_n) \supset Pt_1...t_n \rightarrow \Delta}{\Gamma \rightarrow \Delta}$$

10.5.2 Soundness of the System $G_=$

The following lemma is easily shown.

Lemma 10.5.1 If a sequent is $G_=$-provable then it is valid. $\square$

10.5.3 Completeness of the System $G_=$

It is not difficult to adapt the proofs of Section 5.6 to obtain the following completeness theorem.

Theorem 10.5.1 (Gödel's extended completeness theorem for $G_=$) Let **L** be a many-sorted first-order language with equality. A sequent (even infinite) is valid iff it is $G_=$-provable. $\square$

10.5.4 Reduction of Many-Sorted Logic to One-Sorted Logic

Although many-sorted first-order logic is very convenient, it is not an essential extension of standard one-sorted first-order logic, in the sense that there is a translation of many-sorted logic into one-sorted logic. Such a translation is described in Enderton, 1972, and the reader is referred to it for details. The essential idea to convert a many-sorted language **L** into a one-sorted language **L'** is to add domain predicate symbols D_s, one for each sort, and to modify quantified formulae recursively as follows:

Every formula A of the form $\forall x : sB$ (or $\exists x : sB$) is converted to the formula $A' = \forall x(D_s(x) \supset B')$, where B' is the result of converting B.

We also define the set MS to be the set of all one-sorted formulae of the form:

(1) $\exists x D_s(x)$, for every sort s, and

(2) $\forall x_1...\forall x_n(D_{s_1}(x_1) \wedge ... \wedge D_{s_n}(x_n) \supset D_s(f(x_1, ..., x_n)))$, for every function symbol f of rank $(s_1...s_n, s)$.

Then, the following lemma can be shown.

Lemma 10.5.2 Given a many-sorted first-order language $\mathbf{L}$, for every set T of many-sorted formulae in $\mathbf{L}$ and for every many-sorted formula A in $\mathbf{L}$,

$$T \vdash A \quad \text{iff}$$
$$T' \cup MS \vdash A', \quad \text{in the translated one-sorted language } \mathbf{L}',$$

where T' is the set of formulae in T translated into one-sorted formulae, and A' is the one-sorted translation of A. $\square$

Lemma 10.5.2 can be used to transfer results from one-sorted logic to many-sorted logic. In particular, the compactness theorem, model existence theorem, and Löwenheim-Skolem theorem hold in many-sorted first-order logic.

PROBLEMS

10.4.1. Prove lemma 10.4.1.

10.4.2. Define many-sorted Hintikka sets for languages without equality, and prove lemma 5.4.5 for the many-sorted case.

10.4.3. Prove the completeness theorem (theorem 10.4.1) for many-sorted logic without equality.

10.5.1. Prove lemma 10.5.1.

10.5.2. Generalize the other theorems of Section 5.5 to the many-sorted case (compactness, model existence, Löwenheim-Skolem).

10.5.3. Define many-sorted Hintikka sets for languages with equality, and prove lemma 5.6.1 for the many-sorted case.

10.5.3. Prove the completeness theorem (theorem 10.5.1) for many-sorted logic with equality.

10.5.4. Generalize the other theorems of Section 5.6 to the many-sorted case (compactness, model existence, Löwenheim-Skolem).

10.5.5. Prove lemma 10.5.2.

10.6 Decision Procedures Based on Congruence Closure

In this section, we show that there is an algorithm for deciding whether a quantifier-free formula without predicate symbols in a many-sorted language with equality is valid, using a method due to Nelson and Oppen (Nelson and Oppen, 1980). Then, we show how this algorithm can easily be extended to deal with arbitrary quantifier-free formulae in a many-sorted language with equality.

10.6.1 Decision Procedure for Quantifier-free Formulae Without Predicate Symbols

First, we state the following lemma whose proof is left as an exercise.

Lemma 10.6.1 Given any quantifier-free formula A in a many-sorted first-order language **L**, a formula B in disjunctive normal form such that $A \equiv B$ is valid can be constructed. $\square$

Using lemma 10.6.1, observe that a quantifier-free formula A is valid iff the disjunctive normal form B of $\neg A$ is unsatisfiable. But a formula $B = C_1 \vee \dots \vee C_n$ in disjunctive normal form is unsatisfiable iff every conjunct C_i is unsatisfiable. Hence, in order to have an algorithm for deciding validity of quantifier-free formulae, it is enough to have an algorithm for deciding whether a conjunction of literals is unsatisfiable.

If the language **L** has no equality symbols, the problem is trivial since a conjunction of literals is unsatisfiable iff it contains some atomic formula $Pt_1...t_n$ and its negation. Otherwise, we follow a method due to Nelson and Oppen (Nelson and Oppen, 1980). First, we assume that the language **L** does not have predicate symbols. Then, every conjunct C consists of equations and of negations of equations:

$$t_1 \doteq_{s_1} u_1 \wedge \dots \wedge t_m \doteq_{s_m} u_m \wedge \neg r_1 \doteq_{s'_1} v_1 \wedge \dots \wedge \neg r_n \doteq_{s'_n} v_n$$

We give a method inspired from Nelson and Oppen for deciding whether a conjunction C as above is unsatisfiable. For this, we define the concept of a congruence on a graph.

10.6.2 Congruence Closure on a Graph

First, we define the notion of a labeled graph. We are considering graphs in which for every node u, the set of immediate successors is ordered. Also, every node is labeled with a function symbol from a many-sorted ranked alphabet, and the labeling satisfies a condition similar to the condition imposed on many-sorted terms.

Definition 10.6.1 A *finite labeled graph* G is a quadruple $(V, \Sigma, \Lambda, \delta)$, where:

V is an S-indexed family of finite sets V_s of *nodes* (or *vertices*);

Σ is an S-sorted ranked alphabet;

$\Lambda : V \rightarrow \Sigma$ is a *labeling function* assigning a symbol $\Lambda(v)$ in Σ to every node v in V;

$$\delta : \{(v, i) \mid v \in V, 1 \leq i \leq n, \ r(\Lambda(v)) = (u_1...u_n, s)\} \rightarrow V,$$

is a function called the *successor function*.

The functions Λ and δ also satisfy the following properties: For every node v in V_s, the rank of the symbol $\Lambda(v)$ labeling v is of the form $(u_1...u_n, s)$, and for every node $\delta(v, i)$, the sort of the symbol $\Lambda(\delta(v, i))$ labeling $\delta(v, i)$ is equal to u_i.

For every node v, $\delta(v, i)$ is called the *i-th successor* of v, and is also denoted as $v[i]$. Note that for a node v such that $r(\Lambda(v)) = (e, s)$, δ is not defined. Such a node is called a *terminal node*, or *leaf*. Given a node u, the set P_u of *predecessors* of u is the set $\{v \in V \mid \delta(v, i) = u, for\ some\ i\}$. A node u such that $P_u = \emptyset$ is called an *initial node*, or *root*. A pair (v, i) as in the definition of δ is also called an *edge*.

Next, we define a certain kind of equivalence relation on a graph called a congruence. A congruence is an equivalence relation closed under a certain form of implication.

Definition 10.6.2 Given a (finite) graph $G = (V, \Sigma, \Lambda, \delta)$, an S-indexed family R of relations R_s over V_s is a *G-congruence* (for short, a congruence) iff:

(1) Each R_s is an equivalence relation;

(2) For every pair (u, v) of nodes in V^2, if $\Lambda(u) = \Lambda(v)$ and $r(\Lambda(u)) = (e, s)$ then uR_sv, else if $\Lambda(u) = \Lambda(v)$, $r(\Lambda(u)) = (s_1...s_n, s)$, and for every i, $1 \le i \le n$, $u[i]R_{s_i}v[i]$, then uR_sv.

In particular, note that any two terminal nodes labeled with the same symbol of arity e are congruent.

Graphically, if u and v are two nodes labeled with the same symbol f of rank $(s_1...s_n, s)$, if $u[1], ..., u[n]$ are the successors of u and $v[1], ..., v[n]$ are the successors of v,

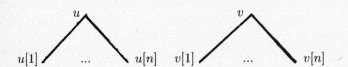

if $u[i]$ and $v[i]$ are equivalent for all i, $1 \le i \le n$, then u and v are also equivalent. Hence, we have a kind of backward closure.

We will prove shortly that given any finite graph and S-indexed family R of relations on G, there is a smallest congruence on G containing R, called the *congruence closure* of R. First, we show how the congruence closure concept can be used to give an algorithm for deciding unsatisfiability.

10.6.3 The Graph Associated With a Conjunction

The key to the algorithm for deciding unsatisfiability is the computation of a certain congruence over a finite graph induced by C and defined below.

Definition 10.6.3 Given a conjunction C of the form

$$t_1 \doteq_{s_1} u_1 \wedge \ldots \wedge t_m \doteq_{s_m} u_m \wedge \neg r_1 \doteq_{s'_1} v_1 \wedge \ldots \wedge \neg r_n \doteq_{s'_n} v_n,$$

let $TERM(C)$ be the set of all subterms of terms occurring in the conjunction C, including the terms t_i, u_i, r_j, v_j themselves. Let $S(C)$ be the set of sorts of all terms in $TERM(C)$. For every sort s in $S(C)$, let $TERM(C)_s$ be the set of all terms of sort s in $TERM(C)$. Note that by definition, each set $TERM(C)_s$ is nonempty. Let Σ be the $S(C)$-ranked alphabet consisting of all constants, function symbols, and variables occurring in $TERM(C)$. The graph $G(C) = (TERM(C), \Sigma, \Lambda, \delta)$ is defined as follows:

For every node t in $TERM(C)$, if t is either a variable or a constant then $\Lambda(t) = t$, else t is of the form $f y_1 \ldots y_k$ and $\Lambda(t) = f$;

For every node t in $TERM(C)$, if t is of the form $f y_1 \ldots y_k$, then for every i, $1 \le i \le k$, $\delta(t, i) = y_i$, else t is a constant or a variable and it is a terminal node of $G(C)$.

Finally, let $E = \{(t_1, u_1), \ldots, (t_m, u_m)\}$ be the set of pairs of terms occurring in the positive (nonnegated) equations in the conjunction C.

EXAMPLE 10.6.1

Consider the alphabet in which $S = \{i, s\}$, $\Sigma = \{f, g, a, b, c\}$, with $r(f) = (is, s)$, $r(g) = (si, s)$, $r(a) = i$, $r(b) = r(c) = s$. Let C be the conjunction

$$f(a, b) \doteq c \wedge \neg g(f(a, b), a) \doteq g(c, a).$$

Then, $TERM(C)_i = \{a\}$, $TERM(C)_s = \{b, c, f(a, b), g(c, a), g(f(a, b), a)\}$, and $E = \{(f(a, b), c)\}$. The graph $G(C)$ is the following:

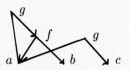

The key to the decision procedure is that the least congruence on $G(C)$ containing E exists, and that there is an algorithm for computing it. Indeed, assume that this least congruence $\overset{*}{\longleftrightarrow}_E$ containing E (called the *congruence closure of* E) exists and has been computed. Then, we have the following result.

Lemma 10.6.2 Given a conjunction C of the form

$$t_1 \doteq_{s_1} u_1 \wedge \ldots \wedge t_m \doteq_{s_m} u_m \wedge \neg r_1 \doteq_{s'_1} v_1 \wedge \ldots \wedge \neg r_n \doteq_{s'_n} v_n$$

of equations and negations of equations, if $\overset{*}{\longleftrightarrow}_E$ is the congruence closure on $G(C)$ of the relation $E = \{(t_1, u_1), ..., (t_m, u_m)\}$, then

$$C \text{ is unsatisfiable iff for some } j, 1 \leq j \leq n, \quad r_j \overset{*}{\longleftrightarrow}_E v_j.$$

Proof: First, we prove that the $S(C)$-indexed family R of relations R_s on $TERM(C)$ defined such that

$$t R_s u \quad \text{iff} \quad E \models t \doteq_s u,$$

is a congruence on $G(C)$ containing E. It is obvious that

$$E \models t_i \doteq_{s_i} u_i$$

for every (t_i, u_i) in E, $1 \leq i \leq m$, and so

$$t_i R_{s_i} u_i.$$

For every two subterms of the form $f y_1 ... y_k$ and $f z_1 ... z_k$ such that f is of rank $(w_1 ... w_k, s)$, if for every i, $1 \leq i \leq k$,

$$E \models y_i \doteq_{w_i} z_i$$

then by the definition of the semantics of equality,

$$E \models f(y_1, ..., y_k) \doteq_s f(z_1, ..., z_k).$$

Hence, R is a congruence on $G(C)$ containing E. Since $\overset{*}{\longleftrightarrow}_E$ is the least congruence on $G(C)$ containing E,

$$\text{if} \quad r_j \overset{*}{\longleftrightarrow}_E v_j, \quad \text{then} \quad E \models r_j \doteq_{s'_j} v_j.$$

But then, since any model satisfying C satisfies E, both $r_j \doteq_{s'_j} v_j$ and $\neg r_j \doteq_{s'_j} v_j$ would be satisfied, a contradiction. We conclude that if for some j, $1 \leq j \leq n$,

$$r_j \overset{*}{\longleftrightarrow}_E v_j,$$

then C is unsatisfiable. Conversely, assuming that there is no j, $1 \leq j \leq n$, such that

$$r_j \overset{*}{\longleftrightarrow}_E v_j,$$

we shall construct a model **M** of C. First, we make the $S(C)$-indexed family $TERM(C)$ into a many-sorted Σ-algebra **C** as follows:

For each sort s in $S(C)$, each constant or variable t of sort s is interpreted as the term t itself.

For every function symbol f in Σ of rank $(w_1...w_k, s)$, for every k terms $y_1, ..., y_k$ in $TERM(C)$, each y_i being of sort w_i, $1 \leq i \leq k$,

$$f_C(y_1, ..., y_k) = \begin{cases} fy_1...y_k & \text{if } fy_1...y_k \in TERM(C)_s; \\ fz_1...z_k & \text{if there are terms } z_1, ..., z_k \text{ such that,} \\ & y_i \overset{*}{\longleftrightarrow}_E z_i, \text{ and } fz_1...z_k \in TERM(C)_s; \\ t_0 & \text{otherwise, where } t_0 \text{ is some arbitrary term} \\ & \text{chosen in } TERM(C)_s. \end{cases}$$

Next, we prove that $\overset{*}{\longleftrightarrow}_E$ is a congruence on **C**. Indeed, for every function symbol f in Σ of rank $(w_1...w_k, s)$, for every k pairs of terms $(y_1, z_1), ..., (y_k, z_k)$, with y_i, z_i of sort w_i, $1 \leq i \leq k$, if

$$y_i \overset{*}{\longleftrightarrow}_E z_i,$$

then:

(1) If $fy_1...y_k$ and $fz_1...z_k$ are both in $TERM(C)$, then

$$f_C(y_1, ..., y_k) = fy_1...y_k, \quad \text{and} \quad f_C(z_1, ..., z_k) = fz_1...z_k,$$

and since $\overset{*}{\longleftrightarrow}_E$ is a congruence on $G(C)$,

$$fy_1...y_k \overset{*}{\longleftrightarrow}_E fz_1...z_k;$$

(2) If $fy_1...y_k$ is not in $TERM(C)$, but $fz_1...z_k$ is in $TERM(C)$, then

$$f_C(z_1, ..., z_k) = fz_1...z_k \quad \text{and} \quad f_C(y_1, ..., y_k) = fz_1...z_k,$$

and we conclude using the reflexivity of $\overset{*}{\longleftrightarrow}_E$;

(3) If neither $fy_1...y_k$ nor $fz_1...z_k$ is in $TERM(C)$, then

$$f_C(y_1, ..., y_k) = f_C(z_1, ..., z_k) = t_0$$

for some chosen term t_0 in $TERM(C)$, and we conclude using the reflexivity of $\overset{*}{\longleftrightarrow}_E$.

Let **M** be the quotient of the algebra **C** by the congruence $\overset{*}{\longleftrightarrow}_E$, as defined in Subsection 2.5.8. Let v be the assignment defined as follows: For every variable x occurring in some term in $TERM(C)$, $v(x)$ is the congruence class $[x]$ of x.

We claim that for every term t in $TERM(C)$,

$$t_M[v] = [t],$$

the congruence class of t. This is easily shown by induction and is left as an exercise. But then, C is satisfied by v in $\mathbf{M}$. Indeed, for every (t_i, u_i) in E,

$$[t_i] = [u_i],$$

and so

$$\mathbf{M} \models (t_i \doteq_{s_i} u_i)[v],$$

and for every j, $1 \leq j \leq n$, since we have assumed that the congruence classes $[r_j]$ and $[v_j]$ are unequal,

$$\mathbf{M} \models (\neg r_j \doteq_{s'_j} v_j)[v].$$

□

The above lemma provides a decision procedure for quantifier-free formulae without predicate symbols. It only remains to prove that $\overset{*}{\longleftrightarrow}_E$ exists and to give an algorithm for computing it. First, we give an example.

EXAMPLE 10.6.2

Recall the conjunction C of example 10.6.1:

$$f(a, b) \doteq c \wedge \neg g(f(a, b), a) \doteq g(c, a).$$

Then, $TERM(C)_i = \{a\}$, $TERM(C)_s = \{b, c, f(a, b), g(c, a), g(f(a, b), a)\}$, and it is not difficult to verify that the congruence closure of the relation $E = \{(f(a, b), c)\}$ has the following congruence classes: $\{a\}, \{b\}$, $\{f(a, b), c\}$ and $\{g(f(a, b), a), g(c, a)\}$. A possible candidate for the algebra $\mathbf{C}$ is given by the following table:

	f	g
(a, b)	$f(a, b)$	
(b, a)		b
(c, a)		$g(c, a)$
(a, c)	b	
$(f(a, b), a)$		$g(f(a, b), a)$
$(a, f(a, b))$	b	
$(a, g(c, a))$	b	
$(g(c, a), a)$		b
$(a, g(f(a, b), a))$	b	
$(g(f(a, b), a), a)$		b

By lemma 10.6.2, C is unsatisfiable.

10.6.4 Existence of the Congruence Closure

We now prove that the congruence closure of a relation R on a finite graph G exists.

Lemma 10.6.3 (Existence of the congruence closure) Given a finite graph $G = (V, \Sigma, \Lambda, \delta)$ and a relation R on V, there is a smallest congruence $\xleftrightarrow{*}_R$ on G containing R.

  Proof: We define a sequence R^i of S-indexed families of relations inductively as follows: For every sort $s \in S$,

$$R_s^0 = R_s \cup \{(u, v) \in V^2 \mid \Lambda(u) = \Lambda(v), \text{and } r(\Lambda(u)) = (e, s)\} \cup \{(u, u) \mid u \in V\};$$

For every sort $s \in S$,

$$\begin{aligned}
R_s^{i+1} = R_s^i &\cup \{(v, u) \in V^2 \mid (u, v) \in R_s^i\} \\
&\cup \{(u, w) \in V^2 \mid \exists v \in V, \ (u, v) \in R_s^i \text{ and } (v, w) \in R_s^i\} \\
&\cup \{(u, v) \in V^2 \mid \Lambda(u) = \Lambda(v), r(\Lambda(u)) = (s_1...s_n, s), \\
&\qquad \text{and } u[j] R_{s_j}^i v[j], \ 1 \le j \le n\}.
\end{aligned}$$

Let

$$(\xleftrightarrow{*}_R)_s = \bigcup R_s^i.$$

  It is easily shown by induction that every congruence on G containing R contains every R^i, and that $\xleftrightarrow{*}_R$ is a congruence on G. Hence, $\xleftrightarrow{*}_R$ is the least congruence on G containing R. $\square$

  Since the graph G is finite, there must exist some integer i such that $R^i = R^{i+1}$. Hence, the congruence closure of R is also computable. We shall give later a better algorithm due to Nelson and Oppen. But first, we show how the method of Subsection 10.6.3 can be used to give an algorithm for deciding the unsatisfiability of a conjunction of literals over a many-sorted language with equality (and function, constant, and predicate symbols).

10.6.5 Decision Procedure for Quantifier-free Formulae

The crucial observation that allows us to adapt the congruence closure method is the following:

  For any atomic formula $Pt_1...t_k$, if $\top$ represents the logical constant always interpreted as **T**, then $Pt_1...t_k$ is logically equivalent to $(Pt_1...t_k \equiv \top)$, in the sense that
$$Pt_1...t_k \equiv (Pt_1...t_k \equiv \top)$$
is valid.

But then, this means that $\equiv$ behaves semantically exactly as the identity relation on BOOL. Hence, we can treat $\equiv$ as the equality symbol $\doteq_{bool}$ of sort *bool*, and interpret it as the identity on BOOL.

Hence, every conjunction C of literals is equivalent to a conjunction C' of equations and negations of equations, such that every atomic formula $Pt_1...t_k$ in C is replaced by the equation $Pt_1...t_k \equiv \top$, and every formula $\neg Pt_1...t_k$ in C is replaced by $\neg(Pt_1...t_k \equiv \top)$.

Then, as in Subsection 10.6.3, we can build the graph $G(C')$ associated with C', treating $\top$ as a constant of sort *bool*, and treating every predicate symbol P of arity $u_1...u_k$ as a function symbol of rank $(u_1...u_k, bool)$. We then have the following lemma generalizing lemma 10.6.2.

Lemma 10.6.4 Let C' be the conjunction obtained from a conjunction of literals obtained by changing literals of the form $Pt_1...t_k$ or $\neg Pt_1...t_k$ into equations as explained above. If C' is of the form

$$t_1 \doteq_{s_1} u_1 \wedge ... \wedge t_m \doteq_{s_m} u_m \wedge \neg r_1 \doteq_{s'_1} v_1 \wedge ... \wedge \neg r_n \doteq_{s'_n} v_n$$

and if $\overset{*}{\longleftrightarrow}_E$ is the congruence closure on $G(C)$ of the relation $E = \{(t_1, u_1), ..., (t_m, u_m)\}$, then

$$C \quad \text{is unsatisfiable iff for some } j,\ 1 \leq j \leq n, \quad r_j \overset{*}{\longleftrightarrow}_E v_j.$$

Proof: We define $TERM(C)$ as in Subsection 10.6.3, except that $S(C)$ also contains the sort *bool*, and we have a set of terms of sort *bool*. First, we prove that the $S(C)$-indexed family R of relations R_s on $TERM(C)$ defined such that

$$tR_s u \quad \text{iff} \quad E \models t \doteq_s u,$$

is a congruence on $G(C)$ containing E. For function symbols of sort $s \neq bool$, the proof is identical to the proof of lemma 10.6.2. For every two subterms of the form $Py_1...y_k$ and $Pz_1...z_k$ such that P is of rank $(w_1...w_k, bool)$, if for every $i,\ 1 \leq i \leq k$,

$$E \models y_i \doteq_{w_i} z_i,$$

then by the definition of the semantics of equality symbols,

$$E \models P(y_1, ..., y_k) \equiv P(z_1, ..., z_k).$$

Hence, R is a congruence on $G(C)$ containing E. Now there are two cases.

(i) If $r_j \overset{*}{\longleftrightarrow}_E v_j$ and the sort of r_j and v_j is not *bool*, we conclude as in the proof of lemma 10.6.2.

(ii) Otherwise, r_j is some atomic formula $Py_1...y_k$ and v_j is the constant $\top$. In that case,

$$E \models r_j.$$

But then, since any model satisfying C satisfies E, both r_j and $\neg r_j$ would be satisfied, a contradiction. We conclude that if for some j, $1 \leq j \leq n$,

$$r_j \overset{*}{\longleftrightarrow}_E v_j,$$

then C is unsatisfiable. Conversely, assuming that there is no j, $1 \leq j \leq n$, such that

$$r_j \overset{*}{\longleftrightarrow}_E v_j,$$

we shall construct a model **C** of C. First, we make the $S(C)$-indexed family $TERM(C)$ into a many-sorted Σ-algebra **C** as in the proof of lemma 10.6.2. The new case is the case of symbols of sort *bool*. For every predicate symbol P of rank $(w_1...w_k, bool)$, for every k terms $y_1, ..., y_k \in TERM(C)$, each y_i being of sort w_i, $1 \leq i \leq k$,

$$P_{\mathbf{C}}(y_1, ..., y_k) = \begin{cases} \mathbf{T} & \text{if } Py_1...y_k \overset{*}{\longleftrightarrow}_E \mathsf{T}; \\ \mathbf{F} & \text{otherwise.} \end{cases}$$

Next, we prove that $\overset{*}{\longleftrightarrow}_E$ is a congruence on **C**. The case of symbols of sort $\neq bool$ has already been treated in the proof of lemma 10.6.2, and we only need to consider the case of predicate symbols. For every predicate symbol P of rank $(w_1...w_k, bool)$, for every k pairs of terms $(y_1, z_1), ..., (y_k, z_k)$, with y_i, z_i of sort w_i, $1 \leq i \leq k$, if

$$y_i \overset{*}{\longleftrightarrow}_E z_i,$$

then:

(1) If $Py_1...y_k \overset{*}{\longleftrightarrow}_E \mathsf{T}$, then since $y_i \overset{*}{\longleftrightarrow}_E z_i$, and $\overset{*}{\longleftrightarrow}_E$ is a congruence, we have

$$Pz_1...z_k \overset{*}{\longleftrightarrow}_E \mathsf{T},$$

and so

$$P_{\mathbf{C}}(y_1, ..., y_k) = \mathbf{T} \quad \text{and} \quad P_{\mathbf{C}}(z_1, ..., z_k) = \mathbf{T};$$

(2) If $Pz_1...z_k \overset{*}{\longleftrightarrow}_E \mathsf{T}$, this case is similar to case (1).

(3) Otherwise, both

$$P_{\mathbf{C}}(y_1, ..., y_k) = \mathbf{F} \quad \text{and} \quad P_{\mathbf{C}}(z_1, ..., z_k) = \mathbf{F}.$$

Let **M** be the quotient of the algebra **C** by the congruence $\overset{*}{\longleftrightarrow}_E$, as defined in Subsection 2.5.8. Let v be the assignment defined as follows: For every variable x occurring in some term in $TERM(C)$, $v(x)$ is the congruence class $[x]$ of x.

As in the proof of lemma 10.6.2, it is easily shown that for every term t in $TERM(C)$,

$$t_{\mathbf{M}}[v] = [t],$$

the congruence class of t. But then, C is satisfied by v in $\mathbf{M}$. The case of equations not involving predicate symbols in treated as in the proof of lemma 10.6.2. Clearly, for every equation $Py_1...y_k \equiv \top$ in C', by the definition of $\mathbf{M}$,

$$\mathbf{M} \models Py_1...y_k.$$

Since we have assumed that for every negation $\neg(Py_1...y_k \equiv \top)$ in C', it is not the case that

$$Py_1...y_n \overset{*}{\longleftrightarrow}_E \top,$$

by the definition of $\mathbf{M}$,

$$\mathbf{M} \models \neg Py_1...y_k.$$

This concludes the proof. $\square$ The above lemma provides a decision procedure for quantifier-free formulae.

EXAMPLE 10.6.3

Consider the following conjunction C over a (one-sorted) first-order language:

$$f(f(f(a))) \doteq a \wedge f(f(f(f(f(a))))) \doteq a \wedge P(a) \wedge \neg P(f(a)),$$

where f is a unary function symbol and P a unary predicate. First, C is converted to C':

$$f(f(f(a))) \doteq a \wedge f(f(f(f(f(a))))) \doteq a \wedge P(a) \equiv \top \wedge \neg P(f(a)) \equiv \top.$$

The graph $G(C)$ corresponding to C' is the following:

Initially, $R = \{(V1,V4),(V1,V6),(V8,V7)\}$, and let R' be its congruence closure. Since $(V1,V4)$ is in R, $(V2,V5)$ is in R'. Since $(V2,V5)$ is in R', $(V3,V6)$ is in R'. Since both $(V1,V6)$ and $(V3,V6)$ are in R', $(V1,V3)$ is in R'. Hence, $(V2,V4)$ is in R'. So the nodes $V1,V2,V3,V4,V5,V6$ are all equivalent under R'. But then, since $(V1,V2)$ is in R', $(V8,V9)$ is in R', and since $(V8,V7)$ is in R, $(V9,V7)$ is in R'. Consequently, $P(f(a))$ is equivalent to $\top$, and C' is unsatisfiable.

10.6.6 Computing the Congruence Closure

We conclude this section by giving an algorithm due to Nelson and Oppen
(Nelson and Oppen, 1980) for computing the congruence closure R' of a relation R on a graph G. This algorithm uses a procedure $MERGE$ that, given
a graph G, a congruence R on G, and a pair (u, v) of nodes, computes the
congruence closure of $R \cup \{(u, v)\}$. An equivalence relation is represented by
its corresponding partition; that is, the set of its equivalence classes. Two
procedures for operating on partitions are assumed to be available: $UNION$
and $FIND$.

$UNION(u, v)$ combines the equivalence classes of nodes u and v into a
single class. $FIND(u)$ returns a unique name associated with the equivalence
class of node u.

The recursive procedure $MERGE(R, u, v)$ makes use of the function
$CONGRUENT(R, u, v)$ that determines whether two nodes u, v are congruent.

Definition 10.6.4 (Oppen and Nelson's congruence closure algorithm)

Function $CONGRUENT$

```
function CONGRUENT(R': congruence; u, v: node): boolean;
   var flag: boolean; i, n: integer;
   begin
     if Λ(u) = Λ(v) then
       let n = |w| where r(Λ(u)) = (w, s);
       flag := true;
       for i := 1 to n do
         if FIND(u[i]) <> FIND(v[i]) then
           flag := false
         endif
       endfor;
       CONGRUENT := flag
     else
       CONGRUENT := false
     endif
   end
```

Procedure $MERGE$

```
procedure MERGE(var R': congruence; u, v: node);
   var X, Y: set-of-nodes; x, y: node;
   begin
     if FIND(u) <> FIND(v) then
       X := the union of the sets Px of predecessors of all
            nodes x in [u], the equivalence class of u;
```

$Y :=$ the union of the sets P_y of predecessors of all
    nodes y in $[v]$, the equivalence class of v;
$UNION(u, v)$;
for each pair (x, y) such that $x \in X$ **and** $y \in Y$ **do**
  if $FIND(x) <> FIND(y)$ **and** $CONGRUENT(R', x, y)$
  then
    $MERGE(R', x, y)$
  endif
endfor
  endif
end

In order to compute the congruence closure R' of a relation $R = \{(u_1, v_1), ..., (u_n, v_n)\}$ on a graph G, we use the following algorithm, which computes R' incrementally. It is assumed that R' is a global variable.

```
program closure;
    var R': relation;
    function CONGRUENT(R': congruence; u, v: node): boolean;
    procedure MERGE(var R': congruence; u, v: node);
    begin
      input(G, R);
      R' := Id; {the identity relation on the set V of nodes}
      for each (u_i, v_i) in R do
        MERGE(R', u_i, v_i)
      endfor
    end
```

To prove the correctness of the above algorithm, we need the following lemma.

Lemma 10.6.5 (Correctness of the congruence closure algorithm) Given a finite graph G, a congruence R on G, and any pair (u, v) of nodes in G, let R_1 be the least equivalence relation containing $R \cup \{(u, v)\}$, and R_3 be the congruence closure of $R_1 \cup R_2$, where

$$R_2 = \{(x, y) \in X \times Y \mid CONGRUENT(R_1, x, y) = \textbf{true}\},$$

with

$X =$ the union of the sets P_x of predecessors of all nodes x in $[u]$, the equivalence class of u (modulo R_1), and

$Y =$ the union of the sets P_y of predecessors of all nodes y in $[v]$, the equivalence class of v (modulo R_1).

Then the relation R_3 is the congruence closure of $R \cup \{(u, v)\}$.

Proof: First, since R_3 is the congruence closure of $R_1 \cup (X \times Y)$ and R_1 contains $R \cup \{(u,v)\}$, R_3 is a congruence containing $R \cup \{(u,v)\}$. To show that it is the smallest one, observe that for any congruence R' containing $R \cup \{(u,v)\}$, by the definition of a congruence, R' has to contain all pairs in R_2, as well as all pairs in R and (u,v). Hence, any such congruence R' contains R_3. $\square$

Using lemma 10.6.5, it is easy to justify that if $MERGE(R,u,v)$ terminates, then it computes the congruence closure of $R \cup \{(u,v)\}$. The only fact that remains to be checked is that the procedure terminates. But note that $MERGE(R',u,v)$ calls $UNION(u,v)$ iff u and v are not already equivalent, before calling $MERGE(R',x,y)$ recursively iff x and y are not equivalent. Hence, every time $MERGE$ is called recursively, the number of inequivalent nodes decreases, which guarantees termination. Then, the correctness of the algorithm closure is straigtforward. $\square$

The complexity of the procedure $MERGE$ has been analyzed in Nelson and Oppen, 1980. Nelson and Oppen showed that if G has m edges and G has no isolated nodes, in which case the number of nodes n is $O(m)$, then the algorithm can be implemented to run in $O(m^2)$-time. Downey, Sethi, and Tarjan (Downey, 1980) have also studied the problem of congruence closure, and have given a faster algorithm running in $O(mlog(m))$-time.

It should also be noted that there is a dual version of the congruence closure problem, the *unification closure problem*, and also a symmetric version of the problem, which have been investigated in Oppen, 1980a, and Downey, 1980. Both have applications to decision problems for certain classes of formulae. For instance, in Oppen, 1980a, the symmetric congruence closure is used to give a decision procedure for the theory of recursively defined data structures, and in Nelson and Oppen, 1980, the congruence closure is used to give a decision problem for the theory of list structures. Other applications of the concept of congruence closure are also found in Oppen, 1980b; Nelson and Oppen, 1979; and Shostak, 1984b.

The dual concept of the congruence closure is defined as follows. An equivalence relation R on a graph G is a *unification closure* iff, for every pair (u,v) of nodes in V^2, whenever uR_sv then:

(1) Either $\Lambda(u) = \Lambda(v)$, or one of $\Lambda(u)$, $\Lambda(v)$ has arity e;

(2) If $\Lambda(u) = \Lambda(v)$ and $r(\Lambda(u)) = (s_1...s_n, s)$, then for every i, $1 \le i \le n$, $u[i]R_{s_i}v[i]$.

Graphically, if u and v are two nodes labeled with the same symbol f of rank $(s_1...s_n, s)$, if $u[1], ..., u[n]$ are the successors of u and $v[1], ..., v[n]$ are the successors of v,

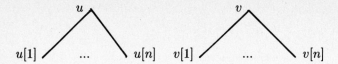

if u and v are equivalent then $u[i]$ and $v[i]$ are equivalent for all $i, 1 \leq n \leq n$. Hence, we have a kind of forward closure. In contrast with the congruence closure, the least unification closure containing a given relation R does not necessarily exist, because condition (1) may fail (see the problems).

The unification closure problem has applications to the unification of trees, and to the equivalence of deterministic finite automata. There is a linear-time unification closure algorithm due to Paterson and Wegman (Paterson and Wegman, 1978) when a certain acyclicity condition is satisfied, and in the general case, there is an $O(m\alpha(m))$-time algorithm, where α is a functional inverse of Ackermann's function. For details, the reader is referred to Downey, 1980. An $O(m\alpha(m))$-time unification algorithm is also given in Huet, 1976.

PROBLEMS

10.6.1. Prove lemma 10.6.1.

10.6.2. Give the details of the proof of lemma 10.6.3.

10.6.3. Use the method of Subsection 10.6.3 to show that the following formula is valid:
$$x \doteq y \supset f(x, y) \doteq f(y, x)$$

10.6.4. Use the method of Subsection 10.6.3 to show that the following formula is valid:
$$f(f(f(a))) \doteq a \wedge f(f(a)) \doteq a \supset f(a) \doteq a$$

10.6.5. Use the method of Subsection 10.6.5 to show that the following formula is valid:
$$x \doteq y \wedge g(f(x, y)) \doteq h(f(x, y)) \wedge P(g(f(x, y)) \supset P(h(f(y, x))))$$

10.6.6. Use the method of Subsection 10.6.5 to show that the following formula is not valid:
$$x \doteq y \wedge g(z) \doteq h(z) \wedge P(g(f(x, y))) \supset P(h(f(y, x)))$$

* **10.6.7.** An equivalence relation R on a graph G is a *unification closure* iff, for every pair (u, v) of nodes in V^2, whenever $uR_s v$ then:

(1) Either $\Lambda(u) = \Lambda(v)$, or one of $\Lambda(u)$, $\Lambda(v)$ has arity e;

(2) If $\Lambda(u) = \Lambda(v)$ and $r(\Lambda(u)) = (s_1...s_n, s)$, then for every i, $1 \leq i \leq n$, $u[i]R_{s_i}v[i]$.

(a) Given an arbitrary relation R_0 on a graph G, give an example showing that the smallest unification closure containing R_0 does not necessarily exists, because condition (1) may fail.

(b) Using the idea of Subsection 10.6.4, show that there is an algorithm for deciding whether the smallest unification closure containing a relation R_0 on G exists, and if so, for computing it.

$*$ **10.6.8.** In order to test whether two trees t_1 and t_2 are unifiable, we can compute the unification closure of the relation $\{t_1, t_2\}$ on the graph $G(t_1, t_2)$ constructed from t_1 and t_2 as follows:

(i) The set set of nodes of $G(t_1, t_2)$ is the set of all subterms of t_1 and t_2.

(ii) Every subterm that is either a constant or a variable is a terminal node labeled with that symbol.

(iii) For every subterm of the form $fy_1...y_k$, the label is f, and there is an edge from $fy_1...y_k$ to y_i, for each i, $1 \leq i \leq k$.

Let R be the least unification closure containing the relation $\{t_1, t_2\}$ on the graph $G(t_1, t_2)$, if it exits. A new graph $G(t_1, t_2)/R$ can be constructed as follows:

(i) The nodes of $G(t_1, t_2)/R$ are the equivalence classes of R.

(ii) There is an edge from a class C to a class C' iff there is an edge in $G(t_1, t_2)$ from some node y in class C to some node z in class C'.

Prove that t_1 and t_2 are unifiable iff the unification closure R exists and the graph $G(t_1, t_2)/R$ is acyclic. Show that a unifying substitution (if it exists) is given by all pairs (x, t) such that x and t are in a same equivalence class, and x is a variable.

(A cycle in a graph is a sequence $v_1, ..., v_k$, of nodes such that $v_1 = v_k$, and there is an edge from v_i to v_{i+1}, $1 \leq i \leq k-1$. A graph is acyclic iff it does not have any cycle.)

$*$ **10.6.9.** Let C be a quantifier-free formula of the form

$$(s_1 \doteq t_1) \wedge ... \wedge (s_n \doteq t_n) \supset (s \doteq t).$$

Let $COM(C)$ be the conjunction of all formulae of the form

$$\forall x \forall y (f(x, y) \doteq f(y, x)),$$

for every binary function symbol f in C.

Show that the decision procedure provided by the congruence closure algorithm can be adpated to decide whether formulae of the form $COM(C) \supset C$ are valid.

Hint: Make the following modification in building the graph $G(C)$: For every term ft_1t_2, create a node labeled ft_1t_2 having two ordered successors:

A terminal node labeled f, and a node labeled with $v(ft_1t_2)$, such that $v(ft_1t_2)$ has two unordered successors labeled with t_1 and t_2.

Also modify the definition of a congruence, so that for any two nodes u and v labeled with $v(ft_1t_2)$, where f is a binary function symbol, if either

(i) $u[1]$ is congruent to $v[1]$ and $u[2]$ is congruent to $v[2]$, or

(ii) $u[1]$ is congruent to $v[2]$ and $u[2]$ is congruent to $v[1]$,

then u and v are congruent.

Notes and Suggestions for Further Reading

Many-sorted first-order logic is now used quite extensively in computer science. Its main uses are to the definition of abstract data types and to programming with rewrite rules. Brief presentations of many-sorted first-order logic can be found in Enderton, 1972, and Monk, 1976. A pioneering paper appears to be Wang, 1952.

The literature on abstact data types and rewrite rules is now extensive. A good introduction to abstract data types can be found in the survey paper by Goguen,Thatcher,Wagner, and Wright, in Yeh, 1978. For an introduction to rewrite rules, the reader is referred to the survey paper by Huet and Oppen, in Book, 1980, and to Huet, 1980.

Congruence closure algorithms were discovered independently by Nelson and Oppen (Nelson and Oppen, 1980, Oppen, 1980a), and Downey, Sethi, and Tarjan (Downey, 1980). The extension given in Subsection 10.6.4 appears to be new. Problems 10.6.7 and 10.6.8 are inspired from Paterson and Wegman, 1978, where a linear-time algorithm is given, and problem 10.6.9 is inspired from Downey, 1980, where fast algorithms are given.

Appendix
and
References

2.4 Algebras

We have already encountered the concept of an algebra in example 2.3.1 and
in the section containing theorem 2.3.1. An algebra is simply a pair $< A, F >$
consisting of a nonempty set A together with a collection F of functions also
called operations, each function in F being of the form $f : A^n \to A$, for some
natural number $n > 0$. When working with algebras, one is often dealing with
algebras having a *common structure*, in the sense that for any two algebras
$< A, F >$ and $< B, G >$, there is a function $d : F \to G$ from the set of
functions F to the set of functions G. A more convenient way to indicate
common structure is to define in advance the set Σ of *function symbols* as a
ranked alphabet used to name operators used in these algebras. Then, given
an algebra $< A, F >$, each function name f receives an *interpretation* $I(f)$
which is a function in F. In other words, the set F of functions is defined by
an interpretation $I : \Sigma \to F$ assigning a function of rank n to every function
symbol of rank n in Σ. Given any two algebras $< A, I : \Sigma \to F >$ and
$< B, J : \Sigma \to G >$, the mapping d from F to G indicating common structure
is the function such that $d(I(f)) = J(f)$, for every function name f in Σ.
This leads to the following formal definition.

2.4.1 Definition of an Algebra

Given a ranked alphabet Σ, a Σ-*algebra* **A** is a pair $< A, I >$ where A is a nonempty set called the *carrier*, and I is an *interpretation function* assigning functions to the function symbols as follows:

(i) Each symbol f in Σ of rank $n > 0$ is interpreted as a function $I(f) : A^n \rightarrow A$;

(ii) Each constant c in Σ is interpreted as an element $I(c)$ in A.

The following abbreviations will also be used: $I(f)$ will be denoted as $f_{\mathbf{A}}$ and $I(c)$ as $c_{\mathbf{A}}$.

Roughly speaking, the ranked alphabet describes the syntax, and the interpretation function I describes the semantics.

EXAMPLE 2.4.1

Let A be any (nonempty) set and let 2^A denote the *power-set* of A, that is, the set of all subsets of A. Let $\Sigma = \{0, 1, -, +, *\}$, where $0,1$ are constants, that is of rank 0, $-$ has rank 1, and $+$ and $*$ have rank 2. If we define the interpretation function I such that $I(0) = \emptyset$ (the empty set), $I(1) = A$, $I(-) = set\ complementation$, $I(+) = set\ union$, and $I(*) = set\ intersection$, we have the algebra $\mathcal{B}_A = < 2^A, I >$. Such an algebra is a *boolean algebra of sets in the universe A*.

EXAMPLE 2.4.2

Let Σ be a ranked alphabet. Recall that the set CT_Σ denotes the set of all finite or infinite Σ-trees. Every function symbol f of rank $n > 0$ defines the function $\overline{f} : CT_\Sigma^n \rightarrow CT_\Sigma$ as follows: for all $t_1, t_2, ..., t_n \in CT_\Sigma$, $\overline{f}(t_1, t_2, ..., t_n)$ is the tree denoted by $ft_1t_2...t_n$ and whose graph is the set of pairs

$$\{(e, f)\} \cup \bigcup_{i=1}^{i=n} \{(iu, t_i(u)) \mid u \in dom(t_i)\}.$$

The tree $ft_1...t_n$ is the tree with f at the root and t_i as the subtree at address i. Let I be the interpretation function such that for every $f \in \Sigma_n$, $n > 0$, $I(f) = \overline{f}$. The pair (CT_Σ, I) is a Σ-algebra. Similarly, (T_Σ, J) is a Σ-algebra for the interpretation J such that for every $f \in \Sigma_n$, $n > 0$, $J(f)$ is the restriction of $\overline{f}$ to finite trees. For simplicity of notation, the algebra (CT_Σ, I) is denoted by CT_Σ and (T_Σ, J) by T_Σ.

The notion of a function preserving algebraic structure is defined below

2.4.2 Homomorphisms

Given two Σ-algebras $\mathbf{A}$ and $\mathbf{B}$, a function $h : A \rightarrow B$ is a *homomorphism* iff:

(i) For every function symbol f of rank $n > 0$, for every $(a_1, ..., a_n) \in A^n$,

$$h(f_{\mathbf{A}}(a_1, ..., a_n)) = f_{\mathbf{B}}(h(a_1), ..., h(a_n));$$

(ii) For every constant c, $h(c_\mathbf{A}) = c_\mathbf{B}$.

If we define $A^0 = \{e\}$ and the function $h^n : A^n \to B^n$ by $h(a_1, ..., a_n) = (h(a_1), ..., h(a_n))$, then the fact that the function h is a homomorphism is expressed by the commutativity of the following diagram:

$$\begin{array}{ccc} A^n & \xrightarrow{f_\mathbf{A}} & A \\ h^n \downarrow & & \downarrow h \\ B^n & \xrightarrow{f_\mathbf{B}} & B \end{array}$$

We say that a homomorphism $h : \mathbf{A} \to \mathbf{B}$ is an *isomorphism* iff there is a homomorphism $g : \mathbf{B} \to \mathbf{A}$ such that $h \circ g = I_A$ and $g \circ h = I_B$. Note that if h is an isomorphism, then it is a bijection.

Inductive sets correspond to subalgebras. This concept is defined as follows.

2.4.3 Subalgebras

An algebra $\mathbf{B} =< B, J >$ is a *subalgebra* of an algebra $\mathbf{A} =< A, I >$ (with the same ranked alphabet Σ) iff:

(1) B is a subset of A;

(2) For every constant c in Σ, $c_\mathbf{B} = c_\mathbf{A}$, and for every function symbol f of rank $n > 0$, $f_\mathbf{B} : B^n \to B$ is the restriction of $f_\mathbf{A} : A^n \to A$.

The fact that $\mathbf{B}$ is an algebra implies that B is *closed under the operations*; that is, for every function symbol f of rank $n > 0$ in Σ, for all $b_1, ..., b_n \in B$, $f_\mathbf{B}(b_1, ..., b_n) \in B$. Now, inductive closures correspond to least subalgebras.

2.4.4 Least Subalgebra Generated by a Subset

Given an algebra $\mathbf{A}$ and a subset X of A, the inductive closure of X in $\mathbf{A}$ is the *least subalgebra of* $\mathbf{A}$ *containing* X. It is also called the least subalgebra of $\mathbf{A}$ generated by X. This algebra can be defined as in lemma 2.3.1. Let $([X]_i)_{i \geq 0}$ be the sequence of subsets of A defined by induction as follows:

$$[X]_0 = X \cup \{c_\mathbf{A} \mid c \text{ is a constant in } \Sigma\};$$
$$[X]_{i+1} = [X]_i \cup \{f_\mathbf{A}(a_1, ..., a_n) \mid a_1, ..., a_n \in [X]_i, f \in \Sigma_n, n \geq 1\}.$$

Let

$$[X] = \bigcup_{i \geq 0} [X]_i.$$

One can verify easily that $[X]$ is closed under the operations of $\mathbf{A}$. Hence, $[X]$ together with the restriction of the operators to $[X]$ is a subalgebra of $\mathbf{A}$.

Let [**X**] denote this subalgebra. The following lemma is easily proved (using a proof similar to that of lemma 2.3.1).

Lemma 2.4.1 Given any algebra **A** and any subset X of A, [**X**] is the least subalgebra of **A** containing X.

Important note: The carrier of an algebra is always *nonempty*. To avoid having the carrier $[X]$ empty, we will make the assumption that $[X]_0 \neq \emptyset$ (either X is nonempty or there are constant symbols).

Finally, the notion of free generation is generalized as follows.

2.4.5 Subalgebras Freely Generated by a Set X

We say that the algebra [**X**] is *freely generated by X in* **A** iff the following conditions hold:

(1) For every f (not a constant) in Σ, the restriction of the function $f_{\mathbf{A}} : A^m \to A$ to $[X]^m$ is injective.

(2) For every $f_{\mathbf{A}} : A^m \to A$, $g_{\mathbf{A}} : A^n \to A$ with $f, g \in \Sigma$, $f([X]^m)$ is disjoint from $g([X]^n)$ whenever $f \neq g$ (and $c_{\mathbf{A}} \neq d_{\mathbf{A}}$ for constants $c \neq d$).

(3) For every $f_{\mathbf{A}} : A^n \to A$ with $f \in \Sigma$ and every $(x_1, ..., x_n) \in [X]^n$, $f_{\mathbf{A}}(x_1, ..., x_n) \notin X$.

As in lemma 2.3.3, it can be shown that for every $(x_1, ..., x_n) \in [X]_i^n - [X]_{i-1}^n$, $f_{\mathbf{A}}(x_1, ..., x_n) \notin [X]_i$, $(i \geq 0$, with $[X]_{-1} = \emptyset)$.

We have the following version of theorem 2.3.1.

Theorem 2.4.1 (Unique homomorphic extension theorem) Let **A** and **B** be two Σ-algebras, X a subset of A, let [**X**] be the least subalgebra of **A** containing X, and assume that [**X**] is freely generated by X. For every function $h : X \to B$, there is a unique homomorphism $\widehat{h} : [\mathbf{X}] \to \mathbf{B}$ such that:

(1) For all $x \in X$, $\widehat{h}(x) = h(x)$;

For every function symbol f of rank $n > 0$, for all $(x_1, ..., x_n) \in [X]^n$,

(2) $\widehat{h}(f_{\mathbf{A}}(x_1, ..., x_n)) = f_{\mathbf{B}}(\widehat{h}(x_1), ..., \widehat{h}(x_n))$, and $\widehat{h}(c_{\mathbf{A}}) = c_{\mathbf{B}}$, for each constant c.

This is also expressed by the following diagram.

$$
\begin{array}{ccc}
X & \longrightarrow & [\mathbf{X}] \\
& h \searrow & \downarrow \widehat{h} \\
& & \mathbf{B}
\end{array}
$$

The following lemma shows the importance of the algebra of finite trees.

Lemma 2.4.2 For every ranked alphabet Σ, for every set X, if $X \cup \Sigma_0 \neq \emptyset$, then the Σ-algebra $T_\Sigma(X)$ of finite trees over the ranked alphabet $\Sigma \cup X$ obtained by adjoining the set X to Σ_0 (the set of constants in Σ) is freely generated by X.

Proof: First, it is easy to show from the definitions that for every tree t such that $depth(t) > 0$,

$$t = \overline{f}(t/1, ..., t/n) \quad \text{and} \quad depth(t/i) < depth(t), \ 1 \le i \le n,$$

where $t(e) = f$ (the label of the root of t), $n = r(f)$ (the rank of f) and t/i is the "i-th subtree" of t, $1 \le i \le n$. Using this property, we prove by induction on the depth of trees that every tree in $T_\Sigma(X)$ belongs to the inductive closure of X. Since a tree of depth 0 is a one-node tree labeled with either a constant or an element of X, the base case of the induction holds. Assume by induction that every tree of depth at most k belongs to X_k, the k-th stage of the inductive closure of X. Since every tree of depth $k + 1$ can be written as $t = \overline{f}(t/1, ..., t/n)$ where every subtree t/i has depth at most k, the induction hypothesis applies. Hence, each t/i belongs to X_k. But then $\overline{f}(t/1, ..., t/n) = t$ belongs to X_{k+1}. This concludes the induction showing that $T_\Sigma(X)$ is a subset of X_+. For every $f \in \Sigma_n$, $n > 0$, by the definition of $\overline{f}$, if $t_1, ..., t_n$ are finite trees, $\overline{f}(t_1, ..., t_n)$ is a finite tree (because its domain is a finite union of finite domains). Hence, every X_k is a set of finite trees, and thus a subset of $T_\Sigma(X)$. But then X_+ is a subset of $T_\Sigma(X)$ and $T_\Sigma(X) = X_+$. Note also that for any two trees t and t' (even infinite), $t = t'$ if and only if, either

(1) $t = \overline{f}(t_1, ..., t_m) = t'$, for some unique $f \in \Sigma_m$, $(m > 0)$, and some unique trees $t_1, ..., t_m \in CT_\Sigma(X)$, or

(2) $t = a = t'$, for some unique $a \in X \cup \Sigma_0$.

Then, it is clear that each function $\overline{f}$ is injective, that $range(\overline{f})$ and $range(\overline{g})$ are disjoint whenever $f \neq g$, and that for every f of rank $n > 0$, $\overline{f}(t_1, ..., t_n)$ is never a one-node tree labeled with a constant. Hence, conditions (1),(2),(3) for free generation are satisfied. $\square$

In Chapter 5, when proving the completeness theorem for first-order logic, it is necessary to define the quotient of an algebra by a type of equivalence relation. Actually, in order to define an algebraic structure on the set of equivalence classes, we need a stronger concept known as a *congruence relation*.

2.4.6 Congruences

Given a Σ-algebra **A**, a *congruence* $\cong$ on A is an equivalence relation on the carrier A satisfying the following conditions: For every function symbol f of rank $n > 0$ in Σ, for all $x_1, ..., x_n, y_1, ..., y_n \in A$, if $x_i \cong y_i$ for all $i, 1 \le i \le n$, then

$$f_\mathbf{A}(x_1, ..., x_n) \cong f_\mathbf{A}(y_1, ..., y_n).$$

The equivalence class of x modulo $\cong$ is denoted by $[x]_\cong$, or more simply by $[x]$ or $\hat{x}$.

Given any function symbol f of rank $n > 0$ in Σ and any n subsets $B_1, ..., B_n$ of A, we define

$$f_\mathbf{A}(B_1, ..., B_n) = \{f_\mathbf{A}(b_1, ..., b_n) \mid b_i \in B_i, 1 \le i \le n\}.$$

If $\cong$ is a congruence on A, for any $a_1, ..., a_n \in A$, $f_\mathbf{A}([a_1], ..., [a_n])$ is a subset of some unique equivalence class which is in fact $[f_\mathbf{A}(a_1, ..., a_n)]$. Hence, we can define a structure of Σ-algebra on the set $A/\cong$ of equivalence classes.

2.4.7 Quotient Algebras

Given a Σ-algebra $\mathbf{A}$ and a congruence $\cong$ on A, the *quotient algebra* $\mathbf{A}/\cong$ has the set $A/\cong$ of equivalence classes modulo $\cong$ as its carrier, and its operations are defined as follows: For every function symbol f of rank $n > 0$ in Σ, for all $[a_1], ..., [a_n] \in A/\cong$,

$$f_{\mathbf{A}/\cong}([a_1], ..., [a_n]) = [f_\mathbf{A}(a_1, ..., a_n)],$$

and for every constant c,

$$c_{\mathbf{A}/\cong} = [c].$$

One can easily verify that the function $h_\cong : A \to A/\cong$ such that $h_\cong(x) = [x]$ is a homomorphism from $\mathbf{A}$ to $\mathbf{A}/\cong$.

EXAMPLE 2.4.3

Let $\Sigma = \{0, 1, -, +, *\}$ be the ranked alphabet of example 2.4.1. Let A be any nonempty set and let $T_\Sigma(2^A)$ be the tree algebra freely generated by 2^A. By lemma 2.4.2 and theorem 2.4.1, the identity function $Id : 2^A \to 2^A$ extends to a unique homomorphism $h : T_\Sigma(2^A) \to \mathcal{B}_A$, where the range of the function h is the boolean algebra $\mathcal{B}_A$ (and not just its carrier 2^A as in the function Id). Let $\cong$ be the relation defined on $T_\Sigma(2^A)$ such that:

$$t_1 \cong t_2 \quad \text{if and only if} \quad h(t_1) = h(t_2).$$

For example, if $A = \{a, b, c, d\}$, for the trees t_1 and t_2 given by the terms $t_1 = +(\{a\}, *(\{b, c, d\}, \{a, b, c\}))$ and $t_2 = +(\{a, b\}, \{c\})$, we have

$$h(t_1) = h(t_2) = \{a, b, c\}.$$

It can be verified that $\cong$ is a congruence, and that the quotient algebra $T_\Sigma(2^A)/\cong$ is isomorphic to $\mathcal{B}_A$.

2.5 Many-Sorted Algebras

For many computer science applications, and for the definition of data types in particular, it is convenient to generalize algebras by allowing domains and operations of different types (also called sorts). A convenient way to do so is to introduce the concept of a *many-sorted algebra*. In Chapter 10, a generalization of first-order logic known as many-sorted first-order logic is also presented. Since the semantics of this logic is based on many-sorted algebras, we present in this section some basic material on many-sorted algebra.

Let S be a set of *sorts* (or *types*). Typically, S consists of types in a programming language (such as *integer*, *real*, *boolean*, *character*, etc.).

2.5.1 S-Ranked Alphabets

An *S-ranked alphabet* is pair (Σ, r) consisting of a set Σ together with a function $r : \Sigma \rightarrow S^* \times S$ assigning a *rank* (u, s) to each symbol f in Σ.

The string u in S^* is the *arity* of f and s is the *sort* (or *type*) of f. If $u = s_1...s_n$, $(n \geq 1)$, a symbol f of rank (u, s) is to be interpreted as an operation taking arguments, the i-th argument being of type s_i and yielding a result of type s. A symbol of rank (e, s) (when u is the empty string) is called a *constant of sort s*. For simplicity, a ranked alphabet (Σ, r) is often denoted by Σ.

2.5.2 Definition of a Many-Sorted Algebra

Given an S-ranked alphabet Σ, a *many-sorted Σ-algebra* $\mathbf{A}$ is a pair $< A, I >$, where $A = (A_s)_{s \in S}$ is an S-indexed family of nonempty sets, each A_s being called a *carrier of sort s*, and I is an *interpretation function* assigning functions to the function symbols as follows:

(i) Each symbol f of rank (u, s) where $u = s_1...s_n$ is interpreted as a function $I(f) : A_{s_1} \times ... \times A_{s_n} \rightarrow A_s$;

(ii) Each constant c of sort s is interpreted as an element $I(c)$ in A_s.

The following abbreviations will also be used: $I(f)$ will be denoted as $f_\mathbf{A}$ and $I(c)$ as $c_\mathbf{A}$; If $u = s_1...s_n$, we let $A^u = A_{s_1} \times ... \times A_{s_n}$ and $A^e = \{e\}$. (Since there is a bijection between A and $A \times \{e\}$, A and $A \times \{e\}$ will be identified.) Given an S-indexed family $h = (h_s)_{s \in S}$ of functions $h_s : A_s \rightarrow B_s$, the function $h^u : A^u \rightarrow B^u$ is defined so that, for all $(a_1, ..., a_n) \in A^u$, $h^u(a_1, ..., a_n) = (h_{s_1}(a_1), ..., h_{s_n}(a_n))$.

EXAMPLE 2.5.1

The algebra of *stacks of natural numbers* is defined as follows. Let $S = \{\textbf{int}, \textbf{stack}\}$, $\Sigma = \{\Lambda, ERROR, POP, PUSH, TOP\}$, where Λ is a constant of sort **stack**, $ERROR$ is a constant of sort **int**, $PUSH$ is a function symbol of rank (**int.stack,stack**), POP a function symbol of rank (**stack,stack**) and TOP a function symbol of rank (**stack,int**). The carrier of sort **int** is $\mathbf{N} \cup \{error\}$, and the carrier of sort **stack**, the set of functions of the form $X : [n] \to \mathbf{N} \cup \{error\}$, where $n \in \mathbf{N}$. When $n = 0$, the unique function from the empty set ($[0]$) is denoted by Λ and is called the empty stack. The constant $ERROR$ is interpreted as the element *error*. Given any stack $X : [n] \to \mathbf{N} \cup \{error\}$ and any element $a \in \mathbf{N} \cup \{error\}$, $PUSH(a, X)$ is the stack $X' : [n+1] \to \mathbf{N} \cup \{error\}$ such that $X'(k) = X(k)$ for all k, $1 \le k \le n$, and $X'(n + 1) = a$; $TOP(X)$ is the top element $X(n)$ of X if $n > 0$, *error* if $n = 0$; $POP(X)$ is the empty stack Λ if either X is the empty stack or $n = 1$, or the stack $X' : [n-1] \to \mathbf{N} \cup \{error\}$ such that $X'(k) = X(k)$ for all k, $1 \le k \le n-1$ if $n > 1$.

Note: This formalization of a stack is not perfectly faithful because $TOP(X) = error$ does not necessarily imply that $X = \Lambda$. However, it is good enough as an example.

2.5.3 Homomorphisms

Given two many-sorted Σ-algebras **A** and **B**, an S-indexed family $h = (h_s)_{s \in S}$ of functions $h_s : A_s \to B_s$ is a *homomorphism* iff:

(i) For every function symbol f or rank (u, s) with $u = s_1...s_n$, for every $(a_1, ..., a_n) \in A^u$,

$$h_s(f_\mathbf{A}(a_1, ..., a_n)) = f_\mathbf{B}(h_{s_1}(a_1), ..., h_{s_n}(a_n));$$

(ii) For every sort s, for every constant c of sort s, $h_s(c_\mathbf{A}) = c_\mathbf{B}$.

These conditions can be represented by the following commutative diagram.

$$
\begin{array}{ccc}
A^u & \xrightarrow{f_\mathbf{A}} & A_s \\
h^u \downarrow & & \downarrow h \\
B^u & \xrightarrow[f_\mathbf{B}]{} & B_s
\end{array}
$$

2.5.4 Subalgebras

An algebra $\mathbf{B} = <B, J>$ is a *subalgebra* of an algebra $\mathbf{A} = <A, I>$ (with the same ranked alphabet Σ) iff:

(1) Each B_s is a subset of A_s, for every sort $s \in S$;

(2) For every sort $s \in S$, for every constant c of sort s, $c_{\mathbf{B}} = c_{\mathbf{A}}$, and for every function symbol f of rank (u, s) with $u = s_1...s_n$, $f_{\mathbf{B}} : B^u \rightarrow B_s$ is the restriction of $f_{\mathbf{A}} : A^u \rightarrow A_s$.

The fact that $\mathbf{B}$ is an algebra implies that B is *closed under the operations*; that is, for every function symbol f of rank (u, s) with $u = s_1...s_n$, for every $(b_1, ..., b_n) \in B^u$, $f_{\mathbf{A}}(b_1, ..., b_n)$ is in B_s, and for every constant c of sort s, $c_{\mathbf{A}}$ is in B_s.

2.5.5 Least Subalgebras

Given a Σ-algebra $\mathbf{A}$, let $X = (X_s)_{s \in S}$ be an S-indexed family with every X_s a subset of A_s. As in the one-sorted case, the least subalgebra of $\mathbf{A}$ containing X can be characterized by a bottom-up definition. We define this algebra $[\mathbf{X}]$ as follows.

The sequence of S-indexed families of sets $([X]_i)$, $(i \geq 0)$ is defined by induction: For every sort s,

$$[X_s]_0 = X_s \cup \{c_{\mathbf{A}} \mid c \text{ a constant of sort } s\},$$
$$[X_s]_{i+1} = [X_s]_i \cup \{f_{\mathbf{A}}(x_1, ..., x_n) \mid r(f) = (u, s), (x_1, ..., x_n) \in ([X]_i)^u\},$$

with $u = s_1...s_n$, $n \geq 1$.

The carrier of sort s of the algebra $[\mathbf{X}]$ is

$$\bigcup_{i \geq 0} [X_s]_i.$$

Important note: The carriers of an algebra are always *nonempty*. To avoid having any carrier $[X_s]$ empty, we will make the assumption that for every sort s, $[X_s]_0 \neq \emptyset$. This will be satisfied if either X_s is nonempty or there are constants of sort s. A more general condition can be given. Call a sort s *nonvoid* if either there is some constant of sort s, or there is some function symbol f of rank $(s_1...s_n, s)$ such that $s_1, ..., s_n$ are all non-void $(n \geq 1)$. Then, $[X]_s$ is nonempty if and only if either $X_s \neq \emptyset$ or s is non-void.

We also have the following induction principle.

Induction Principle for Least Subalgebras

If $[\mathbf{X}]$ is the least subalgebra of $\mathbf{A}$ containing X, for every subfamily Y of $[\mathbf{X}]$, if Y contains X and is closed under the operations of $\mathbf{A}$ (and contains $\{c_{\mathbf{A}} \mid c \text{ is a constant}\}$) then $\mathbf{Y} = [\mathbf{X}]$.

2.5.6 Freely Generated Subalgebras

We say that $[\mathbf{X}]$ is *freely generated by* X *in* $\mathbf{A}$ iff the following conditions hold:

(1) For every f (not a constant) in Σ, the restriction of the function $f_{\mathbf{A}} : A^u \to A_s$ to $[X]^u$ is injective.

(2) For every $f_{\mathbf{A}} : A^u \to A_s$, $g_{\mathbf{A}} : A^v \to A_{s'}$ with $f, g \in \Sigma$, $f([X]^u)$ is disjoint from $g([X]^v)$ whenever $f \neq g$ (and $c_{\mathbf{A}} \neq d_{\mathbf{A}}$ for constants $c \neq d$).

(3) For every $f_{\mathbf{A}} : A^u \to A_s$ with $f \in \Sigma$ and every $(x_1, ..., x_n) \in [X]^u$, $f_{\mathbf{A}}(x_1, ..., x_n) \notin X_s$.

As in lemma 2.3.3 , it can be shown that for every $(x_1, ..., x_n)$ in $[X]_i^u - [X]_{i-1}^u$, $f_{\mathbf{A}}(x_1, ..., x_n) \notin [X_s]_i$, ($i \geq 0$, with $[X_s]_{-1} = \emptyset$).

We have the following generalization of theorem 2.4.1.

Theorem 2.5.1 (Unique homomorphic extension theorem) Let $\mathbf{A}$ and $\mathbf{B}$ be two many-sorted Σ-algebras, X an S-indexed family of subsets of A, let $[\mathbf{X}]$ be the least subalgebra of $\mathbf{A}$ containing X, and assume that $[\mathbf{X}]$ is freely generated by X. For every S-indexed family $h : X \to B$ of functions $h_s : X_s \to B_s$, there is a unique homomorphism $\widehat{h} : [\mathbf{X}] \to \mathbf{B}$ such that:

(1) For all $x \in X_s$, $\widehat{h}_s(x) = h_s(x)$ for every sort s;

For every function symbol f of rank (u, s), with $u = s_1...s_n$, for all $(x_1, ..., x_n) \in [X]^u$,

(2) $\widehat{h}_s(f_{\mathbf{A}}(x_1, ..., x_n)) = f_{\mathbf{B}}(\widehat{h}_{s_1}(x_1), ..., \widehat{h}_{s_n}(x_n))$, and $\widehat{h}_s(c_{\mathbf{A}}) = c_{\mathbf{B}}$ for a constant c of sort s.

$$X \longrightarrow [\mathbf{X}]$$
$$h \searrow \quad \downarrow \widehat{h}$$
$$\mathbf{B}$$

2.5.7 Congruences

Given a Σ-algebra $\mathbf{A}$, a *congruence* $\cong$ on A is an S-indexed family $(\cong_s)_{s \in S}$ of relations, each $\cong_s$ being an equivalence relation on the carrier A_s, and satisfying the following conditions: For every function symbol f of rank (u, s), with $u = s_1...s_n$, for all $(x_1, ..., x_n)$ and $(y_1, ..., y_n)$ in A^u,

if $x_i \cong_{s_i} y_i$ for all i, $1 \leq i \leq n$, then

$$f_{\mathbf{A}}(x_1, ..., x_n) \cong_s f_{\mathbf{A}}(y_1, ..., y_n).$$

The equivalence class of x modulo $\cong_s$ is denoted by $[x]_{\cong_s}$, or more simply by $[x]_s$ or $\widehat{x}_s$.

Given any function symbol f of rank (u, s), with $u = s_1...s_n$, and any n subsets $B_1, ..., B_n$ such that B_i is a subset of A_{s_i}, we define

$$f_{\mathbf{A}}(B_1, ..., B_n) = \{f_{\mathbf{A}}(b_1, ..., b_n) \mid (b_1, ..., b_n) \in A^u\}.$$

If $\cong$ if a congruence on A, for any $(a_1, ..., a_n) \in A^u$, $f_{\mathbf{A}}([a_1]_{s_1}, ..., [a_n]_{s_n})$ is a subset of some unique equivalence class which is in fact $[f_{\mathbf{A}}(a_1, ..., a_n)]_s$. Hence, we can define a structure of Σ-algebra on the S-indexed family $A/\cong$ of sets of equivalence classes.

2.5.8 Quotient Algebras

Given a Σ-algebra $\mathbf{A}$ and a congruence $\cong$ on A, the *quotient algebra* $\mathbf{A}/\cong$ has the S-indexed family $A/\cong$ of sets of equivalence classes modulo $\cong_s$ as its carriers, and its operations are defined as follows: For every function symbol f of rank (u, s), with $u = s_1...s_n$, for all $([a_1]_{s_1}, ..., [a_n]_{s_n}) \in (A/\cong)^u$,

$$f_{\mathbf{A}/\cong}([a_1]_{s_1}, ..., [a_n]_{s_n}) = [f_{\mathbf{A}}(a_1, ..., a_n)]_s,$$

and for every constant c of sort s,

$$c_{\mathbf{A}/\cong_s} = [c]_s.$$

The S-indexed family $h_\cong$ of functions $h_{\cong_s} : A_s \to A/\cong_s$ such that $h_{\cong_s}(x) = [x]_{\cong_s}$ is a homomorphism from $\mathbf{A}$ to $\mathbf{A}/\cong$.

Finally, many-sorted trees are defined as follows.

2.5.9 Many-Sorted Trees

Given a many-sorted alphabet Σ (with set S of sorts), a Σ-*tree of sort* s is any function $t : D \to \Sigma$ where D is a tree domain denoted by $dom(t)$ and t satisfies the following conditions:

1) The root of t is labeled with a symbol $t(e)$ in Σ of sort s.

2) For every node $u \in dom(t)$, if $\{i \mid ui \in dom(t)\} = [n]$, for each ui, $i \in [n]$, if $t(ui)$ is a symbol of sort u_i, then $t(u)$ is a symbol of rank (u, v) with $u = u_1...u_n$, for some $v \in S$.

The set of all finite trees of sort s is denoted by T_Σ^s, and the set of all finite trees by T_Σ. Given an S-indexed family $X = (X_s)_{s \in S}$, we can form the sets of trees $T_\Sigma^s(X_s)$ obtained by adjoining each set X_s to the set of constants of sort s. $T_\Sigma(X)$ is a Σ-algebra, and lemma 2.4.2 generalizes as follows.

Lemma 2.5.1 For every many-sorted Σ-algebra $\mathbf{A}$ and S-indexed family $X = (X_s)_{s \in S}$, the Σ-algebra $T_\Sigma(X)$ of finite trees over the ranked alphabet obtained by adjoining each set X_s to the set of constants of sort s in Σ is freely generated by X.

EXAMPLE 2.5.2

Referring to example 2.5.1, let Σ be the ranked alphabet of the algebra $\mathbf{A}$ of stacks. Let $T_\Sigma(\mathbf{N})$ be the tree algebra freely generated by the pair of sets $(\emptyset, \mathbf{N})$. The identity function on $(\emptyset, \mathbf{N})$ extends to a unique homormophism h from $T_\Sigma(\mathbf{N})$ to $\mathbf{A}$. Define the relations $\cong_{\mathbf{int}}$ and $\cong_{\mathbf{stack}}$ on $T_\Sigma(\mathbf{N})$ as follows: For all t_1, t_2 of sort **stack**,

$$t_1 \cong_{\mathbf{int}} t_2 \text{ iff } h(t_1) = h(t_2),$$

and for all t_1, t_2 of sort **int**,

$$t_1 \cong_{\mathbf{stack}} t_2 \text{ iff } h(t_1) = h(t_2).$$

One can check that $\cong$ is a congruence, and that $T_\Sigma(\mathbf{N})/\cong$ is isomorphic to $\mathbf{A}$. One can also check that the following holds for all trees X of sort **stack** and all trees a of sort **int**:

$$POP(PUSH(a, X)) \cong_{\mathbf{stack}} X,$$
$$POP(\Lambda) \cong_{\mathbf{stack}} \Lambda,$$
$$TOP(PUSH(a, X)) \cong_{\mathbf{int}} a,$$
$$TOP(\Lambda) \cong_{\mathbf{int}} ERROR.$$

The reader is referred to Cohn, 1981, or Gratzer, 1979, for a complete exposition of universal algebra. For more details on many-sorted algebras, the reader is referred to the article by Goguen, Thatcher, Wagner and Wright in Yeh, 1978, or the survey article by Huet and Oppen, in Book, 1980.

PROBLEMS

2.4.1. Let $\mathbf{A}$ and $\mathbf{B}$ two Σ-algebras and X a subset of A. Assume that $\mathbf{A}$ is the least subalgebra generated by X. Show that if h_1 and h_2 are any two homomorphisms from $\mathbf{A}$ to $\mathbf{B}$ such that h_1 and h_2 agree on X (that is, $h_1(x) = h_2(x)$ for all $x \in X$), then $h_1 = h_2$.

2.4.2. Let $h : \mathbf{A} \to \mathbf{B}$ be a homomorphism of Σ-algebras.

(a) Given any subalgebra $\mathbf{X}$ of $\mathbf{A}$, prove that $h(X)$ is a subalgebra of $\mathbf{B}$ (denoted by $h(\mathbf{X})$).

(b) Given any subalgebra $\mathbf{Y}$ of $\mathbf{B}$, prove that $h^{-1}(Y)$ is a subalgebra of $\mathbf{A}$ (denoted by $h^{-1}(\mathbf{Y})$).

2.4.3. Let $h : \mathbf{A} \to \mathbf{B}$ be a homomorphism of Σ-algebras. Let $\cong$ be the relation defined on A such that, for all $x, y \in A$,

$$x \cong y \quad \text{if and only if} \quad h(x) = h(y).$$

Prove that $\cong$ is a congruence on A, and that $h(\mathbf{A})$ is isomorphic to $\mathbf{A}/\cong$.

2.4.4. Prove that for every Σ-algebra **A**, there is some tree algebra $T_\Sigma(X)$ freely generated by some set X and some congruence $\cong$ on $T_\Sigma(X)$ such that $T_\Sigma(X)/\cong$ is isomorphic to **A**.

2.4.5. Let **A** be a Σ-algebra, X a subset of A, and assume that $[\mathbf{X}] = \mathbf{A}$, that is, X generates **A**.

Prove that if for every Σ-algebra **B** and function $h : X \to \mathbf{B}$ there is a unique homomorphism $\hat{h} : \mathbf{A} \to \mathbf{B}$ extending h, then **A** is freely generated by X.

* **2.4.6.** Given a Σ-algebra **A** and any relation R on A, prove that there is a least congruence $\cong$ containing R.

2.5.1. Do problem 2.4.1 for many-sorted algebras.

2.5.2. Do problem 2.4.2 for many-sorted algebras.

2.5.3. Do problem 2.4.3 for many-sorted algebras.

2.5.4. Do problem 2.4.4 for many-sorted algebras.

2.5.5. Do problem 2.4.5 for many-sorted algebras.

* **2.5.6.** Do problem 2.4.6 for many-sorted algebras.

* **2.5.7.** Referring to example 2.5.2, prove that the quotient algebra $T_\Sigma(\mathbf{N})/\cong$ is isomorphic to the stack algebra **A**.

* **2.5.8.** Prove that the least congruence containing the relation R defined below is the congruence $\cong$ of problem 2.5.7. The relation R is defined such that, for all trees X of sort **stack** and all trees a of sort **int**:

$$POP(PUSH(a, X)) \; R_{\mathbf{stack}} \; X,$$
$$POP(\Lambda) \; R_{\mathbf{stack}} \; \Lambda,$$
$$TOP(PUSH(a, X)) \; R_{\mathbf{int}} \; a,$$
$$TOP(\Lambda) \; R_{\mathbf{int}} \; ERROR.$$

This problem shows that the stack algebra is isomorphic to the quotient of the tree algebra $T_\Sigma(\mathbf{N})$ by the least congruence $\cong$ containing the above relation.

REFERENCES

Aho, Alfred V., and Ullman, Jeffrey D. 1979. *Principles of Compiler Design*. Reading, MA: Addison Wesley.

Anderson, R. 1970. Completeness Results for E-resolution. *Proceedings AFIPS 1970*, Spring Joint Computer Conference, Vol. 36. Montvale, NJ: AFIPS Press, 653-656.

Andrews, Peter B. 1970. Resolution in Type Theory. *Journal of Symbolic Logic* 36(3), 414-432.

Andrews, Peter B. 1981. Theorem Proving via General Matings. *J.ACM* 28(2), 193-214.

Apt, Krzysztof R., and VanEmden, M. H. 1982. Contributions to the Theory of Logic Programming. *J.ACM* 29(3), 841-862.

Barwise, Jon. 1977. *Handbook of Mathematical Logic*. Studies in Logic, Vol. 90. New York: Elsevier North-Holland.

Bell, J. L., and Slomson, A. B. 1974. *Models and Ultraproducts*. Amsterdam: Elsevier North-Holland.

Birkhoff, Garrett. 1973. *Lattice Theory*. American Mathematical Society Colloquium Publications, Vol. 25.

Book, Ronald. 1980. *Formal Language Theory*. New York: Academic Press.

Boyer, R. S., and Moore, J. S. 1979. *A Computational Logic*. New York: Academic Press.

Bundy, Alan. 1983. *The Computer Modelling of Mathematical Reasoning*. New York: Academic Press.

Campbell, J. A. 1984. *Implementations of PROLOG*. Chichester, England: Ellis & Horwood.

Chang, C. C., and Keisler, H. J. 1973. *Model Theory*. Studies in Logic, Vol. 73. Amsterdam: Elsevier North-Holland.

Chang, C., and Lee, R. C. 1973. *Symbolic Logic and Mechanical Theorem Proving*. New York: Academic Press.

Clocksin, William F., and Mellish, Christopher S. 1981. *Programming in PROLOG*. New York: Springer Verlag.

Cohn, Paul M. 1981. *Universal Algebra*. Hingham, MA: Reidel Publishing Company.

Cook, Stephen A. 1971. The Complexity of Theorem-Proving Procedures. *Proceedings of the Third Annual ACM Symposium on the Theory of Computing*, pp. 151-158, Association for Computing Machinery, New York.

Cook, Stephen A., and Reckhow, Robert A. 1979. The Relative Efficiency of Propositional Proof Systems. *Journal of Symbolic Logic* 44(1), 33-50.

Davis, Martin D. 1963. Eliminating the Irrelevant from Mechanical Proofs. *Proceedings of a Symposium on Applied Mathematics*, Vol. XV, Providence, RI, 15-30.

Davis, Martin D., and Putnam, H. 1960. A Computing Procedure for Quantification Theory. *J.ACM* 7, 210-215.

Davis, Martin D. and Weyuker, Elaine J. 1983. *Computability, Complexity and Languages*. New York: Academic Press.

Dowling, William F., and Gallier, Jean H. 1984. Linear-time Algorithms for Testing the Satisfiability of Propositional Horn Formulae. *Journal of Logic Programming* 3, 267-284.

Downey, Peter J., Sethi, Ravi, and Tarjan, Endre R. 1980. Variations on the Common Subexpressions Problem. *J.ACM* 27(4), 758-771.

Dreben, B., and Goldfarb, W. D. 1979. *The Decision Problem*. Reading, MA: Addison-Wesley.

Enderton, Herbert B. 1972. *A Mathematical Introduction to Logic*. New York: Academic Press.

Enderton, Herbert B. 1977. *Elements of Set Theory*. New York: Academic Press.

Garey, M. R., and Johnson, D. S. 1979. *Computers and Intractability: A Guide to the Theory of NP-Completeness*. New York: Freeman.

Gorn, Saul. 1965. Explicit Definitions and Linguistic Dominoes. In John Hart and Satoru Takasu, eds., *Systems and Computer Science*. Toronto, Canada: University of Toronto Press.

Gorn, Saul. 1984. Data Representation and Lexical Calculi. *Information Processing and Management* 20(1-2), 151-174.

Gratzer, G. 1979. *Universal Algebra*. New York: Springer Verlag.

Halmos, Paul R. 1974. *Lectures on Boolean Algebras*. New York: Springer Verlag.

Henkin, L., Monk, J. D., and Tarski, A. 1971. *Cylindric Algebras*. Studies in Logic, Vol. 64, New York: Elsevier North-Holland.

Herbrand, Jacques. 1971. *Logical Writing*. Hingham, MA: Reidel Publishing Company.

Huet, Gérard. 1973. A Mechanization of Type Theory. *Proceedings of the Third International Joint Conference on Artificial Intelligence*, Stanford, CA, 139-146.

Huet, Gérard. 1976. Résolution d'Equations dans les Languages d'Ordre 1,2,...ω. Doctoral Thesis, Université de Paris VII.

Huet, Gérard. 1980. Confluent Reductions: Abstract Properties and Applications to Term Rewriting Systems. *J.ACM* 27(4), 797-821.

Joyner, William H. 1974. Automatic Theorem Proving and The Decision Problem. Ph.D thesis, Harvard University, Cambridge, MA.

Kapur, D., Krishnamoorthy, M. S., and Narendran, P. 1982. A New Linear Algorithm for Unification. General Electric, Report no. 82CRD-100, New York.

Kleene, Stephen C. 1952. *Introduction to Metamathematics*. Amsterdam: Elsevier North-Holland.

Kleene, Stephen C. 1967. *Mathematical Logic*. New York: Wiley Interscience.

Knuth, Donald. 1968. *The Art of Computer Programming, Vol. 1: Fundamental Algorithms*. Reading, MA: Addison Wesley.

Kowalski, Robert A. 1979. *Logic for Problem Solving*. New York: Elsevier North-Holland.

Kowalski, Robert A., and Kuehner, D. 1970. Linear Resolution with Selection Function. *Artificial Intelligence* 2, 227-260.

Kowalski, Robert A., and Van Emden, M. H. 1976. The Semantics of Predicate Logic as a Programming Language. *J.ACM* 23(4), 733-742.

Kuratowski, K., and Mostowski, A. 1976. *Set Theory*. Studies in Logic, Vol. 86. New York: Elsevier North-Holland.

Levy, A. 1979. *Basic Set Theory*. Berlin, Heidelberg, New York: Springer-Verlag,

Lewis, H. R. 1979. *Unsolvable Classes of Quantificational Formulas*. Reading, MA: Addison-Wesley.

Lewis, H. R., and Papadimitriou, C. H. 1981. *Elements of the Theory of Computation*. Englewood Cliffs, NY: Prentice-Hall.

Lloyd, J. W. 1984. *Foundations of Logic Programming*. New York: Springer-Verlag.

Loveland, Donald W. 1979. *Automated Theorem Proving: A Logical Basis*. New York: Elsevier North-Holland.

Machtey, Michael, and Young, Paul. 1978. *An Introduction to the General Theory of Algorithms*. New York: Elsevier, North-Holland.

Manna, Zohar. 1974. *Mathematical Theory of Computation*. New York: McGraw-Hill.

Manna, Zohar, and Waldinger, Richard. 1985. *The Logical Basis for Computer Programming*. Reading, MA: Addison Wesley.

Miller, Dale. 1984. Herbrand's Theorem in Higher-Order Logic. Technical Report, Department of Computer and Information Science, University of Pennsylvania, Philadelphia, PA 19104.

Monk, J. D. 1976. *Mathematical Logic*. G.T.M Vol. 37. New York: Springer-Verlag.

Morris, J. B. 1969. E-resolution: Extension of Resolution to Include the Equality Relation. *Proceedings International Joint Conference on Artificial Intelligence*, Washington D.C.

Nelson, G., and Oppen, D. C. 1979. Simplification by Cooperating Decision Procedures. *TOPLAS* 1(2), 245-257.

Nelson G., and Oppen, D. C. 1980. Fast Decision Procedures Based on Congruence Closure. *J. ACM* 27(2), 356-364.

Oppen, D. C. 1980a. Reasoning About Recursively Defined Data Structures. *J. ACM* 27(3), 403-411.

Oppen, D. C. 1980b. Complexity, Convexity and Combinations of Theories. *Theoretical Computer Science* 12(3), 291-302.

Paterson, M. S., and Wegman, M. N. 1978. Linear Unification. *Journal of Computer and System Sciences* 16(2), 158-167.

Pietrzykowski, Tomasz. 1973. A Complete Mechanization of Second-Order Type Theory. *J.ACM* 20, 333-364.

Prawitz, Dag. 1960. An Improved Proof Procedure. *Theoria* 26, 102-139.

Prawitz, Dag. 1965. *Natural Deduction*. Stockholm: Almquist & Wiskell.

Robinson, J. A. 1965. A Machine Oriented Logic Based on the Resolution Principle. *J.ACM* 12(1), 23-41.

Robinson, J. A. 1979. *Logic: Form and Function*. New York: Elsevier North-Holland.

Robinson, G. A., and Wos, L. 1969. Paramodulation and Theorem-Proving in First-order Logic with Equality. *Machine Intelligence*, Vol. 4, 135-150.

Rogers, H. R. 1967. *Theory of Recursive Functions and Effective Computability*. New York: McGraw-Hill.

Schoenfield, J. R. 1967. *Mathematical Logic*. Reading, MA: Addison Wesley.

Shostak, Robert. E. 1984a. *7th International Conference on Automated Deduction*. Lecture Notes in Computer Science, Vol. 170. New York: Springer Verlag.

Shostak, Robert. E. 1984b. Deciding Combinations of Theories. *J.ACM* 31(1), 1-12.

Siekman J., and Wrightson, G. 1983. *Automation of Reasoning*. New York: Springer Verlag.

Smorynski, C. 1983. "Big" News From Archimedes to Friedman. *Notices of the American Mathematical Society* 30(3).

Smullyan, R. M. 1968. *First-order Logic*. Berlin, Heidelberg, New York: Springer-Verlag.

Statman, R. 1979. Lower Bounds on Herbrand's Theorem. *Proceedings of the American Mathematical Society* 75(1), 104-107.

Suppes, P. 1972. *Axiomatic Set Theory*. New York: Dover.

Szabo, M. E. 1970. *The Collected Papers of Gerhard Gentzen*. Studies in Logic. New York: Elsevier North-Holland.

Tait, W. W. 1968. Normal Derivability in Classical Logic. In *The Syntax and Semantics of Infinitary Languages*, J. Barwise, ed., Lecture Notes in Mathematics. New York: Springer Verlag.

Takeuti, G. 1975. *Proof Theory*. Studies in Logic, Vol. 81. Amsterdam: Elsevier North-Holland.

Tarski, A., Mostowski, A., and Robinson, R. M. 1971. *Undecidable Theories*. Studies in Logic. Amsterdam: Elsevier North-Holland.

Van Dalen, D. 1980. *Logic and Structure*. Berlin, Heidelberg, New York: Springer-Verlag.

Van Heijenoort, Jean. 1967. *From Frege to Gödel*. Cambridge, MA: Harvard University Press.

Wang, Hao. 1952. Logic of Many-Sorted Theories. *Journal of Symbolic Logic* 17(2), 105-116.

Yeh, Raymond. 1978. *Current Trends in Programming Methodology*, Vol. IV. Englewood Cliffs, NJ: Prentice Hall.

Index of Symbols

Index of Definitions

Subject Index